T0221034

Mathematics of Economics and Business

Knowledge of mathematical methods has become a prerequisite for all students who wish to understand current economic and business literature. This book covers all the major topics required to gain a firm grounding in the subject, such as sequences, series, application in finance, functions, differentiations, differential and difference equations, optimizations with and without constraints, integrations and much more.

Written in an easy and accessible style with precise definitions and theorems, *Mathematics of Economics and Business* contains exercises and worked examples, as well as economic applications. This book will provide the reader with a comprehensive understanding of the mathematical models and tools used in both economics and business.

Frank Werner is Extraordinary Professor of Mathematics at Otto-von-Guericke University in Magdeburg, Germany.

Yuri N. Sotskov is Professor at the United Institute of Informatics Problems, National Academy of Science of Belarus, Minsk.

Mathematics of Economics and Business

Frank Werner and Yuri N. Sotskov

Routledge
Taylor & Francis Group

LONDON AND NEW YORK

First published 2006 by Routledge

Published 2017 by Routledge
2 Park Square, Milton Park, Abingdon, Oxon OX14 4RN
711 Third Avenue, New York, NY 10017, USA

Routledge is an imprint of the Taylor & Francis Group, an informa business

Copyright © 2006 Frank Werner and Yuri N. Sotskov

Typeset in Times New Roman by
Newgen Imaging Systems (P) Ltd, Chennai, India

The Open Access version of this book, available at www.tandfebooks.com,
has been made available under a Creative Commons Attribution-Non
Commercial-No Derivatives 4.0 license.

British Library Cataloguing in Publication Data
A catalogue record for this book is available from the British Library

Library of Congress Cataloging in Publication Data
A catalog record for this title has been requested

ISBN13: 9-78-0-415-33280-4 (hbk)
ISBN13: 9-78-0-415-33281-1 (pbk)

Contents

Preface

Today, a firm understanding of mathematics is essential for any serious student of economics. Students of economics need nowadays several important mathematical tools. These include calculus for functions of one or several variables as well as a basic understanding of optimization with and without constraints, e.g. linear programming plays an important role in optimizing production programs. Linear algebra is used in economic theory and econometrics. Students in other areas of economics can benefit for instance from some knowledge about differential and difference equations or mathematical problems arising in finance. The more complex economics becomes, the more deep mathematics is required and used. Today economists consider mathematics as the most important tool of economics and business. This book is not a book on mathematical economics, but a book on higher-level mathematics for economists.

Experience shows that students who enter a university and specialize in economics vary enormously in the range of their mathematical skills and aptitudes. Since mathematics is often a requirement for specialist studies in economics, we felt a need to provide as much elementary material as possible in order to give those students with weaker mathematical backgrounds the chance to get started. Using this book may depend on the skills of readers and their purposes. The book starts with very basic mathematical principles. Therefore, we have included some material that covers several topics of mathematics in school (e.g. fractions, powers, roots and logarithms in Chapter 1 or functions of a real variable in Chapter 3). So the reader can judge whether or not he (she) is sufficiently familiar with mathematics to be able to skip some of the sections or chapters.

Studying mathematics is very difficult for most students of economics and business. However, nowadays it is indeed necessary to know a lot of results of higher mathematics to understand the current economic literature and to use modern economic tools in practical economics and business. With this in mind, we wrote the book as simply as possible. On the other hand, we have presented the mathematical results strongly correct and complete, as is necessary in mathematics. The material is appropriately ordered according to mathematical requirements (while courses, e.g. in macroeconomics, often start with advanced topics such as constrained optimization for functions of several variables). On the one hand, previous results are used by later results in the text. On the other hand, current results in a chapter make it clear why previous results were included in the book.

The book is written for non-mathematicians (or rather, for those people who only want to use mathematical tools in their practice). It intends to support students in learning the basic mathematical methods that have become indispensable for a proper understanding of the current economic literature. Therefore, the book contains a lot of worked examples and economic applications. It also contains many illustrations and figures to simplify the mathematical techniques used and show how mathematical results may be used in economics and business. Some of these examples have been taken from former examinations (at the Otto-von-Guericke University of Magdeburg), and many of the exercises given at the end of each chapter have been used in the tutorials for a long time. In this book, we do not show how the mathematical results have been obtained and proved, but we show how they may be used in real-life economics and business. Therefore, proofs of theorems have been skipped (with a few exceptions) so that the volume of the book does not substantially exceed 500 pages, but in spite of the relatively short length the book includes the main mathematical subjects useful for practical economics and an efficient business.

The book should serve not only as a textbook for a course on mathematical methods for students, but also as a reference book for business people who need to use higher-level mathematics to increase profits. (Of course, one will not increase profit by solving e.g. a differential equation, but one can understand why somebody has increased profits after modelling a real process and finding a solution for it.) One of the purposes of this book is to introduce the reader to the most important mathematical methods used in current economic literature. We also provide an introduction to the close relationship between mathematical methods and problems arising in the economy. However, we have included only such economic applications as do not require an advanced knowledge of economic disciplines, since mathematics is usually taught in the first year of studies at university.

The reader needs only knowledge of elementary mathematics from secondary school to understand and use the results of the book, i.e. the content is self-sufficient for understanding. For a deeper understanding of higher mathematics used in economics, we also suggest a small selection of German and English textbooks and lecture materials listed in the literature section at the end of the book. Some of these books have been written at a comparable mathematical level (e.g. Opitz, *Mathematik*; Simon and Blume, *Mathematics for Economists*; Sydsaeter and Hammond, *Mathematics for Economic Analysis*) while others are more elementary in style (e.g. Misrahi and Sullivan, *Mathematics* and *Finite Mathematics*; Ohse, *Mathematik für Wirtschaftswissenschaftler*; Rosser, *Basic Mathematics for Economists*). The booklets (Schulz, *Mathematik für wirtschaftswissenchaftliche Studiengänge*; Werner, *Mathematics for Students of Economics and Management*) contain most important definitions, theorems of a one-year lecture course in mathematics for economists in a compact form and they sketch some basic algorithms taught in the mathematics classes for economists at the Otto-von-Guericke University of Magdeburg during recent decades. Bronstein and Semandjajew, *Taschenbuch der Mathematik*, and Eichholz and Vilkner, *Taschenbuch der Wirtschaftsmathematik*, are well-known handbooks of mathematics for students. Varian, *Intermediate Microeconomics*, is a standard textbook of intermediate microeconomics, where various economic applications of mathematics can be found.

The book is based on a two-semester course with four hours of lectures per week which the first author has given at the Otto-von-Guericke University of Magdeburg within the last ten years. The authors are indebted to many people in the writing of the book. First of all, the authors would like to thank Dr Iris Paasche, who was responsible for the tutorials from the beginning of this course in Magdeburg. She contributed many suggestions for including exercises and for improvements of the contents and, last but not least, she prepared the answers to the exercises. Moreover, the authors are grateful to Dr Günther Schulz for his support and valuable suggestions which were based on his wealth of experience in teaching students of economics and management at the Otto-von-Guericke University of Magdeburg for more than twenty years. The authors are grateful to both colleagues for their contributions.

The authors also thank Ms Natalja Leshchenko of the United Institute of Informatics Problems of the National Academy of Sciences of Belarus for reading the whole manuscript (and carefully checking the examples) and Mr Georgij Andreev of the same institute for preparing a substantial number of the figures. Moreover, many students of the International Study Programme of Economics and Management at the Otto-von-Guericke University of Magdeburg have read single chapters and contributed useful suggestions, particularly the students from the course starting in October 2002. In addition, the authors would like to express their gratitude to the former Ph.D. students Dr Nadezhda Sotskova and Dr Volker Lauff, who carefully prepared in the early stages a part of the book formerly used as printed manuscript in LATEX and who made a lot of constructive suggestions for improvements.

Although both authors have taught in English at universities for many years and during that time have published more than 100 research papers in English, we are nevertheless not native speakers. So we apologize for all the linguistic weaknesses (and hope there are not too many). Of course, for all remaining mathematical and stylistic mistakes we take full responsibility, and we will be grateful for any further comments and suggestions for improvements by readers for inclusion in future editions (e-mail address for correspondence: frank.werner@mathematik.uni-magdeburg.de). Furthermore, we are grateful to Routledge for their pleasant cooperation during the preparation of the book. The authors wish all readers success in studying mathematics.

We dedicate the book to our parents Hannelore Werner, Willi Werner, Maja N. Sotskova and Nazar F. Sotskov.

F.W.
Y.N.S.

Abbreviations

p.a.	per annum
NPV	net present value
resp.	respectively
rad	radian
l	litre
m	metre
cm	centimetre
km	kilometre
s	second
EUR	euro
LPP	linear programming problem
s.t.	subject to
bv	basic variables of a system of linear equations
nbv	non-basic variables of a system of linear equations

Notations

$\overline{A}$	negation of proposition A
$A \wedge B$	conjunction of propositions A and B
$A \vee B$	disjunction of propositions A and B
$A \Longrightarrow B$	implication (if A then B)
$A \Longleftrightarrow B$	equivalence of propositions A and B
$\bigwedge\limits_{x} A(x)$	universal proposition
$\bigvee\limits_{x} A(x)$	existential proposition
$a \in A$	a is an element of set A
$b \notin A$	b is not an element of set A
$\varnothing$	empty set
$|A|$	cardinality of a set A (if A is a finite set, then $|A|$ is equal to the number of elements in set A), the same notation is used for the determinant of a square matrix A
$P(A)$	power set of set A
$A \subseteq B$	set A is a subset of set B
$A \cup B$	union of sets A and B
$A \cap B$	intersection of sets A and B
$A \setminus B$	difference of sets A and B
$A \times B$	Cartesian product of sets A and B
$\bigtimes\limits_{i=1}^{n} A_i$	Cartesian product of sets $A_1, A_2, \ldots, A_n$
A^n	Cartesian product $\bigtimes\limits_{i=1}^{n} A_i$, where $A_1 = A_2 = \ldots = A_n = A$
$n!$	n factorial: $n! = 1 \cdot 2 \cdot \ldots \cdot (n-1) \cdot n$
$\binom{n}{k}$	binomial coefficient: $$\binom{n}{k} = \frac{n!}{k! \cdot (n-k)!}$$
$n = 1, 2, \ldots, k$	equalities $n = 1, n = 2, \ldots, n = k$
$\mathbb{N}$	set of all natural numbers: $\mathbb{N} = \{1, 2, 3, \ldots\}$
$\mathbb{N}_0$	union of set $\mathbb{N}$ with number zero: $\mathbb{N}_0 = \mathbb{N} \cup \{0\}$
$\mathbb{Z}$	union of set $\mathbb{N}_0$ with the set of all negative integers

$\mathbb{Q}$	set of all rational numbers, i.e. set of all fractions p/q with $p \in \mathbb{Z}$ and $q \in \mathbb{N}$
$\mathbb{R}$	set of all real numbers
$\mathbb{R}_+$	set of all non-negative real numbers
(a, b)	open interval between a and b
$[a, b]$	closed interval between a and b
$\pm$	denotes two cases of a mathematical term: the first one with sign $+$ and the second one with sign $-$
$\mp$	denotes two cases of a mathematical term: the first one with sign $-$ and the second one with sign $+$
$\|a\|$	absolute value of number $a \in \mathbb{R}$
$\approx$	sign of approximate equality, e.g. $\sqrt{2} \approx 1.41$
$\neq$	sign 'not equal'
π	irrational number equal to the circle length divided by the diameter length: $\pi \approx 3.14159...$
e	Euler's number: $e \approx 2.71828...$
∞	infinity
$\sqrt{a}$	square root of a
exp	notation used for the exponential function with base e: $y = \exp(x) = e^x$
log	notation used for the logarithm: if $y = \log_a x$, then $a^y = x$
lg	notation used for the logarithm with base 10: $\lg x = \log_{10} x$
ln	notation used for the logarithm with base e: $\ln a = \log_e x$
$\sum$	summation sign: $$\sum_{i=1}^{n} a_i = a_1 + a_2 + \cdots + a_n$$
$\prod$	product sign: $$\prod_{i=1}^{n} a_i = a_1 \cdot a_2 \cdot \ldots \cdot a_n$$
i	imaginary unit: $i^2 = -1$
$\mathbb{C}$	set of all complex numbers: $z = a + bi$, where a and b are real numbers
$\|z\|$	modulus of number $z \in \mathbb{C}$
$\{a_n\}$	sequence: $\{a_n\} = a_1, a_2, a_3, \ldots, a_n, \ldots$
$\{s_n\}$	series, i.e. the sequence of partial sums of a sequence $\{a_n\}$
lim	limit sign
aRb	a is related to b by the binary relation R
$a\not\!Rb$	a is not related to b by the binary relation R
R^{-1}	inverse relation of R
$S \circ R$	composite relation of R and S
$f : A \to B$	mapping or function $f \in A \times B$: f is a binary relation which assigns to $a \in A$ exactly one $b \in B$
$b = f(a)$	b is the image of a assigned by mapping f
f^{-1}	inverse mapping or function of f
$g \circ f$	composite mapping or function of f and g
D_f	domain of a function f of $n \geq 1$ real variables
R_f	range of a function f of $n \geq 1$ real variables
$y = f(x)$	$y \in \mathbb{R}$ is the function value of $x \in \mathbb{R}$, i.e. the value of function f at point x

$\deg P$	degree of polynomial P
$x \to x_0$	x tends to x_0
$x \to x_0 + 0$	x tends to x_0 from the right side
$x \to x_0 - 0$	x tends to x_0 from the left side
f'	derivative of function f
$f'(x), y'(x)$	derivative of function f with $y = f(x)$ at point x
$f''(x), y''(x)$	second derivative of function f with $y = f(x)$ at point x
$dy, \quad df$	differential of function f with $y = f(x)$
$\equiv$	sign of identical equality, e.g. $f(x) \equiv 0$ means that equality $f(x) = 0$ holds for any value x
$\rho_f(x_0)$	proportional rate of change of function f at point x_0
$\varepsilon_f(x_0)$	elasticity of function f at point x_0
$\int$	integral sign
$\mathbb{R}^n$	n-dimensional Euclidean space, i.e. set of all real n-tuples
$\mathbb{R}^n_+$	set of all non-negative real n-tuples
$\mathbf{a}$	vector: ordered n-tuple of real numbers $a_1, a_2, \ldots, a_n$ corresponding to a matrix with one column
$\mathbf{a}^\mathsf{T}$	transposed vector of vector $\mathbf{a}$
$|\mathbf{a}|$	Euclidean length or norm of vector $\mathbf{a}$
$|\mathbf{a} - \mathbf{b}|$	Euclidean distance between vectors $\mathbf{a} \in \mathbb{R}^n$ and $\mathbf{b} \in \mathbb{R}^n$
$\mathbf{a} \perp \mathbf{b}$	means that vectors $\mathbf{a}$ and $\mathbf{b}$ are orthogonal
$\dim V$	dimension of the vector space V
$A_{m,n}$	matrix of order (dimension) $m \times n$
A^T	transpose of matrix A
A^n	nth power of a square matrix A
$\det A, \quad (\text{or } |A|)$	determinant of a matrix A
A^{-1}	inverse matrix of matrix A
$\mathrm{adj}\,(A)$	adjoint of matrix A
$r(A)$	rank of matrix A
$x_1, x_2, \ldots, x_n \geq 0$	denotes the inequalities $x_1 \geq 0, x_2 \geq 0, \ldots, x_n \geq 0$
$R_i \in \{\leq, =, \geq\}$	means that one of these signs hold in the ith constraint of a system of linear inequalities
$z \to \min!$	indicates that the value of function z should become minimal for the desired solution
$z \to \max!$	indicates that the value of function z should become maximal for the desired solution
$f_x(x^0, y^0)$	partial derivative of function f with $z = f(x, y)$ with respect to x at point (x^0, y^0)
$f_{x_i}(\mathbf{x}^0)$	partial derivative of function f with $z = f(x_1, x_2, \ldots, x_n)$ with respect to x_i at point $\mathbf{x}^0 = (x_1^0, x_2^0, \ldots, x_n^0)$
$\mathbf{grad}\ f(\mathbf{x}^0)$	gradient of function f at point $\mathbf{x}^0$
$\rho_{f,x_i}(\mathbf{x}^0)$	partial rate of change of function f with respect to x_i at point $\mathbf{x}^0$
$\varepsilon_{f,x_i}(\mathbf{x}^0)$	partial elasticity of function f with respect to x_i at point $\mathbf{x}^0$
$H_f(\mathbf{x}^0)$	Hessian matrix of function f at point $\mathbf{x}^0$
∎	Q.E.D. (quod erat demonstrandum – 'that which was to be demonstrated')

1 Introduction

In this chapter, an overview on some basic topics in mathematics is given. We summarize elements of logic and basic properties of sets and operations with sets. Some comments on basic combinatorial problems are given. We also include the main important facts concerning number systems and summarize rules for operations with numbers.

1.1 LOGIC AND PROPOSITIONAL CALCULUS

This section deals with basic elements of mathematical logic. In addition to propositions and logical operations, we discuss types of mathematical proofs.

1.1.1 Propositions and their composition

Let us consider the following four statements A, B, C and D.

A Number 126 is divisible by number 3.
B Equality $5 \cdot 11 = 65$ holds.
C Number 11 is a prime number.
D On 1 July 1000 it was raining in Magdeburg.

Obviously, the statements A and C are true. Statement B is false since $5 \cdot 11 = 55$. Statement D is either true or false but today probably nobody knows. For each of the above statements we have only two possibilities concerning their truth values (to be true or to be false). This leads to the notion of a proposition introduced in the following definition.

Definition 1.1   A statement which is either true or false is called a *proposition*.

Remark   For a proposition, there are no other truth values than 'true' (T) or 'false' (F) allowed. Furthermore, a proposition cannot have both truth values 'true' and 'false' (*principle of excluded contradiction*).

Next, we consider logical operations. We introduce the negation of a proposition and connect different propositions. Furthermore, the truth value of such *compound propositions* is investigated.

Definition 1.2 A proposition $\overline{A}$ (read: not A) is called the *negation* of proposition A. Proposition $\overline{A}$ is true if A is false. Proposition $\overline{A}$ is false if A is true.

One can illustrate Definition 1.2 by a so-called *truth table*. According to Definition 1.2, the truth table of the *negation* of proposition A is as follows:

A	T	F
$\overline{A}$	F	T

Considering the negations of the propositions A, B and C, we obtain:

$\overline{A}$ Number 126 is not divisible by the number 3.
$\overline{B}$ Equality $5 \cdot 11 = 65$ does not hold.
$\overline{C}$ The number 11 is not a prime number.

Propositions $\overline{A}$ and $\overline{C}$ are false and $\overline{B}$ is true.

Definition 1.3 The proposition $A \wedge B$ (read: A and B) is called a *conjunction*. Proposition $A \wedge B$ is true only if propositions A and B are both true. Otherwise, $A \wedge B$ is false.

According to Definition 1.3, the truth table of the *conjunction* $A \wedge B$ is as follows:

A	T	T	F	F
B	T	F	T	F
$A \wedge B$	T	F	F	F

Definition 1.4 The proposition $A \vee B$ (read: A or B) is called a *disjunction*. Proposition $A \vee B$ is false only if propositions A and B are both false. Otherwise, $A \vee B$ is true.

According to Definition 1.4, the truth table of the *disjunction* $A \vee B$ is as follows:

A	T	T	F	F
B	T	F	T	F
$A \vee B$	T	T	T	F

The symbol $\vee$ stands for the 'inclusive or' which allows that both propositions are true (in contrast to the 'exclusive or', where the latter is not possible).

Example 1.1   Consider the propositions M and P.

M In 2003 Magdeburg was the largest city in Germany.
P In 2003 Paris was the capital of France.

Although proposition P is true, the conjunction $M \wedge P$ is false, since Magdeburg was not the largest city in Germany in 2003 (i.e. proposition M is false). However, the disjunction $M \vee P$ is true, since (at least) one of the two propositions is true (namely proposition P).

Definition 1.5   The proposition $A \implies B$ (read: if A then B) is called an *implication*. Only if A is true and B is false, is the proposition $A \implies B$ defined to be false. In all remaining cases, the proposition $A \implies B$ is true.

According to Definition 1.5, the truth table of the *implication* $A \implies B$ is as follows:

A	T	T	F	F
B	T	F	T	F
$A \implies B$	T	F	T	T

For the implication $A \implies B$, proposition A is called the hypothesis and proposition B is called the conclusion. An implication is also known as a *conditional statement*. Next, we give an illustration of the implication. A student says: If the price of the book is at most 20 EUR, I will buy it. This is an implication $A \implies B$ with

A The price of the book is at most 20 EUR.
B The student will buy the book.

In the first case of the four possibilities in the above truth table (second column), the student confirms the validity of the implication $A \implies B$ (due to the low price of no more than 20 EUR, the student will buy the book). In the second case (third column), the implication is false since the price of the book is low but the student will not buy the book. The truth value of an implication is also true if A is false but B is true (fourth column). In our example, this means that it is possible that the student will also buy the book in the case of an unexpectedly high price of more than 20 EUR. (This does not contradict the fact that the student certainly will buy the book for a price lower than or equal to 20 EUR.) In the fourth case (fifth column of the truth table), the high price is the reason that the student will not buy the book. So in all four cases, the definition of the truth value corresponds with our intuition.

Example 1.2   Consider the propositions A and B defined as follows:

A The natural number n is divisible by 6.
B The natural number n is divisible by 3.

We investigate the implication $A \implies B$. Since each of the propositions A and B can be true and false, we have to consider four possibilities.

If n is a multiple of 6 (i.e. $n \in \{6, 12, 18, \ldots\}$), then both A and B are true. According to Definition 1.5, the implication $A \implies B$ is true. If n is a multiple of 3 but not of 6 (i.e. $n \in \{3, 9, 15, \ldots\}$), then A is false but B is true. Therefore, implication $A \implies B$ is true. If n is not a multiple of 3 (i.e. $n \in \{1, 2, 4, 5, 7, 8, 10, \ldots\}$), then both A and B are false, and by Definition 1.5, implication $A \implies B$ is true. It is worth noting that the case where A is true but B is false cannot occur, since no natural number which is divisible by 6 is not divisible by 3.

Remark   For an implication $A \implies B$, one can also say:

(1) *A implies B*;
(2) *from A* it *follows B*;
(3) *A* is *sufficient* for *B*;
(4) *B* is *necessary* for *A*;
(5) *A* is true *only if B* is true;
(6) *if A* is true, then *B* is true.

The latter four formulations are used in connection with the presentation of mathematical theorems and their proof.

Example 1.3   Consider the propositions

H Claudia is happy today.
E Claudia does not have an examination today.

Then the implication $H \implies E$ means: If Claudia is happy today, she does not have an examination today. Therefore, a necessary condition for Claudia to be happy today is that she does not have an examination today.

In the case of the opposite implication $E \implies H$, a sufficient condition for Claudia to be happy today is that she does not have an examination today.

If both implications $H \implies E$ and $E \implies H$ are true, it means that Claudia is happy today if and only if she does not have an examination today.

Definition 1.6   The proposition $A \iff B$ (read: A is equivalent to B) is called *equivalence*. Proposition $A \iff B$ is true if both propositions A and B are true or propositions A and B are both false. Otherwise, proposition $A \iff B$ is false.

According to Definition 1.6, the truth table of the *equivalence* $A \iff B$ is as follows:

A	T	T	F	F
B	T	F	T	F
$A \iff B$	T	F	F	T

Remark   For an equivalence $A \Longleftrightarrow B$, one can also say

(1) A holds *if and only if* B holds;
(2) A is *necessary and sufficient* for B.

For a compound proposition consisting of more than two propositions, there is a hierarchy of the logical operations as follows. The negation of a proposition has the highest priority, then both the conjunction and the disjunction have the second highest priority and finally the implication and the equivalence have the lowest priority. Thus, the proposition

$$\overline{A} \wedge B \Longleftrightarrow C$$

may also be written as

$$\left[(\overline{A}) \wedge B\right] \Longleftrightarrow C.$$

By means of a truth table we can investigate the truth value of arbitrary compound propositions.

Example 1.4   We investigate the truth value of the compound proposition

$$\overline{A \vee B} \Longrightarrow \left(\overline{B} \Longrightarrow \overline{A}\right).$$

One has to consider four possible combinations of the truth values of A and B (each of the propositions A and B can be either true or false):

	T	T	F	F
A				
B	T	F	T	F
$A \vee B$	T	T	T	F
$\overline{A \vee B}$	F	F	F	T
$\overline{A}$	F	F	T	T
$\overline{B}$	F	T	F	T
$\overline{B} \Longrightarrow \overline{A}$	T	F	T	T
$\overline{A \vee B} \Longrightarrow \left(\overline{B} \Longrightarrow \overline{A}\right)$	**T**	**T**	**T**	**T**

In Example 1.4, the implication is always true, independently of the truth values of the individual propositions. This leads to the following definition.

Definition 1.7   A compound proposition which is true independently of the truth values of the individual propositions is called a *tautology*. A compound proposition being always false is called a *contradiction*.

Example 1.5 We investigate whether the implication

$$\left(\overline{A \Longrightarrow B}\right) \Longrightarrow \left(\overline{A} \Longrightarrow \overline{B}\right) \tag{1.1}$$

is a tautology. As in the previous example, we have to consider four combinations of truth values of propositions A and B. This yields the following truth table:

A	T	T	F	F
B	T	F	T	F
$A \Longrightarrow B$	T	F	T	T
$\overline{A \Longrightarrow B}$	F	T	F	F
$\overline{A}$	F	F	T	T
$\overline{B}$	F	T	F	T
$\overline{A} \Longrightarrow \overline{B}$	T	T	F	T
$(\overline{A \Longrightarrow B}) \Longrightarrow (\overline{A} \Longrightarrow \overline{B})$	T	T	T	T

Independently of the truth values of A and B, the truth value of the implication considered is true. Therefore, implication (1.1) is a tautology.

Some further tautologies are presented in the following theorem.

THEOREM 1.1 The following propositions are tautologies:

(1) $A \wedge B \Longleftrightarrow B \wedge A, \qquad A \vee B \Longleftrightarrow B \vee A$
(commutative laws of conjunction and disjunction);
(2) $(A \wedge B) \wedge C \Longleftrightarrow A \wedge (B \wedge C), \qquad (A \vee B) \vee C \Longleftrightarrow A \vee (B \vee C)$
(associative laws of conjunction and disjunction);
(3) $(A \wedge B) \vee C \Longleftrightarrow (A \vee C) \wedge (B \vee C), \qquad (A \vee B) \wedge C \Longleftrightarrow (A \wedge C) \vee (B \wedge C)$
(distributive laws).

Remark The negation of a disjunction of two propositions is a conjunction and, analogously, the negation of a conjunction of two propositions is a disjunction. We get

$$\overline{A \vee B} \Longleftrightarrow \overline{A} \wedge \overline{B} \qquad \text{and} \qquad \overline{A \wedge B} \Longleftrightarrow \overline{A} \vee \overline{B} \qquad \text{(de Morgan's laws)}.$$

PROOF De Morgan's laws can be proved by using truth tables. Let us prove the first equivalence. Since each of the propositions A and B has two possible truth values T and F, we have to consider four combinations of the truth values of A and B:

A	T	T	F	F
B	T	F	T	F
$A \vee B$	T	T	T	F
$\overline{A \vee B}$	F	F	F	T
$\overline{A}$	F	F	T	T
$\overline{B}$	F	T	F	T
$\overline{A} \wedge \overline{B}$	F	F	F	T

If we compare the fourth and last rows, we get identical truth values for each of the four possibilities and therefore the first law of de Morgan has been proved. ∎

The next theorem presents some tautologies which are useful for different types of mathematical proofs.

THEOREM 1.2 The following propositions are tautologies:

(1) $(A \Longleftrightarrow B) \Longleftrightarrow \left[(A \Longrightarrow B) \wedge (B \Longrightarrow A)\right]$;

(2) $(A \Longrightarrow B) \wedge (B \Longrightarrow C) \Longrightarrow (A \Longrightarrow C)$;

(3) $(A \Longrightarrow B) \Longleftrightarrow (\overline{B} \Longrightarrow \overline{A}) \Longleftrightarrow \overline{A \wedge \overline{B}}$.

PROOF We prove only part (1) and have to consider four possible combinations of the truth values of propositions A and B. This yields the following truth table:

A	T	T	F	F
B	T	F	T	F
$A \Longrightarrow B$	T	F	T	T
$B \Longrightarrow A$	T	T	F	T
$(A \Longrightarrow B) \wedge (B \Longrightarrow A)$	**T**	**F**	**F**	**T**
$A \Longleftrightarrow B$	**T**	**F**	**F**	**T**

The latter two rows give identical truth values for all four possibilities and thus it has been proved that the equivalence of part (1) is a tautology. ∎

Part (1) of Theorem 1.2 can be used to prove the logical equivalence of two propositions, i.e. we can prove both implications $A \Longrightarrow B$ and $B \Longrightarrow A$ separately. Part (2) of Theorem 1.2 is known as the *transitivity* property. The equivalences of part (3) of Theorem 1.2 are used later to present different types of mathematical proof.

1.1.2 Universal and existential propositions

In this section, we consider propositions that depend on the value(s) of one or several variables.

Definition 1.8 A proposition depending on one or more variables is called an *open proposition* or a *propositional function*.

We denote by $A(x)$ a proposition A that depends on variable x. Let $A(x)$ be the following open proposition: $x^2 + x - 6 = 0$. For $x = 2$, the proposition $A(2)$ is true since $2^2 + 2 - 6 = 0$. On the other hand, for $x = 0$, the proposition $A(0)$ is false since $0^2 + 0 - 6 \neq 0$.

Next, we introduce the notions of a universal and an existential proposition.

Definition 1.9   A conjunction

$$\bigwedge_{x} A(x)$$

of propositions $A(x)$, where x can take given values, is called a *universal proposition*. The universal proposition is true if the propositions $A(x)$ are true for all x. If at least one of these propositions is false, the universal proposition is false.

Proposition

$$\bigwedge_{x} A(x)$$

may be written in an equivalent form:

$$\forall x: \quad A(x).$$

Definition 1.10   A disjunction

$$\bigvee_{x} A(x)$$

of propositions $A(x)$ is called an *existential proposition*. If at least one x exists such that $A(x)$ is true, the existential proposition is true. If such an x does not exist, the existential proposition is false.

Proposition

$$\bigvee_{x} A(x)$$

may be written in an equivalent form:

$$\exists x: \quad A(x).$$

If variable x can take infinitely many values, the universal and existential propositions are compounded by infinitely many propositions.

Remark   The negation of an existential proposition is always a universal proposition and vice versa. In particular, we get as a generalization of de Morgan's laws that

$$\overline{\bigvee_x A(x)} \iff \bigwedge_x \overline{A(x)} \qquad \text{and} \qquad \overline{\bigwedge_x A(x)} \iff \bigvee_x \overline{A(x)}$$

are tautologies.

Example 1.6   Let x be a real number and

$$A(x): \quad x^2 + 3x - 4 = 0.$$

Then the existential proposition

$$\bigvee_x A(x)$$

is true since there exists a real x for which $x^2 + 3x - 4 = 0$, namely for the numbers $x = 1$ and $x = -4$ the equality $x^2 + 3x - 4 = 0$ is satisfied, i.e. propositions $A(1)$ and $A(-4)$ are true.

Example 1.7   Let x be a real number and consider

$$A(x): \quad x^2 \geq 0.$$

Then the universal proposition

$$\bigwedge_x A(x)$$

is true since the square of any real number is non-negative.

1.1.3   Types of mathematical proof

A mathematical theorem can be formulated as an implication $A \implies B$ (or as several implications) where A represents a proposition or a set of propositions called the *hypothesis* or the *premises* ('what we know') and B represents a proposition or a set of propositions that are called the *conclusions* ('what we want to know'). One can prove such an implication in different ways.

Direct proof

We show a series of implications $C_i \implies C_{i+1}$ for $i = 0, 1, \ldots, n-1$ with $C_0 = A$ and $C_n = B$. By the repeated use of the transitivity property given in part (2) of Theorem 1.2, this means that $A \implies B$ is true. As a consequence, if the premises A and the implication $A \implies B$ are true, then conclusion B must be true, too. Often several cases have to be considered in order to prove an implication. For instance, if a proposition A is equivalent to a disjunction of propositions, i.e.

$$A \iff A_1 \vee A_2 \vee \ldots \vee A_n,$$

one can prove the implication $A \implies B$ by showing the implications

$$A_1 \implies B, \quad A_2 \implies B, \quad \ldots, \quad A_n \implies B.$$

Indirect proof

According to part (3) of Theorem 1.2, instead of proving an implication $A \implies B$ directly, one can prove an implication in two other variants indirectly.

(1) *Proof of contrapositive.*   For $A \implies B$, it is equivalent (see part (3) of Theorem 1.2) to show $\overline{B} \implies \overline{A}$. As a consequence, if the conclusion B does not hold ($\overline{B}$ is true) and $\overline{B} \implies \overline{A}$, then the premises A cannot hold ($\overline{A}$ is true). We also say that $\overline{B} \implies \overline{A}$ is a *contrapositive* of implication $A \implies B$.

(2) *Proof by contradiction.*   This is another variant of an indirect proof. We know (see part (3) of Theorem 1.2) that

$$(A \implies B) \iff \overline{A \wedge \overline{B}}.$$

Thus, we investigate $A \wedge \overline{B}$ which must lead to a *contradiction*, i.e. in the latter case we have shown that proposition $\overline{A \wedge \overline{B}}$ is true.

Next, we illustrate the three variants of a proof mentioned above by a few examples.

Example 1.8   Given are the following propositions A and B:

$$A: x \neq 1.$$

$$B: x^3 + 4x + \frac{1}{x-1} - 3 \neq 2.$$

We prove implication $A \implies B$ by a direct proof.
To this end, we consider two cases, namely propositions A_1 and A_2:

$$A_1: x < 1.$$

$$A_2: x > 1.$$

It is clear that $A \iff A_1 \vee A_2$. Let us prove both implications $A_1 \implies B$ and $A_2 \implies B$.

$A_1 \Longrightarrow B$. Since inequalities $x^3 < 1, 4x < 4$ and $1/(x-1) < 0$ hold for $x < 1$, we obtain (using the rules for adding inequalities, see also rule (5) for inequalities in Chapter 1.4.1) the implication

$$x < 1 \quad \Longrightarrow \quad x^3 + 4x + \frac{1}{x-1} < 5.$$

Moreover,

$$x^3 + 4x + \frac{1}{x-1} < 5 \quad \Longrightarrow \quad x^3 + 4x + \frac{1}{x-1} - 3 \neq 2,$$

i.e. $A_1 \Longrightarrow B$.

$A_2 \Longrightarrow B$. Since inequalities $x^3 > 1, 4x > 4$ and $1/(x-1) > 0$ hold for $x > 1$, we obtain (using the rules for adding inequalities) the implication

$$x > 1 \quad \Longrightarrow \quad x^3 + 4x + \frac{1}{x-1} > 5.$$

Moreover,

$$x^3 + 4x + \frac{1}{x-1} > 5 \quad \Longrightarrow \quad x^3 + 4x + \frac{1}{x-1} - 3 \neq 2,$$

i.e. $A_2 \Longrightarrow B$.

Due to $A \Longleftrightarrow A_1 \vee A_2$, we have proved the implication $A \Longrightarrow B$.

Example 1.9   We prove the implication $A \Longrightarrow B$, where

$A :$ x is a positive real number.

$B :$ $x + \dfrac{16}{x} \geq 8.$

Applying the proof by contradiction, we assume that proposition $A \wedge \overline{B}$ is true, i.e.

$A \wedge \overline{B} :$ x is a positive real number and $x + \dfrac{16}{x} < 8.$

Then we get the following series of implications:

$$\left(x + \frac{16}{x} < 8 \right) \wedge (x > 0) \Longrightarrow x^2 + 16 < 8x$$

$$x^2 + 16 < 8x \Longrightarrow x^2 - 8x + 16 < 0$$

$$x^2 - 8x + 16 < 0 \Longrightarrow (x-4)^2 < 0.$$

Now we have obtained the contradiction $(x-4)^2 < 0$ since the square of any real number is non-negative. Hence, proposition $A \wedge \overline{B}$ is false and therefore the implication $A \Longrightarrow B$ is true. Notice that the first of the three implications above holds due to the assumption $x > 0$.

Example 1.10   We prove $A \Longleftrightarrow B$, where

   $A : x$ is an even natural number.

   $B : x^2$ is an even natural number.

According to Theorem 1.2, part (1), $A \Longleftrightarrow B$ is equivalent to proving both implications $A \Longrightarrow B$ and $B \Longrightarrow A$.
(a) $A \Longrightarrow B$. Let x be an even natural number, i.e. x can be written as $x = 2n$ with n being some natural number. Then

$$x^2 = (2n)^2 = 4n^2 = 2(2n^2),$$

i.e. x^2 has the form $2k$ and is therefore an even natural number too.
(b) $B \Longrightarrow A$. We use the indirect proof by showing $\overline{A} \Longrightarrow \overline{B}$. Assume that x is an odd natural number, i.e. $x = 2n + 1$ with n being some natural number. Then

$$x^2 = (2n + 1)^2 = 4n^2 + 4n + 1 = 2(2n^2 + 2n) + 1.$$

Therefore, x^2 has the form $2k + 1$ and is therefore an odd natural number too, and we have proved implication $\overline{A} \Longrightarrow \overline{B}$ which is equivalent to implication $B \Longrightarrow A$.

Example 1.11   Consider the propositions

   $A : \quad -x^2 + 3x \geq 0 \quad$ and $\quad B : \quad x \geq 0.$

We prove the implication $A \Longrightarrow B$ by all three types of mathematical proofs discussed before.
(1) To give a direct proof, we suppose that $-x^2 + 3x \geq 0$. The latter equation can be rewritten as $3x \geq x^2$. Since $x^2 \geq 0$, we obtain $3x \geq x^2 \geq 0$. From $3x \geq 0$, we get $x \geq 0$, i.e. we have proved $A \Longrightarrow B$.
(2) To prove the contrapositive, we have to show that $\overline{B} \Longrightarrow \overline{A}$. We therefore suppose that $x < 0$. Then $3x < 0$ and $-x^2 + 3x < 0$ since the term $-x^2$ is always non-positive.
(3) For a proof by contradiction, we suppose that $A \wedge \overline{B}$ is true, which corresponds to the following proposition. There exists an x such that

   $-x^2 + 3x \geq 0 \quad$ and $\quad x < 0.$

However, if $x < 0$, we obtain

   $-x^2 + 3x < -x^2 \leq 0.$

The inequality $-x^2 + 3x < 0$ is a contradiction of the above assumption. Therefore, we have proved that proposition $A \wedge \overline{B}$ is true.

We finish this section with a type of proof that can be used for universal propositions of a special type.

Proof by induction

This type of proof can be used for propositions

$$\bigwedge_n A(n),$$

where $n = k, k + 1, \ldots$, and k is a natural number. The proof consists of two steps, namely the initial and the inductive steps:

(1) *Initial step.* We prove that the proposition is true for an initial natural number $n = k$, i.e. $A(k)$ is true.
(2) *Inductive step.* To show a sequence of (infinitely many) implications $A(k) \Longrightarrow A(k + 1), A(k + 1) \Longrightarrow A(k + 2)$, and so on, it is sufficient to prove once the implication

$$A(n) \Longrightarrow A(n + 1)$$

for an arbitrary $n \in \{k, k + 1, \ldots\}$. The hypothesis that $A(n)$ is true for some n is also denoted as *inductive hypothesis*, and we prove in the inductive step that then also $A(n+1)$ must be true. In this way, we can conclude that, since $A(k)$ is true by the initial step, $A(k + 1)$ must be true by the inductive step. Now, since $A(k + 1)$ is true, it follows again by the inductive step that $A(k + 2)$ must be true and so on.

The proof by induction is illustrated by the following two examples.

Example 1.12   We want to prove by induction that

$$A(n): \quad \sum_{i=1}^{n} \frac{1}{i(i + 1)} = \frac{n}{n + 1}$$

is true for any natural number n.
In the initial step, we consider $A(1)$ and obtain

$$A(1): \quad \sum_{i=1}^{1} \frac{1}{i(i + 1)} = \frac{1}{2}$$

which is obviously true. In the inductive step, we have to prove that, if $A(n)$ is true for some natural number $n = k$, then also

$$A(n + 1): \quad \sum_{i=1}^{n+1} \frac{1}{i(i + 1)} = \frac{n + 1}{n + 2}$$

has to be true for the next natural number $n + 1$. We consider the inductive hypothesis $A(n)$ and add $1/[(n + 1)(n + 2)]$ on both sides. This yields

$$\sum_{i=1}^{n} \frac{1}{i(i + 1)} + \frac{1}{(n + 1)(n + 2)} = \frac{n}{n + 1} + \frac{1}{(n + 1)(n + 2)}$$

$$\sum_{i=1}^{n+1} \frac{1}{i(i + 1)} = \frac{n(n + 2) + 1}{(n + 1)(n + 2)}$$

$$\sum_{i=1}^{n+1} \frac{1}{i(i + 1)} = \frac{n^2 + 2n + 1}{(n + 1)(n + 2)}$$

$$\sum_{i=1}^{n+1} \frac{1}{i(i + 1)} = \frac{(n + 1)^2}{(n + 1)(n + 2)}$$

$$\sum_{i=1}^{n+1} \frac{1}{i(i + 1)} = \frac{n + 1}{n + 2}.$$

Thus, we have shown that $A(1)$ holds and that implication $A(n) \Longrightarrow A(n + 1)$ holds for an arbitrary natural number. Therefore, the proposition $A(n)$ is true for all natural numbers n.

A proof by induction can also be used when proving the validity of certain inequalities.

Example 1.13   We prove by induction that

$$A(n): \qquad 2^n > n$$

is true. In the initial step, we obviously get for $n = 1$

$$A(1): \qquad 2^1 = 2 > 1,$$

i.e. $A(1)$ is true. In the inductive step, we have to prove implication $A(n) \Longrightarrow A(n + 1)$ with

$$A(n + 1): \qquad 2^{n+1} > n + 1,$$

i.e. we show that

$$2^n > n \quad \Longrightarrow \quad 2^{n+1} > n + 1.$$

Multiplying both sides of the inductive hypothesis $A(n)$ (which is assumed to be true) by 2 yields

$$2 \cdot 2^n > 2n.$$

From $n \geq 1$, we get

$$2n \geq n + 1.$$

Combining both inequalities gives

$$2^{n+1} > n + 1,$$

and thus we can conclude that $A(n)$ is true for all natural numbers n.

1.2   SETS AND OPERATIONS ON SETS

In this section, we introduce the basic notion of a set and discuss operations on sets.

1.2.1   Basic definitions

A *set* is a fundamental notion of mathematics, and so there is no definition of a set by other basic mathematical notions. A set may be considered as a collection of *distinct* objects which are called the elements of the set. For each object, it can be uniquely decided whether it is an element of the set or not. We write:

$a \in A$: a is an element of set A;
$b \notin A$: b is not an element of set A.

A set can be given by

(1) *enumeration*, i.e. $A = \{a_1, a_2, \ldots, a_n\}$ which means that set A consists of the elements $a_1, a_2, \ldots, a_n$, or
(2) *description*, i.e. $A = \{a \mid$ property P$\}$ which means that set A contains all elements with property P.

Example 1.14   Let $A = \{3, 5, 7, 9, 11\}$. Here set A is given by enumeration and it contains as elements the five numbers 3, 5, 7, 9 and 11. Set A can also be given by description as follows:

$$A = \{x \mid (3 \leq x \leq 11) \wedge (x \text{ is an odd integer})\}.$$

Definition 1.11   A set with a finite number of elements is called a *finite* set. The number of elements of a finite set A is called the *cardinality* of a set A and is denoted by $|A|$.
A set with an infinite number of elements is called an *infinite* set.

Finite sets are always *denumerable* (or *countable*), i.e. their elements can be counted one by one in the sequence $1, 2, 3, \ldots$. Infinite sets are either denumerable (e.g. the set of all even positive integers or the set of all rational numbers) or not denumerable (e.g. the set of all real numbers).

Definition 1.12   A set B is called a *subset* of a set A (in symbols: $B \subseteq A$) if every element of B is also an element of A.

Two sets A and B are said to be *equal* (in symbols: $A = B$) if both inclusions $A \subseteq B$ and $B \subseteq A$ hold.

A set A is called an *empty set* (in symbols: $A = \emptyset$) if A contains no element.

If B is a subset of A, one can alternatively say that set B is contained in set A or that set A includes set B. In order to prove that two sets A and B are equal, we either prove both inclusions $A \subseteq B$ and $B \subseteq A$, or alternatively we can prove that some element x is contained in set A *if and only if* it is contained in set B. The latter can be done by a series of logical equivalences.

Example 1.15   Let $A = \{1, 3, 5\}$. We calculate the number of subsets of set A.

We get the three one-element sets $A_1 = \{1\}$, $A_2 = \{3\}$, $A_3 = \{5\}$, the three two-element sets $A_4 = \{1, 3\}$, $A_5 = \{1, 5\}$, $A_6 = \{3, 5\}$ and the two limiting cases $\emptyset$ and A. Thus we have found eight subsets of set A.

Definition 1.13   The set of all subsets of a set A is called the *power set* of set A and is denoted by $P(A)$. The limiting cases $\emptyset$ and A itself belong to set $P(A)$.

The number of elements of the power set of a finite set is given by the following theorem.

THEOREM 1.3   The cardinality of set $P(A)$ is given by $|P(A)| = 2^{|A|}$.

For the set A given in Example 1.15, we have $|A| = 3$. According to Theorem 1.3, $|P(A)| = 2^3 = 8$ is obtained what we have already found by a direct enumeration.

1.2.2   Operations on sets

In this section, we discuss some operations on sets.

Definition 1.14   The set of all elements which belong either only to a set A or only to a set B or to both sets A and B is called the *union* of the two sets A and B (in symbols $A \cup B$, read: A union B):

$$A \cup B = \{x \mid x \in A \vee x \in B\}.$$

Set $A \cup B$ contains all elements that belong *at least* to one of the sets A and B.

Definition 1.15   The set of all elements belonging to both sets A and B is called the *intersection* of the two sets A and B (in symbols $A \cap B$, read: A intersection B):

$$A \cap B = \{x \mid x \in A \wedge x \in B\}.$$

Two sets A and B are called *disjoint* if $A \cap B = \emptyset$.

Definition 1.16   The set of all elements belonging to a set A but not to a set B is called the *difference set* of A and B (in symbols $A \setminus B$, read: A minus B):

$$A \setminus B = \{x \mid x \in A \wedge x \notin B\}.$$

If $B \subseteq A$, then the set $A \setminus B$ is called the *complement* of B with respect to A (in symbols $\bar{B}_A$).

Definitions 1.14, 1.15 and 1.16 are illustrated by so-called *Venn* diagrams in Figure 1.1, where sets are represented by areas in the plane. The union, intersection and difference of the two sets as well as the complement of a set are given by the dashed areas. For the difference set $A \setminus B$, we have the following property:

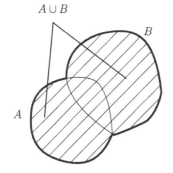

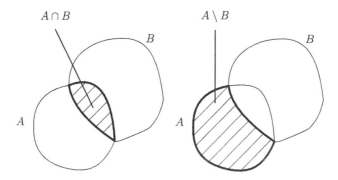

Figure 1.1 Venn diagrams for the union, intersection and difference of sets A and B.

THEOREM 1.4   Let A and B be arbitrary sets. Then:

$$A \setminus B = A \setminus (A \cap B) = (A \cup B) \setminus B.$$

In Theorem 1.4 sets A and B do not need to be finite sets. Theorem 1.4 can be illustrated by the two Venn diagrams given in Figure 1.2.

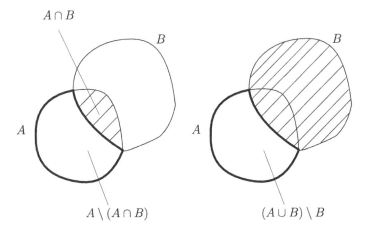

Figure 1.2 Illustration of Theorem 1.4.

Example 1.16   Let

$$A = \{3, 5, 7\} \qquad \text{and} \qquad B = \{2, 3, 4, 7, 8\}.$$

Then

$$A \cup B = \{2, 3, 4, 5, 7, 8\}, \quad A \cap B = \{3, 7\},$$
$$A \setminus B = \{5\}, \qquad\qquad B \setminus A = \{2, 4, 8\}.$$

Example 1.17   Let A, B, C be arbitrary sets. We prove that

$$(A \cup B) \setminus C = (A \setminus C) \cup (B \setminus C).$$

As mentioned before, equality of two sets can be shown by proving that an element x belongs to the first set if and only if it belongs to the second set. By a series of logical equivalences, we obtain

$$x \in (A \cup B) \setminus C \Longleftrightarrow x \in A \cup B \wedge x \notin C$$
$$\Longleftrightarrow (x \in A \vee x \in B) \wedge x \notin C$$
$$\Longleftrightarrow (x \in A \wedge x \notin C) \vee (x \in B \wedge x \notin C)$$
$$\Longleftrightarrow x \in A \setminus C \vee x \in B \setminus C$$
$$\Longleftrightarrow x \in (A \setminus C) \cup (B \setminus C).$$

Thus, we have proved that $x \in (A \cup B) \setminus C$ if and only if $x \in (A \setminus C) \cup (B \setminus C)$. In the above way, we have simultaneously shown that

$$(A \cup B) \setminus C \subseteq (A \setminus C) \cup (B \setminus C)$$

and

$$(A \setminus C) \cup (B \setminus C) \subseteq (A \cup B) \setminus C.$$

The first inclusion is obtained when starting from the left-hand side and considering the implications '$\Longrightarrow$', whereas the other inclusion is obtained when starting from the right-hand side of the last row and considering the implications '$\Longleftarrow$'.

Next, we present some rules for the set operations of intersection and union.

THEOREM 1.5   Let A, B, C, D be arbitrary sets. Then:

(1) $A \cap B = B \cap A, \qquad A \cup B = B \cup A$
    (commutative laws of intersection and union);
(2) $(A \cap B) \cap C = A \cap (B \cap C), \qquad (A \cup B) \cup C = A \cup (B \cup C)$
    (associative laws of intersection and union);
(3) $(A \cap B) \cup C = (A \cup C) \cap (B \cup C), \qquad (A \cup B) \cap C = (A \cap C) \cup (B \cap C)$
    (distributive laws of intersection and union).

As a consequence, we do not need to use parentheses when considering the union or the intersection of three sets due to part (2) of Theorem 1.5.

Example 1.18   We illustrate the first equality of part (3) of Theorem 1.5. Let

$$A = \{3, 4, 5\}, \quad B = \{3, 5, 6, 7\} \quad \text{and} \quad C = \{2, 3\}.$$

For the left-hand side, we obtain

$$(A \cap B) \cup C = \{3, 5\} \cup \{2, 3\} = \{2, 3, 5\},$$

and for the right-hand side, we obtain

$$(A \cup C) \cap (B \cup C) = \{2, 3, 4, 5\} \cap \{2, 3, 5, 6, 7\} = \{2, 3, 5\}.$$

In both cases we obtain the same set, i.e. $(A \cap B) \cup C = (A \cup C) \cap (B \cup C)$.

Remark   There exist relationships between set operations and logical operations described in Chapter 1.1.1. Let us consider the propositions A and B:

$A : a \in A'$;

$B : b \in B'$.

Then:

(1) conjunction $A \wedge B$ corresponds to intersection $A' \cap B'$;
(2) disjunction $A \vee B$ corresponds to union $A' \cup B'$;
(3) implication $A \Rightarrow B$ corresponds to the subset relation (inclusion) $A' \subseteq B'$;
(4) equivalence $A \Leftrightarrow B$ corresponds to set equality $A' = B'$.

The following theorem gives the cardinality of the union and the difference of two sets in the case of *finite sets*.

THEOREM 1.6   Let A and B be two finite sets. Then:

(1) $|A \cup B| = |A| + |B| - |A \cap B|$;
(2) $|A \backslash B| = |A| - |A \cap B| = |A \cup B| - |B|$.

Example 1.19   A car dealer has sold 350 cars during the quarter of a year. Among them, 130 cars have air conditioning, 255 cars have power steering and 110 cars have a navigation system as extras. Furthermore, 75 cars have both power steering and a navigation system, 10 cars have all of these three extras, 10 cars have only a navigation system, and 20 cars have none of these extras. Denote by A the set of cars with air conditioning, by P the set of cars with power steering and by N the set of cars with navigation system. Then

$$|A| = 130, \qquad |P| = 255, \qquad |N| = 110,$$
$$|N \cap P| = 75, \quad |A \cap N \cap P| = 10, \quad |N \setminus (A \cup P)| = 10.$$

Moreover, let C be the set of sold cars. Then

$$|C| = 350 \qquad \text{and} \qquad |(\overline{A \cup N \cup P})_C| = 20$$

i.e.

$$|A \cup P \cup N| = 330.$$

First, the intersection $N \cap P$ can be written as the union of two disjoint sets as follows:

$$N \cap P = \left[(N \cap P) \setminus A \right] \cup \left[A \cap N \cap P \right].$$

Therefore,

$$|N \cap P| = |(N \cap P) \setminus A| + |A \cap N \cap P|$$
$$75 = |(N \cap P) \setminus A| + 10$$

from which we obtain

$$|(N \cap P) \setminus A| = 65,$$

i.e. 65 cars have a navigation system and power steering but no air conditioning. Next, we determine the number of cars having both navigation system and air conditioning but no power steering, i.e. the cardinality of set $(N \cap A) \setminus P$. The set N is the union of disjoint sets of cars having only a navigation system, having a navigation system plus one of the other extras and the cars having all extras. Therefore,

$$|N| = |N \setminus (A \cup P)| + |(A \cap N) \setminus P| + |(N \cap P) \setminus A| + |A \cap N \cap P|$$
$$110 = 10 + |(A \cap N) \setminus P| + 65 + 10$$

from which we obtain

$$|(A \cap N) \setminus P| = 25,$$

i.e. 25 cars have air conditioning and navigation system but no power steering. Next, we determine the number of cars having only air conditioning as an extra, i.e. we determine the cardinality of set $A \setminus (N \cup P)$. Using

$$|N \cup P| = |N| + |P| - |N \cap P|$$
$$= 110 + 255 - 75 = 290,$$

we obtain now from Theorem 1.6

$$|A \setminus (N \cup P)| = |A \cup (N \cup P)| - |N \cup P|$$
$$= 330 - 290 = 40.$$

Now we can determine the number of cars having both air conditioning and power steering but no navigation system, i.e. the cardinality of set $(A \cap P) \setminus N$. Since set A can be written as the union of disjoint sets in an analogous way to set N above, we obtain:

$$|A| = |A \setminus (N \cup P)| + |(A \cap N) \setminus P| + |(A \cap P) \setminus N| + |A \cap N \cap P|$$
$$130 = 40 + 25 + |(A \cap P) \setminus N| + 10$$

from which we obtain

$$|A \cap P) \setminus N| = 55.$$

It remains to determine the number of cars having only power steering as an extra, i.e. the cardinality of set $P \setminus (A \cup N)$. We get

$$|P| = |P \setminus (A \cup N)| + |(A \cap P) \setminus N| + |(N \cap P) \setminus A| + |A \cap N \cap P|$$
$$255 = |P \setminus (A \cup N)| + 55 + 65 + 10$$

from which we obtain

$$|P \setminus (A \cup N)| = 125.$$

The Venn diagram for the corresponding sets is given in Figure 1.3.

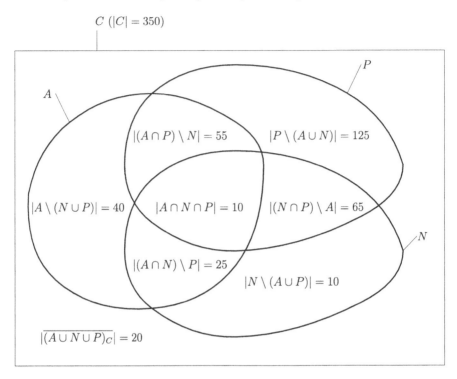

C ($|C| = 350$)

P

A

$|(A \cap P) \setminus N| = 55$

$|P \setminus (A \cup N)| = 125$

$|A \setminus (N \cup P)| = 40$

$|A \cap N \cap P| = 10$

$|(N \cap P) \setminus A| = 65$

N

$|(A \cap N) \setminus P| = 25$

$|N \setminus (A \cup P)| = 10$

$|\overline{(A \cup N \cup P)}_C| = 20$

Figure 1.3 Venn diagram for Example 1.19.

Example 1.20   Among 30 secretaries of several institutes of a German university, each of them speaks at least one of the foreign languages English, French or Russian. There are 11 secretaries speaking only English as a foreign language, 6 secretaries speaking only French and 2 secretaries speaking only Russian. Moreover, it is known that 7 of these secretaries speak exactly two of the three languages and that 21 secretaries speak English.
Denote by E the set of secretaries speaking English, by F the set of secretaries speaking French and by R the set of secretaries speaking Russian. Then

$$|E \cup F \cup R| = 30.$$

Moreover, we know the numbers of secretaries speaking only one of these foreign languages:

$$|E \setminus (F \cup R)| = 11, \quad |F \setminus (E \cup R)| = 6, \quad |R \setminus (E \cup F)| = 2. \tag{1.2}$$

In addition we know that $|E| = 21$. For the cardinality of the set L_1 of secretaries speaking exactly one foreign language we have $|L_1| = 11 + 6 + 2 = 19$. For the cardinality of set L_2 of secretaries speaking exactly two foreign languages, we have to calculate the cardinality of the union of the following sets:

$$|L_2| = \left| \left[(E \cap F) \setminus R \right] \cup \left[(E \cap R) \setminus F \right] \cup \left[(F \cap R) \setminus E \right] \right|$$
$$= \left| (E \cap F) \setminus R \right| + \left| (E \cap R) \setminus F \right| + \left| (F \cap R) \setminus E \right| = 7.$$

Of course, in the last row we can add the cardinality of the three sets since they are pairwise disjoint. To find the number of secretaries speaking all three foreign languages, we have to subtract from the total number the number of secretaries speaking exactly one foreign language and the number of secretaries speaking exactly two foreign languages. We obtain

$$|E \cap F \cap R| = |E \cup F \cup R| - |L_1| - |L_2| = 30 - 19 - 7 = 4.$$

Next, we determine the number of secretaries speaking both French and Russian but not English, i.e. we want to determine the cardinality of set $(F \cap R) \setminus E$. Using $|E| = 21$, we get

$$|(E \cap F) \setminus R| + |(E \cap R) \setminus F| = |E| - |(E \cap F \cap R) \cup [E \setminus (F \cup R)]|$$
$$= |E| - \left[|E \cap F \cap R| + |E \setminus (F \cup R)|\right]$$
$$= 21 - (4 + 11) = 6.$$

Therefore,

$$|(F \cap R) \setminus E| = |L_2| - \left(|(E \cap F) \setminus R| + |(E \cap R) \setminus F|\right) = 7 - 6 = 1.$$

Thus, 1 secretary speaks French and Russian but not English. This solution is illustrated by a Venn diagram in Figure 1.4. With the given information it is not possible to determine the cardinalities of the sets $(E \cap F) \setminus R$ and $(E \cap R) \setminus F$. In order to be able to find these cardinalities, one must know either the number of secretaries speaking French (i.e. $|F|$) or the number of secretaries speaking Russian (i.e. $|R|$). Without this information, we only know that the sum of the cardinalities of these two sets is equal to six (see Figure 1.4).

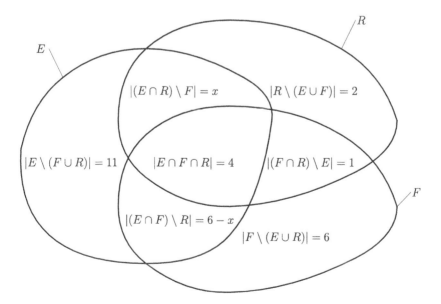

Figure 1.4 Venn diagram for Example 1.20.

Definition 1.17   The set of all ordered pairs (a, b) with $a \in A$ and $b \in B$ is called the *Cartesian product* $A \times B$:

$$A \times B = \{(a, b) \mid a \in A \land b \in B\}.$$

Example 1.21   Let $X = \{(x \mid (1 \leq x \leq 4) \land (x$ is a real number)$)\}$ and $Y = \{y \mid (2 \leq y \leq 3) \land (y$ is a real number)$)\}$. Then the Cartesian product $X \times Y$ is given by

$$X \times Y = \{(x, y) \mid (1 \leq x \leq 4) \land (2 \leq y \leq 3) \land (x, y \text{ are real numbers})\}.$$

The Cartesian product can be illustrated in a coordinate system as follows (see Figure 1.5). The set of all ordered pairs $(x, y) \in X \times Y$ are in one-to-one correspondence to the points in the xy plane whose first coordinate is equal to x and whose second coordinate is equal to y.

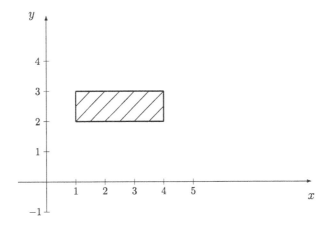

Figure 1.5 Cartesian product $X \times Y$ in Example 1.21.

Example 1.22   We prove the following equality:

$$(A_1 \cup A_2) \times B = (A_1 \times B) \cup (A_2 \times B).$$

By logical equivalences we obtain

$$\begin{aligned}
(a, b) \in (A_1 \cup A_2) \times B &\Longleftrightarrow a \in A_1 \cup A_2 \land b \in B \\
&\Longleftrightarrow (a \in A_1 \lor a \in A_2) \land b \in B \\
&\Longleftrightarrow (a \in A_1 \land b \in B) \lor (a \in A_2 \land b \in B) \\
&\Longleftrightarrow (a, b) \in A_1 \times B \lor (a, b) \in A_2 \times B \\
&\Longleftrightarrow (a, b) \in (A_1 \times B) \cup (A_2 \times B).
\end{aligned}$$

We have shown that an ordered pair (a, b) is contained in set $(A_1 \cup A_2) \times B$ if and only if it is contained in set $(A_1 \times B) \cup (A_2 \times B)$. Therefore the corresponding sets are equal.

Example 1.23 Let $A = \{2,3\}$ and $B = \{3,4,5\}$. Then

$$A \times B = \{(2,3),(2,4),(2,5),(3,3),(3,4),(3,5)\}$$

and

$$B \times A = \{(3,2),(3,3),(4,2),(4,3),(5,2),(5,3)\}.$$

Example 1.23 shows that in general we have $A \times B \neq B \times A$. However, as the following theorem shows, the cardinalities of both Cartesian products are equal provided that A and B are finite sets.

THEOREM 1.7   Let A,B be finite sets with $|A| = n$ and $|B| = m$. Then:

$$|A \times B| = |B \times A| = |A| \cdot |B| = n \cdot m.$$

Now we generalize the Cartesian product to the case of more than two sets.

Definition 1.18   Let $A_1, A_2, \ldots, A_n$ be sets. The set

$$\underset{i=1}{\overset{n}{\times}} A_i = A_1 \times A_2 \times \cdots \times A_n$$

$$= \{(a_1, a_2, \ldots, a_n) \mid a_1 \in A_1 \wedge a_2 \in A_2 \wedge \cdots \wedge a_n \in A_n\}$$

is called the *Cartesian product* of the sets $A_1, A_2, \ldots, A_n$.
An element $(a_1, a_2, \ldots, a_n) \in \underset{i=1}{\overset{n}{\times}} A_i$ is called *ordered n-tuple*.

For $A_1 = A_2 = \ldots = A_n = A$, we also write $\underset{i=1}{\overset{n}{\times}} A_i = A^n$.

Example 1.24   Let $A = \{3,4\}$ and $B = \{10,12,14,16\}$ be given. Then the Cartesian product $A^2 \times B$ is defined as follows:

$$A^2 \times B = \{(x,y,z) \mid x,y \in A, z \in B\}.$$

In generalization of Theorem 1.7, we obtain $|A^2 \times B| = |A|^2 \cdot |B| = 2^2 \cdot 4 = 16$.
Accordingly, if $A = \{1,2\}$, $B = \{b_1, b_2\}$ and $C = \{0,2,4\}$, we get $|A \times B \times C| = |A| \cdot |B| \cdot |C| = 2 \cdot 2 \cdot 3 = 12$, i.e. the Cartesian product $A \times B \times C$ contains 12 elements:

$$A \times B \times C = \{(1,b_1,0),(1,b_1,2),(1,b_1,4),(1,b_2,0),(1,b_2,2),(1,b_2,4),$$
$$(2,b_1,0),(2,b_1,2),(2,b_1,4),(2,b_2,0),(2,b_2,2),(2,b_2,4)\}.$$

1.3   COMBINATORICS

In this section, we summarize some basics on combinatorics. In particular, we investigate two questions:

(1) How many possibilities exist of sequencing the elements of a set?
(2) How many possibilities exist of selecting a certain number of elements from a set?

Let us start with the determination of the number of possible sequences formed with a set of elements. To this end, we first introduce the notion of a permutation.

> **Definition 1.19**   Let $M = \{a_1, a_2, \ldots, a_n\}$. Any sequence $(a_{p_1}, a_{p_2}, \ldots, a_{p_n})$ of all elements of set M is called a *permutation*.

In order to determine the number of permutations, we introduce $n!$ (read: n factorial) which is defined as follows: $n! = 1 \cdot 2 \cdot \ldots \cdot (n-1) \cdot n$ for $n \geq 1$. For $n = 0$, we define $0! = 1$.

THEOREM 1.8   Let a set M consisting of $n \geq 1$ elements be given. Then there exist $P(n) = n!$ permutations.

The latter theorem can easily be proved by induction.

Example 1.25   We enumerate all permutations of the elements of set $M = \{1, 2, 3\}$. We can form $P(3) = 3! = 6$ sequences of the elements from the set M : $(1, 2, 3)$, $(1, 3, 2)$, $(2, 1, 3)$, $(2, 3, 1)$, $(3, 1, 2)$, $(3, 2, 1)$.

Example 1.26   Assume that 7 jobs have to be processed on a single machine and that all job sequences are feasible. Then there exist $P(7) = 7! = 1 \cdot 2 \cdot \ldots \cdot 7 = 5,040$ feasible job sequences.

Example 1.27   Given 6 cities, how many possibilities exist of organizing a tour starting from one city visiting each of the remaining cities exactly once and to return to the initial city? Assume that 1 city is the starting point, all remaining 5 cities can be visited in arbitrary order. Therefore, there are

$$P(5) = 5! = 1 \cdot 2 \cdot \ldots \cdot 5 = 120$$

possible tours. (Here it is assumed that it is a different tour when taking the opposite sequence of the cities.)

If there are some non-distinguishable elements the number of permutations of such elements reduces in comparison with $P(n)$. The following theorem gives the number of permutations for the latter case.

THEOREM 1.9   Let n elements consisting of groups of $n_1, n_2, \ldots, n_r$ non-distinguishable (identical) elements with $n = n_1 + n_2 + \cdots + n_r$ be given. Then there exist

$$P(n; \, n_1, n_2, \ldots, n_r) = \frac{n!}{n_1! \cdot n_2! \cdot \ldots \cdot n_r!}$$

permutations.

Example 1.28   How many distinct numbers with nine digits exist which contain three times digit 1, two times digit 2 and four times digit 3? Due to Theorem 1.9, there are

$$P(9; \, 3, 2, 4) = \frac{9!}{3! \cdot 2! \cdot 4!} = \frac{5 \cdot 6 \cdot 7 \cdot 8 \cdot 9}{6 \cdot 2} = 1,260$$

distinct numbers with these properties.

Example 1.29   In a school, a teacher wishes to put 13 textbooks of three types (mathematics, physics, and chemistry textbooks) on a shelf. How many possibilities exist of arranging the 13 books on a shelf when there are 4 copies of a mathematics textbook, 6 copies of a physics textbook and 3 copies of a chemistry textbook? The problem is to find the number of possible permutations with non-distinguishable (copies of the same textbook) elements.

For the problem under consideration, we have $n = 13, n_1 = 4, n_2 = 6$ and $n_3 = 3$. Thus, due to Theorem 1.9, there are

$$P(13; \, 4, 6, 3) = \frac{13!}{4! \cdot 6! \cdot 3!} = \frac{7 \cdot 8 \cdot 9 \cdot 10 \cdot 11 \cdot 12 \cdot 13}{4! \cdot 3!} = 60,060$$

possibilities of arranging the books on the shelf.

Next, we investigate the question of how many possibilities exist of selecting a certain number of elements from some basic set when the order in which the elements are chosen is important. We distinguish the cases where a repeated selection of elements is allowed and forbidden, respectively. First, we introduce binomial coefficients and present some properties for them.

Definition 1.20   For $0 \leq k \leq n$ with k, n being integer, the term

$$\binom{n}{k} = \frac{n!}{k! \cdot (n-k)!}$$

is called a *binomial coefficient* (read: from n choose k). For $k > n$, we define

$$\binom{n}{k} = 0.$$

For instance, we get

$$\binom{10}{4} = \frac{10!}{4! \cdot 6!} = \frac{1 \cdot 2 \cdot 3 \cdot 4 \cdot 5 \cdot 6 \cdot 7 \cdot 8 \cdot 9 \cdot 10}{(1 \cdot 2 \cdot 3 \cdot 4) \cdot (1 \cdot 2 \cdot 3 \cdot 4 \cdot 5 \cdot 6)} = \frac{7 \cdot 8 \cdot 9 \cdot 10}{1 \cdot 2 \cdot 3 \cdot 4} = 210.$$

In order to determine the binomial coefficient $\binom{n}{k}$ with $n > k$, we have to compute the quotient of the product of the k largest integers no greater than n, i.e. $(n-k+1) \cdot (n-k+2) \cdot \ldots \cdot n$, divided by the product of the first k integers (i.e. $1 \cdot 2 \cdot \ldots \cdot k$). The following theorem gives basic properties of binomial coefficients.

THEOREM 1.10   Let $0 \leq k \leq n$ with k and n being integers. Then:

(1) $\quad \binom{n}{0} = \binom{n}{n} = 1, \qquad \binom{n}{k} = \binom{n}{n-k};$

(2) $\quad \binom{n}{k} + \binom{n}{k+1} = \binom{n+1}{k+1}.$

The first property in Theorem 1.10 describes a symmetry property of the binomial coefficients and the second can be interpreted as an addition theorem for binomial coefficients. Applying the binomial coefficient, we get the following theorem:

THEOREM 1.11   Let a, b be real numbers and n be a natural number. Then:

$$(a+b)^n = a^n + \binom{n}{1}a^{n-1}b + \binom{n}{2}a^{n-2}b^2 + \ldots + \binom{n}{n-1}ab^{n-1} + b^n$$

$$= \sum_{i=0}^{n} \binom{n}{i} a^{n-i} b^i.$$

Remark   Using $a = b = 1$ in Theorem 1.11, we get the following special case:

$$\sum_{k=0}^{n} \binom{n}{k} = \binom{n}{0} + \binom{n}{1} + \ldots + \binom{n}{n} = 2^n.$$

The above comment confirms the previous result that the power set of a set having n elements has cardinality 2^n. The coefficients in Theorem 1.11 can be determined by means of *Pascal's triangle:*

$$
\begin{array}{ll}
(a+b)^0 & \quad\quad\quad\quad\quad 1 \\
(a+b)^1 & \quad\quad\quad\quad 1 \quad 1 \\
(a+b)^2 & \quad\quad\quad 1 \quad 2 \quad 1 \\
(a+b)^3 & \quad\quad 1 \quad 3 \quad 3 \quad 1 \\
(a+b)^4 & \quad 1 \quad 4 \quad 6 \quad 4 \quad 1 \\
\vdots & \quad\quad\quad\quad\quad \vdots
\end{array}
$$

$$
(a+b)^n \quad \binom{n}{0} \; \binom{n}{1} \; \cdots \; \binom{n}{k} \; \binom{n}{k+1} \; \cdots \; \binom{n}{n-1} \; \binom{n}{n}
$$

$$
(a+b)^{n+1} \quad \binom{n+1}{0} \; \binom{n+1}{1} \; \cdots \; \binom{n+1}{k+1} \; \cdots \; \binom{n+1}{n} \; \binom{n+1}{n+1}
$$

In Pascal's triangle, each inner number is obtained by adding the two numbers in the row above which are standing immediately to the left and to the right of the number considered (see equality (2) in Theorem 1.10). For instance, number 6 in the middle of the row for $(a+b)^4$ is obtained by adding number 3 and number 3 again in the row of $(a+b)^3$ (see the numbers in bold face in rows four and five of Pascal's triangle above). Moreover, the numbers in each row are symmetric (symmetry property). In the case of $n = 2$, we obtain from Theorem 1.11 and Pascal's triangle the well-known binomial formula

$$
(a+b)^2 = a^2 + 2ab + b^2.
$$

Example 1.30   We determine the fifth power of $2x + y$. Applying Pascal's triangle, we obtain

$$
(2x+y)^5 = (2x)^5 + 5(2x)^4 y + 10(2x)^3 y^2 + 10(2x)^2 y^3 + 5(2x)y^4 + y^5
$$
$$
= 32x^5 + 80x^4 y + 80x^3 y^2 + 40x^2 y^3 + 10xy^4 + y^5.
$$

In order to get a formula for $(a - b)^n$, we have to replace b by $-b$ in Theorem 1.11, which leads to a minus sign in each second term (i.e. in each term with an odd power of b). We illustrate the computation by the following example.

Example 1.31   We determine the third power of $x - 2y$. From Theorem 1.11, we obtain

$$
(x-2y)^3 = \left[x + (-2y) \right]^3
$$
$$
= x^3 - 3x^2(2y) + 3x(2y)^2 - (2y)^3
$$
$$
= x^3 - 6x^2 y + 12xy^2 - 8y^3.
$$

Next, we consider selections of certain elements from a set starting with *ordered* selections, where the order in which the elements are chosen from the set is important.

THEOREM 1.12 The number of possible selections of k elements from n elements with consideration of the sequence (i.e. the order of the elements), each of them denoted as a *variation*, is equal to

(1) $V(k,n) = \dfrac{n!}{(n-k)!}$ if repeated selection is forbidden (i.e. every element may occur only once in each selection);

(2) $\overline{V}(k,n) = n^k$ if repeated selection is allowed (i.e. an element may occur arbitrarily often in a selection).

Example 1.32 A travelling salesman who should visit 15 cities exactly once may visit only 4 cities on one day. How many possibilities exist of forming a 4-city subtour (i.e. the order in which the cities are visited is important).

This is a problem of determining the number of variations when repeated selection is forbidden. From Theorem 1.12, we obtain

$$V(4,15) = \frac{15!}{(15-4)!} = \frac{15!}{11!} = 12 \cdot 13 \cdot 14 \cdot 15 = 32,760$$

different 4-city subtours.

Example 1.33 A system of switches consists of 9 elements each of which can be in position 'on' or 'off'. We determine the number of distinct constellations of the system and obtain

$$\overline{V}(9,2) = 2^9 = 512.$$

Next, we consider the special case when the order in which the elements are chosen is unimportant, i.e. we consider selections of *unordered* elements.

THEOREM 1.13 The number of possible selections of k elements from n elements without consideration of the sequence (i.e. the order of elements), each of them denoted as a *combination*, is equal to

(1) $C(k,n) = \dbinom{n}{k}$ if repeated selection of the same element is forbidden;

(2) $\overline{C}(k,n) = \dbinom{n+k-1}{k}$ if repeated selection of the same element is allowed.

Example 1.34 In one of the German lottery games, 6 of 49 numbers are drawn in arbitrary order. There exist

$$C(6,49) = \binom{49}{6} = \frac{49!}{6! \cdot 43!} = \frac{44 \cdot 45 \cdot \ldots \cdot 49}{1 \cdot 2 \cdot \ldots \cdot 6} = 13,983,816$$

possibilities of drawing 6 numbers if repeated selection is forbidden. If we consider this game with repeated selection (i.e. when a number that has been selected must be put back and can be drawn again), then there are

$$\bar{C}(6,49) = \binom{54}{6} = \frac{54!}{6! \cdot 48!} = \frac{49 \cdot 50 \cdot \ldots \cdot 54}{1 \cdot 2 \cdot \ldots \cdot 6} = 25,827,165$$

possibilities of selecting 6 (not necessarily different) numbers.

Example 1.35 At a Faculty of Economics, 16 professors, 22 students and 7 members of the technical staff take part in an election. A commission is to be elected that includes 8 professors, 4 students and 2 members of the technical staff. How many ways exist of constituting the commission?

There are

$$C(16,8) = \binom{16}{8} = \frac{16!}{8! \cdot 8!} = \frac{9 \cdot 10 \cdot 11 \cdot 12 \cdot 13 \cdot 14 \cdot 15 \cdot 16}{1 \cdot 2 \cdot 3 \cdot 4 \cdot 5 \cdot 6 \cdot 7 \cdot 8} = 12,870$$

ways of electing 8 out of 16 professors,

$$C(22,4) = \binom{22}{4} = \frac{22!}{4! \cdot 18!} = \frac{19 \cdot 20 \cdot 21 \cdot 22}{1 \cdot 2 \cdot 3 \cdot 4} = 7,315$$

ways of electing 4 out of 22 students and

$$C(7,2) = \binom{7}{2} = \frac{7!}{2! \cdot 5!} = \frac{6 \cdot 7}{2} = 21$$

ways of electing 2 out of 7 members of the technical staff. Since any selection of the professors can be combined with any selection of the students and the technical staff, there are

$$C(18,8) \cdot C(22,4) \cdot C(7,2) = 12,870 \cdot 7,315 \cdot 21 = 1,977,025,050$$

ways of constituting such a commission.

Example 1.36 From a delivery containing 28 spare parts, 5 of which are faulty, a sample of 8 parts is taken. How many samples that contain at least one faulty part are possible? From Theorem 1.13, the number of samples with at least one faulty part is given by

$$C(28,8) - C(23,8) = \binom{28}{8} - \binom{23}{8} = 3,108,105 - 490,314 = 2,617,791.$$

Here we have subtracted the number of samples without any faulty spare part from the total number of possible samples.

1.4   REAL NUMBERS AND COMPLEX NUMBERS

This section deals with number systems. In the first subsection, we summarize the most important facts and rules concerning numbers. Then we extend the set of real numbers by introducing so-called complex numbers.

1.4.1   Real numbers

The infinite set $\mathbb{N} = \{1, 2, 3, \ldots\}$ is called the set of natural numbers. If we union this set with number zero, we obtain the set $\mathbb{N}_0 = \mathbb{N} \cup \{0\}$. The natural numbers can be represented on a straight line known as a number line (see Figure 1.6).

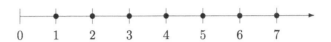

$$0 \quad 1 \quad 2 \quad 3 \quad 4 \quad 5 \quad 6 \quad 7$$

Figure 1.6 Natural numbers on the number line.

We can perform the operations of addition and multiplication within the set $\mathbb{N}$ of natural numbers, i.e. for $a, b \in \mathbb{N}$, we get that $a + b$ belongs to set $\mathbb{N}$ and $a \cdot b$ belongs to set $\mathbb{N}$. In other words, the sum and product of any two natural numbers are again a natural number. However, the difference and quotient of two natural numbers are not necessarily a natural number.

The first extension which we perform is to add the set of negative integers $\{-1, -2, -3, \ldots\}$ to set $\mathbb{N}_0$ which yields the set $\mathbb{Z}$ of integers, i.e.

$$\mathbb{Z} = \mathbb{N}_0 \cup \{-1, -2, -3, \ldots\}.$$

This allows us to perform the three operations of addition, subtraction and multiplication within the set of integers, i.e. for $a, b \in \mathbb{Z}$, we get that $a + b$ belongs to set $\mathbb{Z}$, $a - b$ belongs to set $\mathbb{Z}$ and $a \cdot b$ belongs to set $\mathbb{Z}$.

To be able to perform the division of two integers, we introduce the set of all fractions p/q with $p \in \mathbb{Z}, q \in \mathbb{N}$. The union of the integers and the fractions is denoted as set $\mathbb{Q}$ of rational numbers, i.e.

$$\mathbb{Q} = \mathbb{Z} \cup \left\{ \frac{p}{q} \mid p \in \mathbb{Z},\, q \in \mathbb{N} \right\}.$$

Now all the four basic operations of addition, subtraction, multiplication and division (except by zero) can be performed within the set $\mathbb{Q}$ of rational numbers.

Consider next the equation $x^2 = 2$. This equation cannot be solved within the set of rational numbers, i.e. there exists no rational number p/q such that $(p/q)^2 = 2$. This leads to the extension of the set $\mathbb{Q}$ of rational numbers by the *irrational numbers*. These are numbers which cannot be written as the quotient of two integers. There are infinitely many irrational numbers, e.g.

$$\sqrt{2} \approx 1.41421, \quad \sqrt{3} \approx 1.73205, \quad e \approx 2.71828 \quad \text{and} \quad \pi \approx 3.14159.$$

Irrational numbers are characterized by decimal expansions that never end and by the fact that their digits have no repeating pattern, i.e. no irrational number can be presented as a periodic decimal number.

The union of the set $\mathbb{Q}$ of rational numbers and the set of all irrational numbers is denoted as set $\mathbb{R}$ of real numbers. We get the following property: there is a one-to-one correspondence between real numbers and points on the number line, i.e. any real number corresponds to a point on the number line and vice versa.

The stepwise extension of the set $\mathbb{N}$ of natural numbers to the set $\mathbb{R}$ of real numbers is illustrated in Figure 1.7. Next, we summarize some basic rules for working with the set of real numbers.

Properties of real numbers $(a, b, c \in \mathbb{R})$

(1) $a + b = b + a$ (commutative law of addition);
(2) there exists a number, $0 \in \mathbb{R}$, such that for all a

$$a + 0 = 0 + a = a;$$

(3) for all a, b there exists a number, $x \in \mathbb{R}$, with

$$a + x = x + a = b;$$

(4) $a + (b + c) = (a + b) + c$ (associative law of addition);
(5) $a \cdot b = b \cdot a$ (commutative law of multiplication);
(6) there exists a number, $1 \in \mathbb{R}$, such that for all a

$$a \cdot 1 = 1 \cdot a = a;$$

(7) for all a, b with $a \neq 0$ there exists a real number, $x \in \mathbb{R}$, such that

$$a \cdot x = x \cdot a = b;$$

(8) $(a \cdot b) \cdot c = a \cdot (b \cdot c)$ (associative law of multiplication);
(9) $(a + b) \cdot c = a \cdot c + b \cdot c$ (distributive law).

For special subsets of real numbers, we also use the interval notation. In particular, we use

$$[a, b] = \{x \in \mathbb{R} \mid a \leq x \leq b\};$$
$$(a, b) = \{x \in \mathbb{R} \mid a < x < b\};$$
$$[a, b) = \{x \in \mathbb{R} \mid a \leq x < b\};$$
$$(a, b] = \{x \in \mathbb{R} \mid a < x \leq b\}.$$

The interval $[a, b]$ is called a closed interval whereas (a, b) is called an open interval. Accordingly, the intervals $(a, b]$ and $[a, b)$ are called half-open (left-open and right-open,

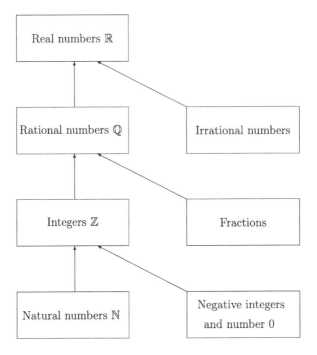

Figure 1.7 Real numbers and subsets.

respectively) intervals. The set $\mathbb{R}$ of real numbers can also be described by the interval $(-\infty, \infty)$. The set of all non-negative real numbers is denoted as $\mathbb{R}_+$, i.e. $\mathbb{R}_+ = [0, \infty)$.

Example 1.37   Often a sum or difference T of different mathematical terms is given and we wish to find a product representation if possible. Let for instance the equality

$$T = 2ax - 12ay + 6bx - 36by$$

be given. Factoring out $2a$ in the first two terms and $6b$ in the third and fourth terms, we get

$$T = 2a \cdot (x - 6y) + 6b \cdot (x - 6y).$$

Now, factoring out $x - 6y$, we get

$$T = (x - 6y) \cdot (2a + 6b)$$
$$= 2 \cdot (x - 6y) \cdot (a + 3b).$$

We now summarize some well-known rules for working with fractions.

Rules for fractions $(a, c \in \mathbb{R},\ b, d \in \mathbb{R} \setminus \{0\})$

(1) $\dfrac{a}{b} \pm \dfrac{c}{d} = \dfrac{ad \pm bc}{bd}$;

(2) $\dfrac{a}{b} \cdot \dfrac{c}{d} = \dfrac{a\,c}{b\,d}$;

(3) $\dfrac{a}{b} : \dfrac{c}{d} = \dfrac{a\,d}{b\,c}$ $(c \neq 0)$.

For a fraction, we can also write a/b (in particular we use the latter form within the text) or $a : b$. According to rule (1), fractions have to have the same denominator before adding or subtracting them. The product of two fractions is obtained by multiplying both numerators and both denominators (rule (2)). The quotient of two fractions is obtained by multiplying the fraction in the numerator by the reciprocal value of the fraction in the denominator (rule (3)). As an application of these rules, we consider the following examples.

Example 1.38   We compute

$$z = \frac{17}{45} - \frac{5}{10} + \frac{7}{15}.$$

Applying rule (2), we have to find a common denominator for all fractions. Instead of taking the product $45 \cdot 10 \cdot 15 = 6,750$, we can use the smallest common multiple of 45, 10 and 15, which is equal to 90, and as a result we obtain

$$z = \frac{17 \cdot 2 - 5 \cdot 9 + 7 \cdot 6}{90} = \frac{34 - 45 + 42}{90} = \frac{31}{90}.$$

Example 1.39   Let

$$\frac{\dfrac{a}{a-x} - \dfrac{x}{a+x}}{\dfrac{a}{a+x} + \dfrac{x}{a-x}} = x,$$

provided that $x \neq a$ and $x \neq -a$. The latter conditions are sufficient to guarantee that no denominator in the above equation is equal to zero. Transforming the left-hand side, we get

$$\frac{\dfrac{a(a+x) - x(a-x)}{(a-x)(a+x)}}{\dfrac{a(a-x) + x(a+x)}{(a+x)(a-x)}} = \frac{a^2 + ax - ax + x^2}{a^2 - ax + ax + x^2} = 1$$

from which we obtain $x = 1$ as the only solution (if $a \neq 1$ and $a \neq -1$).

Example 1.40   Assume that the fuel consumption of a car is 6.5 l per 100 km. How many kilometres can the car go if the tank of this car is filled up with 41 l?
One can solve this problem after establishing a proportion:

$$6.5 : 100 = 41 : x,$$

i.e. if 6.5 l are consumed per 100 km, 41 l are consumed per x km. After cross-multiplying, we get

$$6.5x = 41 \cdot 100$$

from which we obtain

$$x \approx 630.77 \text{ km},$$

i.e. the car can go (approximately) 630 km with 41 l.

In the following, we survey some rules for working with inequalities.

Rules for inequalities $(a, b, c, d \in \mathbb{R})$

(1) If $a < b$, then $a + c < b + c$ and $a - c < b - c$;
(2) if $a < b$ and $b < c$, then $a < c$;
(3) if $a < b$ and $c > 0$, then

$$ac < bc, \quad \frac{a}{c} < \frac{b}{c};$$

(4) if $a < b$ and $c < 0$, then

$$ac > bc, \quad \frac{a}{c} > \frac{b}{c};$$

(5) if $a < b$ and $c < d$, then $a + c < b + d$;
(6) if $0 < a < b$, then

$$\frac{1}{a} > \frac{1}{b};$$

(7) if $a^2 \leq b$ and $b > 0$, then

$$a \geq -\sqrt{b} \quad \wedge \quad a \leq \sqrt{b} \qquad \text{(or correspondingly, } -\sqrt{b} \leq a \leq \sqrt{b}\text{)}.$$

Rule (1) indicates that we can add or subtract any number to an existing inequality without changing the inequality sign. Rule (2) gives a transitivity property (i.e. one can 'connect' several inequalities with the same inequality sign). Rule (3) says that the inequality sign does not change when multiplying (or dividing) both sides of an inequality by a positive number, but the inequality sign changes when multiplying (or dividing) both sides by a negative number (see rule (4)). In particular, if $c = -1$, then it follows from $a < b$ that $-a > -b$. Rule (5) expresses that one can add inequalities with the same inequality sign. However, we cannot state a corresponding rule for the subtraction of two inequalities with the same inequality sign. Rule (6) states that, if we consider the reciprocal values of two positive numbers, the inequality sign is changed. Finally, rule (7) is helpful for solving quadratic inequalities.

Example 1.41 We determine all real numbers satisfying the inequality

$$2x - 4 \leq 5x + 1.$$

Putting all terms depending on x on the left-hand side and the remaining terms on the right-hand side, we get

$$2x - 5x \leq 1 + 4$$

which corresponds to

$$-3x \leq 5.$$

After dividing both sides by -3 (i.e. the inequality sign changes), we obtain

$$x \geq \frac{5}{3}.$$

Thus, the set of all real numbers from the interval $[5/3, \infty)$ satisfies the given inequality.

Definition 1.21 Let $a \in \mathbb{R}$. Then

$$|a| = \begin{cases} a & \text{for } a \geq 0 \\ -a & \text{for } a < 0 \end{cases}$$

is called the *absolute value* of a.

From Definition 1.21 it follows that the absolute value $|a|$ is always a non-negative number. (Note that, if $a < 0$, then $-a > 0$.) The absolute value of a real number represents the distance of this number from point zero on the number line. For instance, $|3| = 3$, $|-5| = 5$ and $|0| = 0$. The following theorem gives some properties of absolute values.

THEOREM 1.14 Let $a, b \in \mathbb{R}$ and $c \in \mathbb{R}_+$. Then:

(1) $|-a| = |a|$;
(2) $|a| \leq c \iff -c \leq a \leq c$;
(3) $|a| \geq c \iff (a < -c) \vee (a > c)$;
(4) $\left| |a| - |b| \right| \leq |a + b| \leq |a| + |b|$;
(5) $|a \cdot b| = |a| \cdot |b|$.

Rule (2) expresses that, for instance, we can rewrite inequality

$$|x| \leq 3$$

in an equivalent form as follows:

$$-3 \le x \le 3.$$

Similarly, inequality $|y| > 6$ can be written in an equivalent form as follows:

$$y < -6 \qquad \text{or} \qquad y > 6.$$

The inequalities of part (4) are known as *triangle inequalities*, referring to the geometric properties of a triangle (see also Chapter 6): the length of the third side of a triangle (i.e. $|a + b|$) is always no greater than the sum of the lengths of the other two sides (i.e. $|a| + |b|$) and is always no smaller than the absolute value of the difference of the lengths of the other two sides.

Often, inequalities contain absolute values of certain mathematical terms, and the problem is to find the solution set satisfying this inequality. In such situations, usually certain cases have to be considered. Let us consider the following examples.

Example 1.42   Let the inequality

$$|x - 3| \le 5$$

be given. In order to find the set of all solutions satisfying the above inequality, we have to distinguish the following two cases.

Case 1 $x - 3 \ge 0$ This means that $x \ge 3$. Then $|x - 3| = x - 3$, and we obtain from $x - 3 \le 5$ the set of solutions L_1 in case 1:

$$L_1 = \{x \in \mathbb{R} \mid (x \ge 3) \wedge (x \le 8)\} = [3, 8].$$

Case 2 $x - 3 < 0$ This means that $x < 3$. Then $|x - 3| = -(x - 3) = -x + 3$, and we obtain from $-x + 3 \le 5$ the set of solutions L_2 in case 2:

$$L_2 = \{x \in \mathbb{R} \mid (x < 3) \wedge (x \ge -2)\} = [-2, 3).$$

In order to get the set L of all solutions satisfying the given inequality, we have to find the union of both sets L_1 and L_2:

$$L = L_1 \cup L_2 = \{x \in \mathbb{R} \mid x \in [-2, 3) \cup [3, 8]\} = [-2, 8].$$

Example 1.43   Let inequality

$$\frac{|2x - 2|}{x + 3} < 1 \tag{1.3}$$

be given. We determine the set of all real numbers satisfying this inequality. In order to solve it for x, we have to remove the absolute values and the fractions. However, the transformation depends on whether the term $2x - 2$ is positive or negative and whether the denominator is positive or negative. Therefore, we determine the roots of both terms. From $2x - 2 = 0$, it follows that $x = 1$ and from $x + 3 = 0$, it follows that $x = -3$. Therefore, we have to consider the following three cases (where each case leads to a different transformation of inequality (1.3), note also that $x = -3$ has to be excluded while $x = 1$ is allowed): case 1: $x < -3$; case 2: $-3 < x < 1$; case 3: $x \geq 1$.

Case 1 If $x < -3$, inequality (1.3) turns into

$$\frac{2 - 2x}{x + 3} < 1$$

since $|2x - 2| = 2 - 2x$ due to inequality $2x - 2 < 0$ and Definition 1.21. Multiplying both sides by the negative number $x + 3$, we obtain

$$2 - 2x > x + 3$$

and therefore, $3x < -1$, which corresponds to inequality $x < -1/3$. Hence, in case 1 we have the following set L_1 of solutions:

$$L_1 = \left\{ x \in \mathbb{R} \mid (x < -3) \wedge \left(x < -\frac{1}{3} \right) \right\} = \{ x \in \mathbb{R} \mid x < -3 \} = (-\infty, -3).$$

Case 2 If $-3 < x < 1$, inequality (1.3) reads as in case 1:

$$\frac{2 - 2x}{x + 3} < 1.$$

However, since $x + 3 > 0$ in case 2, multiplying both sides by the positive number $x + 3$ gives

$$2 - 2x < x + 3$$

which yields $3x > -1$, and thus after dividing by 3 we get

$$x > -\frac{1}{3}.$$

Thus, in case 2 we get the following set L_2 of solutions:

$$L_2 = \left\{ x \in \mathbb{R} \mid (-3 < x < 1) \wedge \left(x > -\frac{1}{3} \right) \right\} = \left(-\frac{1}{3}, 1 \right).$$

Case 3 If $x \geq 1$, then $|2x - 2| = 2x - 2$ and we have

$$\frac{2x - 2}{x + 3} < 1.$$

Multiplying both sides by the positive number $x + 3$, we get

$$2x - 2 < x + 3$$

and therefore $x < 5$. The set L_3 of solutions is as follows:

$$L_3 = \{ x \in \mathbb{R} \mid (x \geq 1) \wedge (x < 5) \} = [1, 5).$$

The complete set L of solutions of inequality (1.3) is obtained as the union of the three sets L_1, L_2 and L_3, i.e.

$$L = L_1 \cup L_2 \cup L_3 = \left\{ x \in \mathbb{R} \mid x \in (-\infty, -3) \cup \left(-\frac{1}{3}, 5 \right) \right\}.$$

Absolute values are, for instance, helpful when solving quadratic inequalities.

Example 1.44   We determine all real numbers satisfying the inequality

$$-3x^2 + 18x - 15 \geq 0.$$

First, we divide both sides by the negative number -3 which changes the sign in the inequality:

$$x^2 - 6x + 5 \leq 0. \tag{1.4}$$

One way to solve this inequality is to consider the corresponding equation

$$x^2 - 6x + 5 = 0$$

and determine its roots:

$$x_1 = 3 + \sqrt{9 - 5} = 5 \qquad \text{and} \qquad x_2 = 3 - \sqrt{9 - 5} = 1.$$

Now we check for an arbitrary 'trial' value (e.g. number 0) whether the given inequality is satisfied. We obtain

$$0^2 - 6 \cdot 0 + 5 \nleq 0,$$

and therefore all $x \in (-\infty, 1)$ do not satisfy the given inequality (for all these values, we have $x^2 - 6x + 5 > 0$). Since we have obtained two distinct roots, the sign of the term $x^2 - 6x + 5$ changes 'at' the roots x_1 and x_2, and therefore the set of solutions L is given by $L = [1, 5]$. (Note that for $x > 5$, the sign changes again so that the given inequality is not satisfied for these values.)

An alternative way to solve inequality (1.4) is as follows. We rewrite the left-hand side using a complete square in the form $x^2 - 2a + a^2 = (x - a)^2$ which gives

$$x^2 - 6x + 9 - 4 = (x - 3)^2 - 4 \leq 0,$$

or equivalently,

$$(x - 3)^2 \leq 4.$$

If we now take the root on both sides, we get the two inequalities

$$x - 3 \geq -2 \qquad \text{and} \qquad x - 3 \leq 2.$$

Using the definition of the absolute value, we can rewrite the latter condition as

$$|x - 3| \leq 2$$

which corresponds to the set of solutions $L = [1, 5]$.

Let $a^n = b$ with $a, b \in \mathbb{R}_+$ and $n \in \mathbb{N}$, i.e. b is the nth *power* of number a. Then $a = \sqrt[n]{b}$ is referred to as the nth *root* of number b. In the following, we review the most important rules for working with powers and roots.

Rules for powers and roots

Powers $(a, b, m, n \in \mathbb{R})$:

(1) $a^m \cdot a^n = a^{m+n}$;

(2) $a^n \cdot b^n = (a\,b)^n$;

(3) $\dfrac{a^m}{a^n} = a^{m-n}$ $\qquad (a \neq 0)$;

(4) $\dfrac{a^n}{b^n} = \left(\dfrac{a}{b}\right)^n$ $\qquad (b \neq 0)$;

(5) $a^{-n} = \dfrac{1}{a^n}$ $\qquad (a \neq 0)$;

(6) $(a^n)^m = a^{n \cdot m} = (a^m)^n$;

Roots $(a, b \in \mathbb{R}_+, m \in \mathbb{Z}, n \in \mathbb{N})$:

(1) $\sqrt[n]{a \cdot b} = \sqrt[n]{a} \cdot \sqrt[n]{b}$;

(2) $\sqrt[n]{\dfrac{a}{b}} = \dfrac{\sqrt[n]{a}}{\sqrt[n]{b}}$ $\qquad (b \neq 0)$;

(3) $\sqrt[n]{a^m} = a^{m/n}$ $\qquad (a \neq 0 \text{ or } m/n > 0)$.

From the latter equality, we get for $m = 1$ the special case

$$\sqrt[n]{a} = a^{1/n}.$$

Example 1.45 We consider the root equation

$$\sqrt{x} + \sqrt{x + 3} = \sqrt{x + 8}$$

and determine all solutions x. Taking the square on both sides, we obtain

$$(\sqrt{x} + \sqrt{x + 3})^2 = x + 8$$

$$x + 2 \cdot \sqrt{x} \cdot \sqrt{x + 3} + (x + 3) = x + 8,$$

which can be written as

$$2 \cdot \sqrt{x} \cdot \sqrt{x+3} = -x + 5.$$

Taking now again the square on both sides, we obtain

$$4x(x+3) = x^2 - 10x + 25$$

which is equivalent to the quadratic equation

$$3x^2 + 22x - 25 = 0$$

or, equivalently,

$$x^2 + \frac{22}{3}x - \frac{25}{3} = 0.$$

A quadratic equation $x^2 + px + q = 0$ has the two real solutions

$$x_{1,2} = -\frac{p}{2} \pm \sqrt{\left(\frac{p}{2}\right)^2 - q}$$

provided that $(p/2)^2 - q \geq 0$. Hereafter, the sign $\pm$ means $+$ or $-$. Hence, the two solutions of the equation obtained are

$$x_1 = -\frac{11}{3} + \sqrt{\frac{121}{9} + \frac{25}{3}} = -\frac{11}{3} + \frac{14}{3} = 1$$

and

$$x_2 = -\frac{11}{3} - \sqrt{\frac{121}{9} + \frac{25}{3}} = -\frac{11}{3} - \frac{14}{3} = -\frac{25}{3}.$$

In the case of root equations, we have to verify that the values obtained are indeed a solution of the original equation. For $x_1 = 1$, the root equation is satisfied. However, for $x_2 = -25/3$, $\sqrt{x_2}$ is not defined since x_2 is negative. Therefore, $x_1 = 1$ is the only real solution of the given root equation.

Example 1.46   If the denominator of a fraction contains roots, it is often desirable to transform the fraction such that the denominator becomes rational. We illustrate this process of rationalizing the denominator. Let

$$z = \frac{a}{x + \sqrt{y}}.$$

In this case, we multiply both the numerator and the denominator by $x - \sqrt{y}$, and we obtain

$$z = \frac{a(x - \sqrt{y})}{(x + \sqrt{y})(x - \sqrt{y})} = \frac{a(x - \sqrt{y})}{x^2 - (\sqrt{y})^2} = \frac{a(x - \sqrt{y})}{x^2 - y}.$$

Example 1.47   We compute

$$x = \sqrt[4]{a \cdot \sqrt[5]{a^2}} \qquad (a > 0)$$

and obtain

$$x = \sqrt[4]{a \cdot a^{2/5}} = \sqrt[4]{a^{7/5}} = a^{7/(4 \cdot 5)} = a^{7/20} = \sqrt[20]{a^7}.$$

Example 1.48   Let the exponential equation

$$2e^{-2x} - e^{-x} - 1 = 0$$

be given. In order to determine all real solutions x of this equation, we can transform it into a quadratic equation by using the substitution $y = e^{-x}$. Taking into account that

$$e^{-2x} = \left(e^{-x}\right)^2,$$

we get the equation

$$2y^2 - y - 1 = 0$$

which can be rewritten as

$$y^2 - \frac{1}{2}y - \frac{1}{2} = 0.$$

The latter equation has the two solutions

$$y_1 = \frac{1}{4} + \sqrt{\frac{1}{16} + \frac{1}{2}} = \frac{1}{4} + \sqrt{\frac{9}{16}} = \frac{1}{4} + \frac{3}{4} = 1$$

and

$$y_2 = \frac{1}{4} - \sqrt{\frac{1}{16} + \frac{1}{2}} = \frac{1}{4} - \sqrt{\frac{9}{16}} = \frac{1}{4} - \frac{3}{4} = -\frac{1}{2}.$$

Substituting variables back, this gives

$$e^{-x_1} = 1 \qquad \text{and} \qquad e^{-x_2} = -\frac{1}{2}.$$

From the first equality we get the solution $x_1 = -\ln 1 = 0$ while the second equality does not have a solution since e^{-x_2} is always greater than zero. Therefore, $x_1 = 0$ is the only solution of the given exponential equation.

Example 1.49   We consider the equation

$$x^4 + 9(a^2 + x^2) = -a^2 x^2$$

and determine for which values of parameter a real solutions exist. We obtain

$$x^4 + 9a^2 + 9x^2 + a^2 x^2 = x^4 + (9 + a^2)x^2 + 9a^2 = 0.$$

Substituting $z = x^2$, we get a quadratic equation in the variable z:

$$z^2 + (9 + a^2)z + 9a^2 = 0.$$

This equation has the two solutions

$$z_{1,2} = -\frac{9 + a^2}{2} \pm \sqrt{\frac{(9 + a^2)^2}{4} - 9a^2}$$

$$= -\frac{9 + a^2}{2} \pm \sqrt{\frac{81 + 18a^2 + a^4 - 36a^2}{4}}$$

$$= -\frac{9 + a^2}{2} \pm \sqrt{\frac{(9 - a^2)^2}{4}}$$

$$= -\frac{9 + a^2}{2} \pm \frac{9 - a^2}{2}.$$

From the last equation we obtain

$$z_1 = -9 \qquad \text{and} \qquad z_2 = -a^2.$$

For the first solution in z, we get after back substitution $x^2 = -9$, i.e. there is no real solution of this equation. Considering the second solution and substituting back, we get $x^2 = -a^2$ which does not have a real solution for $a \neq 0$. Thus, only for $a = 0$ do we have $x^2 = 0$, i.e. in this case we get the solutions $x_1 = x_2 = 0$.

Definition 1.22   Let $a^x = b$ with $a, b > 0$ and $a \neq 1$. Then

$$x = \log_a b$$

is defined as the *logarithm* of (number) b to the base a.

Thus, the logarithm of b to the base a is the power to which one must raise a to yield b. As a consequence from Definition 1.22, we have

$$a^{\log_a b} = b.$$

Rules for logarithms ($a > 0$, $a \neq 1$, $x > 0$, $y > 0$, $n \in \mathbb{R}$)

(1) $\log_a 1 = 0$; $\log_a a = 1$;

(2) $\log_a (x \cdot y) = \log_a x + \log_a y$;

(3) $\log_a \left(\dfrac{x}{y} \right) = \log_a x - \log_a y$;

(4) $\log_a (x^n) = n \cdot \log_a x$.

It is worth noting that there is no rule for the logarithm of the sum or difference of two numbers x and y. When using a pocket calculator, it is often only possible to use one of the bases e or 10. In the case of bases other than e and 10, it is best to transform the base of logarithms to one of the above bases. Therefore, we now derive a formula which presents a relationship between logarithms to different bases. Let

$$x = \log_a t \qquad \text{and} \qquad y = \log_b t.$$

Solving both equations for t, we get

$$a^x = t = b^y.$$

Taking now the logarithm to base a and applying the rules for logarithms presented above, we obtain

$$\log_a \left(a^x \right) = x \cdot \log_a a = x = \log_a \left(b^y \right) = y \cdot \log_a b.$$

Therefore,

$$\log_a b = \frac{x}{y} = \frac{\log_a t}{\log_b t}.$$

Considering now the left-hand side and the right-hand side from the above equality, we get the change-of-base formula

$$\log_b t = \frac{\log_a t}{\log_a b}.$$

As an example, if we want to determine $\log_5 11$, we can apply the above formula with base $a = e$ and obtain

$$\log_5 11 = \frac{\log_e 11}{\log_e 5} \approx \frac{2.39789527\ldots}{1.60943791\ldots} \approx 1.4899,$$

where both logarithms on the right-hand side can be determined using a pocket calculator. The dots in the numerator and denominator of the last fraction indicate that we dropped the remaining decimal places, and the final result on the right-hand side is rounded to four decimal places. The same result is obtained when taking e.g. base 10, i.e.

$$\log_5 11 = \frac{\log_{10} 11}{\log_{10} 5} \approx \frac{1.04139268\ldots}{0.69897000\ldots} \approx 1.4899.$$

For the special cases of base e and 10, respectively, we use in the following the abbreviations $\log_e b = \ln b$ and $\log_{10} b = \lg b$.

To illustrate the application of the above logarithm rules, we consider some examples.

Example 1.50   Let the logarithmic equation

$$\log_2 \left[3 + \log_4(x + 1) \right] = 2$$

be given, and we solve it for x. Applying Definition 1.22 repeatedly, we get

$$3 + \log_4(x + 1) = 2^2 = 4$$
$$\log_4(x + 1) = 1$$
$$x + 1 = 4^1 = 4$$
$$x = 3$$

i.e. $x = 3$ is the only solution of the given logarithmic equation.

Example 1.51   Let

$$z = \frac{\log_a \left(\dfrac{a^x}{y} \right) \cdot \log_b(y \cdot a^x)}{\log_b a} ; \qquad a > 1, \quad b > 1, \quad y > 0$$

be given. We simplify this logarithmic term as much as possible and obtain

$$z = \frac{\left[\log_a(a^x) - \log_a y \right] \cdot \left[\log_b y + \log_b(a^x) \right]}{\log_b a}$$

$$= (x - \log_a y) \cdot \left(\frac{\log_b y}{\log_b a} + \frac{x \log_b a}{\log_b a} \right)$$

$$= (x - \log_a y) \cdot \left(\frac{\log_b y}{\log_b a} + x \right)$$

$$= (x - \log_a y) \cdot (\log_a y + x)$$

$$= x^2 - \left(\log_a y \right)^2 .$$

Example 1.52   We simplify the term

$$z = \log_x \sqrt[3]{\frac{1}{x^2 y}}$$

by applying rules for logarithms as follows:

$$z = \frac{1}{3} \cdot \log_x \left(\frac{1}{x^2 y} \right)$$
$$= \frac{1}{3} \left(\log_x 1 - 2 \log_x x - \log_x y \right)$$
$$= \frac{1}{3} \cdot \left(0 - 2 \cdot 1 - \log_x y \right)$$
$$= -\frac{1}{3} \cdot \left(2 + \log_x y \right).$$

Example 1.53 We consider the exponential equation

$$\left(\frac{4}{5} \right)^{x-2} = \left(\frac{5}{3} \right)^{2x-3}$$

and solve it for x. Such an equation is called an exponential equation since the variable x occurs only in the exponent. Taking the logarithm to base 10 (alternatively we can also choose any other positive base) on both sides, we obtain

$$(x - 2) \lg \left(\frac{4}{5} \right) = (2x - 3) \lg \left(\frac{5}{3} \right)$$

which can be transformed further as follows:

$$x \lg \left(\frac{4}{5} \right) - 2x \lg \left(\frac{5}{3} \right) = 2 \lg \left(\frac{4}{5} \right) - 3 \lg \left(\frac{5}{3} \right)$$

$$x = \frac{2 \lg \left(\frac{4}{5} \right) - 3 \lg \left(\frac{5}{3} \right)}{\lg \left(\frac{4}{5} \right) - 2 \lg \left(\frac{5}{3} \right)}$$

$$x \approx 1.58963.$$

1.4.2 Complex numbers

The set $\mathbb{R}$ of real numbers is still not sufficient for all purposes. For instance, consider the equation $x^2 + 1 = 0$ or the equation $x^2 = -9$ obtained in Example 1.49. These equations do not have a real solution, and so we wish to extend the set of real numbers such that all roots can be determined within an extended set of numbers.

Let i be the symbol ('number') such that $i^2 = -1$ and we call i the *imaginary unit*. Then we can define the set of complex numbers as follows.

Definition 1.23 The complex number system $\mathbb{C}$ is the set of all symbols $z = a + bi$, where a and b are real numbers. The number $a = \Re(z)$ is called the *real part* and $b = \Im(z)$ is called the *imaginary part* of the complex number z.

The representation $z = a + bi$ is called the *Cartesian form* of the complex number z. The number bi is also known as an *imaginary number*. The set of complex numbers contains the set of all real numbers: if $b = \Im(z)$ is equal to zero, we get the real number $a + 0i = a$. The number $\bar{z} = a - bi$ is the *complex conjugate* of number $z = a + bi$. We are now able to solve any quadratic equation within the set of complex numbers.

Example 1.54   Let $x^2 = -9$. Because $x^2 = 9 \cdot i^2$ we obtain $x_1 = 3i$ and $x_2 = -3i$.

Example 1.55   We consider the quadratic equation

$$x^2 + 4x + 13 = 0$$

and obtain

$$x_1 = -2 + \sqrt{4 - 13} = -2 + \sqrt{-9} = -2 + 3i$$

as well as

$$x_2 = -2 - \sqrt{4 - 13} = -2 - \sqrt{-9} = -2 - 3i.$$

In the latter representations, we assumed that $i = \sqrt{-1}$ by definition.

Definition 1.24   Let $z_1 = a + bi$ and $z_2 = c + di$. The respective operations of *addition, subtraction, multiplication* and *division* are defined as follows:

$$z_1 + z_2 = (a + bi) + (c + di) = (a + c) + (b + d)i,$$
$$z_1 - z_2 = (a + bi) - (c + di) = (a - c) + (b - d)i,$$
$$z_1 \cdot z_2 = (a + bi) \cdot (c + di) = (ac - bd) + (ad + bc)i,$$
$$\frac{z_1}{z_2} = \frac{(a + bi) \cdot (c - di)}{(c + di) \cdot (c - di)} = \frac{(ac + bd) + (bc - ad)i}{c^2 + d^2}.$$

When computing the quotient z_1/z_2, the denominator is transformed into a real number by multiplying both the numerator and the denominator by $\bar{z_2} = c - di$. From the above definitions, we see that the sum, difference, product and quotient of two complex numbers again yield a complex number.

For the power of a complex number we obtain

$$z^n = (a + bi)^n = \underbrace{(a + bi) \cdot (a + bi) \cdot \ldots \cdot (a + bi)}_{n \text{ factors}}$$

For the above computations we can apply Theorem 1.11. However, for large numbers n a more convenient calculation of a power is discussed later.

Example 1.56   Let $z_1 = -2 + 2i$ and $z_2 = 3 + i$. Then:

$$z_1 + z_2 = (-2 + 3) + (2 + 1)i = 1 + 3i,$$

$$z_1 - z_2 = (-2 - 3) + (2 - 1)i = -5 + i,$$

$$z_1 \cdot z_2 = (-6 - 2) + (-2 + 6)i = -8 + 4i,$$

$$\frac{z_1}{z_2} = \frac{-2 + 2i}{3 + i} = \frac{(-2 + 2i) \cdot (3 - i)}{(3 + i) \cdot (3 - i)} = \frac{-4 + 8i}{10} = -\frac{2}{5} + \frac{4}{5}i.$$

We can graph complex numbers in the plane by the so-called *Argand diagram* (see Figure 1.8). The x axis represents the real part of the complex number z, and the y axis represents the imaginary number (i.e. the imaginary part together with the imaginary unit). Then each point in the plane represents a complex number and vice versa. The points on the real axis correspond to the real numbers.

From Figure 1.8, we can derive another representation of a complex number z. Instead of giving the real part a and the imaginary part b, we can characterize a complex number by means of the angle φ between the real axis and the line connecting the origin and point $z = a + bi$ as well as the length of this line. We get

$$a = r \cos \varphi \qquad \text{and} \qquad b = r \sin \varphi.$$

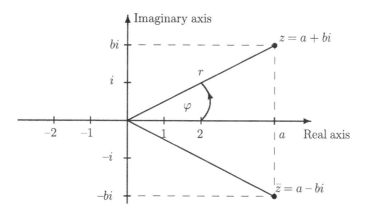

Figure 1.8 Argand diagram.

Therefore, we have

$$z = a + bi = r \cdot (\cos \varphi + i \sin \varphi).$$

The right-hand side representation of number z is called the *polar (or trigonometric) form* of the complex number z. The number

$$r = |z| = \sqrt{a^2 + b^2}$$

is denoted as the *modulus* and φ is denoted as the *argument* of the complex number z. The argument φ is given as an angle either in degrees or in radians.

Relationships between polar and Cartesian coordinates

Let $z = a + bi = r \cdot (\cos\varphi + i\sin\varphi)$. Then

(1) $a = r\cos\varphi,$ $b = r\sin\varphi;$

(2) $r = |z| = \sqrt{a^2 + b^2},$ $\tan\varphi = \dfrac{b}{a}$

$\left(\text{or equivalently } \cos\varphi = \dfrac{a}{r} \quad \text{and} \quad \sin\varphi = \dfrac{b}{r}\right).$

Part (1) of the above relationship can be used for transforming a complex number given in the polar form into the Cartesian form whereas part (2) can be used for transforming a complex number from the Cartesian form into the polar form. We give two examples for transforming a complex number from one form into the other.

Example 1.57   Let $z = -1 + \sqrt{3}i$ be given in Cartesian form. To get the number z in polar form, we determine the modulus r and the argument φ of the complex number z. We obtain

$$r = |z| = \sqrt{(-1)^2 + (\sqrt{3})^2} = \sqrt{4} = 2$$

and

$$\cos\varphi = \frac{a}{r} = -\frac{1}{2}; \qquad \sin\varphi = \frac{b}{r} = \frac{1}{2}\sqrt{3}$$

$$\left(\text{or equivalently, } \tan\varphi = \frac{\sin\varphi}{\cos\varphi} = \frac{b}{a} = -\sqrt{3}\right).$$

For $0° \leq \varphi \leq 360°$, the equation $\cos\varphi = -1/2$ has the two solutions $\varphi_1 = 120°$ and $\varphi_2 = 240°$. Analogously, equation $\sin\varphi = \sqrt{3}/2$ has the two solutions $\varphi_3 = 60°$ and $\varphi_4 = 120°$. Since both equations must be satisfied, we have found $\varphi = 120°$. This result can also be obtained without considering the sine function. The second solution $\varphi_2 = 240°$ of the cosine equation cannot be the argument of number z, since $a = -1 < 0$ and $b = \sqrt{3} > 0$ (i.e. number z is in the second orthant, which means that $90° \leq \varphi \leq 180°$). If we operate with the tangent function, we obtain

$$\tan\varphi = \frac{1}{\sqrt{3}} = \frac{1}{3}\sqrt{3},$$

where two solutions exist: $\varphi_1 = 120°$ and $\varphi_2 = 300°$. However, since we must have $\varphi \in [90°, 180°]$ because $a < 0$ and $b > 0$, we would also obtain $\varphi = 120°$ in this case. We have found number z in polar form:

$$z = 2 \cdot (\cos 120° + i\sin 120°).$$

Example 1.58   Let $z = 2 \cdot (\cos 225° + i \sin 225°)$. Since

$$\cos 225° = -\frac{1}{2}\sqrt{2} \qquad \text{and} \qquad \sin 225° = -\frac{1}{2}\sqrt{2},$$

we get the number z in Cartesian form as follows:

$$z = 2 \cdot \left(-\frac{\sqrt{2}}{2} - i\frac{\sqrt{2}}{2}\right) = -\sqrt{2} - \sqrt{2}\,i,$$

i.e. $a = b = -\sqrt{2}$.

The above transformations require some knowledge about trigonometric functions. Some basic properties of such functions are summarized in Chapter 3.3 when dealing in detail with functions of one real variable.

We summarize some rules for operations with complex numbers in polar form.

Rules for complex numbers in polar form

Let $z_1 = r_1 \cdot (\cos \varphi_1 + i \sin \varphi_1)$ and $z_2 = r_2 \cdot (\cos \varphi_2 + i \sin \varphi_2)$. Then

(1) $z_1 \cdot z_2 = r_1 \cdot r_2 \cdot \left[\cos (\varphi_1 + \varphi_2) + i \sin (\varphi_1 + \varphi_2) \right]$;

(2) $\dfrac{z_1}{z_2} = \dfrac{r_1}{r_2} \cdot \left[\cos (\varphi_1 - \varphi_2) + i \sin (\varphi_1 - \varphi_2) \right].$

We multiply two complex numbers in polar form by multiplying their moduli and adding their arguments. Similarly, we divide two complex numbers in polar form by dividing their moduli and subtracting their arguments. For performing the addition or subtraction of two complex numbers, the Cartesian form is required. Next, we derive a formula for the power of a complex number in polar form.

If $r_1 = r_2 = 1$ and $\varphi_1 = \varphi_2 = \varphi$, then the above formula for $z_1 \cdot z_2$ reads as

$$\left(\cos \varphi + i \sin \varphi \right)^2 = \cos 2\varphi + i \sin 2\varphi.$$

By induction, we can obtain *de Moivre's formula*:

$$(\cos \varphi + i \sin \varphi)^n = \cos n\varphi + i \sin n\varphi, \qquad n = 1, 2, 3, \ldots$$

Using de Moivre's formula, we can easily compute powers of complex numbers. Let $z = r \cdot (\cos \varphi + i \sin \varphi)$ be given in polar form. Then

$$z^n = \left[r \cdot (\cos \varphi + i \sin \varphi) \right]^n = r^n \cdot \left(\cos n\varphi + i \sin n\varphi \right).$$

Example 1.59   We determine $z = (1 - i)^4$. We transform $w = 1 - i$ into the polar form and obtain

$$|w| = \sqrt{1^2 + (-1)^2} = \sqrt{2} \quad \text{and} \quad \cos \varphi = \frac{1}{\sqrt{2}} = \frac{1}{2}\sqrt{2}, \quad \sin \varphi = -\frac{1}{\sqrt{2}} = -\frac{1}{2}\sqrt{2}.$$

Since the cosine value is positive and the sine value is negative, the argument φ is in the interval $[270°, 360°]$. From $\cos 45° = \cos(360° - 45°) = \sqrt{2}/2$, we obtain $\varphi = 360° - 45° = 315°$. This gives us

$$z = \left[\sqrt{2} \cdot \left(\cos 315° + i \sin 315° \right) \right]^4$$
$$= (\sqrt{2})^4 \cdot (\cos 1,260° + i \sin 1,260°).$$

Since

$$\cos x = \cos(x + k \cdot 360°) = \cos(x + 2k\pi) \quad \text{and}$$
$$\sin x = \sin(x + k \cdot 360°) = \sin(x + 2k\pi)$$

for $k \in \mathbb{Z}$, we obtain $1,260° - 3 \cdot 360° = 180°$ and therefore

$$z = 4 \cdot (\cos 180° + i \sin 180°) = 4 \cdot (-1 + i \cdot 0) = -4.$$

The same result is obtained when using the Cartesian form directly:

$$z = (1 - i)^4 = 1 - 4i + 6i^2 - 4i^3 + i^4 = -4.$$

The latter is obtained due to $i^3 = -i$ and $i^4 = 1$.

Definition 1.25   Let $z_0 \in \mathbb{C}$ and $n \in \mathbb{N}$. Every complex number satisfying the equation $z^n = z_0$ is called the *n*th root of z_0, and we write $z = \sqrt[n]{z_0}$.

To derive a formula to determine all the roots of the above equation $z^n = z_0$, let

$$z = r \cdot (\cos \varphi + i \sin \varphi) \qquad \text{and} \qquad z_0 = r_0 \cdot (\cos \varphi_0 + i \sin \varphi_0).$$

Then $z^n = z_0$ can be written as follows:

$$z^n = r^n \cdot (\cos n\varphi + i \sin n\varphi) = r_0 \cdot (\cos \varphi_0 + i \sin \varphi_0) = z_0. \tag{1.5}$$

In equation (1.5), we can compare the moduli and the arguments of both complex numbers, and we obtain

$$r^n = r_0 \qquad \text{and} \qquad n\varphi = \varphi_0 + 2k\pi, \ k \in \mathbb{Z}.$$

Notice that the right equation concerning the arguments is valid due to the periodicity of the trigonometric functions, i.e. $\sin x = \sin(x + 2k\pi)$, $\cos x = \cos(x + 2k\pi)$, where $k \in \mathbb{Z}$. The above equalities for finding r and φ can be rewritten as

$$r = \sqrt[n]{r_0} \qquad \text{and} \qquad \varphi_k = \frac{\varphi_0 + 2k\pi}{n}.$$

Thus, the equation $z^n = z_0 = r_0 \cdot (\cos \varphi_0 + i \sin \varphi_0)$ has the n solutions

$$z_{k+1} = \sqrt[n]{r_0} \cdot \left(\cos \frac{\varphi_0 + 2k\pi}{n} + i \sin \frac{\varphi_0 + 2k\pi}{n} \right), \quad k = 0, 1, 2, \ldots, n-1. \tag{1.6}$$

For $k \geq n$, we would again get consecutively the solutions obtained for $k = 0, 1, 2, \ldots,$ $n-1$. Therefore, we have exactly n roots of the equation $z^n = z_0$. Writing the argument of the complex numbers in degrees instead of radians, equation (1.6) can be written as follows:

$$z_{k+1} = \sqrt[n]{r_0} \cdot \left(\cos \frac{\varphi_0 + k \cdot 360°}{n} + i \sin \frac{\varphi_0 + k \cdot 360°}{n} \right), \quad k = 0, 1, 2, \ldots, n-1.$$

Example 1.60 We determine all roots of $z^4 = 1$. Writing $z = 1 = 1 + 0i$ in polar form yields

$$z^4 = \cos 0° + i \sin 0°.$$

Applying the formula, we obtain

$$z_1 = 1 \cdot (\cos 0° + i \sin 0°) = 1 + 0i = 1,$$

$$z_2 = 1 \cdot \left(\cos \frac{360°}{4} + i \sin \frac{360°}{4} \right) = \cos 90° + i \sin 90° = 0 + 1i = i,$$

$$z_3 = 1 \cdot \left(\cos \frac{720°}{4} + i \sin \frac{720°}{4} \right) = \cos 180° + i \sin 180° = -1 + 0i = -1,$$

$$z_4 = 1 \cdot \left(\cos \frac{1,080°}{4} + i \sin \frac{1,080°}{4} \right) = \cos 270° + i \sin 270° = 0 - 1i = -i.$$

The four solutions are illustrated in an Argand diagram in Figure 1.9. They are located on a circle with the origin as centre and radius one.

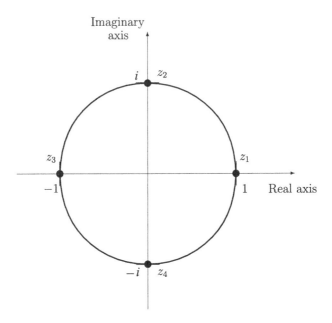

Figure 1.9 Roots $z_1 - z_4$ in Example 1.59.

Example 1.61   Let the equation

$$z^3 + \frac{1-i}{1+i} = \frac{1+i}{1-i}$$

be given and we determine all solutions z. We obtain

$$z^3 = \frac{(1+i)^2 - (1-i)^2}{1-i^2} = \frac{1+2i-1-1+2i+1}{2} = 2i.$$

The modulus of z^3 is therefore equal to two and the argument φ is equal to $90°$ (since $2i$ is on the positive section of the imaginary axis). Thus, we have

$$z^3 = 2 \cdot (\cos 90° + i \sin 90°).$$

Now, we obtain the three solutions

$$z_1 = \sqrt[3]{2} \cdot \left(\cos \frac{90°}{3} + i \sin \frac{90°}{3} \right) = \sqrt[3]{2} \cdot (\cos 30° + i \sin 30°) = \sqrt[3]{2} \cdot \left(\frac{1}{2}\sqrt{3} + \frac{1}{2}i \right),$$

$$z_2 = \sqrt[3]{2} \cdot \left(\cos \frac{90° + 360°}{3} + i \sin \frac{90° + 360°}{3} \right) = \sqrt[3]{2} \cdot (\cos 150° + i \sin 150°)$$

$$= \sqrt[3]{2} \cdot \left(-\frac{1}{2}\sqrt{3} + \frac{1}{2}i \right),$$

$$z_3 = \sqrt[3]{2} \cdot \left(\cos \frac{90° + 720°}{3} + i \sin \frac{90° + 720°}{3} \right) = \sqrt[3]{2} \cdot (\cos 270° + i \sin 270°)$$

$$= \sqrt[3]{2} \cdot (0 - i) = -\sqrt[3]{2}i.$$

The three solutions are illustrated in an Argand diagram in Figure 1.10.

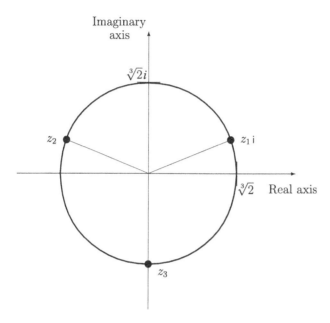

Figure 1.10 Roots $z_1 - z_3$ in Example 1.61.

The following representation for an exponential term with imaginary argument is called *Euler's formula*, where angle φ is presented in radians:

$$e^{i\varphi} = \cos \varphi + i \sin \varphi$$

Using Euler's formula, we obtain the equality

$$z = r \cdot e^{i\varphi} = r \cdot (\cos \varphi + i \sin \varphi).$$

From the latter equality, we can introduce the *exponential form* as the third representation of a complex number:

$$z = r \cdot e^{i\varphi},$$

which is immediately obtained from the polar form, since it uses the same two variables r and φ. (Again angle φ is presented in radians.)

We can summarize the following rules for the product and the quotient of two complex numbers as well as for the power of a complex number using the exponential form.

Rules for complex numbers in exponential form

Let $z_1 = r_1 \cdot (\cos \varphi_1 + i \sin \varphi_1)$, $z_2 = r_2 \cdot (\cos \varphi_2 + i \sin \varphi_2)$, $z = r \cdot (\cos \varphi + i \sin \varphi)$ and $z_0 = r_0 \cdot (\cos \varphi_0 + i \sin \varphi_0)$. Then

(1) $z_1 \cdot z_2 = r_1 \cdot r_2 \cdot e^{i(\varphi_1 + \varphi_2)}$;

(2) $\dfrac{z_1}{z_2} = \dfrac{r_1}{r_2} \cdot e^{i(\varphi_1 - \varphi_2)}$;

(3) $z^n = r^n \cdot e^{in\varphi}$ and

(4) $z_{k+1} = \sqrt[n]{z_0} = \sqrt[n]{r_0} \cdot e^{i(\varphi_0 + 2k\pi)/n}, \quad k = 0, 1, 2, \ldots, n-1.$

Example 1.62   Let

$$z_1 = e^{i\pi/2} \quad \text{and} \quad z_2 = 2 \cdot e^{i\pi/4},$$

i.e.

$$r_1 = 1, \quad r_2 = 2, \quad \varphi_1 = \frac{\pi}{2}, \quad \text{and} \quad \varphi_2 = \frac{\pi}{4}.$$

Then

$$z_1 \cdot z_2 = 2 \cdot e^{i(\pi/2 + \pi/4)} = 2 \cdot e^{3i\pi/4}.$$

Using the exponential form of a complex number, we can easily determine the logarithm of a complex number. Let

$$z = r \cdot e^{i\varphi} = r \cdot e^{(i\varphi + 2k\pi)}, \quad k \in \mathbb{Z}.$$

Taking the logarithm we obtain

$$\ln z = \ln \left[r \cdot e^{i(\varphi + 2k\pi)} \right]$$
$$= \ln r + i(\varphi + 2k\pi), \quad k \in \mathbb{Z}.$$

From the above formula we see that the imaginary part of $\ln z$ is not uniquely determined. For simplicity, one takes the imaginary part using $k = 0$.

Example 1.63   Let $z = -12$, i.e. $z = 12 \cdot e^{i\pi}$, and we determine the logarithm:

$$\ln z = \ln 12 + i \cdot \pi \approx 2.4849 + \pi i.$$

Notice again that in the above representation, the argument φ is given in radians.

From Example 1.63 we see that we can compute the logarithm of a negative number within the set of complex numbers.

EXERCISES

1.1 Which of the following terms are propositions? If they are propositions, what is their truth value?

(a) $2 + 1 = 15$; (b) Please hold the line.
(c) $\int x^2 dx$; (d) $5 < 10$;
(e) 3 is an even number.

1.2 Given are the following propositions

    A   Peter meets Ann.
    B   Peter meets Betty.
    C   Peter meets Cindy.

Express the following propositions by use of conjunction, disjunction and (or) negation of the given propositions:

    D   Peter meets all three girls.
    E   Peter meets only Ann.
    F   Peter meets at least one of the three girls.
    G   Peter meets exactly one of the three girls.
    H   Peter meets no girl.
    I   Peter does not meet all three girls.

1.3 Verify by means of truth tables that

(a) $\overline{A \wedge B} \Longleftrightarrow \overline{A} \vee \overline{B}$; (b) $(A \Rightarrow B) \Longleftrightarrow (\overline{B} \Rightarrow \overline{A})$

are tautologies.

1.4 Find out in which cases A is sufficient for B or A is necessary for B or both.

(a) $A : x \in \mathbb{N}$ is even; (b) $A : x/3 \in \mathbb{N}$;
    $B : x/2 \in \mathbb{N}$; $B : x$ is integer;
(c) $A : x < 4$; (d) $A : x^2 = 16$;
    $B : x^2 < 16$; $B : (x = 4) \vee (x = -4)$;
(e) $A : (x > 0) \wedge (y > 0)$;
    $B : xy > 0$.

1.5 Check whether the following propositions are true for $x \in \mathbb{R}$:

(a) $\bigwedge_x x^2 - 5x + 10 > 0$;
(b) $\bigwedge_x x^2 - 2x > 0$.

Find the negations of the propositions and give their truth values.

1.6 Prove indirectly

$$x + \frac{1}{x} \geq 2$$

with the premise $x > 0$. Use both methods, proof by contradiction and proof of contrapositive.

1.7   Prove by induction:

(a) $\sum_{i=1}^{n}(2i-1) = n^2$;

(b) $\sum_{i=1}^{n} i^3 = \left[\frac{n(n+1)}{2}\right]^2$;

(c)   Let $a > 0$ and $n = 1, 2, \ldots$. Prove the inequality $(1+a)^n \geq 1 + na$.

1.8   Which of the propositions

$$1 \in S; \qquad 0 \in S; \qquad 2 \notin S; \qquad -1 \notin S$$

are true and which are false in each of the following cases?

(a) $S = \{1, 2\}$; (b) $S = \{x \mid x^2 + 2x - 3 = 0\}$;
(c) $S = \{0, 1, 2\} \cup \{-1, 0\}$; (d) $S = \{0, 1, 2\} \cap \{-1, 0\}$.

1.9   Let $A = \{1, 3, 5, 7, 9, 11\}$ and $B = \{1, 2, 3, 7, 8, 9\}$. Find $A \cup B, A \cap B, |A|, |B|, |A \cup B|,$ $|A \cap B|, A \setminus B$ and $|A \setminus B|$.

1.10   Given is $A = \{1, 2\}$. Enumerate all the subsets of the set A. What is the cardinality of set $P(P(A))$?

1.11   Illustrate by means of Venn diagrams:

(a) $A \cup (B \cap C) = (A \cup B) \cap (A \cup C)$;
(b) $(A \cup B) \setminus (A \cap B) = (A \setminus B) \cup (B \setminus A)$.

1.12   Prove that $(A \cup B) \setminus (A \cap B) = (A \setminus B) \cup (B \setminus A)$.

1.13   8 students of a group study mathematics, 13 students of the group do not live in the student hostel, and 17 students study mathematics or do not live in the student hostel. How many students of the group studying mathematics do live in the student hostel?

1.14   Of 1,100 students studying economics at a university, 550 students have a car and 400 students have a PC, and 260 students have neither a car nor a PC. Determine the number of students having a car or a PC, the number of students having both a car and a PC, the number of students having a car but no PC and the number of students having a PC but no car.

1.15   Given are $A = \{1, 2\}, B = \{2, 3\}$ and $C = \{0\}$. Find $A \times A, A \times B, B \times A, A \times C,$ $A \times C \times B$ and $A \times A \times B$.

1.16   Let the sets

$$M_1 = \{x \in \mathbb{R} \mid 1 \leq x \leq 4\}, \quad M_2 = \{y \in \mathbb{R} \mid -2 \leq y \leq 3\},$$
$$M_3 = \{z \in \mathbb{R} \mid 0 \leq z \leq 5\}$$

be given. Find $M_1 \times M_2 \times M_3$. Illustrate this Cartesian product, which corresponds to a set of points in the three-dimensional space, in a rectangular coordinate system.

1.17   There are 12 books on a shelf. How many possibilities for arranging these books exist? How does this number change if three books have to stand side by side?

1.18   A designer wants to arrange 3 blue and 5 red buttons in a row. How many possibilities are there for arranging them?

1.19 Given the set of numbers $\{1, 2, 3, 4, 5\}$, how many three-figure numbers can you select? How does the answer change when all the figures have to be different?

1.20 A group of n people all say 'hello' to each other once. How many 'hellos' can somebody from the group hear?

1.21 In the vicinity of a holiday resort 15 walks are to be marked with two coloured parallel lines. What is the minimum number of colours required to mark all footpaths if both lines can have the same colour and the sequence of colours cannot be inverted?

1.22 A set of weights consists of weights of 1, 2, 5, 10, 50, 100 and 500 units.

   (a) How many combinations of these weights are possible? (Use combinatorics and consider the cases when first no weight is selected, then exactly one weight is selected, and so on.)
   (b) Using your knowledge about sets, explain that for a set of n weights there are exactly 2^n combinations of weights.
   (c) Prove by induction:

$$\sum_{i=0}^{n} \binom{n}{i} = 2^n.$$

1.23 Simplify the following fractions:

   (a) $\dfrac{4x + 5y}{x^2 + 2xy} - \dfrac{3x - 2y}{4y^2 + 2xy} + \dfrac{x^2 - 15y^2}{3x^2y + 6xy^2}$;

   (b) $\dfrac{\dfrac{a}{a-b} - \dfrac{b}{a+b}}{1 + \dfrac{a^2 + b^2}{a^2 - b^2}}$.

1.24 Solve the following inequalities:

   (a) $5x + 3 \geq -2x - 4$;
   (b) $\dfrac{3}{2x - 4} \leq 2$;
   (c) $\dfrac{x^2 - 1}{x + 1} \leq \dfrac{2}{x}$;
   (d) $\dfrac{x + 6}{x - 2} > x + 1$.

1.25 Solve the following inequalities:

   (a) $|x| < 2$;
   (b) $|x - 3| < 2$;
   (c) $|2x - 3| < 2$;
   (d) $|x - 4| < |x + 4|$;
   (e) $\dfrac{|x - 1|}{2x + 2} \geq 1$.

1.26 Simplify the following terms:

   (a) $\dfrac{4 + 5\sqrt{2}}{2 - 3\sqrt{2}} + \dfrac{11\sqrt{2}}{7}$;
   (b) $\dfrac{24a^3 b^{-5}}{7c^3} : \dfrac{8b^{-4}c^{-4}}{21a^{-3}b^{-5}}$.

1.27 Solve the following equations for x:

   (a) $\sqrt{3x - 9} - \sqrt{2x - 5} = 1$;
   (b) $9^{x^2 - 1} - 36 \cdot 3^{x^2 - 3} + 3 = 0$;
   (c) $2\sqrt[3]{x + 2} + 3 = \sqrt{15 + 3x}$.

1.28 Simplify the following terms:

(a) $a = -\log_3 (\log_3 \sqrt[3]{\sqrt[3]{3}})$;

(b) $x = \dfrac{\log_a N}{\log_{ab} N} - \log_a b$.

1.29 Solve the following equations for x:

(a) $\frac{1}{4} \ln x^5 + 3 \ln \sqrt{x} - 3 \ln \sqrt[4]{x} = 2 (\ln 2 + \ln 3)$;

(b) $x^{2-(\lg x)^2 - \lg x^2} - \frac{1}{x} = 0$;

(c) $\log_3 x + \log_{\sqrt{x}} x - \log_{\frac{1}{3}} x = 6$.

1.30 Find the roots of the following equations and illustrate them in an Argand diagram:

(a) $x^2 - 3x + 9 = 0$;

(b) $x^4 + 13x^2 + 36 = 0$.

1.31 Illustrate the set of all complex numbers z with $|z - i| < 4$ in an Argand diagram.

1.32 Find the sum, difference, product and quotient of the following complex numbers:
$z_1 = 1 + 4i$; $z_2 = -2 + i$.

1.33 Find the Cartesian form of the following complex numbers:

(a) $z = \frac{1}{i}$;

(b) $z = \left(\frac{1+i}{1-i}\right)^2$;

(c) $z = re^{i\varphi}$ with $r = 2\sqrt{3}$ and $\varphi = -2\pi/3$.

1.34 Find the polar and exponential forms of the following complex numbers:

(a) $z = \frac{1}{i}$;

(b) $z = \left(\frac{1+i}{1-i}\right)^2$;

(c) $z = \frac{3}{2} + 3\frac{\sqrt{3}}{2}i$;

(d) $z = \dfrac{2 - i}{3i + (i - 1)^2}$.

1.35 Given the complex numbers $z_1 = 2 - 2i$ and $z_2 = 1 + i$, find z_1^4 and z_1/z_2 by using the polar form. Compare the results with those you get by means of the Cartesian form of the complex numbers.

1.36 Show that i^i is a real number.

1.37 Find the solutions of the following equations:

(a) $z^4 = -8 + 8\sqrt{3}i$;

(b) $z^3 + \frac{5}{8}i = \frac{15}{i}$.

1.38 Find the real numbers a_1 and a_2 so that $z = 4(\cos 40° + i \sin 40°)$ is a solution of the equation

$$z^3 - \frac{a_1(\sqrt{3} + a_2 i)}{5i + 2(1 - i)^2} = 0.$$

2 Sequences; series; finance

This chapter deals first with sequences and series. Sequences play a role e.g. when solving difference equations. Special types of sequences and series are closely related to problems in finance, which we discuss at the end of this chapter.

2.1 SEQUENCES

2.1.1 Basic definitions

> **Definition 2.1** If to each positive integer $n \in \mathbb{N}$, there is assigned a real number a_n, then
>
> $$\{a_n\} = a_1, a_2, a_3, \ldots, a_n, \ldots$$
>
> is called a *sequence*.

The elements $a_1, a_2, a_3, \ldots$ are called the *terms* of the sequence. In particular a_n is denoted as the nth term. There are two ways of defining a sequence:

(1) *explicitly* by giving the nth term of the sequence;
(2) *recursively* by giving the first term(s) of the sequence and a recursive formula for calculating the remaining terms.

This is illustrated by the following example.

Example 2.1   (a) Consider the sequence $\{a_n\}$ with

$$a_n = \frac{n-1}{2n+1}, \qquad n \in \mathbb{N}.$$

We get the terms

$$a_1 = 0, \quad a_2 = \frac{1}{5}, \quad a_3 = \frac{2}{7}, \quad a_4 = \frac{3}{9}, \quad a_5 = \frac{4}{11}, \quad \ldots.$$

In this case, sequence $\{a_n\}$ is given explicitly.

(b) Let sequence $\{b_n\}$ be given by

$$b_1 = 2 \quad \text{and} \quad b_{n+1} = 2b_n - 1, \quad n \in \mathbb{N}.$$

Then we get the terms

$$b_1 = 2, \quad b_2 = 3, \quad b_3 = 5, \quad b_4 = 9, \quad b_5 = 17, \quad \ldots.$$

The latter sequence is given recursively.

(c) Let sequence $\{c_n\}$ be given by

$$c_n = (-1)^n, \quad n \in \mathbb{N}.$$

In this case, we get the terms

$$c_1 = -1, \quad c_2 = 1, \quad c_3 = -1, \quad c_4 = 1, \quad c_5 = -1, \quad \ldots.$$

This is an alternating sequence, where any two successive terms have a different sign.

Arithmetic and geometric sequences

We continue with two special types of sequences.

> **Definition 2.2**   A sequence $\{a_n\}$, where the difference of any two successive terms is constant, is called an *arithmetic sequence*, i.e.
>
> $$a_{n+1} - a_n = d \quad \text{for all } n \in \mathbb{N}, \text{ where } d \text{ is constant.}$$

Thus, the terms of an arithmetic sequence with the first term a_1 follow the pattern

$$a_1, \quad a_1 + d, \quad a_1 + 2d, \quad a_1 + 3d, \quad \ldots, \quad a_1 + (n-1)d, \quad \ldots$$

and we obtain the following explicit formula for the nth term:

$$a_n = a_1 + (n-1)d \quad \text{for all } n \in \mathbb{N}.$$

Example 2.2   A car company produces 750 cars of a certain type in the first month of its production and then in each month the production is increased by 20 cars in comparison with the preceding month. What is the production in the twelfth month? This number is obtained as the term a_{12} of an arithmetic sequence with $a_1 = 750$ and $d = 20$, and we get

$$a_{12} = a_1 + 11 \cdot d = 750 + 11 \cdot 20 = 970,$$

i.e. in the twelfth month, 970 cars of this type have been produced.

Definition 2.3   A sequence $\{a_n\}$, where the ratio of any two successive terms is the same number $q \neq 0$, is called a *geometric sequence*, i.e.

$$\frac{a_{n+1}}{a_n} = q \qquad \text{for all } n \in \mathbb{N}, \text{ where } q \text{ is constant.}$$

Thus, the terms of a geometric sequence with the first term a_1 follow the pattern

$$a_1, \quad a_1 \cdot q, \quad a_1 \cdot q^2, \quad a_1 \cdot q^3, \quad \ldots, \quad a_1 \cdot q^{n-1}, \quad \ldots$$

and we obtain the following explicit formula for the nth term:

$$a_n = a_1 \cdot q^{n-1} \qquad \text{for all } n \in \mathbb{N}.$$

Example 2.3   Consider sequence $\{a_n\}$ with

$$a_n = (-1)^n \cdot 4^{-n}, \qquad n \in \mathbb{N}.$$

The first term is

$$a_1 = (-1)^1 \cdot 4^{-1} = -\frac{1}{4}.$$

Using

$$a_{n+1} = (-1)^{n+1} \cdot 4^{-(n+1)} \qquad \text{and} \qquad a_n = (-1)^n \cdot 4^{-n},$$

we obtain

$$\frac{a_{n+1}}{a_n} = (-1) \cdot 4^{-n-1+n} = -4^{-1} = -\frac{1}{4}.$$

Thus, $\{a_n\}$ is a geometric sequence with common ratio $q = -1/4$.

Example 2.4   A firm produced 20,000 DVD players in its first year 2001. If the production increases every year by 10 per cent, what will be the production in 2009?
The answer is obtained by the ninth term of a sequence $\{a_n\}$ with $q = 1 + 0.1$:

$$a_9 = a_1 \cdot q^8 = 20,000 \cdot 1.1^8 \approx 42,872.$$

Next, we investigate what will be the first year so that under the above assumptions, production will exceed 55,000 DVD players. From

$$a_n = 55,000 = 20,000 \cdot 1.1^{n-1} = a_1 \cdot q^{n-1}$$

we obtain

$$1.1^{n-1} = \frac{55,000}{20,000} = 2.75$$

and

$$n - 1 = \frac{\ln 2.75}{\ln 1.1} \approx \frac{1.0116}{0.0953} \approx 10.61,$$

i.e. $n \approx 11.61$. Thus, 2012 will be the first year with a production of more than 55,000 DVD players.

Next, we introduce some basic notions.

Definition 2.4   A sequence $\{a_n\}$ is called *increasing* (resp. *strictly increasing*) if

$$a_n \leq a_{n+1} \quad (\text{resp.} \quad a_n < a_{n+1}) \qquad \text{for all } n \in \mathbb{N}.$$

Sequence $\{a_n\}$ is called *decreasing* (resp. *strictly decreasing*) if

$$a_n \geq a_{n+1} \quad (\text{resp.} \quad a_n > a_{n+1}) \qquad \text{for all } n \in \mathbb{N}.$$

A sequence $\{a_n\}$ which is (strictly) increasing or decreasing is also denoted as (strictly) *monotone*. When checking a sequence $\{a_n\}$ for monotonicity, we investigate the difference $D_n = a_{n+1} - a_n$ of two successive terms. If $D_n \geq 0$ ($D_n > 0$) for all $n \in \mathbb{N}$, sequence $\{a_n\}$ is increasing (strictly increasing). If $D_n \leq 0$ ($D_n < 0$) for all $n \in \mathbb{N}$, sequence $\{a_n\}$ is decreasing (strictly decreasing).

Example 2.5   We investigate sequence $\{a_n\}$ with

$$a_n = 2(n-1)^2 - n, \qquad n \in \mathbb{N},$$

for monotonicity, i.e. we investigate the difference of two successive terms and obtain

$$a_{n+1} - a_n = 2n^2 - (n+1) - \left[2(n-1)^2 - n \right]$$

$$= 2n^2 - n - 1 - \left(2n^2 - 4n + 2 - n \right)$$

$$= 4n - 3.$$

Since $4n - 3 > 0$ for all $n \in \mathbb{N}$, we get $a_{n+1} > a_n$ for all $n \in \mathbb{N}$. Therefore, sequence $\{a_n\}$ is strictly increasing.

Definition 2.5   A sequence $\{a_n\}$ is called *bounded* if there exists a finite constant C (denoted as bound) such that

$$|a_n| \leq C \qquad \text{for all } n \in \mathbb{N}.$$

Example 2.6   We consider the sequence $\{a_n\}$ with

$$a_n = \frac{(-1)^n \cdot 2}{n^2}, \qquad n \in \mathbb{N},$$

and investigate whether it is bounded. We can estimate $|a_n|$ as follows:

$$|a_n| = \left| \frac{(-1)^n \cdot 2}{n^2} \right| = \frac{2}{n^2} \leq 2 \qquad \text{for all } n \in \mathbb{N}.$$

Therefore, sequence $\{a_n\}$ is bounded. Note that sequence $\{a_n\}$ is not monotone, since the signs of the terms alternate due to factor $(-1)^n$.

2.1.2   Limit of a sequence

Next, we introduce the basic notion of a limit of a sequence.

Definition 2.6   A finite number a is called the *limit* of a sequence $\{a_n\}$ if, for any given $\varepsilon > 0$, there exists an index $n(\varepsilon)$ such that

$$|a_n - a| < \varepsilon \qquad \text{for all } n \geq n(\varepsilon).$$

To indicate that number a is the limit of sequence $\{a_n\}$, we write

$$\lim_{n \to \infty} a_n = a.$$

The notion of the limit a is illustrated in Figure 2.1. Sequence $\{a_n\}$ has the limit a, if there exists some index $n(\varepsilon)$ (depending on ε) such that the absolute value of the difference between the term a_n and the limit a becomes smaller than the given value ε for all terms a_n with $n \geq n(\varepsilon)$, i.e. from some n on, the terms of the sequence $\{a_n\}$ are very close to the limit a. If ε becomes smaller, the corresponding value $n(\varepsilon)$ becomes larger. To illustrate the latter definition, we consider the following example.

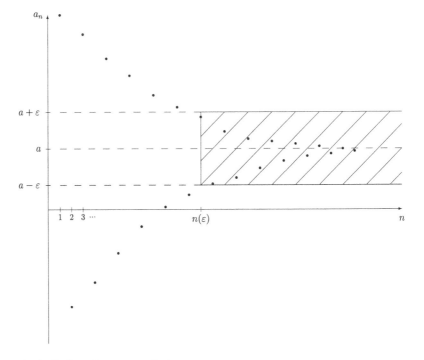

Figure 2.1 The limit a of a sequence $\{a_n\}$.

Example 2.7   Let sequence $\{a_n\}$ be given with

$$a_n = 1 + \frac{1}{n^2}, \qquad n \in \mathbb{N},$$

i.e. we get

$$a_1 = 2, \quad a_2 = \frac{5}{4}, \quad a_3 = \frac{10}{9}, \quad a_4 = \frac{17}{16}, \quad \ldots$$

The terms of sequence $\{a_n\}$ tend to number one. For example, if $\varepsilon = 1/100$, we get

$$|a_n - 1| = \left| \left(1 + \frac{1}{n^2} \right) - 1 \right| = \frac{1}{n^2} < \frac{1}{100},$$

which is satisfied for $n \geq n(\varepsilon) = 11$, i.e. all terms a_n with $n \geq 11$ have an absolute difference from the limit $a = 1$ smaller than $\varepsilon = 1/100$.

If we choose a smaller value of ε, say $\varepsilon = 1/10,000$, we obtain for $n \geq n(\varepsilon) = 101$ only that

$$|a_n - 1| = \frac{1}{n^2} < \frac{1}{10,000},$$

i.e. all terms a_n with $n \geq n(\varepsilon) = 101$ have an absolute difference smaller than $\varepsilon = 1/10,000$ from the limit $a = 1$ of the sequence. Note, however, that independently of the choice of ε, there are infinitely many terms of the sequence in the interval $(a - \varepsilon, a + \varepsilon)$.

The definition of a limit is not appropriate for determining its value. It can be used to confirm that some (guessed) value a is in fact the limit of a certain sequence.

Example 2.8 Consider the sequence $\{a_n\}$ with

$$a_n = \frac{n-2}{2n+3}, \qquad n \in \mathbb{N}.$$

We get

$$a_1 = \frac{1-2}{2 \cdot 1 + 3} = -\frac{1}{5}, \quad a_2 = \frac{2-2}{2 \cdot 2 + 3} = 0, \quad a_3 = \frac{3-2}{2 \cdot 3 + 3} = \frac{1}{9}, \quad \dots$$

For larger values of n, we obtain e.g.

$$a_{50} = \frac{48}{103} \approx 0.466, \quad a_{100} = \frac{98}{203} \approx 0.483, \quad a_{1,000} = \frac{998}{2,003} \approx 0.498,$$

and we guess that $a = 1/2$ is the limit of sequence $\{a_n\}$. To prove it, we apply Definition 2.6 and obtain

$$|a_n - a| = \left| \frac{n-2}{2n+3} - \frac{1}{2} \right| = \left| \frac{2(n-2) - (2n+3)}{2 \cdot (2n+3)} \right| = \left| \frac{2n - 4 - 2n - 3}{4n+6} \right|$$

$$= \left| \frac{-7}{4n+6} \right| = \frac{7}{4n+6} < \varepsilon.$$

By taking the reciprocal values of the latter inequality, we get

$$\frac{4n+6}{7} > \frac{1}{\varepsilon}$$

which can be rewritten as

$$n > \frac{7 - 6\varepsilon}{4\varepsilon},$$

i.e. in dependence on ε we get a number $n = n(\varepsilon)$ such that all terms a_n with $n \geq n(\varepsilon)$ deviate from a by no more than ε. For $\varepsilon = 10^{-2} = 0.01$, we get

$$n > \frac{7 - 6 \cdot 0.01}{4 \cdot 0.01} = 173.5,$$

i.e. we have $n(\varepsilon) = 174$ and from the 174th term on, all terms of the sequence deviate by less than $\varepsilon = 0.01$ from the limit $a = 1/2$. If ε becomes smaller, the number $n(\varepsilon)$ increases, e.g. for $\varepsilon = 10^{-4} = 0.0001$, we get

$$n > \frac{7 - 6 \cdot 0.0001}{4 \cdot 0.0001} \approx 17,498.5,$$

i.e. $n(\varepsilon) = 17,499$.

If a sequence has a limit, then it is said to *converge*. Now we are able to give a first convergence criterion for an arbitrary sequence.

THEOREM 2.1   A bounded and monotone (i.e. increasing or decreasing) sequence $\{a_n\}$ converges.

We illustrate the use of Theorem 2.1 by the following example.

Example 2.9   Consider the recursively defined sequence $\{a_n\}$ with

$$a_1 = 1 \quad\text{and}\quad a_{n+1} = \sqrt{3 \cdot a_n}, \quad n \in \mathbb{N}.$$

First, we prove that this sequence is bounded. All terms are certainly positive, and it is sufficient to prove by induction that $a_n < 3$ for all $n \in \mathbb{N}$. In the initial step, we find that $a_1 = 1 < 3$. Assuming $a_n < 3$ in the inductive step, we obtain

$$a_{n+1} = \sqrt{3 \cdot a_n} < \sqrt{3 \cdot 3} = 3$$

for all $n \in \mathbb{N}$. Therefore, implication

$$a_n < 3 \quad\Longrightarrow\quad a_{n+1} < 3$$

is true. Hence, sequence $\{a_n\}$ is bounded by three.

To show that sequence $\{a_n\}$ is increasing, we may investigate the quotient of two successive terms (notice that $a_n > 0$ for $n \in \mathbb{N}$) and find

$$\frac{a_{n+1}}{a_n} = \frac{\sqrt{3 \cdot a_n}}{a_n} = \sqrt{\frac{3}{a_n}} > 1.$$

The latter inequality is obtained due to $a_n < 3$. Therefore, $a_{n+1}/a_n > 1$ and thus $a_{n+1} > a_n$, i.e. sequence $\{a_n\}$ is strictly increasing. Since sequence $\{a_n\}$ is bounded by three, we have found by Theorem 2.1 that sequence $\{a_n\}$ converges.

Now, we give a few rules for working with limits of sequences.

THEOREM 2.2   Assume that the limits

$$\lim_{n\to\infty} a_n = a \quad\text{and}\quad \lim_{n\to\infty} b_n = b$$

exist. Then the following limits exist, and we obtain:

(1) $\displaystyle\lim_{n\to\infty} (a_n \pm c) = \lim_{n\to\infty} (a_n) \pm c = a \pm c$ *for constant $c \in \mathbb{R}$;*

(2) $\displaystyle\lim_{n\to\infty} (c \cdot a_n) = c \cdot a$ *for constant $c \in \mathbb{R}$;*

(3) $\displaystyle\lim_{n\to\infty} (a_n \pm b_n) = a \pm b$;

(4) $\lim_{n \to \infty} (a_n \cdot b_n) = a \cdot b$;

(5) $\lim_{n \to \infty} \dfrac{a_n}{b_n} = \dfrac{a}{b}$ $\quad (b \neq 0)$.

We illustrate the use of the above rules by the following example.

Example 2.10 Let the sequences $\{a_n\}$ and $\{b_n\}$ with

$$a_n = \frac{2n^2 + 4n - 1}{5n^2 + 10}, \qquad b_n = \frac{3n^2 - 1}{n^3 + 2n}, \qquad n \in \mathbb{N},$$

be given. Then

$$\lim_{n \to \infty} a_n = \lim_{n \to \infty} \frac{2n^2 + 4n - 1}{5n^2 + 10} = \lim_{n \to \infty} \frac{n^2 \left(2 + \frac{4}{n} - \frac{1}{n^2}\right)}{n^2 \left(5 + \frac{10}{n^2}\right)}$$

$$= \frac{\lim\limits_{n \to \infty} 2 + \lim\limits_{n \to \infty} \frac{4}{n} - \lim\limits_{n \to \infty} \frac{1}{n^2}}{\lim\limits_{n \to \infty} 5 + \lim\limits_{n \to \infty} \frac{10}{n^2}} = \frac{2 + 0 - 0}{5 + 0} = \frac{2}{5}.$$

Similarly, we get

$$\lim_{n \to \infty} b_n = \lim_{n \to \infty} \frac{3n^2 - 1}{n^3 + 2n} = \lim_{n \to \infty} \frac{n^2 \left(3 - \frac{1}{n^2}\right)}{n^2 \left(n + \frac{2}{n}\right)}$$

$$= \frac{\lim\limits_{n \to \infty} 3 - \lim\limits_{n \to \infty} \frac{1}{n^2}}{\lim\limits_{n \to \infty} n + \lim\limits_{n \to \infty} \frac{2}{n}} = \frac{3 - 0}{\lim\limits_{n \to \infty} n + 0} = 0.$$

Consider now sequence $\{c_n\}$ with

$$c_n = a_n + b_n, \qquad n \in \mathbb{N}.$$

Applying the given rules for limits, we get

$$\lim_{n \to \infty} c_n = \lim_{n \to \infty} a_n + \lim_{n \to \infty} b_n = \frac{2}{5} + 0 = \frac{2}{5}.$$

As an alternative to the above calculation, we can also first determine the nth term of sequence $\{c_n\}$:

$$c_n = a_n + b_n = \frac{2n^2 + 4n - 1}{5(n^2 + 2)} + \frac{3n^2 - 1}{n(n^2 + 2)}$$

$$= \frac{\frac{1}{5} \cdot (2n^3 + 4n^2 - n) + 3n^2 - 1}{n^3 + 2n} = \frac{\frac{2}{5} n^3 + \frac{19}{5} n^2 - \frac{1}{5} n - 1}{n^3 + 2n}.$$

Determining now

$$\lim_{n\to\infty} c_n$$

we can get the same result as before.

In generalization of the above example, we can get the following result for $p, q \in \mathbb{N}$:

$$\lim_{n\to\infty} \frac{a_p n^p + a_{p-1} n^{p-1} + \ldots + a_1 n + a_0}{b_q n^q + b_{q-1} n^{q-1} + \ldots + b_1 n + b_0} = \begin{cases} 0 & \text{for } p < q \\ \dfrac{a_p}{b_q} & \text{for } p = q \\ \infty & \text{for } p > q. \end{cases}$$

This means that, in order to find the above limit, we have to check only the terms with the largest exponent in the numerator and in the denominator.

Example 2.11   We consider sequence $\{a_n\}$ with

$$a_n = \left(\frac{3n+2}{7n-3}\right)^2 + 2, \qquad n \in \mathbb{N}.$$

Then

$$\lim_{n\to\infty} \left[\left(\frac{3n+2}{7n-3}\right)^2 + 2\right] = \left(\lim_{n\to\infty} \frac{3n+2}{7n-3}\right)^2 + \lim_{n\to\infty} 2$$

$$= \left(\frac{3}{7}\right)^2 + 2 = \frac{9}{49} + 2 = \frac{107}{49}.$$

We finish this section with some frequently used limits.

Some limits

(1) $\lim\limits_{n\to\infty} \dfrac{1}{n} = 0;$

(2) $\lim\limits_{n\to\infty} a^n = 0 \qquad$ for $|a| < 1;$

(3) $\lim\limits_{n\to\infty} \sqrt[n]{c} = 1 \qquad$ for constant $c \in \mathbb{R}, c > 0;$

(4) $\lim\limits_{n\to\infty} \sqrt[n]{n} = 1;$

(5) $\lim\limits_{n\to\infty} \dfrac{a^n}{n!} = 0;$

(6) $\lim\limits_{n\to\infty} \dfrac{n^k}{n!} = 0;$

(7) $\lim\limits_{n\to\infty} \dfrac{n!}{n^n} = 0;$

(8) $\lim\limits_{n\to\infty} \left(1 + \dfrac{1}{n}\right)^n = e.$

From limits (5) up to (7), one can conclude that the corresponding term in the denominator grows faster than the term in the numerator as n tends to infinity.

2.2 SERIES

2.2.1 Partial sums

We start this section with the introduction of the partial sum of some terms of a sequence.

Definition 2.7 Let $\{a_n\}$ be a sequence. Then the sum of the first n terms

$$s_n = a_1 + a_2 + \ldots + a_n = \sum_{k=1}^{n} a_k$$

is called the *nth partial sum* s_n.

Example 2.12 Consider the sequence $\{a_n\}$ with

$$a_n = 3 + (-1)^n \cdot \frac{2}{n}, \qquad n \in \mathbb{N},$$

i.e.

$$a_1 = 3 - 1 \cdot 2 = 1, \quad a_2 = 3 + 1 = 4, \quad a_3 = 3 - \frac{2}{3} = \frac{7}{3}, \quad a_4 = 3 + \frac{1}{2} = \frac{7}{2}, \ldots.$$

Then we get the following partial sums:

$$s_1 = a_1 = 1, \ s_2 = a_1 + a_2 = 1 + 4 = 5, \ s_3 = a_1 + a_2 + a_3 = 1 + 4 + \frac{7}{3} = \frac{22}{3},$$

$$s_4 = a_1 + a_2 + a_3 + a_4 = 1 + 4 + \frac{7}{3} + \frac{7}{2} = \frac{65}{6}, \ldots.$$

For special types of sequences (namely, arithmetic and geomctric sequences), we can easily give the corresponding partial sums as follows.

THEOREM 2.3   The nth partial sum of an arithmetic sequence $\{a_n\}$ with $a_n = a_1 + (n-1) \cdot d$ is given by

$$s_n = \frac{n}{2} \cdot (a_1 + a_n) = \frac{n}{2} \cdot [2a_1 + (n-1)d].$$

The nth partial sum of a geometric sequence $\{a_n\}$ with $a_n = a_1 \cdot q^{n-1}$ and $q \neq 1$ is given by

$$s_n = a_1 \cdot \frac{1 - q^n}{1 - q}.$$

Example 2.2 (continued)   We determine the total car production within the first twelve months of production. To this end, we have to determine the twelfth partial sum s_{12} of an arithmetic sequence with $a_1 = 750$ and $d = 20$. We obtain

$$s_{12} = \frac{n}{2} \cdot (2a_1 + 11 \cdot d) = 6 \cdot (1{,}500 + 11 \cdot 20) = 10{,}320,$$

i.e. the total car production within the first year is equal to 10,320.

Example 2.13   Consider a geometric sequence with $a_1 = 2$ and $q = -4/3$, i.e. the next four terms of the sequence are

$$a_2 = -\frac{8}{3}, \qquad a_3 = \frac{32}{9}, \qquad a_4 = -\frac{128}{27}, \qquad a_5 = \frac{512}{81}.$$

According to Theorem 2.3, we obtain the fifth partial sum as follows:

$$s_5 = a_1 \cdot \frac{1 - q^5}{1 - q} = 2 \cdot \frac{1 - (-\frac{4}{3})^5}{1 - (-\frac{4}{3})} = 2 \cdot \frac{1 - \left(-\frac{1024}{243}\right)}{\frac{7}{3}} = 2 \cdot \frac{1267}{243} \cdot \frac{3}{7} = \frac{362}{81}.$$

Example 2.4 (continued)   We wish to know what will be the total production of DVD players from the beginning of 2001 until the end of 2012. We apply the formula for the partial sum of a geometric sequence given in Theorem 2.3 and obtain

$$s_{12} = a_1 \cdot \frac{1 - q^n}{1 - q} = 20{,}000 \cdot \frac{1 - 1.1^{12}}{1 - 1.1} = \frac{20{,}000}{-0.1} \cdot \left(1 - 1.1^{12}\right) \approx 427{,}686.$$

The firm will produce about 427,686 DVD players in the years from 2001 until 2012.

Example 2.14   Two computer experts, Mr Bit and Mrs Byte, started their jobs on 1 January 2002. During the first year, Mr Bit got a fixed salary of 2,500 EUR every month and his salary rises by 4 per cent every year. Mrs Byte started with an annual salary of 33,000 EUR, and her salary rises by 2.5 per cent every year.

First, we determine the annual salary of Mr Bit and Mrs Byte in 2010. Mr Bit got 30,000 EUR in 2002, and in order to determine his salary in 2010, we have to find the term a_9 of a geometric sequence with $a_1 = 30,000$ and $q = 1.04$:

$$a_9 = 30,000 \cdot 1.04^8 = 41,057.07.$$

To find the salary of Mrs Byte, we have to find the term b_9 of a sequence with $b_1 = 33,000$ and $q = 1.025$:

$$b_9 = 33,000 \cdot 1.025^8 = 40,207.30.$$

In 2010, the annual salary of Mr Bit will be 41,057.07 EUR and the annual salary of Mrs Byte will be 40,207.30 EUR. Next, we wish to know which of them will earn more money over the years from 2002 until the end of 2010. To this end, we have to determine the partial sums of the first nine terms of both sequences. For sequence $\{a_n\}$, we obtain

$$s_9 = a_1 \cdot \frac{1 - q^9}{1 - q} = 30,000 \cdot \frac{1 - 1.04^9}{1 - 1.04} \approx 317,483.86.$$

For sequence $\{b_n\}$, we obtain

$$s_9 = b_1 \cdot \frac{1 - q^9}{1 - q} = 33,000 \cdot \frac{1 - 1.025^9}{1 - 1.025} \approx 328,499.12.$$

Hence, Mrs Byte will earn about 11,000 EUR more than Mr Bit over the nine years.

2.2.2 Series and convergence of series

Definition 2.8 The sequence $\{s_n\}$ of partial sums of a sequence $\{a_n\}$ is called an (*infinite*) *series*.

A series $\{s_n\}$ is said to *converge* if the sequence $\{s_n\}$ has a limit s. This value

$$s = \lim_{n \to \infty} s_n = \lim_{n \to \infty} \sum_{k=1}^{n} a_k = \sum_{k=1}^{\infty} a_k$$

is called the *sum* of the (infinite) series $\{s_n\}$.

If $\{s_n\}$ does not have a limit, the series is said to *diverge*.

Example 2.15 We investigate whether the series $\{s_n\}$ with

$$s_n = \sum_{k=1}^{n} \frac{2}{k^2}, \qquad n \in \mathbb{N},$$

converges. To this end, we apply Theorem 2.1 to the sequence of partial sums $\{s_n\}$. First, we note that this sequence $\{s_n\}$ is strictly increasing, since every term $a_k = 2/k^2$ is positive for all $k \in \mathbb{N}$ and therefore, $s_{n+1} = s_n + a_n > s_n$. It remains to prove that sequence $\{s_n\}$ of the partial sums is also bounded. For $k \geq 2$, we obtain the following estimate of term a_k:

$$a_k = \frac{2}{k^2} \leq \frac{2}{k^2 - k} = \frac{2}{k(k-1)} = 2\left(\frac{1}{k-1} - \frac{1}{k}\right).$$

Then we obtain the following estimate for the nth partial sum s_n:

$$s_n = \sum_{k=1}^{n} \frac{2}{k^2} = 2 + \sum_{k=2}^{n} \frac{2}{k^2} \leq 2 + 2\sum_{k=2}^{n} \left(\frac{1}{k-1} - \frac{1}{k}\right)$$

$$= 2 + 2\left[\left(\frac{1}{2-1} - \frac{1}{2}\right) + \left(\frac{1}{3-1} - \frac{1}{3}\right) + \left(\frac{1}{4-1} - \frac{1}{4}\right) + \ldots + \left(\frac{1}{n-1} - \frac{1}{n}\right)\right]$$

$$= 2 + 2 \cdot \left(1 - \frac{1}{n}\right) \leq 4.$$

The last equality holds since all terms within the brackets are mutually cancelled except the first and last terms. Thus, the strictly increasing sequence $\{s_n\}$ of partial sums is also bounded. From Theorem 2.1, series $\{s_n\}$ with

$$s_n = \sum_{k=1}^{n} \frac{2}{k^2}$$

converges.

THEOREM 2.4 (Necessary convergence criterion)   Let the series $\{s_n\}$ with

$$s_n = \sum_{k=1}^{n} a_k, \qquad n \in \mathbb{N},$$

be convergent. Then

$$\lim_{n \to \infty} a_n = 0.$$

The condition of Theorem 2.4 is *not* sufficient. For instance, let us consider the so-called harmonic series $\{s_n\}$ with

$$s_n = \sum_{k=1}^{n} \frac{1}{k}, \qquad n \in \mathbb{N}.$$

Next, we show that this series does not converge although

$$\lim_{n \to \infty} a_n = \lim_{n \to \infty} \frac{1}{n} = 0.$$

Using the fact that for $n \geq 4$, $n \in \mathbb{N}$, there exists a number $k \in \mathbb{N}$ such that $2^{k+1} \leq n < 2^{k+2}$, we obtain the following estimate for the nth partial sum s_n:

$$s_n = 1 + \frac{1}{2} + \frac{1}{3} + \cdots + \frac{1}{2^{k+1}} + \cdots + \frac{1}{n}$$

$$= 1 + \frac{1}{2} + \left(\frac{1}{3} + \frac{1}{4}\right) + \left(\frac{1}{5} + \cdots + \frac{1}{8}\right) + \left(\frac{1}{2^k + 1} + \cdots + \frac{1}{2^{k+1}}\right) + \cdots + \frac{1}{n}$$

$$> 1 + \frac{1}{2} + 2 \cdot \frac{1}{4} + 4 \cdot \frac{1}{8} + \cdots + 2^k \cdot \frac{1}{2^{k+1}} = \underbrace{1 + \frac{1}{2} + \frac{1}{2} + \cdots + \frac{1}{2}}_{k+1 \text{ summands}}$$

Consequently, we get

$$s_n > \frac{2}{2} + \frac{k+1}{2} = \frac{k+3}{2}.$$

Since for any (arbitrarily large) $k \in \mathbb{N}$, there exists a number $n \in \mathbb{N}$ such that s_n can become arbitrarily large, the partial sum s_n is not bounded. Thus, the sequence of the partial sums $\{s_n\}$ does not converge.

Definition 2.9 Let $\{a_n\}$ be a geometric sequence. The series $\{s_n\}$ with

$$s_n = \sum_{k=1}^{n} a_k = \sum_{k=1}^{n} a_1 \cdot q^{k-1}$$

is called a *geometric* series.

For a geometric series, we can easily decide whether it converges or not by means of the following theorem.

THEOREM 2.5 A geometric series $\{s_n\}$ with $a_1 \neq 0$ converges for $|q| < 1$, and we obtain

$$\lim_{n \to \infty} s_n = \lim_{n \to \infty} \sum_{k=1}^{n} a_1 \cdot q^{k-1} = \lim_{n \to \infty} a_1 \cdot \frac{1 - q^n}{1 - q} = \frac{a_1}{1 - q}.$$

For $|q| \geq 1$, series $\{s_n\}$ diverges.

Example 2.16 Consider the series $\{s_n\}$ with

$$s_n = \sum_{k=1}^{n} 2 \cdot \left(-\frac{1}{2}\right)^{k-1}, \qquad n \in \mathbb{N}.$$

This is a geometric series with $a_1 = 2$ and $q = -1/2$. Applying Theorem 2.5 we find that this series converges, since $|q| = 1/2 < 1$, and we obtain

$$\lim_{n \to \infty} s_n = \frac{a_1}{1-q} = \frac{2}{1 - \left(-\frac{1}{2}\right)} = \frac{2}{\frac{3}{2}} = \frac{4}{3}.$$

A first general criterion for checking convergence of a series has already been discussed in connection with sequences (see Theorem 2.1). If the sequence of the partial sums is bounded and monotone, the series $\{s_n\}$ certainly converges. We consider a similar example to Example 2.15, which however uses Theorem 2.5 for estimating the terms s_n.

Example 2.17   We consider the series $\{s_n\}$ with

$$s_n = \sum_{k=1}^{n} \frac{2}{(k-1)!}, \qquad n \in \mathbb{N}.$$

Determining the first partial sums, we obtain

$$s_1 = \frac{2}{0!} = 2, \; s_2 = s_1 + \frac{2}{1!} = 2 + 2 = 4, \; s_3 = s_2 + \frac{2}{2!} = 4 + 1 = 5,$$

$$s_4 = s_3 + \frac{2}{3!} = 5 + \frac{2}{6} = \frac{16}{3}, \dots$$

Since

$$a_k = \frac{2}{(k-1)!} > 0 \qquad \text{for all } k \in \mathbb{N},$$

sequence $\{s_n\}$ of the partial sums is strictly increasing. To check whether sequence $\{s_n\}$ is bounded, we first obtain for $k \geq 3$:

$$a_k = \frac{2}{(k-1)!} = \frac{2}{1 \cdot 2 \cdot 3 \cdot \dots \cdot (k-1)} \leq \frac{2}{1 \cdot \underbrace{2 \cdot 2 \cdot \dots \cdot 2}_{k-2 \text{ factors}}} = \frac{1}{2^{k-3}}.$$

Denoting $a_k^* = 1/2^{k-3}, k \geq 3$, the following estimate holds for $n \geq 3$:

$$s_n \leq a_1 + a_2 + a_3^* + \dots + a_n^*$$

$$= 2 + \frac{2}{1!} + \sum_{k=3}^{n} \frac{1}{2^{k-3}} = 4 + \sum_{k=1}^{n-2} \frac{1}{2^{k-1}}$$

$$= 4 + \sum_{k=1}^{n-2} \left(\frac{1}{2}\right)^{k-1} \leq 4 + \sum_{k=1}^{\infty} \left(\frac{1}{2}\right)^{k-1} = 4 + \frac{1}{1 - \frac{1}{2}} = 6,$$

i.e. each partial sum s_n with $n \geq 3$ is no greater than six (notice that $s_1 = 2$ and $s_2 = 4$). In the above estimate of s_n, we have used the sum of a geometric series given in Theorem 2.5. Thus, sequence $\{s_n\}$ is strictly increasing and bounded. From Theorem 2.1, the series $\{s_n\}$ with

$$s_n = \sum_{k=1}^{n} \frac{2}{(k-1)!}$$

converges.

Definition 2.10   A series $\{s_n\}$ with

$$s_n = \sum_{k=1}^{n} a_k$$

is called *alternating* if the signs of the terms a_k alternate.

We can give the following convergence criterion for an alternating series.

THEOREM 2.6 (Leibniz's criterion)   Let $\{s_n\}$ with

$$s_n = \sum_{k=1}^{n} a_k$$

be an alternating series. If the sequence $\{|a_k|\}$ of the absolute values is decreasing and

$$\lim_{k \to \infty} a_k = 0,$$

series $\{s_n\}$ converges.

Example 2.18   We consider the series $\{s_n\}$ with

$$s_n = \sum_{k=1}^{n} a_k = \sum_{k=1}^{n} \frac{(-1)^{k+1} \cdot 3}{k}, \qquad n \in \mathbb{N}.$$

This is an alternating series with

$$s_1 = a_1 = 3, \ s_2 = s_1 + a_2 = 3 - \frac{3}{2} = \frac{3}{2}, \ s_3 = s_2 + a_3 = \frac{3}{2} + 1 = \frac{5}{2},$$

$$s_4 = s_3 + a_4 = \frac{5}{2} - \frac{3}{4} = \frac{7}{4}, \dots$$

First, due to

$$|a_{k+1}| - |a_k| = \frac{3}{k+1} - \frac{3}{k} = \frac{3k - 3(k+1)}{k(k+1)} = -\frac{3}{k(k+1)} < 0,$$

sequence $\{|a_k|\}$ is strictly decreasing. Furthermore,

$$\lim_{k \to \infty} \frac{(-1)^{k+1} \cdot 3}{k} = 0,$$

and thus by Theorem 2.6, series $\{s_n\}$ converges.

We now present two convergence criteria which can be applied to arbitrary series.

THEOREM 2.7 (quotient criterion)   Let $\{s_n\}$ be a series with

$$s_n = \sum_{k=1}^{n} a_k.$$

Then:

(1) If

$$Q = \lim_{k \to \infty} \left| \frac{a_{k+1}}{a_k} \right| < 1,$$

series $\{s_n\}$ converges.
(2) If $Q > 1$, series $\{s_n\}$ diverges.
(3) For $Q = 1$, a decision is not possible by means of the quotient criterion.

Example 2.19   We investigate whether the series $\{s_n\}$ with

$$s_n = \sum_{k=1}^{n} \frac{k}{4^k}, \qquad n \in \mathbb{N},$$

converges. Applying the quotient criterion, we get with $a_k = k/4^k$

$$\left| \frac{a_{k+1}}{a_k} \right| = \frac{(k+1) \cdot 4^k}{4^{k+1} \cdot k} = \frac{k+1}{4k}.$$

Therefore,

$$Q = \lim_{k \to \infty} \left| \frac{a_{k+1}}{a_k} \right| = \lim_{k \to \infty} \frac{k+1}{4k} = \frac{1}{4} < 1.$$

Thus, the given series converges.

Example 2.20   Consider the series $\{s_n\}$ with

$$s_n = \sum_{k=1}^{n} \frac{3k^3}{2^k}, \qquad n \in \mathbb{N}.$$

Using the quotient criterion, we get

$$\left|\frac{a_{k+1}}{a_k}\right| = \frac{3(k+1)^3 \cdot 2^k}{2^{k+1} \cdot 3k^3} = \frac{1}{2} \cdot \frac{k^3 + 3k^2 + 3k + 1}{k^3}$$

and

$$Q = \lim_{k \to \infty} \left|\frac{a_{k+1}}{a_k}\right| = \frac{1}{2}.$$

Thus, the series considered converges.

THEOREM 2.8 (root criterion)   Let $\{s_n\}$ be a series with

$$s_n = \sum_{k=1}^{n} a_k.$$

Then:

(1) If

$$R = \lim_{k \to \infty} \sqrt[k]{|a_k|} < 1,$$

series $\{s_n\}$ converges.
(2) If $R > 1$, series $\{s_n\}$ diverges.
(3) For $R = 1$, a decision is not possible by means of the root criterion.

Example 2.21   Let the series $\{s_n\}$ with

$$s_n = \sum_{k=1}^{n} \frac{k}{2^k}, \qquad n \in \mathbb{N},$$

be given. In order to investigate it for convergence or divergence, we apply the root criterion and obtain

$$\sqrt[k]{|a_k|} = \sqrt[k]{a_k} = \sqrt[k]{\frac{k}{2^k}} = \frac{1}{2} \cdot \sqrt[k]{k}.$$

Further,

$$R = \lim_{k \to \infty} \sqrt[k]{|a_k|} = \lim_{k \to \infty} \left(\frac{1}{2} \cdot \sqrt[k]{k}\right) = \frac{1}{2} \cdot \lim_{k \to \infty} \sqrt[k]{k} = \frac{1}{2} \cdot 1 = \frac{1}{2}.$$

The latter result is obtained using the given limit (see limit (4) at the end of Chapter 2.1.2)

$$\lim_{k \to \infty} \sqrt[k]{k} = 1.$$

Hence, the series considered converges due to $R < 1$.

Example 2.22   Consider the series $\{s_n\}$ with

$$s_n = \sum_{k=1}^{n} \left(\frac{k+1}{k}\right)^{k^2} \cdot 2^{-k}, \qquad n \in \mathbb{N}.$$

To check $\{s_n\}$ for convergence, we apply the root criterion and obtain

$$R = \lim_{k \to \infty} \sqrt[k]{|a_k|} = \lim_{k \to \infty} \sqrt[k]{\left(\frac{k+1}{k}\right)^{k^2} \cdot 2^{-k}}$$

$$= \lim_{k \to \infty} \sqrt[k]{\left(1 + \frac{1}{k}\right)^{k^2}} \cdot \left(\frac{1}{2}\right)^{k} = \frac{1}{2} \cdot \lim_{k \to \infty} \left(1 + \frac{1}{k}\right)^{k} = \frac{1}{2} \cdot e > 1.$$

The latter equality holds due to limit (8) given at the end of Chapter 2.1.2. Hence, since $R > 1$, series $\{s_n\}$ diverges.

2.3   FINANCE

In this section, we discuss some problems arising in finance which are an application of sequences and series in economics. Mainly we use geometric and arithmetic sequences and partial sums of such sequences. In the first two subsections, we investigate how one payment develops over time and then how several periodic payments develop over time. Then we consider some applications of these considerations such as loan repayments, evaluation of investment projects and depreciation of machinery.

2.3.1   Simple interest and compound interest

We start with some notions. Let P denote the *principal*, i.e. it is the total amount of money borrowed (e.g. by an individual from a bank in the form of a loan) or invested (e.g. by an individual at a bank in the form of a savings account). *Interest* can be interpreted as money paid for the use of money. The *rate of interest* is the amount charged for the use of the principal for a given length of time, usually on a yearly (or per annum, abbreviated p.a.) basis, given either as a percentage (p per cent) or as a decimal i, i.e.

$$i = \frac{p}{100} \,\widehat{=}\, p\,\%$$

If we use the notation i for the rate of interest, then we always consider a decimal value in the rest of this chapter, and we assume that $i > 0$.

In the following, we consider two variants of interest payments. *Simple interest* is interest computed on the principal for the entire period it is borrowed (or invested). In the case of simple interest, it is assumed that this interest is not reinvested together with the original capital. If a principal P is invested at a simple interest rate of i per annum (where i is a decimal) for a period of n years, the *interest charge* I_n is given by

$$I_n = P \cdot i \cdot n.$$

The *amount* A owed at the end of n years is the sum of the principal borrowed and the interest charge:

$$A = P + I_n = P + P \cdot i \cdot n = P \cdot (1 + i \cdot n).$$

Simple interest is usually applied when interest is paid over a period of less than one year.

If a lender deducts the interest from the amount of the loan at the time the loan is made, the loan is said to be *discounted*. In this case, the borrower receives the amount P (also denoted as proceeds) given by

$$P = A - A \cdot i \cdot n = A \cdot (1 - i \cdot n),$$

where $A \cdot i \cdot n$ is the discount and A is the amount to be repaid at the end of the period of time (also known as the maturity value). We emphasize that in the case of a discounted loan, interest has to be paid on the total amount A (not on the proceeds P) at the beginning of the period.

Next, we assume that at the end of each year, the interest which is due at this time is added to the principal so that the interest computed for the next year is based on this new amount (of old principal plus interest). This is known as *compound interest*. Let A_k be the amount accrued on the principal at the end of year k. Then we obtain the following:

$$A_1 = P + P \cdot i = P \cdot (1 + i),$$

$$A_2 = A_1 + A_1 \cdot i = A_1 \cdot (1 + i) = P \cdot (1 + i)^2,$$

$$A_3 = A_2 + A_2 \cdot i = A_2 \cdot (1 + i) = P \cdot (1 + i)^3,$$

and so on. The amount A_n accrued on a principal P after n years at a rate of interest i per annum is given by

$$A_n = P \cdot (1 + i)^n = P \cdot q^n. \tag{2.1}$$

The term $q = 1 + i$ is also called the *accumulation factor*. Note that the amounts $A_k, k = 1, 2, \ldots$, are the terms of a geometric sequence with the common ratio $q = 1 + i$. It is worth noting that in the case of setting $A_0 = P$, it is a geometric sequence starting with $k = 0$ in contrast to our considerations in Chapter 2.1, where sequences began always with the first term a_1. (However, when beginning with A_0, one has to take into account that A_n is already the $(n + 1)$th term of the sequence.) The above formula (2.1) can be used to solve the following four basic types of problems:

(1) Given the values P, i and n, we can immediately determine the final amount A_n after n years using formula (2.1).
(2) Given the final amount A_n, the interest rate i and the number of years n, we determine the principal P required now such that it turns after n years into the amount A_n when the rate of interest is i. From formula (2.1) we obtain

$$P = \frac{A_n}{(1 + i)^n}.$$

(3) Given the values n, P and A_n, we determine the accumulation factor q and the rate of interest i, respectively, which turns principal P after n years into the final amount A_n. From formula (2.1) we obtain

$$\frac{A_n}{P} = (1+i)^n. \tag{2.2}$$

Solving for i, we get

$$\sqrt[n]{\frac{A_n}{P}} = 1+i$$

$$i = \sqrt[n]{\frac{A_n}{P}} - 1.$$

(4) Given the values A_n, P and i, we determine the time n necessary to turn principal P at a given rate of interest i into a final amount A_n. Solving formula (2.2) for n, we obtain

$$\ln \frac{A_n}{P} = n \cdot \ln(1+i)$$

$$n = \frac{\ln A_n - \ln P}{\ln(1+i)}.$$

We give a few examples to illustrate the above basic problems.

Example 2.23   What principal is required now so that after 6 years at a rate of interest of 5 per cent p.a. the final amount is 20,000 EUR?
This is a problem of type (2) with $n = 6, i = 0.05$ and $A_6 = 20,000$. We obtain:

$$P = \frac{A_6}{(1+i)^6} = \frac{20,000}{(1+0.05)^6} = 14,924.31 \text{ EUR}$$

i.e. the principal required now is equal to 14,924.31 EUR.

Example 2.24   How long will it take to double a principal of 20,000 EUR when the interest rate is 6 per cent p.a.?
This is a problem of type (4). Using $P = 20,000, A_n = 40,000$ and $i = 0.06$, we obtain

$$n = \frac{\ln 40,000 - \ln 20,000}{\ln 1.06} \approx 11.9$$

i.e. it takes about 11.9 years to double the principal.

Often, we do not have an annual period for interest payments, i.e. compounding takes place several times per year, e.g.

(1) *semi-annually* – there are 2 interest payments per year, namely after every six months;
(2) *quarterly* – there are 4 payments per year, namely one after every three months;
(3) *monthly* – there are 12 payments, namely one per month;

(4) *daily* – compounding takes place 360 times per year (in this section, we always count 12 months with 30 days each).

Note that in all cases, it is assumed that interest payment is made after *equal length* periods. In the case of m interest payments per year, the rate of interest per payment period is equal to i/m, where i is as before the per annum rate of interest as a decimal, and the number of interest payments within a period of n years is equal to $n \cdot m$. Denoting by $A_{n,m}$ the amount at the end of n years with m interest payments per year, formula (2.1) changes into

$$A_{n,m} = P \cdot \left(1 + \frac{i}{m}\right)^{n \cdot m}. \tag{2.3}$$

Example 2.25   Mrs Rich wishes to invest 15,000 EUR. The bank offers a rate of interest of 6 per cent p.a.

We compute the resulting amount of money at the end of four years when compounding takes place annually, quarterly and monthly. We use formulas (2.1) and (2.3) with $P = 15,000$ EUR and $i = 0.06$.

In the case of annual compounding, we get

$$A_4 = P \cdot (1 + i)^4 = 15,000 \cdot (1 + 0.06)^4 = 18,937.15 \text{ EUR.}$$

Hereafter, we use the equality sign although all amounts are rounded to two decimal places. When compounding takes place four times per year, we compute $A_{4,4}$ and obtain

$$A_{4,4} = P \cdot \left(1 + \frac{i}{4}\right)^{4 \cdot 4} = 15,000 \cdot 1.015^{16} = 19,034.78 \text{ EUR.}$$

Finally, when compounding takes place monthly, we compute $A_{4,12}$ and obtain

$$A_{4,12} = P \cdot \left(1 + \frac{i}{12}\right)^{4 \cdot 12} = 15,000 \cdot 1.005^{48} = 19,057.34 \text{ EUR.}$$

If the compounding takes place several times per year or if the interest rate is not constant over time, one is often looking for the *effective rate of interest* i_{eff}. This is the rate of interest necessary to turn the principal into the final amount when compounding takes place annually. In the case of m interest payments per year at a per annum rate i of interest, there is an interest payment of i/m per interest payment period, and we obtain from formula (2.3)

$$(1 + i_{\text{eff}}) = \left(1 + \frac{i}{m}\right)^m$$

$$i_{\text{eff}} = \left(1 + \frac{i}{m}\right)^m - 1.$$

Similarly, if there are different per annum rates of interest $i_1, i_2, \ldots, i_n$ with an annual compounding, we obtain from formula (2.1)

$$(1 + i_{\text{eff}})^n = (1 + i_1) \cdot (1 + i_2) \cdot \ldots \cdot (1 + i_n)$$

$$i_{\text{eff}} = \sqrt[n]{(1 + i_1) \cdot (1 + i_2) \cdot \ldots \cdot (1 + i_n)} - 1.$$

Example 2.26 A bank offers a rate of interest of 6 per cent p.a. compounded quarterly. What is the corresponding effective rate of interest i_{eff}?
Since $m = 4$, the rate of interest per period is

$$\frac{i}{4} = 0.015,$$

and we obtain

$$i_{\text{eff}} = (1.015)^4 - 1 = 0.06136355$$

i.e. the offer from the bank corresponds to an effective rate of interest of 6.14 per cent p.a.

Finally we briefly discuss a mixture of simple and compound interest. This is used when interest is paid not only for complete interest periods. We illustrate this by the following example.

Example 2.27 Assume that Mr Happy gives 10,000 EUR to a bank on 15 November 2001 and that he would like to have money back on 20 February 2005. The bank offers a constant rate of interest of 4 per cent p.a. compounded annually.
We can apply the formula for compound interest for the years 2002, 2003 and 2004, but for the parts of 2001 and 2005, we apply the formula for simple interest. Remembering our assumption that every month consists of 30 days, the bank pays interest for 45 days in 2001 and for 50 days in 2005.
Therefore, the amount at the end is given by

$$A = 10,000 \cdot \left(1 + 0.04 \cdot \frac{45}{360}\right) \cdot (1 + 0.04)^3 \cdot \left(1 + 0.04 \cdot \frac{50}{360}\right)$$

$$= 11,367.69 \text{ EUR}.$$

In the above calculations, simple interest is computed for the 45 days from 16 November to 31 December 2001 (i.e. the principal is multiplied by $1 + 0.04 \cdot 45/360$) and for the 50 days from 1 January to 20 February 2005 (i.e. the amount at the end of 2004 is multiplied by $1 + 0.04 \cdot 50/360$). For the three intermediate years from 2002 to 2004, we apply the compound interest formula, which leads to the factor $(1 + 0.04)^3$.

2.3.2   Periodic payments

In the previous subsection, we considered the computation of the future value of *one* investment when a fixed amount of money is deposited in an account that pays interest compounded periodically. In many situations, there are periodic payments (i.e. deposits or withdrawals) and the question is which amount is accrued (or left over) after a number of payment periods. Such situations occur, for example, in connection with annual life insurance premiums, monthly deposits at a bank, loan repayments, sinking funds and so on.

The notion *annuity* is used in the following for a sequence of (usually *equal*) *periodic payments*. Here we consider only the case when *payment periods and the periods for interest payments coincide*. Moreover, the interest is always credited at the end of a payment period.

First, we consider a so-called *ordinary annuity*, where the payments (deposits) are made at the same time the interest is credited, namely at the end of the period. We mark all values by the superscript 'E' which stands for 'end of the period'.

Annual payments

Again, we first consider the case of *annual payments r*. The amount of an annuity is the sum of all payments (deposits) made plus all interest accumulated. Next, we compute the amount of an annuity A_n^E, i.e. the total amount after n annual payment periods. Denoting by D_k the amount resulting from the kth deposit at the end of n years, we get

$$A_n^E = D_1 + D_2 + \ldots + D_n = r \cdot (1+i)^{n-1} + r \cdot (1+i)^{n-2} + \ldots + r \cdot (1+i) + r$$

$$= r \cdot \left[1 + (1+i) + \ldots + (1+i)^{n-1} \right].$$

The term in brackets on the right-hand side represents the partial sum s_n of a geometric sequence with $a_1 = 1$ and $q = 1+i$. Note that the kth deposit accumulates the value D_k, where the payments are compounded over $n - k$ years. We use the formula for the partial sum of the first n terms of a geometric sequence and obtain

$$1 + (1+i) + \cdots + (1+i)^{n-1} = \frac{1 - (1+i)^n}{1 - (1+i)} = \frac{1 - (1+i)^n}{-i} = \frac{(1+i)^n - 1}{i}.$$

Thus, we have shown the correctness of the following theorem.

THEOREM 2.9 (amount of an annuity)   If an amount r is deposited at the end of each year at a per annum rate of interest i, the amount A_n^E of the annuity after n annual payments is

$$A_n^E = r \cdot \frac{(1+i)^n - 1}{i}. \tag{2.4}$$

The present value of an annuity is the amount of an annuity discounted over the whole time period, i.e. it is the amount of money needed now so that after depositing this amount for a period of n years at a per annum rate of interest of i, the amount of an annuity A_n^E results. We get

$$P_n^{\mathrm{E}} = \frac{1}{(1+i)^n} \cdot A_n^{\mathrm{E}} = \frac{1}{(1+i)^n} \cdot r \cdot \frac{(1+i)^n - 1}{i}$$

$$= r \cdot \frac{(1+i)^{-n} \cdot [(1+i)^n - 1]}{i} = r \cdot \frac{1 - (1+i)^{-n}}{i}.$$

In connection with periodic withdrawals the present value P_n^{E} has an important meaning. It is the amount of money needed now so that, if it is invested at a per annum rate of interest i, n equal money amounts r can be withdrawn at the end of each year without any money left over at the end.

To illustrate this, let us argue that the sum of the present values of the individual withdrawals is indeed equal to P_n^{E}. Assume that the withdrawals are made at the end of each year and let PV_k be the present value of the kth withdrawal. Then

$$PV_k = r \cdot \frac{1}{(1+i)^k}.$$

Summing up, we get

$$PV = PV_1 + PV_2 + \cdots + PV_n$$

$$= r \cdot \frac{1}{1+i} + r \cdot \frac{1}{(1+i)^2} + \cdots + r \cdot \frac{1}{(1+i)^n}$$

$$= r \cdot \frac{1}{(1+i)^n} \cdot \left[1 + (1+i) + \cdots + (1+i)^{n-1}\right]$$

$$= r \cdot (1+i)^{-n} \cdot \frac{1 - (1+i)^n}{1 - (1+i)} = r \cdot \frac{(1+i)^{-n} - 1}{-i}$$

$$= r \cdot \frac{1 - (1+i)^{-n}}{i} = P_n^{\mathrm{E}},$$

i.e. the sum of the present values of all the n withdrawals corresponds to P_n^{E}. Therefore, we have proved the following theorem.

THEOREM 2.10 (present value of an annuity) If r represents the withdrawal at the end of each of n years, the present value P_n^{E} of an annuity at a per annum rate of interest i is

$$P_n^{\mathrm{E}} = r \cdot \frac{1 - (1+i)^{-n}}{i}. \tag{2.5}$$

As a consequence, we do not need to distinguish between periodic deposits and withdrawals in the sense that formulas follow the same pattern. When considering periodic deposits, we are usually interested in the amount of an annuity A_n^{E}, and when considering periodic withdrawals we are interested in the present value of an annuity P_n^{E}.

Next, we discuss the case when the payments are made at the beginning of the period while interest is still credited at the end. We mark the corresponding values by the superscript 'B' which stands for 'beginning of the period'. Notice that this is not an ordinary annuity.

First, we discuss the modification of the formula for the amount of the annuity A_n^{B}. The difference from the previous case is that for each deposit, interest is credited for one more period, i.e. the kth deposit accumulates interest over $n - k + 1$ years. Let D_k be the amount at the end of n years which results from the kth deposit. Then $D_k = r \cdot (1+i)^{n-k+1}$. Therefore, we get

$$A_n^{\mathrm{B}} = D_{n-1} + D_{n-2} + \ldots + D_1 = r \cdot (1+i) + r \cdot (1+i)^2 + \ldots + r \cdot (1+i)^n$$

$$= r \cdot (1+i) \cdot \left[1 + (1+i) + (1+i)^2 + \ldots + (1+i)^{n-2} + (1+i)^{n-1} \right]$$

$$= r \cdot (1+i) \cdot \frac{(1+i)^n - 1}{i}.$$

By the above computations we have found that $A_n^{\mathrm{B}} = (1+i) \cdot A_n^{\mathrm{E}}$ which reflects the fact that for each deposit, interest is credited for one more year in comparison with deposits at the end of each year.

Correspondingly, the present value P_n^{B} is obtained by discounting the amount of an annuity for a period of n years, i.e.

$$P_n^{\mathrm{B}} = \frac{1}{(1+i)^n} \cdot A_n^{\mathrm{B}} = r \cdot (1+i) \cdot \frac{1 - (1+i)^{-n}}{i}.$$

We now discuss some basic types of problems connected with periodic payments (deposits or withdrawals). Here we consider only the case when payments are made at the end of each year and interest is credited at the same time. In the above formulas we have used the values $P_n^{\mathrm{E}}, A_n^{\mathrm{E}}, r, i$ and n. Among these values, at least two (or even three) must be given for each problem type, and we are looking for one of the remaining values.

(1) Given the per annum rate of interest i and the number of years n, we are looking for the amount of the annuity A_n^{E}. This problem arises when annual deposits are made in an account, rate of interest as well as the time period are known in advance, and we wish to determine the amount in the account at the end of n years. In this case, we can immediately apply formula (2.4).

(2) Given the values r, i and n, we are looking for the present value of the annuity P_n^{E}. This type of problem arises when periodic withdrawals are made, and we wish to know the current amount required to guarantee the n withdrawals such that no money is left over at the end when the time period and rate of interest are known. In this case, we can immediately apply formula (2.5).

(3) Given the values A_n^{E}, i and n, we are looking for the annual payment r to get the final amount A_n^{E} provided that the time period of the periodic payments and the rate of interest are known. From formula (2.4), we obtain

$$r = A_n^{\mathrm{E}} \cdot \frac{i}{(1+i)^n - 1}.$$

(4) Given the values A_n^{E}, r and i, we are looking for the time period n which is required to accumulate the final amount A_n^{E} by periodic payments of an amount r at a given constant rate of interest i. In this case, we obtain

$$A_n^{\mathrm{E}} \cdot \frac{i}{r} = (1+i)^n - 1$$

which can also be written as

$$(1+i)^n = 1 + A_n^E \cdot \frac{i}{r}.$$

Taking the natural logarithm on both sides, we obtain

$$\ln(1+i)^n = \ln\left(1 + A_n^E \cdot \frac{i}{r}\right).$$

Solving the latter equation for n and using the equality $\ln(1+i)^n = n \cdot \ln(1+i)$, we get

$$n = \frac{\ln\left(1 + A_n^E \cdot \frac{i}{r}\right)}{\ln(1+i)}. \tag{2.6}$$

(5) Given the values A_n^E, n and r, we wish to determine the accumulation factor q resp. the per annum rate of interest $i = q - 1$. This is, from a mathematical point of view, a more complicated problem. First, from formula (2.4), we obtain

$$A_n^E \cdot i = r \cdot \left[(1+i)^n - 1\right].$$

This yields

$$A_n^E \cdot (1+i) - A_n^E = r \cdot (1+i)^n - r$$

$$r \cdot (1+i)^n - A_n^E \cdot (1+i) + (A_n^E - r) = 0$$

$$(1+i)^n - \frac{A_n^E}{r} \cdot (1+i) + \left(\frac{A_n^E}{r} - 1\right) = 0.$$

Now, the problem corresponds to finding the zeroes of a polynomial of degree n in the variable $q = 1 + i$ and in general, we cannot solve this equation for q. Therefore, we have to apply approximation methods for finding a root of this equation greater than one (remember that only values $q = 1 + i > 1$ are of interest). In Chapter 4, we discuss some procedures for finding a zero of a function numerically. However, there is another difficulty for big values of n: possibly there is no uniquely determined solution $q > 1$.

Similarly to problems of type (3), (4) and (5), we can formulate the corresponding problems, where the amount of an annuity A_n^E has been replaced by the present value of an annuity P_n^E. As an example, for the problem of type (3), this gives the equality

$$r = P_n^E \cdot \frac{i}{1 - (1+i)^{-n}}. \tag{2.7}$$

Analogously, one can formulate the corresponding problems for the case when the payments are made at the beginning of the years. This is left as an exercise for the reader. We illustrate the solution of the basic problems mentioned above by some examples.

Example 2.28   What is the principal required now in order to be able to withdraw 3,000 EUR at the end of each of ten years (beginning at the end of the first year) so that no money is left over after these ten years provided that a rate of interest of 5 per cent p.a. is offered?

This question is the same as asking for the present value of an annuity P_{10}^E with $r = 3,000$ and $i = 0.05$. We obtain

$$P_{10}^E = r \cdot \frac{1 - (1+i)^{-10}}{i} = 3,000 \cdot \frac{1 - (1+0.05)^{-10}}{0.05} = 23,165.20 \text{ EUR},$$

i.e. the principal required now to ensure an annual payment of 3,000 EUR at the end of each year over a period of ten years is 23,165.20 EUR.

Example 2.29 Mrs Thrifty intends to accumulate an amount of 25,000 EUR until the end of a period of 12 years. If she wants to make annual payments at the end of each year and the bank offers a rate of interest of 6.25 per cent p.a., what is the required annual payment r?
This is a problem of type (3) with $n = 12$, $A_{12}^E = 25,000$ and $i = 0.0625$, and so we obtain

$$r = A_{12}^E \cdot \frac{i}{(1+i)^{12} - 1} = 25,000 \cdot \frac{0.0625}{(1+0.0625)^{12} - 1} = 1,460.43 \text{ EUR},$$

i.e. the annual amount that Mrs Thrifty has to put into an account at the end of each of 12 years to reach her goal is 1,460.43 EUR.

Example 2.30 Assume that Kevin has 2,000 EUR left over at the end of each year and that he puts this amount into a savings account. How long does Kevin need in order to accumulate 20,000 EUR in his account provided that the rate of interest is 5.5 per cent p.a.?
This is a problem of type (4), and so we apply formula (2.6) with $r = 2,000$ EUR, $A_n^E = 20,000$ EUR and $i = 0.055$. This yields

$$n = \frac{\ln\left(1 + A_n^E \cdot \frac{i}{r}\right)}{\ln(1+i)} = \frac{\ln\left(1 + 20,000 \cdot \frac{0.055}{2,000}\right)}{\ln(1 + 0.055)} \approx 8.185,$$

i.e. after approximately 8.2 years Kevin has accumulated 20,000 EUR in his account.

Several payments per year

Next, we briefly discuss the modifications if there are several payment and interest periods per year. Remember that we assume that payment periods are identical to interest periods. As before, denote by m the number of payment periods per year, and $A_{n,m}$ and $P_{n,m}$ (either with superscript 'E' or 'B') denote the amount of the annuity resp. the present value of the annuity provided that the payments go over n years with m payment periods per year. In this case, the amount of the annuity is given by

$$A_{n,m}^E = r \cdot \frac{\left(1 + \frac{i}{m}\right)^{n \cdot m} - 1}{\frac{i}{m}}.$$

Furthermore, the present value $P_{n,m}^E$ of the annuity is given by

$$P_{n,m}^E = r \cdot \frac{1 - \left(1 + \frac{i}{m}\right)^{-n \cdot m}}{\frac{i}{m}}.$$

The corresponding values for the case when payments are made at the beginning of each period are obtained as

$$A_{n,m}^B = r \cdot \left(1 + \frac{i}{m}\right) \cdot \frac{\left(1 + \frac{i}{m}\right)^{n \cdot m} - 1}{\frac{i}{m}}$$

and

$$P_{n,m}^B = r \cdot \left(1 + \frac{i}{m}\right) \cdot \frac{1 - \left(1 + \frac{i}{m}\right)^{-n \cdot m}}{\frac{i}{m}}.$$

Example 2.31   Claudia wants to put aside 100 EUR at the end of each month for a period of seven years. The bank offers a rate of interest of 3.5 per cent p.a. compounded monthly. What is the amount in Claudia's account after seven years?

This is a problem of type (1), and Claudia needs to determine the amount of an annuity $A_{7,12}^E$, where payments are made at the end of each period. Using $r = 100$ EUR and $i = 0.035$, she obtains

$$A_{7,12}^E = r \cdot \frac{\left(1 + \frac{i}{12}\right)^{7 \cdot 12} - 1}{\frac{i}{12}} = 100 \cdot \frac{\left(1 + \frac{0.035}{12}\right)^{84} - 1}{\frac{0.035}{12}} = 9{,}502.83 \text{ EUR},$$

i.e. after seven years, there are 9,502.83 EUR in Claudia's account.

2.3.3   Loan repayments, redemption tables

One application of periodic payments is loan repayments. First, we again consider annual payments and later we briefly discuss the modifications in the case of several payments per year. We use the following notations for year k:

D_k current amount of the loan (i.e. the debt) at the end of year $k, k = 1, 2, \ldots, n$ (more precisely, after the payment at the end of year k), moreover let D_0 be the initial amount of the loan;

I_k interest payment on the loan at the end of the kth year, $k = 1, 2, \ldots, n$;

R_k amortization instalment at the end of the kth year (i.e. the repayment on the loan), $k = 1, 2, \ldots, n$;

A_k annuity of the kth year, i.e. the amount that is paid at the end of year $k, A_k = R_k + I_k$, $k = 1, 2, \ldots, n$.

A loan with a fixed rate of interest is said to be *amortized* if both principal (i.e. the amount of the loan) and interest are paid by a sequence of payments made over equal time periods.

We consider the following two variants of loan repayments:

(1) repayment by equal annuities (i.e. in each period a constant amount of amortization instalment and interest is paid);

(2) repayment by equal amortization instalments (i.e. the repayments on the loan are identical in all years, but different payments are made over the years, since interest payments vary and therefore annuities are different).

Equal annuities

In this case, we assume that the annuity is the same in each year, i.e.

$$A_k = I_k + R_k = A, \qquad \text{where } A \text{ is constant.}$$

First, we mention that the present value of all payments for paying off a loan corresponds to the present value P_n^E of periodic payments at the end of each year. Therefore, finding the annuity A corresponds to the problem of finding the annual payment if a rate of interest i, the duration n of paying off the loan and the present value of an annuity are known, and payments are made at the end of each year. Thus, it corresponds to a problem of type (3) in the previous subsection, and we can apply formula (2.7) with $P_n^E = D_0$ and $r = A$:

$$A = D_0 \cdot \frac{i}{1 - (1 + i)^{-n}} \tag{2.8}$$

The current amount D_k of the loan at the end of year k can be obtained as follows. First, if no payments would be made, the amount of the loan at the end of k years would be equal to $D_k^* = D_0 \cdot (1 + i)^k$ according to formula (2.1). Second, the annual payments at the end of each year result after k years in the amount A_k^E of an annuity with payment A at a rate of interest i for k years according to formula (2.4). Then the remaining amount of the loan at the end of year k (i.e. after the payment in year k) is given by the difference of the two amounts D_k^* and A_k^E, i.e.

$$D_k = D_k^* - A_k^E = D_0 \cdot (1 + i)^k - A \cdot \frac{(1 + i)^k - 1}{i}. \tag{2.9}$$

In formula (2.9), we can replace the initial amount of the loan D_0 using formula (2.5) with $r = A$ and obtain

$$D_k = A \cdot \frac{1 - (1 + i)^{-n}}{i} \cdot (1 + i)^k - A \cdot \frac{(1 + i)^k - 1}{i}$$

$$= \frac{A}{i} \cdot \left[(1 + i)^k - (1 + i)^{-n+k} - (1 + i)^k + 1 \right]$$

which gives

$$D_k = A \cdot \frac{1 - (1 + i)^{-n+k}}{i}. \tag{2.10}$$

For $k = 0$, formula (2.10) has the same structure as formula (2.5) for finding the present value of an annuity P_n^E. However, the following comment has to be added. When computing the annuity A according to formula (2.8), in practice this value is rounded to two decimal places. This leads to the situation that the annuity of the last year is usually a bit larger or smaller than the rounded value. Therefore, usually formula (2.9) is applied to the rounded value A, and formula (2.10) is only exact when the same annuity is paid in all years. (Alternatively, one can use a slightly different annuity in the last period so that $D_n = 0$ is obtained.)

The interest payment I_k at the end of year k can be obtained as follows by taking into account that interest has to be paid on the current amount D_{k-1} of the loan at the end of year $k - 1$:

$$I_k = D_{k-1} \cdot i = \left[D_0 \cdot (1+i)^{k-1} - A \cdot \frac{(1+i)^{k-1} - 1}{i} \right] \cdot i$$

$$= D_0 \cdot i \cdot (1+i)^{k-1} - A \cdot (1+i)^{k-1} + A = A - (A - D_0 \cdot i) \cdot (1+i)^{k-1}.$$

Because $I_k = A - R_k$, we immediately get the repayment R_k on the loan in year k from the last formula:

$$R_k = (A - D_0 \cdot i) \cdot (1+i)^{k-1}.$$

From the formula before last we again see that $I_1 = D_0 \cdot i$. As a conclusion, the amortization instalments $\{R_k\}$ form a geometric sequence with the first term

$$R_1 = A - D_0 \cdot i$$

and the quotient $q = 1 + i$, i.e. we have

$$R_k = R_1 \cdot (1+i)^{k-1}.$$

It is worth noting that in the case of paying in each year exactly the same annuity, the formulas for determining I_k and R_k can be simplified as follows:

$$I_k = D_{k-1} \cdot i = A \cdot (1 - (1+i)^{k-1-n}),$$

$$R_k = A - I_k = A \cdot (1+i)^{k-1-n}.$$

It should be mentioned again that, since all values are rounded to two decimal places, there may be slight rounding differences depending on the formula used.

In the case of m payments per year and m interest payments, we have to replace the per annum rate of interest i again by the rate of interest i/m per period and the exponent n (which is equal to the number of compoundings) by $n \cdot m$. This yields an annuity A'_m to be paid in each of the m payment periods per year

$$A'_m = D_0 \cdot \frac{\frac{i}{m}}{1 - \left(1 + \frac{i}{m}\right)^{-n \cdot m}}.$$

Similarly, the two replacements concerning the interest per period and the number of payment periods have to be made in the other formulas.

Equal amortization instalments

In this case, we assume that the repayment on the loan is the same in each year, i.e.

$$R_k = R, \qquad \text{where } R \text{ is constant.}$$

This means that in each year the repayment on the loan is

$$R = \frac{D_0}{n}.$$

Therefore, the remaining amount of the loan at the end of year k is given by

$$D_k = D_0 - k \cdot \frac{D_0}{n} = D_0 \cdot \left(1 - \frac{k}{n}\right).$$

The amounts $\{D_k\}$ form an arithmetic sequence with difference $-D_0/n$. In period k, interest has to be paid on the current amount of the loan at the end of period k, i.e.

$$I_k = D_{k-1} \cdot i = D_0 \cdot \left(1 - \frac{k-1}{n}\right) \cdot i.$$

Thus, the annuity A_k of year k is given by

$$A_k = R_k + I_k = \frac{D_0}{n} + D_{k-1} \cdot i$$

$$= \frac{D_0}{n} + D_0 \cdot \left(1 - \frac{k-1}{n}\right) \cdot i = \frac{D_0}{n} \cdot [1 + (n - k + 1) \cdot i].$$

In the case of m payment and interest periods per year, we have to replace i by i/m and n by $n \cdot m$. A summary of the payments and the present values of the loan can be given in a *redemption* or *amortization table*. It contains for each period k:

(1) the current amount of the loan at the beginning (D_{k-1}) and at the end (D_k) of the period;
(2) the interest payment I_k;
(3) the amortization instalment R_k;
(4) the annuity (sum of interest payment plus amortization instalment), i.e. $A_k = I_k + R_k$.

We illustrate the two types of loan repayments by the following examples.

Example 2.32 Tom intends to amortize a loan of 10,000 EUR at a rate of interest of 7 per cent p.a. in six years. He decides to make annual equal payments at the end of each year. Using $D_0 = 10,000$ EUR, $i = 0.07$ and $n = 6$, the annual annuity is obtained as

$$A = D_0 \cdot \frac{i}{1 - (1 + i)^{-n}} = 10,000 \cdot \frac{0.07}{1 - (1 + 0.07)^{-6}} = 2,097.96 \text{ EUR.}$$

For the first year (i.e. $k = 1$), we get

$$I_1 = D_0 \cdot i = 10,000 \cdot 0.07 = 700.00 \text{ EUR},$$

$$R_1 = A - I_1 = 2,097.96 - 700.00 = 1,397.96 \text{ EUR},$$

$$D_1 = D_0 - R_1 = 10,000 - 1,397.96 = 8,602.04 \text{ EUR}.$$

For the second year (i.e. $k = 2$), we get

$$I_2 = D_1 \cdot i = 8,602.04 \cdot 0.07 = 602.14 \text{ EUR},$$

$$R_2 = A - I_2 = 2,097.96 - 602.14 = 1,495.82 \text{ EUR},$$

$$D_2 = D_1 - R_2 = 8,602.04 - 1,495.82 = 7,106.22 \text{ EUR}.$$

Notice that the amount D_2 can also be directly determined by using formula (2.9) (without computing D_1 first). A complete redemption table is given as Table 2.1. It is worth noting that in the last year, the annual annuity is one cent lower than in the other years which is caused by rounding the value of A to two decimal places. Applying the given formulas, we obtain $R_6 = 1,960.71$ EUR but $D_5 = 1,960.70$ EUR. Therefore, one cent less is paid in year six.

Table 2.1 Redemption table for Example 2.32 (equal annuities, annual payments)

Period (year)	Current amount of the loan at the beginning	Interest	Amortization instalment	Annuity	Amount of the loan at the end
k	D_{k-1}	I_k	R_k	A	D_k
1	10,000.00	700.00	1,397.96	2,097.96	8,602.04
2	8,602.04	602.14	1,495.82	2,097.96	7,106.22
3	7,106.22	497.44	1,600.52	2,097.96	5,505.70
4	5,505.70	385.40	1,712.56	2,097.96	3,793.14
5	3,793.14	265.52	1,832.44	2,097.96	1,960.70
6	1,960.70	137.25	1,960.70	2,097.95	0.00

Example 2.33   The Happy family have raised a mortgage on their house. They can repay the loan of 120,000 EUR at a rate of interest of 8 per cent p.a. over a period of 10 years. They decide to pay quarterly rates. This means that there are $n \cdot m = 10 \cdot 4 = 40$ payment periods. The quarterly annuity is obtained as

$$A'_4 = D_0 \cdot \frac{\frac{i}{m}}{1 - \left(1 + \frac{i}{m}\right)^{-n \cdot m}}$$

$$= 120,000 \cdot \frac{\frac{0.08}{4}}{1 - \left(1 + \frac{0.08}{4}\right)^{-10 \cdot 4}} = 120,000 \cdot \frac{0.02}{1 - 1.02^{-40}} = 4,386.69 \text{ EUR}.$$

Table 2.2 is a redemption table that includes the corresponding values after each year (i.e. the corresponding amounts for periods $4, 8, \ldots 40$) to give an overview on the payments. As an

Table 2.2 Redemption table for Example 2.33 (equal annuities, quarterly payments)

Period (quarter)	Amount of the loan at the beginning	Interest	Amortization instalment	Annuity	Amount of the loan at the end
$4k$	$D_{k,3}$	$I_{k,4}$	$R_{k,4}$	A'_4	$D_{k,4}$
4	113,919.93	2,278.40	2,108.29	4,386.69	111,811.64
8	105,230.38	2,104.61	2,282.08	4,386.69	102,948.30
12	95,824.54	1,916.49	2,470.20	4,386.69	93,354.34
16	85,643.34	1,712.87	2,673.82	4,386.69	82,989,52
20	74,622.89	1,492.46	2,894.23	4,386.69	71,728.66
24	62,694.00	1,253.88	3,132.81	4,386.69	59,561.19
28	49,781.78	995.64	3,391.05	4,386.69	46,390.73
32	35,805.19	716.10	3,670.59	4,386.69	32,134.60
36	20,676.47	413.53	3,973.16	4,386.69	16,703.31
40	4,300.66	86.01	4,300.66	4,386.67	0.00

illustration, we compute some of the values given in Table 2.2. It is worth emphasizing that we have to adapt the formulas given for the annual payments. At the end of the second year, 8 payments have been made. Denoting by the index pair (k, m) the mth period in the kth year, we get for period 8:

$$R_{2,4} = R_{1,1} \cdot \left(1 + \frac{i}{4}\right)^{2 \cdot 4 - 1} = \left(A'_4 - D_0 \cdot \frac{i}{4}\right) \cdot \left(1 + \frac{i}{4}\right)^{2 \cdot 4 - 1}$$

$$= (4,386.69 - 120,000 \cdot 0.02) \cdot 1.02^7 = 1,986.69 \cdot 1.02^7 = 2,282.08 \text{ EUR},$$

$$I_{2,4} = A'_4 - R_{2,4} = 4,386.69 - 2,282.08 = 2,104.61 \text{ EUR},$$

$$D_{2,4} = D_0 \cdot \left(1 + \frac{i}{4}\right)^{2 \cdot 4} - A'_4 \cdot \frac{\left(1 + \frac{i}{4}\right)^{2 \cdot 4} - 1}{\frac{i}{4}}$$

$$= 120,000 \cdot 1.02^8 - 4,386.69 \cdot \frac{1.02^8 - 1}{0.02} = 102,948.30 \text{ EUR}.$$

Similarly, at the end of the fifth year, 20 payments have been made. We get for period 20:

$$R_{5,4} = R_{1,1} \cdot \left(1 + \frac{i}{4}\right)^{5 \cdot 4 - 1} = 1,986.69 \cdot 1.02^{19} = 2,894.23 \text{ EUR},$$

$$I_{5,4} = A'_4 - R_{5,4} = 4,386.69 - 2,894.23 = 1,492.46 \text{ EUR},$$

$$D_{5,4} = D_0 \cdot \left(1 + \frac{i}{4}\right)^{5 \cdot 4} - A'_4 \cdot \frac{\left(1 + \frac{i}{4}\right)^{5 \cdot 4} - 1}{\frac{i}{4}}$$

$$= 120,000 \cdot 1.02^{20} - 4,386.69 \cdot \frac{1.02^{20} - 1}{0.02} = 71,728.66 \text{ EUR}.$$

Notice that for the computation of $R_{5,4}$, amount $R_{1,1}$ has been taken from the previous computation of amount $R_{2,4}$. Again, the annuity of the last period is a bit smaller than the annuities in the previous periods since the amount of the loan at the end of period 39 is

$$D_{10,3} = D_0 \cdot \left(1 + \frac{i}{4}\right)^{39} - A_4' \cdot \frac{\left(1 + \frac{i}{4}\right)^{39} - 1}{\frac{i}{4}}$$

$$= 120,000 \cdot 1.02^{39} - 4,386.69 \cdot \frac{1.02^{39} - 1}{0.02} = 4,300.66 \text{ EUR},$$

and the interest payment in period 40 on the remaining loan is $I_{10,4} = D_{10,3} \cdot i/4 = 86.01$ EUR.

Example 2.34 The Young family intend to repay a loan of 9,000 EUR over a period of six years by equal amortization instalments at the end of each year. The rate of interest on the loan is 7.5 per cent p.a. The complete redemption table is given as Table 2.3. We again compute only some of the values given in Table 2.3. For the second year, we get

$$R_2 = R = \frac{D_0}{n} = \frac{9,000}{6} = 1,500 \text{ EUR},$$

$$I_2 = D_0 \cdot \left(1 - \frac{1}{n}\right) \cdot i = 9,000 \cdot \left(1 - \frac{1}{6}\right) \cdot 0.075 = 562.50 \text{ EUR},$$

$$D_2 = D_0 \left(1 - \frac{2}{n}\right) = 9,000 \cdot \left(1 - \frac{2}{6}\right) = 6,000 \text{ EUR}.$$

Reminding ourselves that $R_4 = R$, we get for the fourth year

$$I_4 = D_0 \cdot \left(1 - \frac{3}{n}\right) \cdot i = 9,000 \cdot \left(1 - \frac{3}{6}\right) \cdot 0.075 = 337.50 \text{ EUR},$$

$$D_4 = D_0 \cdot \left(1 - \frac{4}{n}\right) = 9,000 \cdot \left(1 - \frac{4}{6}\right) = 3,000 \text{ EUR}.$$

Table 2.3 Redemption table for Example 2.34 (equal amortization instalments)

Period (year)	Amount of the loan at the beginning	Interest	Amortization instalment	Annuity	Amount of the loan at the end
k	D_{k-1}	I_k	R_k	A_k	D_k
1	9,000.00	675.00	1,500.00	2,175.00	7,500.00
2	7,500.00	562.50	1,500.00	2,062.50	6,000.00
3	6,000.00	450.00	1,500.00	1,950.00	4,500.00
4	4,500.00	337.50	1,500.00	1,837.50	3,000.00
5	3,000.00	225.00	1,500.00	1,725.00	1,500.00
6	1,500.00	112.50	1,500.00	1,612.50	0.00

2.3.4 Investment projects

The considerations about periodic payments are also useful for evaluating a planned investment project or for comparing several alternatives. In this subsection, we discuss the following
two methods:

(1) method of net present value;
(2) method of internal rate of return.

In both methods, the expected costs and returns are considered for a couple of years. For
further considerations, we introduce the following notations:

n	considered time period (duration) of the project in years;
C_k	(expected) costs connected with the project in year k $(k = 0, 1, \ldots, n)$;
R_k	(expected) returns in year k $(k = 0, 1, \ldots, n)$, usually $R_0 = 0$;
$B_k = R_k - C_k$	(expected) balance in year k $(k = 0, 1, \ldots, n)$;
i	per annum rate of interest (at which the company can invest money otherwise).

In the following, we always assume that the balances B_k for year k are taken into account at
the end of the corresponding year. Moreover, we assume that the rate of interest i is constant
over the years. However, using the comments with variable rates of interest (see part about
compound interest), one can immediately adapt the given formulas to this case.

Method of net present value

The idea of this method is to compare the present values of the expected balances over the
years. The net present value *NPV* is defined as

$$NPV = \sum_{k=0}^{n} B_k \cdot (1 + i)^{-k}$$

$$= B_0 + B_1 \cdot (1 + i)^{-1} + B_2 \cdot (1 + i)^{-2} + \ldots + B_n \cdot (1 + i)^{-n}. \tag{2.11}$$

In the above formula, $B_k \cdot (1 + i)^{-k}$ gives the present value of the balance B_k (i.e. arising at
the end of year k and discounted to the time of the beginning of the project). Factor $(1+i)^{-k}$
is the *reduction factor* applied in year k for finding the present value of the kth balance.
Finally, the net present value is the sum of the present values of all the expected balances
within the considered time period of n years.

If the net present value *NPV* is positive, the project is *advantageous*. In the case of a negative
net present value, the project is *not recommendable*. If several investment possibilities are
compared, the variant with the highest net present value has to be chosen, provided that this
value is greater than zero.

Example 2.35 A company can open a project which is going over a period of six years.
The expected costs and returns are given in the second and third columns of Table 2.4.
Moreover, the rate of interest at which the company can invest money otherwise is equal

Table 2.4 Net present value for the investment project in Example 2.35

Year k	Costs C_k	Returns R_k	Balance B_k	Reduction factor $(1+i)^{-k}$	Present value of balance B_k
0	160,000	0	−160,000	1.000000000	−160,000.00
1	10,000	45,000	35,000	0.914731192	32,110.09
2	12,000	60,000	48,000	0.841679993	40,400.64
3	15,000	67,000	52,000	0.772183480	40,153.54
4	16,000	74,000	58,000	0.708425211	41,088.66
5	25,000	55,000	30,000	0.649931386	17,888.02
				NPV:	11,490.02

to 8 per cent p.a. We apply the method of net present value to evaluate whether the intended project is recommendable. In order to determine *NPV* according to formula (2.11), we compute first the balance for each year (column 4) and then the discounted present value of each balance (column 6). Moreover, the reduction factor $(1+i)^{-k}$ which has to be applied for the corresponding year is given (column 5). The sum of the present values ($NPV = 11,490.02$, see last row) is positive. Therefore, the company decides to implement the project.

Method of internal rate of return

This method assumes that the sum of the present values of the costs over the years is equal to the sum of the present values of the returns. In other words, it is assumed that the net present value is equal to zero. From this condition the rate of interest i (i.e. the internal rate of return) is determined and compared with the expected minimum rentability r. If $i > r$, then the project is *advantageous*, since it has a higher internal rate of return. In the case of $i < r$, the project is *rejected*. If several possible investment projects have to be compared, the project with the highest internal rate of return is chosen, provided that $i > r$.

Example 2.36   A project which is going over two years has the (expected) costs C_k and returns R_k at the end of year k given in Table 2.5. Moreover, the company expects a minimum rentability of 7.5 per cent p.a.
We apply the method of internal rate of return. From $NPV = 0$, we obtain

$$-35,000 + 17,500 \cdot (1+i)^{-1} + 22,000 \cdot (1+i)^{-2} = 0$$

which yields after multiplying by $-(1+i)^2$ the equation

$$35,000 \cdot (1+i)^2 - 17,500 \cdot (1+i) - 22,000 = 0$$

or equivalently

$$(1+i)^2 - \frac{1}{2} \cdot (1+i) - \frac{22}{35} = 0.$$

Table 2.5 Data for the investment
project in Example 2.36

Year k	Costs C_k	Returns R_k	Balance B_k
0	35,000	0	−35,000
1	6,000	23,500	17,500
2	9,000	31,000	22,000

This is a quadratic equation in the variable $q = 1 + i$, and we obtain the following two solutions for q:

$$q_1 = \frac{1}{4} + \sqrt{\frac{1}{16} + \frac{22}{35}} \approx 0.25 + 0.83130706 \approx 1.08131,$$

$$q_2 \approx 0.25 - 0.83130706 \approx -0.58131.$$

From $q_1 \approx 1.08131$ we obtain approximately the rate of interest $p = 100 \cdot (q_1 - 1) \approx 8.131$ per cent p.a. The second solution q_2 is negative and therefore not of interest, since it leads to a negative rate of interest. Because $i \approx 0.0813 > r = 0.075$, the project appears to be worthwhile.

Finally, we consider an example comparing both methods.

Example 2.37 An enterprise has the possibility to realize two projects A and B. Both projects are going over two years and the expected costs and returns are given in columns 2 and 3 of Table 2.6. Alternatively the firm can invest money at a rate of interest of 8 per cent p.a.

We apply both methods to decide whether one of the projects can be chosen and if so which of the projects is better. Based on the balance B_k for the corresponding year k (column 4), in column 5 the reduction factors for computing the present values of the balances for each of both projects are given (column 6). Summing up the present values of the balances, we get a net present value of 1,242.80 EUR for project A and a net present value of −754.46 EUR for project B. According to the method of net present value, project A is chosen, since only project A has a positive net present value. When applying the method of internal rate of return, we get for project A from $NPV = 0$:

$$-36,000 \cdot (1 + i)^2 + 18,000 \cdot (1 + i) + 24,000 = 0$$

which can be written as

$$(1 + i)^2 - \frac{1}{2} \cdot (1 + i) - \frac{2}{3} = 0.$$

This quadratic solution has two solutions in the variable $q = 1 + i$, namely

$$q_1 = \frac{1}{4} + \sqrt{\frac{3 + 32}{48}} \approx \frac{1}{4} + 0.853912563 \approx 1.103912563$$

Table 2.6 Data for projects A and B given in Example 2.37

Year k	Costs C_k	Returns R_k	Balance B_k	Reduction factor $(1+i)^{-k}$	Present value of balance B_k
			Project A		
0	36,000	0	$-36,000$	1.000000000	$-36,000.00$
1	4,000	22,000	18,000	0.925925925	16,666.67
2	7,000	31,000	24,000	0.857338820	20,576.13
				NPV	1,242.80
			Project B		
0	75,000	0	$-75,000$	1.000000000	$-75,000.00$
1	12,000	32,000	20,000	0.925925925	18,518.52
2	16,000	81,000	65,000	0.857338820	55,727.02
				NPV	-754.46

and

$$q_2 \approx \frac{1}{4} - 0.853912563 < 0.$$

Since the second solution leads to a negative rate of interest, we obtain from the first solution the internal rate of return $i \approx 0.1039$, or equivalently $p \approx 10.39$ per cent. Correspondingly, for project B we get from $NPV = 0$

$$-75,000 \cdot (1+i)^2 + 20,000 \cdot (1+i) + 65,000 = 0$$

which yields the quadratic equation

$$(1+i)^2 - \frac{4}{15} \cdot (1+i) - \frac{13}{15} = 0.$$

We obtain the two solutions in the variable $q = 1 + i$

$$q_1 = \frac{2}{15} + \sqrt{\frac{4+195}{225}} \approx \frac{2}{15} + 0.940449065 \approx 1.073782398$$

and

$$q_2 \approx \frac{2}{15} - 0.940449065 \approx -0.807115731 < 0.$$

The first solution yields the internal rate of return of $i \approx 0.0738$, or correspondingly 7.38 per cent.

Using the method of internal rate of return, we use $r = 0.08$ as minimum rentability. We have found that project A leads to an internal rate of return of 10.39 per cent while project B yields an internal rate of return of only 7.38 per cent. By the method of internal rate of return, project A is also preferred, since only project A has an internal rate higher than the minimum rentability of 8 per cent. So both methods yield the same recommendation concerning the possible realization of one of the projects.

2.3.5 Depreciation

The book value of a machine or an industrial facility decreases over the years. Next, we discuss two variants of depreciation on machinery, namely *linear depreciation* and *degressive depreciation*.

Linear depreciation

In the case of linear depreciation, the depreciation rate D per year is constant. Denoting by B_0 the initial book value (i.e. the cost of the machine or facility) and by B_n the final value after the useful life of the machinery, then the depreciation rate D per year is obtained as

$$D = \frac{B_0 - B_n}{n},$$

and we obtain the book value B_k after k years as follows:

$$B_k = B_{k-1} - D = B_1 - (k-1) \cdot D.$$

The book values $\{B_k\}$ form an arithmetic sequence with difference $-D$.

Example 2.38 A facility has been bought for 50,000 EUR. The useful life is six years, and the final book value is 8,000 EUR. Therefore, the annual depreciation rate is

$$D = \frac{B_n - B_0}{n} = \frac{50,000 - 8,000}{6} = 7,000 \text{ EUR.}$$

We obtain the depreciation table given as Table 2.7.

Table 2.7 Depreciation table for Example 2.38 (linear depreciation)

Year k	Book value at the beginning B_{k-1}	Depreciation amount D_k	Book value at the end B_k
1	50,000	7,000	43,000
2	43,000	7,000	36,000
3	36,000	7,000	29,000
4	29,000	7,000	22,000
5	22,000	7,000	15,000
6	15,000	7,000	8,000

Degressive depreciation

We discuss two types of degressive depreciation, namely arithmetic-degressive depreciation and geometric-degressive depreciation. We begin with the *arithmetic-degressive depreciation*. This type of depreciation is characterized by a linearly decreasing depreciation amount, i.e.

$$D_k = D_{k-1} - d.$$

Here we consider only the special case of $D_n = d$, i.e. the last depreciation amount is equal to the annual reduction in the depreciation amount (*digital depreciation*). The depreciation amounts $\{D_k\}$ form an arithmetic sequence with the first term

$$D_1 = n \cdot d$$

and the difference $-d$. The kth depreciation amount in dependence on D_1 and d is given by

$$D_k = D_1 - (k-1) \cdot d.$$

The book value B_k at the end of year k is given by

$$B_k = B_0 - \sum_{i=1}^{k} D_i = B_0 - \frac{k}{2} \cdot (D_1 + D_k)$$

$$= B_0 - \frac{k}{2} \cdot [n \cdot d + (n - k + 1) \cdot d] = B_0 - \frac{k}{2} \cdot (2n - k + 1) \cdot d.$$

Next, we deal with *geometric-degressive depreciation*. This type of depreciation is characterized by a constant depreciation percentage p with $i = p/100$ in comparison with the current book value, i.e. the depreciation amount in year k is given by

$$D_k = B_{k-1} \cdot i, \quad k = 1, 2, \ldots, n.$$

In this case, we get the book value

$$B_k = B_0 \cdot (1 - i)^k \tag{2.12}$$

at the end of year k. Thus, the book values $\{B_k\}$ form a geometric sequence, and the depreciation rate in terms of B_0 and i for the kth year is given by

$$D_k = B_{k-1} \cdot i = B_0 \cdot (1 - i)^{k-1} \cdot i.$$

Typically, the initial book value B_0 (i.e. the cost of the machine or a facility) and the book value at the end of the useful life B_n are given. Then the depreciation percentage p resp. i are obtained from formula (2.12) for $k = n$ as follows:

$$p = 100 \cdot i \qquad \text{with} \qquad i = 1 - \sqrt[n]{\frac{B_n}{B_0}}.$$

Notice that in the case of geometric-degressive depreciation, a final book value of $B_n = 0$ is not possible.

Example 2.39   A company buys a machine for 40,000 EUR and intends to use it for 8 years. The final book value B_8 is 6,000 EUR. Value i is obtained as

$$i = 1 - \sqrt[8]{\frac{B_8}{B_0}} = 1 - \sqrt[8]{\frac{6,000}{40,000}} \approx 0.211119161.$$

We present a depreciation table for the case of geometric-degressive depreciation as Table 2.8. For instance, the values for the third year are obtained as follows:

$$B_2 = B_0 \cdot (1 - i)^2 \approx 40,000 \cdot 0.788880838^2 = 24,893.32 \text{ EUR},$$

$$D_3 = B_0 \cdot (1 - i)^2 \cdot i = 5,255.46 \text{ EUR},$$

$$B_3 = B_0 \cdot (1 - i)^3 = 19,637.86 \text{ EUR}.$$

Table 2.8 Depreciation table for Example 2.39 (geometric-degressive depreciation)

Year k	Book value at the beginning B_{k-1}	Depreciation amount D_k	Book value at the end B_k
1	40,000.00	8,444.77	31,555.23
2	31,555.23	6,661.91	24,893.32
3	24,893.32	5,255.46	19,637.86
4	19,637.86	4,145.93	15,491.93
5	15,491.93	3,270.64	12,221.29
6	12,221.29	2,580.15	9,641.14
7	9,641.14	2,035.43	7,605.71
8	7,605.71	1,605.71	6,000.00

EXERCISES

2.1 (a) An arithmetic sequence $\{a_n\}$ has a first term $a_1 = 15$ and difference $d = 8$. Find the term a_{101}.

 (b) For an arithmetic sequence, the terms $a_8 = 21$ and $a_{10} = 25$ are known. Find difference d and the terms a_1 and a_n.

2.2 The sequence $\{a_n\}$ has

$$a_n = 3 + \frac{n}{2}, \qquad n \in \mathbb{N}.$$

 (a) Is sequence $\{a_n\}$ decreasing or increasing?
 (b) Is $\{a_n\}$ a bounded sequence?
 (c) Find a recursive formula for this sequence.

2.3 (a) Let $\{a_n\}$ be a geometric sequence with the ratio of successive terms $q = -2/3$ and the term $a_7 = 64/243$. Find the first term a_1. Which of the terms is the first with an absolute value less than 0.01?

    (b)   A geometric sequence has the terms $a_2 = 6$ and $a_7 = 2/81$. Find the first term a_1 and the ratio of successive terms q.

2.4   Given are the sequences

$$\{a_n\} = \left\{\frac{3}{n} - 5\right\}, \quad \{b_n\} = \left\{\frac{7 - 2n}{n^2}\right\} \quad \text{and} \quad \{c_n\} = \left\{\frac{2^n}{n!}\right\}, \quad n \in \mathbb{N}.$$

Are these sequences monotone and bounded? Find the limits of the sequences if they exist.

2.5   Find the limits of the following sequences if they exist ($n \in \mathbb{N}$).

    (a) $\{a_n\} = \left\{\dfrac{2n^n}{(n+1)^n}\right\};$ (b) $\{b_n\} = \left\{\dfrac{an^4 + n^3}{3n^3 + 4n}\right\}, \quad a \in \mathbb{R};$

    (c) $\{c_n\}$ with $c_n = c_{n-1}/2$. Check it for $c_1 = 1$ and $c_1 = 4$.

2.6   A textile company starts in its first year with an output of 2,000 shirts and 1,000 trousers. The production of shirts increases by 500 shirts every year, and the number of trousers rises by 20 per cent every year.

    (a)   How many shirts and trousers are produced in the second, third and tenth years?
    (b)   What is the total output of shirts and trousers after 15 years? (Use given formulas.)

2.7   Find the limit $s = \lim\limits_{n \to \infty} s_n$ for the following series $\{s_n\}$:

    (a) $s_n = \displaystyle\sum_{k=1}^{n} \frac{(-2)^{k+1}}{5^k};$ (b) $s_n = \displaystyle\sum_{k=1}^{n} \frac{1}{k(k+1)}$

    (Hint: find a formula for s_n.)

2.8   Check with respect to the real numbers x whether the following series $\{s_n\}$ converge:

    (a) $s_n = \displaystyle\sum_{k=1}^{n} \frac{x^k}{k};$ (b) $s_n = \displaystyle\sum_{k=1}^{n} \frac{k^2}{x^k}.$

2.9   Consider the following series $\{s_n\}$ with:

    (a) $s_n = \displaystyle\sum_{k=1}^{n} (-1)^k \cdot \frac{4(k+2)}{k^2};$ (b) $s_n = \displaystyle\sum_{k=1}^{n} \frac{k^2}{k!};$

    (c) $s_n = \displaystyle\sum_{k=1}^{n} \left(\frac{k}{k+1}\right)^{k^2};$ (d) $s_n = \displaystyle\sum_{k=1}^{n} \left[(-1)^k + \frac{1}{k}\right].$

Find the first four terms of the series and calculate the partial sums s_1, s_2, s_3 and s_4. Check whether the series converge.

2.10   Which is the best investment for a family wishing to invest 10,000 EUR for 10 years:

    (a)   an interest rate of 6 per cent annually for 10 years;
    (b)   an interest rate of 7 per cent for the first 5 years and 5 per cent for the next 5 years;
    (c)   an interest rate of 7.5 per cent for the first 2 years, then 6 per cent for the next 4 years and 5.5 per cent for the last 4 years?

2.11 A grandfather wants to give his grandson 100,000 EUR at his thirtieth birthday. How much does he need to invest on the grandson's eighteenth birthday if the bank offers a rate of interest of 6 per cent p.a.?

2.12 What interest rate will turn 2,000 EUR into 3,582 EUR after ten years?

2.13 Tom invests 10,000 EUR. The bank offers a rate of interest of 5 per cent p.a.

(a) What is the amount after 10 years when interest is compounded annually, quarterly and monthly, respectively?

(b) Find the effective rate of interest i_{eff} for quarterly and monthly compounding.

2.14 Ann has given 5,000 EUR to a bank on 20 October 2003, and she wants to get the money back on 20 April 2006. What is the amount at the end if the bank has credited interest of 3 per cent p.a. compounded annually?

2.15 You decide to put away 200 EUR every month. Bank A offers simple interest of 6 per cent per annum, and bank B offers 0.48 per cent per month compounded monthly. After making 12 deposits, how much money would you have at the end of one year in banks A and B, respectively?

(a) The deposits are made at the same time the interest is credited (i.e. at the end of each month).

(b) The deposits are made at the beginning of each month and the interest is credited at the end.

2.16 A man wants to buy a car for 24,000 EUR. He decides to pay a first instalment of 25 per cent and the remainder with a bank loan under the following conditions. The borrowed sum will be paid back in one payment at the end of three years. During these 36 months interest is to be paid monthly at a per annum rate of 4.8 per cent. At the same time a sinking fund has to be set up in order to repay the loan. The man has to make monthly deposits into his sinking fund, which earns 3 per cent interest per annum compounded monthly.

(a) Find the sinking fund deposit.

(b) How much money does the man need each month to pay the sum of deposit and interest?

2.17 What is the annual deposit required to pay out 10,000 EUR after five years? Interest of 5 per cent per annum is compounded annually.

2.18 (a) Find the present value of an annuity at a rate of 5 per cent p.a. and with a withdrawal of 6,000 EUR at the end of each of 15 years.

(b) How does the present value change if the withdrawal at the end of each month is 500 EUR, assuming an ordinary annuity at 5 per cent p.a.?

2.19 On 1 January 2015, Peter will retire. He has paid an instalment of 20,000 EUR on 1 January 2000 and after that, he makes annual payments of 2,000 EUR every year up to 2014 on 1 January. The bank offers 5 per cent p.a. over the whole period.

(a) What is the amount on 1 January 2015?

(b) Beginning with 31 January 2015, Peter wants to make a monthly withdrawal of 500 EUR. What is the present value of the annuity after ten years? Interest is compounded monthly at a rate of 5 per cent p.a.

2.20 A family has raised a mortgage of 150,000 EUR on its house. It can amortize the balance at 8.5 per cent p.a. for ten years.

(a) What are the annual payments? What is the total payment? Give a complete redemption table with the current amount of the loan at the beginning and at the end of each year, interest payment and amortization instalment for all the 10 years.

(b) Find the total payment if the family repays 15,000 EUR plus the interest on the current amount of the loan every year (equal amortization instalments).

(c) How many years does the family repay when the annual annuity is $A = 18,000$ EUR?

2.21 A project which is going over two years has the expected costs and returns at the end of year k (in EUR) given in the following table:

Year	Costs	Returns
k	C_k	R_k
0	40,000	0
1	5,000	18,000
2	6,000	35,000

(a) Let the rate of interest be $i = 5$ per cent p.a. Check by means of the method of net present value and by the method of internal rate of return whether the project should go ahead.

(b) What are the percentage rates of interest (use only integers) for the project being advantageous?

(c) In the third year, the costs for the project will be $C_3 = 6,000$ EUR and the returns $R_3 = 36,000$ EUR. Is it now profitable to implement the project with $i = 5$ per cent p.a.?

2.22 A transport company has bought a bus for 100,000 EUR. After eight years, the bus is sold to another company for 44,000 EUR. Compare depreciation amounts and book values at the end of each of the eight years when the company uses

(a) linear depreciation,
(b) arithmetic-degressive depreciation,
(c) geometric-degressive depreciation.

3 Relations; mappings; functions of a real variable

In this chapter, we deal with relations and a specification, namely mappings. Furthermore we introduce special mappings that are functions of one real variable, and we survey the most important types of functions of a real variable.

3.1 RELATIONS

A *relation* indicates a relationship between certain objects. A *binary relation* R from a set A to a set B assigns to each pair $(a, b) \in A \times B$ one of the two propositions:

(1) a is related to b by R, in symbols: aRb;
(2) a is not related to b by R, in symbols: $a\cancel{R}b$.

Definition 3.1   A *(binary) relation* R from a set A to a set B is a subset of the Cartesian product $A \times B$, i.e. $R \subseteq A \times B$.

We describe a relation R in one of the two following forms: $(a, b) \in R \iff aRb$ or $R = \{(a, b) \in A \times B \mid aRb\}$. If R is a relation from set A to set A, we also say that R is a relation on A.

Example 3.1   Let us consider a relation $R \subseteq \mathbb{Z} \times \mathbb{Z}$ on $\mathbb{Z}$ as follows. An integer $i \in \mathbb{Z}$ is related to integer $j \in \mathbb{Z}$ by R if and only if they have the same remainder when divided by 5. This means that e.g. 7 is related to 17 by R (i.e. $7R17$) since both numbers give the remainder 2 when dividing by 5. Similarly, 9 is related to 4 by R (for both numbers, we get the remainder 4 when dividing them by 5), while 6 is not related to 13 (since 6 gives remainder 1, but 13 gives remainder 3) and 0 is not related to 23 (since 0 gives remainder 0, but 23 gives remainder 3). Moreover, relation R has the following two properties. First, every integer a is related to itself (i.e. aRa, such a relation is called reflexive), and relation R is symmetric, i.e. it always follows from aRb that bRa.

Next, we give the notions of an inverse and a composite relation.

Definition 3.2   Let $R \subseteq A \times B$ be a binary relation. Then the relation

$$R^{-1} = \{(b,a) \in B \times A \mid (a,b) \in R\} \subseteq B \times A$$

(read: R inverse) is the *inverse relation* of R.

The following definition treats the case when it is possible to 'concatenate' two relations.

Definition 3.3   Let A, B, C be sets and $R \subseteq A \times B$, $S \subseteq B \times C$ be binary relations. Then

$$S \circ R = \{(a,c) \in A \times C \mid \text{there exists some } b \in B \text{ such that } (a,b) \in R \wedge (b,c) \in S\}$$

is called the *composite relation* or *composition* of R and S.

The composition of two relations is illustrated in Figure 3.1. The element $a \in A$ is related to the element $c \in C$ by the composite relation $S \circ R$, if there exists (at least) one $b \in B$ such that simultaneously a is related to b by R and b is related to c by S. We emphasize that the sequence of the relations in the composition is important, since the relations are described by *ordered pairs*. Therefore, in general $S \circ R \neq R \circ S$ may hold, and one can interpret Definition 3.3 in such a way that first relation R and then relation S are used.

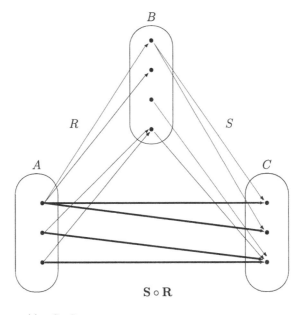

Figure 3.1 The composition $S \circ R$.

The following theorem gives a relationship between inverse and composite relations.

THEOREM 3.1 Let A, B, C be sets and $R \subseteq A \times B$ and $S \subseteq B \times C$ be binary relations. Then

$$(S \circ R)^{-1} = R^{-1} \circ S^{-1}.$$

Theorem 3.1 is illustrated by the following example.

Example 3.2 Consider the relations

$$R = \{(x,y) \in \mathbb{R}^2 \mid y = e^x\} \qquad \text{and} \qquad S = \{(x,y) \in \mathbb{R}^2 \mid x - y \le 2\}.$$

We apply Definition 3.3 and obtain for the possible compositions of both relations

$$
\begin{aligned}
S \circ R &= \{(x,y) \in \mathbb{R}^2 \mid \exists z \in \mathbb{R} \text{ with } z = e^x \wedge z - y \le 2\} \\
&= \{(x,y) \in \mathbb{R}^2 \mid e^x - y \le 2\} \\
&= \{(x,y) \in \mathbb{R}^2 \mid y \ge e^x - 2\}
\end{aligned}
$$

and

$$
\begin{aligned}
R \circ S &= \{(x,y) \in \mathbb{R}^2 \mid \exists z \in \mathbb{R} \text{ with } x - z \le 2 \wedge y = e^z\} \\
&= \{(x,y) \in \mathbb{R}^2 \mid y > 0 \wedge z = \ln y \wedge x - \ln y \le 2\} \\
&= \{(x,y) \in \mathbb{R}^2 \mid y > 0 \wedge \ln y \ge x - 2\} \\
&= \{(x,y) \in \mathbb{R}^2 \mid y \ge e^{x-2}\}.
\end{aligned}
$$

The graph of both composite relations is given in Figure 3.2. Moreover, we get

$$(S \circ R)^{-1} = \{(y,x) \in \mathbb{R}^2 \mid y \ge e^x - 2\},$$
$$(R \circ S)^{-1} = \{(y,x) \in \mathbb{R}^2 \mid y \ge e^{x-2}\}.$$

We now illustrate the validity of Theorem 3.1. Using

$$R^{-1} = \{(y,x) \in \mathbb{R}^2 \mid y = e^x\},$$
$$S^{-1} = \{(y,x) \in \mathbb{R}^2 \mid x - y \le 2\},$$

we obtain

$$
\begin{aligned}
R^{-1} \circ S^{-1} &= \{(y,x) \in \mathbb{R}^2 \mid \exists z \in \mathbb{R} \text{ with } z - y \le 2 \wedge z = e^x\} \\
&= \{(y,x) \in \mathbb{R}^2 \mid e^x - y \le 2\} \\
&= \{(y,x) \in \mathbb{R}^2 \mid y \ge e^x - 2\}
\end{aligned}
$$

and

$$
\begin{aligned}
S^{-1} \circ R^{-1} &= \{(y,x) \in \mathbb{R}^2 \mid \exists z \in \mathbb{R} \text{ with } y = e^z \wedge x - z \le 2\} \\
&= \{(y,x) \in \mathbb{R}^2 \mid y > 0 \wedge x - \ln y \le 2\} \\
&= \{(y,x) \in \mathbb{R}^2 \mid y > 0 \wedge \ln y \ge x - 2\} \\
&= \{(y,x) \in \mathbb{R}^2 \mid y \ge e^{x-2}\}.
\end{aligned}
$$

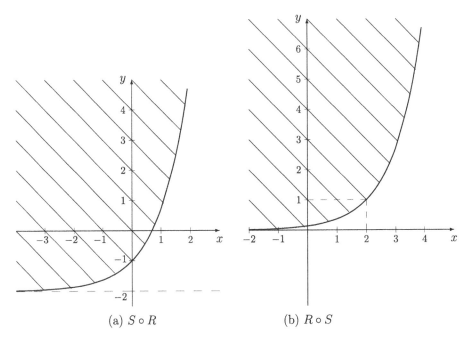

(a) $S \circ R$ (b) $R \circ S$

Figure 3.2 The compositions $S \circ R$ and $R \circ S$ in Example 3.2.

3.2   MAPPINGS

Next, we introduce one of the central concepts in mathematics.

Definition 3.4   Let A, B be sets. A relation $f \subseteq A \times B$, which assigns to every $a \in A$ exactly one $b \in B$, is called a *mapping* (or a *function*) from set A into set B. We write

$$f : A \to B$$

or, if the mapping is considered elementwise:

$$a \in A \mapsto f(a) = b \in B.$$

The set A is called the *domain* of the mapping, and the set B is called the *target*. For each $a \in A, f(a) = b$ is called the *image* of a. The set of images of all elements of the domain is called the *range* $f(A) \subseteq B$ of the mapping.

A mapping associates with each element of A a *unique* element of B; the range $f(A)$ may be a subset of B (see Figure 3.3). It follows from the above definition that a mapping is always

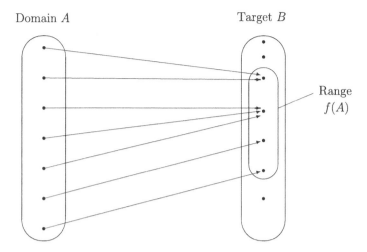

Figure 3.3 Mapping $f : A \to B$.

a relation, namely a particular relation R such that, for each $a \in A$, there is *exactly* one pair $(a, b) \in A \times B$ with $(a, b) \in R$. On the other hand, not every relation constitutes a mapping, e.g. the relation

$$R = \{(x, y) \in \mathbb{R}_+ \times \mathbb{R} \mid x = y^2\}$$

is not a mapping. Indeed, for each $x \in \mathbb{R}_+ \backslash \{0\}$, there exist two values $y_1 = -\sqrt{x}$ and $y_2 = \sqrt{x}$ such that $(x, y_1) \in R$ and $(x, y_2) \in R$ (see Figure 3.4).

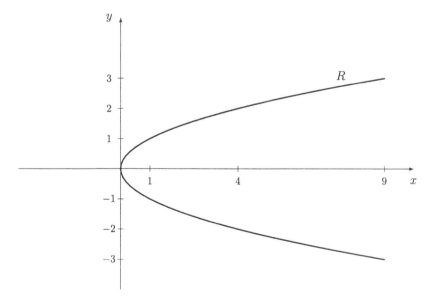

Figure 3.4 Relation $R = \{(x, y) \in \mathbb{R}_+ \times \mathbb{R} \mid x = y^2\}$.

Example 3.3   The German Automobile Association ADAC annually registers the breakdowns of most types of cars in Germany. For the year 2002, the following values (number of breakdowns per 10,000 cars of this type) for 10 types of medium-size cars with an age between four and six years have been obtained (see ADAC journal, May 2003): Toyota Avensis/Carina (TOY): 7.6, Mercedes SLK (MSLK): 9.1, BMW Z3 (BMWZ): 9.9, Mazda 626 (MAZ): 10.2, Mitsubishi Carisma (MIT): 12.0, Nissan Primera (NIS): 13.8, Audi A4/S4 (AUDI): 14.0, VW Passat (VW): 16.1, Mercedes C-class (MC): 17.5, BMW Series 3 (BMW3): 17.8. We should emphasize that we have presented the results only for the ten most reliable cars in this report among the medium-size cars.

Assigning to each car type the rank (i.e. the most reliable car gets rank one, the car in second place gets rank two and so on), we get a mapping $f : D_f \rightarrow \mathbb{N}$ with $D_f = \{TOY, MSLK, BMWZ, MAZ, MIT, NIS, AUDI, VW, MC, BMW3\}$. Mapping f is given by $f : \{(TOY, 1), (MSLK, 2), (BMWZ, 3), (MAZ, 4), (MIT, 5), (NIS, 6), (AUDI, 7), (VW, 8), (MC, 9), (BMW3, 10)\}$.

We continue with some properties of mappings.

Definition 3.5   A mapping f from A into B is called *surjective* if $f(A) = B$, i.e. for each $b \in B$, there exists (at least) an $a \in A$ such that $f(a) = b$.

A surjective mapping is also called an *onto*-mapping. This means that every element in set B is the image of some element(s) in set A.

Definition 3.6   A mapping f from A into B is called *injective* if for all $a_1, a_2 \in A$ the following implication holds:

$$a_1 \neq a_2 \Rightarrow f(a_1) \neq f(a_2)$$

(or equivalently, $f(a_1) = f(a_2) \Rightarrow a_1 = a_2$).

This means that no two elements of set A are assigned to the same element of set B.

Definition 3.7   A mapping f from A into B is called *bijective* if f is surjective and injective.

A bijective mapping is also called a *one-to-one* mapping. The notions of a surjective, an injective and a bijective mapping are illustrated in Figure 3.5.

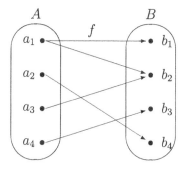

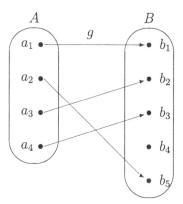

(a) Relation f is
not a mapping

(b) An injective but not
surjective mapping g

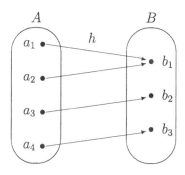

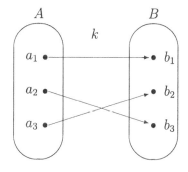

(c) A surjective but not
injective mapping h

(d) A bijective mapping k

Figure 3.5 Relations and different types of mappings.

Example 3.4   Consider the following mappings:

$f : \mathbb{R} \rightarrow \mathbb{R}$ with $f(x) = x^2$

$g : \mathbb{N} \rightarrow \mathbb{N}$ with $g(x) = x^2$

$h : \mathbb{R} \rightarrow \mathbb{R}_+$ with $h(x) = x^2$

$k : \mathbb{R}_+ \rightarrow \mathbb{R}_+$ with $k(x) = x^2$

Although we have the same rule for assigning the image $f(x)$ to some element x for all four mappings, we get different properties of the mappings f, g, h and k. Mapping f is not injective since e.g. for $x_1 = -1$ and $x_2 = 1$, we get $f(x_1) = f(x_2) = 1$. Mapping f is also not surjective since not every $y \in \mathbb{R}$ is the image of some $x \in D_f = \mathbb{R}$. For instance, for $y = -1$, there does not exist an $x \in \mathbb{R}$ such that $x^2 = y$.

Considering mapping g, we see that g is injective. For any two different integers x_1 and x_2, obviously the squares x_1^2 and x_2^2 are different, i.e. no two different natural numbers have the same image. However, mapping g is still not surjective since only squares of natural numbers occur as image $f(x)$. For instance, for $y = 2$, there does not exist a natural number x such that $x^2 = y$.

Mapping h is surjective since for each $y \in \mathbb{R}_+$, there exists (at least) one $x \in \mathbb{R}$ such that $f(x) = y$, namely for $x_1 = \sqrt{y}$ (and for $x_2 = -\sqrt{y}$, respectively), we get

$$f(x_1) = (\sqrt{y})^2 = y \quad (f(x_2) = (-\sqrt{y})^2 = y).$$

Mapping h is not injective for the same reason that mapping f was not injective. Considering now mapping k, it is injective since any two different non-negative real numbers x_1 and x_2 have different squares: $x_1^2 \neq x_2^2$. Mapping k is also surjective since for any $y \in \mathbb{R}_+$, there exists an $x = \sqrt{y} \geq 0$, such that

$$f(x) = f(\sqrt{y}) = (\sqrt{y})^2 = y.$$

Therefore, mapping k is bijective.

Next, we define an inverse mapping and a composite mapping.

Definition 3.8   Let $f : A \to B$ be a bijective mapping. Then the mapping

$$f^{-1} : B \to A$$

(read: f inverse) which assigns to any $b \in B$ that $a \in A$ with $b = f(a)$, is called the *inverse mapping* of mapping f.

Definition 3.9   Let $f : A \to B$ and $g : C \to D$ be mappings with $f(A) \subseteq C$. Then the mapping

$$g \circ f : A \to D,$$

which assigns to any $a \in A$ a unique image $(g \circ f)(a) = g(f(a)) \in D$, is called the *composite mapping* or *composition* of f and g.

The above definition reinforces the concept of a composite relation for the case of mappings. This means that for some $a \in D_f$, mapping f is applied first, to give the image $f(a)$. This image is assumed to be an element of the domain of mapping g, and we apply mapping g to $f(a)$, yielding the image $g(f(a))$. It is worth emphasizing that, analogously to compositions of relations, the order of the sequence of the mappings in the notation $g \circ f$ is important, i.e. 'g is applied after f'.

Example 3.5 Consider the mappings $f : A \rightarrow A$ given by

$$\{(1,3), (2,5), (3,3), (4, 1), (5,2)\}$$

and $g : A \rightarrow A$ given by

$$\{(1,4), (2, 1), (3, 1), (4, 2), (5, 3)\}.$$

For the ranges of the mappings we get

$$f(A) = \{1,2,3,5\} \qquad \text{and} \qquad g(A) = \{1,2,3,4\}.$$

We consider the composite mappings $f \circ g$ and $g \circ f$ and obtain

$$
\begin{aligned}
&(f \circ g)(1) = f(g(1)) = f(4) = 1, \quad (f \circ g)(2) = f(1) = 3, \quad (f \circ g)(3) = f(1) = 3, \\
&(f \circ g)(4) = f(2) = 5, \quad\quad\quad\quad\; (f \circ g)(5) = f(3) = 3, \\
&(g \circ f)(1) = g(f(1)) = g(3) = 1, \quad (g \circ f)(2) = g(5) = 3, \quad (g \circ f)(3) = g(3) = 1, \\
&(g \circ f)(4) = g(1) = 4, \quad\quad\quad\quad\; (g \circ f)(5) = g(2) = 1.
\end{aligned}
$$

The composition $g \circ f$ is illustrated in Figure 3.6.

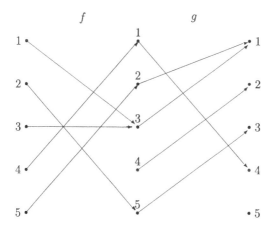

Figure 3.6 The composite mapping $g \circ f$ in Example 3.5.

For composite mappings $g \circ f$, we have the following properties.

THEOREM 3.2 Let $f : A \rightarrow B$ and $g : B \rightarrow C$ be mappings. Then:

(1) If mappings f and g are surjective, mapping $g \circ f$ is surjective.
(2) If mappings f and g are injective, mapping $g \circ f$ is injective.
(3) If mappings f and g are bijective, mapping $g \circ f$ is bijective.

Example 3.6 Consider the following mappings:

$$f : \mathbb{R} \to \mathbb{R} \text{ with } f(x) = 3x + 5;$$
$$g : \mathbb{R} \to \mathbb{R} \text{ with } g(x) = e^x + 1.$$

First, we show that both mappings f and g are injective. Indeed, for mapping f, it follows from $x_1 \neq x_2$ that $3x_1 + 5 \neq 3x_2 + 5$. Analogously, for mapping g, it follows from $x_1 \neq x_2$ that $e^{x_1} + 1 \neq e^{x_2} + 1$. Then mapping $g \circ f$ with

$$(g \circ f)(x) = g(f(x)) = g(3x + 5) = e^{3x+5} + 1$$

is injective as well due to part (2) of Theorem 3.2.

We can formulate a similar property for mappings as given in Theorem 3.1 for relations.

THEOREM 3.3 Let $f : A \to B$ and $g : B \to C$ be bijective mappings. Then the inverse mapping $(f \circ g)^{-1}$ exists, and we get

$$(f \circ g)^{-1} = g^{-1} \circ f^{-1}.$$

Finally, we introduce an identical mapping as follows.

Definition 3.10 A mapping $I : A \to A$ with $I(a) = a$ is called an *identical mapping*.

If for a mapping $f : A \to B$ the inverse mapping $f^{-1} : B \to A$ exists, then

$$f \circ f^{-1} = f^{-1} \circ f = I,$$

where I is the identical mapping. This means that for each $a \in A$, we get

$$f(f^{-1}(a)) = f^{-1}(f(a)) = a.$$

3.3 FUNCTIONS OF A REAL VARIABLE

In this section, we consider mappings or functions $f : A \to B$ with special sets A and B, namely both sets are either equal to the whole set $\mathbb{R}$ of real numbers or equal to a subset of $\mathbb{R}$.

3.3.1   Basic notions

> **Definition 3.11**   A mapping f that assigns to any real number $x \in D_f \subseteq \mathbb{R}$ a unique
> real number y is called a *function of a real variable*. The set D_f is called the *domain*
> of function f, and the set
>
> $$R_f = f(D_f) = \{y \mid y = f(x),\ x \in D_f\}$$
>
> is called the *range* of function f.

We write $y = f(x), x \in D_f$. Variable x is called the *independent variable* or *argument*, and y
is called the *dependent variable*. The real number y is the *image* or the *function value* of x,
i.e. the value of function f at point x. The domain and range of a function are illustrated in
Figure 3.7.

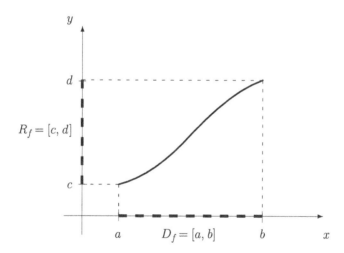

Figure 3.7 Domain and range of a function $f : D_f \rightarrow R_f$.

We also write

$$f : D_f \rightarrow \mathbb{R} \qquad \text{or} \qquad f : D_f \rightarrow R_f.$$

Alternatively we can also use the notation

$$f : \{(x,y) \mid y = f(x),\ x \in D_f\}.$$

Hereafter, the symbol f denotes the function, and $f(x)$ denotes the function value of a certain
real number x belonging to the domain D_f of function f. To indicate that function f depends
on one variable x, one can also write $f = f(x)$ without causing confusion.

A function $f : D_f \rightarrow R_f$ can be represented analytically, graphically or in special cases by
a table presenting for each point $x \in D_f$ the corresponding function value $y = f(x) \in R_f$.

We can use the following analytical representations:

(1) *explicit representation:* $y = f(x)$;
(2) *implicit representation:* $F(x, y) = 0$;
(3) *parametric representation:* $x = g(t)$, $y = h(t)$; $t \in D_g = D_h$.

In the latter representation, t is called a *parameter* which may take certain real values. For each value $t \in D_g = D_h$, we get a point (x, y) with $y = f(x)$.

Example 3.7   Let function $f : \mathbb{R} \to \mathbb{R}$ be given in explicit form by

$$y = f(x) = x^3 + 3x + 2.$$

The corresponding implicit representation is as follows:

$$F(x, y) = x^3 + 3x + 2 - y = 0.$$

We emphasize that, from a given implicit or parametric representation, it is not immediately clear whether it is indeed a function (or only a relation).

Example 3.8   Consider the parametric representation

$$x = r \sin t, \qquad y = r \cos t; \qquad t \in [0, 2\pi]. \tag{3.1}$$

Taking the square of both equations and summing up, we get

$$x^2 + y^2 = r^2 \cdot (\sin^2 t + \cos^2 t) = r^2$$

which describes a circle with radius r and point $(0, 0)$ as centre. Each point $x \in (-r, r)$ is related to two values $y_1 = \sqrt{r^2 - x^2}$ and $y_2 = -\sqrt{r^2 - x^2}$. Therefore equation (3.1) characterizes only a relation.

The concept of bijective mappings can be immediately applied to a function of a real variable and we obtain: if a function $f : D_f \to R_f, D_f \subseteq \mathbb{R}$, is bijective, then there exists a function f^{-1} which assigns to each real number $y \in R_f$ a unique real value $x \in D_f$ with

$$x = f^{-1}(y).$$

Here it should be mentioned that, if $R_f \subseteq \mathbb{R}$ but $R_f \neq \mathbb{R}$ (in this case we say that R_f is a proper subset of $\mathbb{R}$), the inverse function of $f : D_f \to \mathbb{R}$ would not exist, since f would not be surjective in that case. Nevertheless, such a formulation is often used and, in order to determine f^{-1}, one must find the range R_f (in order to have a surjective mapping) or, equivalently, the domain $D_{f^{-1}}$ of the inverse function of f.

We write $f^{-1} : R_f \to \mathbb{R}$ or $R_f \to D_f$. Alternatively, we can use the notation

$$f^{-1} = \{(y,x) \mid y = f(x), x \in D_f\}.$$

Since x usually denotes the independent variable, we can exchange variables x and y and write

$$y = f^{-1}(x)$$

for the inverse function of function f with $y = f(x)$. The graph of the inverse function f^{-1} is given by the mirror image of the graph of function f with respect to the line $y = x$. The graphs of both functions are illustrated in Figure 3.8. (In this case, function f is a linear function and the inverse function f^{-1} is linear as well.)

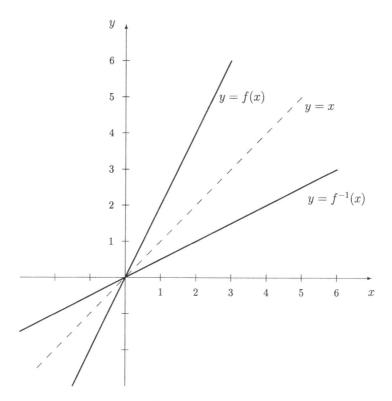

Figure 3.8 Graphs of functions f and f^{-1}.

Example 3.9 Let function $f : \mathbb{R}_+ \to [4, \infty)$ with

$$y = f(x) = x^3 + 4$$

be given. To determine the inverse function f^{-1}, we solve the above equation for x and obtain

$$x^3 = y - 4$$

which yields

$$x = \sqrt[3]{y - 4}.$$

Interchanging both variables, we obtain the inverse function

$$y = f^{-1}(x) = \sqrt[3]{x - 4}.$$

The domain of the inverse function f^{-1} is the range of function f, and the range of function f^{-1} is the domain of function f, i.e.

$$D_{f^{-1}} = [4, \infty) \qquad \text{and} \qquad R_{f^{-1}} = \mathbb{R}_+.$$

Analogously to our previous considerations in this chapter, we define composite functions of a real variable

$$g \circ f : D_f \to \mathbb{R}$$

with $y = g(f(x)), x \in D_f$. Function f is called the *inside function* and function g is called the *outside function*. If there exists the inverse function f^{-1} for a function f, we have

$$f^{-1} \circ f = f \circ f^{-1} = I,$$

i.e.

$$y = f^{-1}(f(x)) = f(f^{-1}(x)) = x,$$

where I is the identity function. We illustrate the concept of a composite function by the following examples.

Example 3.10   Let functions $f : \mathbb{R} \to \mathbb{R}$ and $g : \mathbb{R} \to \mathbb{R}$ with

$$f(x) = 3x - 2 \qquad \text{and} \qquad g(x) = x^2 + x - 1$$

be given. We determine both composite functions $f \circ g$ and $g \circ f$ and obtain

$$(f \circ g)(x) = f(g(x)) = f(x^2 + x + 1) = 3(x^2 + x - 1) - 2 = 3x^2 + 3x - 5$$

and

$$(g \circ f)(x) = g(f(x)) = g(3x - 2) = (3x - 2)^2 + (3x - 2) - 1 = 9x^2 - 9x + 1.$$

Both compositions are defined, since the range of the inside function is either set $\mathbb{R}$ (function f) or a subset (function g) while in both cases the outside function is defined for all real numbers.

Example 3.11 Given are the functions $f : D_f \to \mathbb{R}$ and $g : D_g \to \mathbb{R}$ with

$$f(x) = \sqrt{x+1}, \quad D_f = [-1, \infty)$$

and

$$g(x) = \sin 3x, \quad D_g = \mathbb{R}.$$

For the composite functions we obtain

$$(f \circ g)(x) = f(g(x)) = f(\sin 3x) = \sqrt{\sin 3x + 1}$$

and

$$(g \circ f)(x) = g(f(x)) = g(\sqrt{x+1}) = \sin(3\sqrt{x+1}).$$

Composition $f \circ g$ is defined since the range $R_g = [-1, 1]$ is a subset of the domain D_f, and composition $g \circ f$ is defined since the range $R_f = [0, \infty)$ is a subset of the domain D_g.

3.3.2 Properties of functions

Next, we discuss some properties of functions. We start with an increasing and a decreasing function.

Definition 3.12 A function $f : D_f \to \mathbb{R}$ is *increasing* (resp. *strictly increasing*) on an interval $I \subseteq D_f$ if, for any choice of x_1 and x_2 in I with $x_1 < x_2$, inequality

$$f(x_1) \le f(x_2) \qquad (\text{resp.}\, f(x_1) < f(x_2))$$

holds.

Obviously, a strictly increasing function is a special case of an increasing function. The latter is also called a *non-decreasing function* in order to emphasize that there may be one or several subinterval(s) of I, where function f is constant.

Definition 3.13 A function $f : D_f \to \mathbb{R}$ is *decreasing* (resp. *strictly decreasing*) on an interval $I \subseteq D_f$ if, for any choice of x_1 and x_2 in I with $x_1 < x_2$, inequality

$$f(x_1) \ge f(x_2) \qquad (\text{resp.}\, f(x_1) > f(x_2))$$

holds.

A decreasing function is also called a *non-increasing function*. An illustration is given in Figure 3.9, where function f is strictly increasing on the interval I_1, but strictly decreasing on the interval I_2.

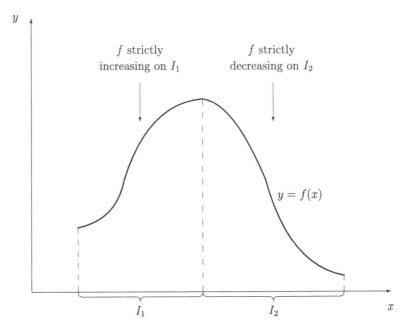

Figure 3.9 Monotonicity of function $f : D_f \to R$.

We note that a function $f : D_f \to W_f$, which is strictly monotone on the domain (i.e. either strictly increasing on D_f or strictly decreasing on D_f, is bijective and therefore, the inverse function $f^{-1} : W_f \to D_f$ exists in this case and is strictly monotone, too.

Example 3.12   We consider function $f : D_f \to \mathbb{R}$ with

$$f(x) = \frac{2}{x+1}.$$

This function is defined for all $x \neq -1$, i.e. $D_f = \mathbb{R} \setminus \{-1\}$. First, we consider the interval $I_1 = (-1, \infty)$. Let $x_1, x_2 \in I_1$ with $x_1 < x_2$. We get $0 < x_1 + 1 < x_2 + 1$ and thus

$$f(x_1) = \frac{2}{x_1 + 1} > \frac{2}{x_2 + 1} = f(x_2).$$

Therefore, function f is strictly decreasing on the interval I_1.
Consider now the interval $I_2 = (-\infty, -1)$ and let $x_1 < x_2 < -1$. In this case, we first get $x_1 + 1 < x_2 + 1 < 0$ and then

$$f(x_1) = \frac{2}{x_1 + 1} < \frac{2}{x_2 + 1} = f(x_2).$$

Therefore, function f is strictly increasing on the interval $(-\infty, -1)$. The graph of function f is given in Figure 3.10.

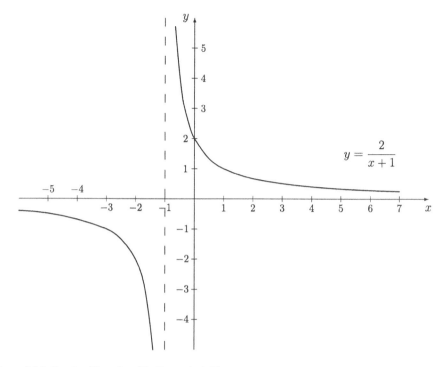

Figure 3.10 Graph of function f in Example 3.12.

Definition 3.14 A function $f : D_f \to \mathbb{R}$ is said to be *bounded from below (from above)* if there exists a constant C such that

$$f(x) \geq C \qquad (\text{resp.} f(x) \leq C)$$

for all $x \in D_f$. Function f is said to be *bounded* if $f(x)$ is bounded from below and from above, i.e.

$$|f(x)| \leq C$$

for all $x \in D_f$.

Example 3.13 We consider function $f : \mathbb{R} \to \mathbb{R}$ with

$$y = f(x) = e^x - 2.$$

Function f is bounded from below since $f(x) \geq -2$ for all $x \in \mathbb{R}$. However, function f is not bounded from above since $f(x)$ can become arbitrarily large when x becomes large. Therefore, function f is not bounded.

Example 3.14   We consider function $f : \mathbb{R} \to \mathbb{R}$ with

$$y = f(x) = 3 + \sin 2x.$$

First, the function values of $\sin 2x$ are in the interval $[-1, 1]$. Therefore, the values of function f are in the interval $[2, 4]$. Consequently, for $C = 4$ we get $|f(x)| \le C$ for all $x \in \mathbb{R}$ and thus, function f is bounded.

Definition 3.15   A function $f : D_f \to \mathbb{R}$ is called *even* (or *symmetric*) if

$$f(-x) = f(x).$$

Function f is called *odd* (or *antisymmetric*) if

$$f(-x) = -f(x).$$

In both cases, the domain D_f has to be symmetric with respect to the origin of the coordinate system.

An even function is symmetric to the y-axis. An odd function is symmetric with respect to the origin of the coordinate system as a point. It is worth noting that a function f is not necessarily either even or odd.

Example 3.15   We consider function $f : \mathbb{R} \setminus \{0\} \to \mathbb{R}$ with

$$f(x) = 4x^3 - 2x + \frac{1}{x}.$$

We determine $f(-x)$ and obtain

$$f(-x) = 4(-x)^3 - 2(-x) + \frac{1}{-x} = -4x^3 + 2x - \frac{1}{x}$$

$$= -\left(4x^3 - 2x + \frac{1}{x}\right) = -f(x).$$

Thus, function f is an odd function.

Example 3.16   Let function $f : \mathbb{R} \to \mathbb{R}$ with

$$f(x) = 3x^6 + x^2$$

be given. Then we obtain

$$f(-x) = 3(-x)^6 + (-x)^2 = 3x^6 + x^2 = f(x).$$

Hence, function f is an even function.

Definition 3.16 A function $f : D_f \to \mathbb{R}$ is called *periodic* if there exists a number T such that for all $x \in D_f$ with $x + T \in D_f$, equality $f(x + T) = f(x)$ holds. The smallest real number T with the above property is called the *period* of function f.

Definition 3.17 A function $f : D_f \to \mathbb{R}$ is called *convex* (*concave*) on an interval $I \subseteq D_f$ if, for any choice of x_1 and x_2 in I and $0 \le \lambda \le 1$, inequality

$$f(\lambda x_1 + (1 - \lambda)x_2) \le \lambda f(x_1) + (1 - \lambda)f(x_2) \tag{3.2}$$

$$\left(\text{resp.} f(\lambda x_1 + (1 - \lambda)x_2) \ge \lambda f(x_1) + (1 - \lambda)f(x_2) \right) \tag{3.3}$$

holds.

If for $0 < \lambda < 1$ and $x_1 \ne x_2$, the sign $<$ holds in inequality (3.2) (resp. the sign $>$ in inequality (3.3)), function f is called *strictly convex* (*strictly concave*).

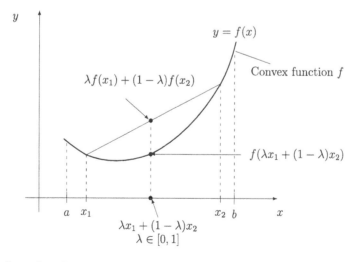

Figure 3.11 Illustration of a convex function $f : D_f \to \mathbb{R}$.

The definition of a convex function is illustrated in Figure 3.11. Function f is convex on an interval I if for any choice of two points x_1 and x_2 from the interval and for any intermediate point x from the interval $[x_1, x_2]$, which can be written as $x = \lambda x_1 + (1 - \lambda)x_2$, the function value of this point (i.e. $f(\lambda x_1 + (1 - \lambda)x_2)$) is not greater than the function value of the straight line through the points $(x_1, f(x_1))$ and $(x_2, f(x_2))$ at the intermediate point x. The latter function value can be written as $\lambda f(x_1) + (1 - \lambda)f(x_2)$.

Checking whether a function is convex or concave by application of Definition 3.17 can be rather difficult. In Chapter 4, we discuss criteria for applying differential calculus, which are easier to use. Here we consider, for illustration only, the following simple example.

Example 3.17   Let function $f : \mathbb{R} \to \mathbb{R}$ with

$$f(x) = ax + b$$

be given. Using formula (3.2) of the definition of a convex function, we obtain the following equivalences:

$$
\begin{aligned}
& f(\lambda x_1 + (1 - \lambda)x_2) && \le && \lambda f(x_1) + (1 - \lambda)f(x_2) \\
\Longleftrightarrow \quad & a \cdot [\lambda x_1 + (1 - \lambda)x_2] + b && \le && \lambda \cdot (ax_1 + b) + (1 - \lambda) \cdot (ax_2 + b) \\
\Longleftrightarrow \quad & a\lambda x_1 + a(1 - \lambda)x_2 + b && \le && \lambda ax_1 + \lambda b + (1 - \lambda)ax_2 + b - \lambda b \\
\Longleftrightarrow \quad & 0 && \le && 0.
\end{aligned}
$$

Since the latter inequality is obviously true, the first of the above inequalities is satisfied too and hence, function f is convex. We mention that function f is also concave (all inequality signs above can be replaced by the sign $\ge$). So a linear function is obviously both convex and concave, but it is neither strictly convex nor strictly concave.

3.3.3   Elementary types of functions

Next, we briefly discuss some types of functions and summarize their main important properties. We start with polynomials and some of their special cases.

Polynomials and rational functions

Definition 3.18   The function $P_n : \mathbb{R} \to \mathbb{R}$ with

$$y = P_n(x) = a_n x^n + a_{n-1}x^{n-1} + \ldots + a_2 x^2 + a_1 x + a_0$$

with $a_n \ne 0$ is called a *polynomial function* (or *polynomial*) *of degree n*. The numbers $a_0, a_1, \ldots, a_n$ are called the *coefficients* of the polynomial.

Depending on the value of n which specifies the highest occurring power of x, we mention the following subclasses of polynomial functions:

$$
\begin{array}{lll}
n = 0: & y = P_0(x) = a_0 & \text{constant function} \\
n = 1: & y = P_1(x) = a_1 x + a_0 & \text{linear function} \\
n = 2: & y = P_2(x) = a_2 x^2 + a_1 x + a_0 & \text{quadratic function} \\
n = 3: & y = P_3(x) = a_3 x^3 + a_2 x^2 + a_1 x + a_0 & \text{cubic function}
\end{array}
$$

We illustrate the special cases of $n = 1$ and $n = 2$.

Case n = 1   In this case, the graph is a (straight) *line* which is uniquely defined by any two distinct points P_1 and P_2 on this line. Assume that P_1 has the coordinates (x_1, y_1) and P_2 has the coordinates (x_2, y_2), then the above parameter a_1 is given by

$$a_1 = \frac{y_2 - y_1}{x_2 - x_1}$$

and is called the *slope* of the line. The parameter a_0 gives the y-coordinate of the intersection of the line with the y-axis. These considerations are illustrated in Figure 3.12 for $a_0 > 0$ and $a_1 > 0$. The point-slope formula of a line passing through point P_1 with coordinates (x_1, y_1) and having a slope a_1 is given by

$$y - y_1 = a_1 \cdot (x - x_1).$$

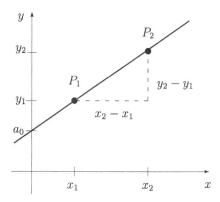

Figure 3.12 A line with positive slope a_1.

Example 3.18   Assume that for some product, the quantity of demand D depends on the price p as follows:

$$D = f(p) = 1,600 - \frac{1}{2} \cdot p.$$

The above equation describes the relationship between the price p of a product and the resulting demand D, i.e. it describes the amount of the product that customers are willing to purchase at different prices. The price p is the independent variable and demand D is the dependent variable. We say that f is the *demand function*. Since both price and demand can be assumed to be non-negative, we get the domain and range as follows:

$$D_f = \{p \in \mathbb{R} \mid 0 \leq p \leq 3,200\} \qquad \text{and} \qquad R_f = \{D \in \mathbb{R} \mid 0 \leq D \leq 1,600\}.$$

If we want to give the price in dependence on the demand, we have to find the inverse function f^{-1} by solving the above equation for p. We obtain

$$\frac{1}{2} \cdot p = 1,600 - D$$

which corresponds to

$$p = f^{-1}(D) = 2 \cdot (1,600 - D) = 3,200 - 2D.$$

Here we did not interchange the variables since 'D' stands for demand and 'p' stands for price. Obviously, we get $D_{f^{-1}} = R_f$ and $R_{f^{-1}} = D_f$.

Case $n = 2$ The graph of a quadratic function is called a *parabola*. If $a_2 > 0$, then the parabola is open from above (*upward parabola*) while, if $a_2 < 0$, then the parabola is open from below (*downward parabola*). A quadratic function with $a_2 > 0$ is a strictly convex function while a quadratic function with $a_2 < 0$ is a strictly concave function. The case of a downward parabola is illustrated in Figure 3.13. Point P is called the *vertex* (or *apex*), and its coordinates can be determined by rewriting the quadratic function in the following form:

$$y - y_0 = a_2 \cdot (x - x_0)^2, \tag{3.4}$$

where (x_0, y_0) are the coordinates of point P. We get

$$x_0 = -\frac{a_1}{2a_2} \qquad \text{and} \qquad y_0 = -\frac{a_1^2}{4a_2} + a_0.$$

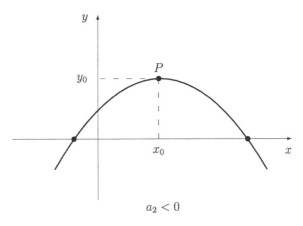

Figure 3.13 A downward parabola $y = a_2 \cdot (x - x_0)^2 + y_0$.

Points x with $y = f(x) = a_2 x^2 + a_1 x + a_0 = 0$ are called the *zeroes* or *roots* of the quadratic function. In the case of $a_1^2 \geq 4a_2 a_0$, a quadratic function has the two real zeroes

$$x_{1,2} = \frac{-a_1 \pm \sqrt{a_1^2 - 4a_2 a_0}}{2a_2}. \tag{3.5}$$

In the case of $a_1^2 = 4a_2 a_0$, we get a zero $x_1 = x_2$ occurring twice. In the case of $a_1^2 < 4a_2 a_0$, there exist two zeroes which are both complex numbers. If a quadratic function is given in normal form $y = x^2 + px + q$, formula (3.5) for finding the zeroes simplifies to

$$x_{1,2} = -\frac{p}{2} \pm \sqrt{\frac{p^2}{4} - q}.$$

Example 3.19 Consider the quadratic function $P_2 : \mathbb{R} \to \mathbb{R}$ with

$$y = P_2(x) = 2x^2 - 4x + 4.$$

First, we write the right-hand side as a complete square and obtain

$$y = 2 \cdot (x^2 - 2x + 2) = 2 \cdot (x - 1)^2 + 2,$$

which can be rewritten as

$$y - 2 = 2 \cdot (x - 1)^2.$$

From the latter representation we see that the graph of function f is an upward parabola with the apex $P = (1, 2)$ (see equation (3.4)). We investigate whether the inverse function f^{-1} exists. We try to eliminate variable x and obtain

$$\frac{y - 2}{2} = (x - 1)^2.$$

If we now take the square root on both sides, we obtain two values on the right-hand side, i.e.

$$\sqrt{\frac{y - 2}{2}} = \pm(x - 1)$$

which yields the following two terms:

$$x_1 = 1 + \sqrt{\frac{y - 2}{2}} \quad \text{and} \quad x_2 = 1 - \sqrt{\frac{y - 2}{2}}.$$

Therefore, the inverse function f^{-1} does not exist since the given function f is not injective.

If we restrict the domain of function f to the interval $D_f = [1, \infty)$, then the inverse function would exist since in this case, only solution x_1 is obtained when eliminating variable x. Thus, we can now interchange both variables and obtain the inverse function f^{-1} with

$$y = f^{-1}(x) = 1 + \sqrt{\frac{x - 2}{2}}, \qquad D_{f^{-1}} = [2, \infty).$$

The graph of function f and the graph of the inverse function f^{-1} for the case when the domain of function f is restricted to $D_f = [1, \infty)$ are given in Figure 3.14.

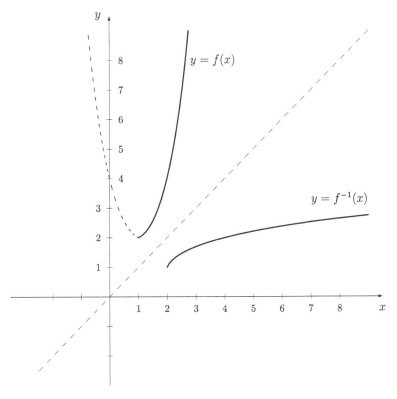

Figure 3.14 Graphs of functions f and f^{-1} in Example 3.19.

Next, we consider the general polynomial P_n of degree n with

$$P_n(x) = a_n x^n + a_{n-1} x^{n-1} + \cdots + a_1 x + a_0, \qquad a_n \neq 0.$$

In the following, we give two properties of general polynomials and start with the fundamental theorem of algebra.

THEOREM 3.4 (fundamental theorem of algebra)   Any polynomial P_n of degree n can be written as a product of polynomials of first or second degree.

Another important property of polynomials is given in the following remainder theorem. Let $\deg P$ denote the *degree* of polynomial P.

THEOREM 3.5 (remainder theorem)   Let $P : \mathbb{R} \to \mathbb{R}$ and $Q : \mathbb{R} \to \mathbb{R}$ be polynomials with $\deg P \geq \deg Q$. Then there always exist unique polynomials $S : \mathbb{R} \to \mathbb{R}$ and $R : \mathbb{R} \to \mathbb{R}$ such that

$$P = S \cdot Q + R, \tag{3.6}$$

where polynomial R is called the remainder and $\deg R < \deg Q$.

If $R(x) \equiv 0$, then we say either that polynomial Q is a factor of polynomial P or that polynomial P is divisible by polynomial Q.

Consider the special case $Q(x) = x - x_0$, i.e. polynomial Q is a linear function and we have deg $Q(x) = 1$. Then it follows from Theorem 3.5 that deg $R = 0$, i.e. $R(x) = r$ is constant. This means that we can write polynomial P in the form

$$P(x) = S(x) \cdot (x - x_0) + r.$$

Consequently, for $x = x_0$ we get the function value $P(x_0) = r$. Furthermore, we have the following equivalence:

$$P(x) = S(x) \cdot (x - x_0) \Longleftrightarrow P(x_0) = 0. \tag{3.7}$$

We can summarize the above considerations as follows. If x_0 is a zero of a polynomial P_n with deg $P_n = n$ (i.e. a root of the equation $P_n(x) = 0$), then, according to the comments above, this polynomial can be written as

$$P_n(x) = (x - x_0) \cdot S_{n-1}(x),$$

where S_{n-1} is a polynomial with deg $S_{n-1} = n - 1$. On the one hand, it follows from equivalence (3.7) that polynomial P_n can have at most n different zeroes. On the other hand, it follows from Theorem 3.4 that polynomial P_n has *exactly* n (real or complex) zeroes. If the polynomial P_n has the complex zero $x = a + bi$, then the complex conjugate $x = a - bi$ is a zero of polynomial P_n, too. Therefore, polynomial P_n has the factors

$$\left[x - (a + bi)\right] \cdot \left[x - (a - bi)\right] = x^2 - 2ax + (a^2 + b^2),$$

which is a polynomial of degree two. So we can confirm Theorem 3.4. A real zero x_0 leads to a polynomial $x - x_0$ of first degree and complex zeroes $a + bi$ and $a - bi$ to a polynomial $x^2 - 2ax + (a^2 + b^2)$ of second degree in the product representation of polynomial P_n. As a consequence of the above considerations, we can conclude for example that a polynomial P_3 of degree three has either three real zeroes or only one real zero (and two complex zeroes).

It may happen that certain factors $x - x_i$ occur repeatedly in the above representation. This leads to the definition of the multiplicity of a zero.

Definition 3.19 Let $f = P_n : \mathbb{R} \to \mathbb{R}$ be a polynomial of degree n with

$$P_n(x) = (x - x_0)^k \cdot S_{n-k}(x) \qquad \text{and} \qquad S_{n-k}(x_0) \neq 0.$$

Then x_0 is called a *zero* (or *root*) of *multiplicity k* of polynomial P_n.

If x_0 is a zero of odd multiplicity, then the sign of function f changes 'at' point x_0, i.e. there exists an interval about x_0 such that function f has positive values to the left of x_0 and negative values to the right of x_0 or vice versa. In contrast, if x_0 is a zero of even multiplicity, there is an interval about x_0 such that the sign of function f is the same on both sides of point x_0.

Next, we investigate the question of how to compute the function value of a polynomial easily. To this end, we introduce *Horner's scheme*.

Calculation of function values of polynomials using Horner's scheme

$$P_n(x_0) = a_n x_0^n + a_{n-1} x_0^{n-1} + \cdots + a_2 x_0^2 + a_1 x_0 + a_0$$
$$= \{\cdots \underbrace{\underbrace{[(a_n x_0 + a_{n-1})}_{A_{n-1}} x_0 + a_{n-2}] x_0 + \cdots + a_1\}}_{A_{n-2}} x_0 + a_0$$
$$\underbrace{}_{A_1}$$

The above computations can be easily performed using the following scheme and starting from the left.

Horner's scheme

	a_n	a_{n-1}	a_{n-2}	$\cdots$	a_2	a_1	a_0
		+	+		+	+	+
$x = x_0$		$a_n x_0$	$A_{n-1} x_0$	$\cdots$	$A_3 x_0$	$A_2 x_0$	$A_1 x_0$
	$\nearrow$	$\nearrow$	$\nearrow$	$\nearrow$	$\nearrow$	$\nearrow$	$\nearrow$
	a_n	A_{n-1}	A_{n-2}	$\cdots$	A_2	A_1	$A_0 = \underline{P_n(x_0)}$

The advantage of Horner's scheme is that, in order to compute the function value of a polynomial, only additions and multiplications have to be performed, but it is not necessary to compute powers.

Example 3.20   We consider the polynomial $P_5 : \mathbb{R} \to \mathbb{R}$ with

$$P_5(x) = 2x^5 + 4x^4 + 8x^3 - 4x^2 - 10x.$$

Since $a_0 = 0$, we get the zero $x_0 = 0$ and dividing $P_4(x)$ by $x - x_0 = x$ yields the polynomial P_4 with

$$P_4(x) = 2x^4 + 4x^3 + 8x^2 - 4x - 10.$$

Checking now the function values for $x_2 = 1$ and $x_3 = -1$ successively by Horner's scheme, we obtain

	2	4	8	-4	-10
$x = x_2 = 1$		2	6	14	10
	2	6	14	10	$\underline{0}$
$x = x_3 = -1$		-2	-4	-10	
	2	4	10	$\underline{0}$	

In the above scheme, we have dropped the arrows and the plus signs for simplicity. We have found that both values $x_2 = 1$ and $x_3 = -1$ are zeroes of the polynomial $P_4(x)$. From the last row we see that the quadratic equation $2x^2 + 4x + 10 = 0$ has to be considered to determine the remaining zeroes of polynomial P_4. We obtain

$$x_{4,5} = -1 \pm \sqrt{1-5}$$

which yields the complex zeroes

$$x_4 = -1 + 2i \qquad \text{and} \qquad x_5 = -1 - 2i.$$

So we have found all five (real and complex) zeroes of polynomial $P_5(x)$, and we get the factorized polynomial

$$P_5(x) = 2x(x - 1)(x + 1)(x^2 + 2x + 5).$$

Example 3.21   We consider the polynomial $P_5 : \mathbb{R} \to \mathbb{R}$ with

$$P_5(x) = x^5 - 5x^4 + 40x^2 - 80x + 48$$

and determine the multiplicity of the (guessed) zero $x = 2$. We obtain:

	1	−5	0	40	−80	48
$x = 2$		2	−6	−12	56	−48
	1	−3	−6	28	−24	0
$x = 2$		2	−2	−16	24	
	1	−1	−8	12	0	
$x = 2$		2	2	−12		
	1	1	−6	0		
$x = 2$		2	6			
	1	3	0			

Thus, $x = 2$ is a zero of multiplicity four, and the polynomial P_5 can be written as a product as follows:

$$P_5(x) = (x - 2)^4 \cdot (x + 3).$$

The last factor is obtained from the remaining polynomial S_1 with $S_1(x) = x + 3$ which has the zero $x = -3$.

In general, it is difficult to find all zeroes of a polynomial of a higher degree. In Chapter 4, we discuss numerical procedures for finding the zeroes approximately. The following theorem gives some relationships between the coefficients of a polynomial and the zeroes.

THEOREM 3.6 (Vieta's theorem)   Let polynomial $P_n : \mathbb{R} \to \mathbb{R}$ with

$$P_n(x) = 1 \cdot x^n + a_{n-1}x^{n-1} + \cdots + a_2x^2 + a_1x + a_0$$

be given and let $x_1, x_2, \cdots, x_n$ be the zeroes of polynomial P_n. Then:

$$
\begin{aligned}
x_1 + x_2 + \cdots + x_n &= -a_{n-1} \\
x_1 x_2 + x_1 x_3 + \cdots + x_{n-1} x_n &= a_{n-2} \\
x_1 x_2 x_3 + x_1 x_2 x_4 + \cdots + x_{n-2} x_{n-1} x_n &= -a_{n-3} \\
&\ \ \vdots \\
x_1 x_2 x_3 \cdots x_n &= (-1)^n a_0
\end{aligned}
$$

In the special case of a quadratic function $P_2(x) = x^2 + px + q$, Vieta's theorem yields the two equations

$$x_1 + x_2 = -p \qquad \text{and} \qquad x_1 x_2 = q.$$

Example 3.22   Let the polynomial $P_3 : \mathbb{R} \to \mathbb{R}$ with

$$P_3(x) = x^3 - 2x^2 - 5x + 6$$

be given. When looking for integer zeroes, it follows from the last equality of Theorem 3.6 that these zeroes must be a divisor of $a_0 = 6$. Therefore, only $x = \pm 1$, $x = \pm 2$ and $x = \pm 3$ are candidates for an integer zero. Checking the function values of P_3 at these points e.g. by Horner's scheme, we obtain the zeroes

$$x_1 = 1, \qquad x_2 = -2 \qquad \text{and} \qquad x_3 = 3$$

and thus the factorized polynomial

$$P_3 = (x - 1) \cdot (x + 2) \cdot (x - 3).$$

Finally, we consider the quotient of two polynomials, P and Q.

Definition 3.20   A function $T : D_T \to \mathbb{R}$ with $T(x) = P(x)/Q(x) = (P/Q)(x)$ is called a *rational function*. The rational function T is called *proper* if $\deg P < \deg Q$ and *improper* if $\deg P \geq \deg Q$.

The domain D_T of function T is given by all real numbers for which the polynomial in the denominator is different from zero, i.e. $D_T = \{x \in \mathbb{R} \mid Q(x) \neq 0\}$.

Example 3.23   Let the cost C of producing x units of a product be equal to

$$C(x) = ax^2 + bx + c, \qquad x \geq 0.$$

Here $c > 0$ describes the fixed cost of production, and a and b are real parameters such that C is a non-negative and strictly increasing function. Then the average cost $C_a(x)$, measuring the cost per unit produced, given by

$$C_a(x) = \frac{ax^2 + bx + c}{x}$$

is an improper rational function which can be rewritten as

$$C_a(x) = ax + b + \frac{c}{x}.$$

By means of *polynomial division*, any improper rational function can be written as the sum of a polynomial and a proper rational function. If $\deg P \geq \deg Q$, then equation (3.6) can be written as

$$\frac{P(x)}{Q(x)} = S(x) + \frac{R(x)}{Q(x)}$$

provided that x is not a zero of polynomial Q. To get the latter representation, we consider both polynomials P and Q in decreasing order of their powers and divide in each step the first term of P by the first term of Q. The result is multiplied by Q and the product is then subtracted from P. This yields a new polynomial P_1 having a smaller degree than polynomial P_n. Now the first term (with the largest exponent) of polynomial P_1 is divided by the first term of polynomial Q, the resulting term is multiplied by polynomial Q and subtracted from P_1, yielding polynomial P_2 and so on. The procedure stops if some resulting polynomial P_i has a smaller degree than polynomial Q. This procedure is illustrated by the following example.

Example 3.24　Let polynomials $P : \mathbb{R} \to \mathbb{R}$ and $Q : \mathbb{R} \to \mathbb{R}$ with

$$P(x) = 2x^4 - 3x^3 + 2x^2 - x + 4 \qquad \text{and} \qquad Q(x) = x^2 - x + 3$$

be given. The function $T : D_T \to \mathbb{R}$ with $T(x) = (P/Q)(x)$ is improper, and by polynomial division we obtain:

$$
\begin{array}{l}
(2x^4 - 3x^3 + 2x^2 - x + 4) \ : \ (x^2 - x + 3) = 2x^2 - x - 5 + \frac{-3x+19}{x^2-x+3} \\
-(2x^4 - 2x^3 + 6x^2) \\
\hline
\quad\quad -x^3 - 4x^2 - x + 4 \\
\quad\quad -(-x^3 + x^2 - 3x) \\
\hline
\quad\quad\quad\quad -5x^2 + 2x + 4 \\
\quad\quad\quad\quad -(-5x^2 + 5x - 15) \\
\hline
\quad\quad\quad\quad\quad\quad -3x + 19
\end{array}
$$

In this case, we have written function T as the sum of the polynomial S of degree two with $S(x) = 2x^2 - x - 5$ and the proper rational function R/Q with

$$(R/Q)(x) = (-3x + 19)/(x^2 - x + 3).$$

Power functions

Definition 3.21   The function $f : D_f \to \mathbb{R}$ with

$$f(x) = x^r, \qquad r \in \mathbb{R},$$

is called a *power function*. In dependence on the value of r, the domain D_f and the range R_f are given as follows:

(1) $r \in \{1, 3, 5, \ldots\} \subseteq \mathbb{N} : D_f = (-\infty, \infty), R_f = (-\infty, \infty);$
(2) $r \in \{2, 4, 6, \ldots\} \subseteq \mathbb{N} : D_f = (-\infty, \infty), R_f = [0, \infty);$
(3) $r \in \mathbb{R} \setminus \mathbb{N} : D_f = (0, \infty), R_f = (0, \infty).$

In case (1), function f is strictly increasing, unbounded and odd. Function f is strictly concave on the interval $(-\infty, 0]$ and strictly convex on the interval $[0, \infty)$. In case (2), function f is strictly decreasing on the interval $(-\infty, 0]$, strictly increasing on the interval $[0, \infty)$, bounded from below, even and strictly convex. Since function f is not injective in this case (note that $f(-x) = -f(x)$), the inverse function does not exist in this case (if one restricts the domain to the interval $[0, \infty)$, then function f is bijective and the inverse function exists). In case (3) with $r > 0$, function f is strictly increasing and bounded from below. Moreover, if $r > 1$, function f is strictly convex and, if $0 < r < 1$, function f is strictly concave. Case (3) with positive value r includes as a special case so-called *root functions*, where $r = 1/n$ with $n \in \mathbb{N}$, i.e.

$$f(x) = x^{1/n} = \sqrt[n]{x}.$$

In the case of a root function, number zero belongs to the domain and range of f, i.e. $D_f = R_f = [0, \infty)$. In case (3) with $r < 0$, function f is strictly decreasing, bounded from below and strictly convex.

Obviously, the inverse function f^{-1} of a power function f with $f(x) = x^r$ exists and is equal to $y = x^{1/r}$. The graph of some power functions for special values of r are given in Figure 3.15. If $r \in \mathbb{Q}$, we say that power function f is an *algebraic function*.

Example 3.25   Let function $f : (0, \infty) \to (0, \infty)$ with

$$f(x) = x^{2.5}$$

be given. It is a power function with the rational exponent $r = 2.5$. This function is bijective, and so the inverse function exists. Solving for variable x, we obtain

$$x = y^{1/2.5} = y^{0.4}.$$

Exchanging both variables x and y, we get the inverse function f^{-1} with

$$y = x^{0.4}, \qquad D_{f^{-1}} = R_f = (0, \infty).$$

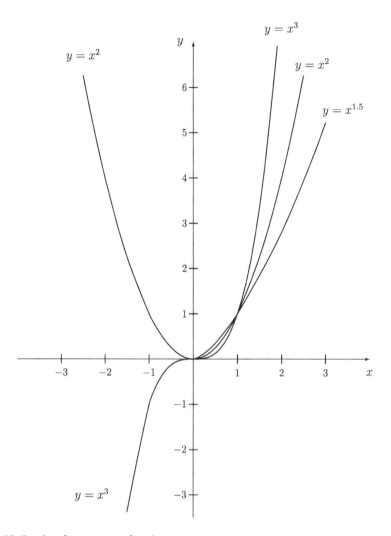

Figure 3.15 Graphs of some power functions.

Exponential and logarithmic functions

Exponential and logarithmic functions play an important role in many economic applications, e.g. growth and decline processes.

Definition 3.22   The function $f : \mathbb{R} \rightarrow (0, \infty)$ with

$$f(x) = a^x, \quad a > 0, \ a \neq 1,$$

is called an *exponential function*.

For all positive values of a, the graph of function f goes through point $(0, 1)$. The graph of the exponential function for three values of a is given in Figure 3.16. For all values of a, the exponential function is only bounded from below (and therefore unbounded). If $a > 1$, the exponential function is strictly increasing, while for $0 < a < 1$ the exponential function is strictly decreasing. All exponential functions are strictly convex. We note that for an exponential function with base e, the notation $y = e^x$ is sometimes also written as $y = \exp(x)$.

Since an exponential function f is bijective for all values $a > 0$ with $a \neq 1$ on the domain D_f, for each of these values of a the inverse function exists. The inverse function of the exponential function is the logarithmic function defined as follows.

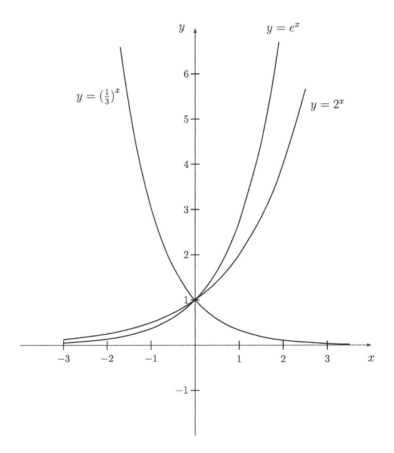

Figure 3.16 Graphs of some exponential functions.

Definition 3.23   The function $f : (0, \infty) \to \mathbb{R}$ with

$$f(x) = \log_a x, \qquad a > 0, \ a \neq 1,$$

is called a *logarithmic function*.

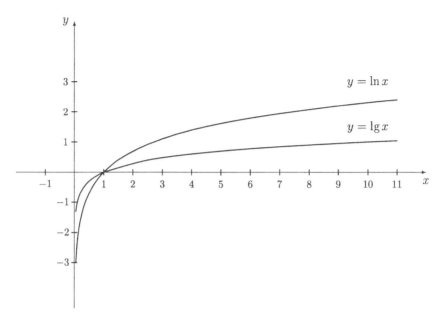

Figure 3.17 Graphs of some logarithmic functions.

The graphs of the logarithmic functions with the two frequently used bases $a = e$ and $a = 10$ are given in Figure 3.17. Since the logarithmic functions are the inverse functions of the corresponding exponential functions, all logarithmic functions go through the point (1,0). Logarithmic functions with a base $a > 1$ are strictly increasing, and logarithmic functions with a base a, where $0 < a < 1$, are strictly decreasing. All logarithmic functions are unbounded. Logarithmic functions with a base $a > 1$ are strictly concave while logarithmic functions with a base a, where $0 < a < 1$, are strictly convex. It is worth emphasizing that both logarithmic and exponential functions are only defined for positive values of the base, where a base equal to one is excluded.

Example 3.26 We consider function $f : D_f \to \mathbb{R}$ with

$$y = f(x) = \ln(3x + 4).$$

Since the logarithmic function is defined only for positive arguments, the inequality $3x+4 > 0$ yields the domain

$$D_f = \left\{ x \in \mathbb{R} \mid x > -\frac{4}{3} \right\}.$$

The logarithmic function has the range $R_f = \mathbb{R}$, and therefore function f also has the range $R_f = \mathbb{R}$. Since the logarithmic function is bijective, function f is bijective as well, and thus the inverse function f^{-1} exists. We obtain

$$y = \ln(3x + 4)$$
$$e^y = 3x + 4$$
$$x = \frac{1}{3} \cdot \left(e^y - 4\right).$$

Exchanging now variables x and y, we have found the inverse function f^{-1} given by

$$y = \frac{1}{3} \cdot (e^x - 4)$$

with the domain $D_{f^{-1}} = R_f = \mathbb{R}$.

Trigonometric functions

Trigonometric (or circular) functions are defined by means of a circle with the origin as centre and radius r (see Figure 3.18). Let P be the point on the circle with the coordinates $(r, 0)$ and assume that this point moves now counterclockwise round the circle. For a certain angle x, we get a point P on the circle with the coordinates (h, v). Then the sine and cosine functions are defined as follows.

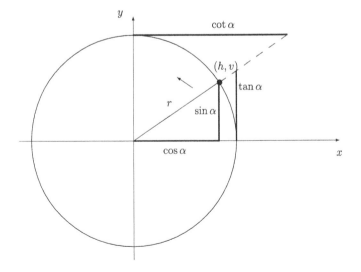

Figure 3.18 The definition of the trigonometric functions.

Definition 3.24 The function $f : \mathbb{R} \to [1, 1]$ with

$$f(x) = \sin x = \frac{v}{r}$$

is called the *sine function*. The function $f : \mathbb{R} \to [1, 1]$ with

$$f(x) = \cos x = \frac{h}{r}$$

is called the *cosine function*.

The graphs of the sine and cosine functions are given in Figure 3.19 (a). The sine and cosine functions are bounded and periodic functions with a period of 2π, i.e. we have for $k \in \mathbb{Z}$:

$$\sin(x + 2k\pi) = \sin x \qquad \text{and} \qquad \cos(x + 2k\pi) = \cos x.$$

The sine function is odd, while the cosine function is even. These functions are often used to model economic cycles.

Remark   Usually, angles are measured in degrees. Particularly in calculus, they are often measured in so-called *radians* (abbreviated rad). An angle of $360°$ corresponds to 2π rad, i.e. we get for instance

$$\sin 360° = \sin 2\pi \text{ rad}.$$

It is worth noting that, using radians, the trigonometric functions have set $\mathbb{R}$ or a subset of $\mathbb{R}$ as domain and due to this, the abbreviation rad is often skipped.

When using powers of trigonometric functions, we also write e.g. $\sin^2 x$ instead of $(\sin x)^2$ or $\cos^3(2x)$ instead of $(\cos 2x)^3$. By means of the sine and the cosine functions, we can introduce two further trigonometric functions as follows.

Definition 3.25   The function $f : D_f \to \mathbb{R}$ with

$$f(x) = \tan x = \frac{\sin x}{\cos x} = \frac{v}{h} \quad \text{and} \quad D_f = \left\{ x \in \mathbb{R} \mid x \neq \frac{\pi}{2} + k\pi, \ k \in \mathbb{Z} \right\}$$

is called the *tangent function*. The function $f : D_f \to \mathbb{R}$ with

$$f(x) = \cot x = \frac{\cos x}{\sin x} = \frac{h}{v} \quad \text{and} \quad D_f = \{ x \in \mathbb{R} \mid x \neq k\pi, \ k \in \mathbb{Z} \}$$

is called the *cotangent function*.

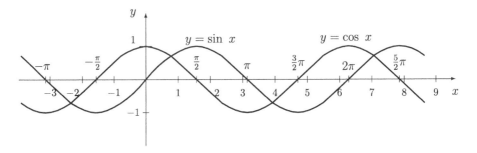

(a) Functions $y = \sin\ x$ and $y = \cos\ x$

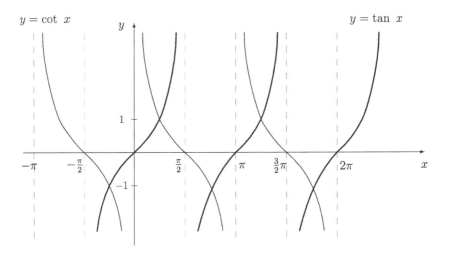

(b) Functions $y = \tan\ x$ and $y = \cot\ x$

Figure 3.19 Graphs of the trigonometric functions.

The graphs of the tangent and cotangent functions are given in Figure 3.19 (b). The tangent and cotangent functions are periodic functions with a period of π, i.e. we have for $k \in \mathbb{Z}$:

$$\tan(x + k\pi) = \tan x \qquad \text{and} \qquad \cot(x + k\pi) = \cot x.$$

Both functions are unbounded and odd.

In the following, we review some basic properties of trigonometric functions.

Some properties of trigonometric functions

(1) $\sin(x \pm y) = \sin x \cos y \pm \cos x \sin y$;

(2) $\cos(x \pm y) = \cos x \cos y \mp \sin x \sin y$;

(3) $\tan(x \pm y) = \dfrac{\tan x \pm \tan y}{1 \mp \tan x \tan y}$;

(4) $\cot(x \pm y) = \dfrac{\cot x \cot y \mp 1}{\cot y \pm \cot x};$

(5) $\sin x \pm \sin y = 2 \cdot \sin \dfrac{x \pm y}{2} \cdot \cos \dfrac{x \mp y}{2};$

(6) $\cos x + \cos y = 2 \cdot \cos \dfrac{x+y}{2} \cdot \cos \dfrac{x-y}{2};$

$\cos x - \cos y = -2 \cdot \sin \dfrac{x+y}{2} \cdot \sin \dfrac{x-y}{2};$

(7) $\sin\left(x \pm \dfrac{\pi}{2}\right) = \pm \cos x, \qquad \cos\left(x \pm \dfrac{\pi}{2}\right) = \mp \sin x;$

(8) $\sin^2 x + \cos^2 x = 1;$

(9) $1 + \tan^2 x = \dfrac{1}{\cos^2 x}.$

Properties (1) up to (4) are known as *addition theorems* of the trigonometric functions. For the special case of $x = y$, properties (1) and (2) turn into

$$\sin 2x = 2 \cdot \sin x \cdot \cos x \qquad \text{and} \qquad \cos 2x = \cos^2 x - \sin^2 x. \tag{3.8}$$

Property (8) is also denoted as a *Pythagorean theorem* for trigonometric functions. As an illustration, we prove the identity given in Property (9) and obtain

$$1 + \tan^2 x = 1 + \frac{\sin^2 x}{\cos^2 x} = \frac{\cos^2 x + \sin^2 x}{\cos^2 x} = \frac{1}{\cos^2 x}.$$

The last equality follows from property (8) above.

Since all four trigonometric functions are not bijective mappings, we have to restrict their domain in order to be able to define inverse functions. If we solve equality $y = \sin x$ for x, we write $x = \arcsin y$. The term $\arcsin y$ gives the angle x whose sine value is y. Similarly, we use the symbol 'arc' in front of the other trigonometric functions when solving for x. Now we can define the arc functions as follows.

Definition 3.26 The function

$$f : [-1, 1] \to \left[-\frac{\pi}{2}, \frac{\pi}{2}\right] \qquad \text{with} \quad f(x) = \arcsin x$$

is called the *arcsine function*. The function

$$f : [-1, 1] \to [0, \pi] \qquad \text{with} \quad f(x) = \arccos x$$

is called the *arccosine function*. The function

$$f : (-\infty, \infty) \to \left(-\frac{\pi}{2}, \frac{\pi}{2}\right) \qquad \text{with} \quad f(x) = \arctan x$$

is called the *arctangent* function. The function

$$f : (-\infty, \infty) \to (0, \pi) \qquad \text{with} \quad f(x) = \text{arccot}\, x$$

is called the *arccotangent function*.

In Definition 3.26, the ranges of the arc functions give the domains of the original trigonometric functions such that they represent bijective mappings and an inverse function can
be defined. The graphs of the arc functions are given in Figure 3.20. All the arc functions
are bounded. The arcsine and arctangent functions are strictly increasing and odd while the
arccosine and arccotangent functions are strictly decreasing. Moreover, the arcsine and arccotangent functions are strictly concave for $x \leq 0$ and they are strictly convex for $x \geq 0$.
The arccosine and arctangent functions are strictly convex for $x \leq 0$ and strictly concave for
$x \geq 0$ (where in all cases, x has to belong to the domain of the corresponding function).

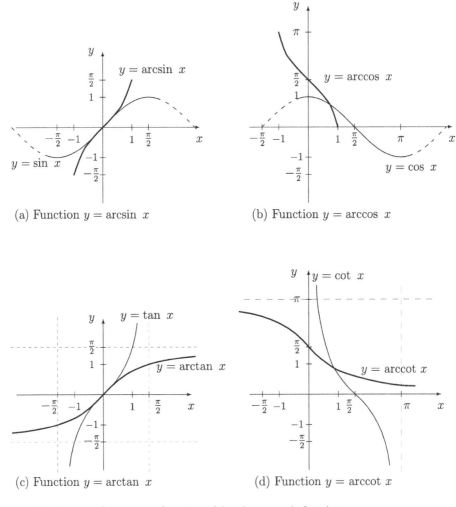

(a) Function $y = \arcsin x$ (b) Function $y = \arccos x$

(c) Function $y = \arctan x$ (d) Function $y = \mathrm{arccot}\, x$

Figure 3.20 Graphs of the inverse functions of the trigonometric functions.

Overview on elementary functions

In Table 3.1, we summarize the domains and ranges of the elementary functions.

Table 3.1 Domains and ranges of elementary functions

Function f with $y = f(x)$		D_f	R_f
$y = c$	c is constant	$-\infty < x < \infty$	$y = c$
$y = x^n$	$n = 2, 3, \dots$	$-\infty < x < \infty$	$-\infty < y < \infty$
$y = x^\alpha$	$\alpha \in \mathbb{R}$	$0 < x < \infty$	$0 < y < \infty$
$y = \sqrt[n]{x}$	$n = 2, 3, \dots$	$0 \leq x < \infty$	$0 \leq y < \infty$
$y = a^x$	$a > 0$	$-\infty < x < \infty$	$0 < y < \infty$
$y = \log_a x$	$a > 0, a \neq 1$	$0 < x < \infty$	$-\infty < y < \infty$
$y = \sin x$		$-\infty < x < \infty$	$-1 \leq y \leq 1$
$y = \cos x$		$-\infty < x < \infty$	$-1 \leq y \leq 1$
$y = \tan x$		$x \neq \dfrac{\pi}{2} + k\pi$	$-\infty < y < \infty$
$y = \cot x$		$x \neq k\pi$	$-\infty < y < \infty$
$y = \arcsin x$		$-1 \leq x \leq 1$	$-\dfrac{\pi}{2} \leq x \leq \dfrac{\pi}{2}$
$y = \arccos x$		$-1 \leq x \leq 1$	$0 \leq x \leq \pi$
$y = \arctan x$		$-\infty < x < \infty$	$-\dfrac{\pi}{2} < x < \dfrac{\pi}{2}$
$y = \text{arccot } x$		$-\infty < x < \infty$	$0 < x < \pi$

EXERCISES

3.1 Let $A = \{1, 2, 3, 4, 5\}$. Consider the relations

$$R = \{(1,2), (1,4), (3,1), (3,4), (3,5), (5,1), (5,4)\} \subseteq A \times A$$

and

$$S = \{(a_1, a_2) \in A \times A \mid a_1 + a_2 = 6\}.$$

(a) Illustrate R and S by graphs and as points in a rectangular coordinate system.
(b) Which of the following propositions are true of relation R:

$$1R2, \quad 2R1, \quad 3\not{R}1, \quad \{2, 3, 4, 5\} = \{a \in A \mid 1Ra\}?$$

(c) Find the inverse relations for R and S.
(d) Check whether the relations R and S are mappings.

3.2   Consider the following relations and find out whether they are mappings. Which of the mappings are surjective, injective, bijective?

(a)   (b)   (c)   (d)

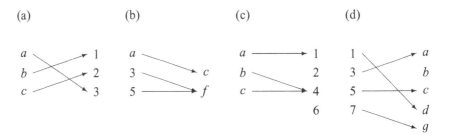

3.3   Let $A = B = \{1,2,3\}$ and $C = \{2,3\}$. Consider the following mappings $f : A \to B$ and $g : C \to A$ with $f(1) = 3, f(2) = 2, f(3) = 1$ and $g(2) = 1, g(3) = 2$.

(a)   Illustrate $f, g, f \circ g$ and $g \circ f$ by graphs if possible.
(b)   Find the domains and the ranges of the given mappings and of the composite mappings.
(c)   What can you say about the properties of these mappings?

3.4   Given is the relation

$$F = \{(x_1,x_2) \in \mathbb{R}^2 \mid |x_2| = x_1 + 2\}.$$

Check whether F or F^{-1} is a mapping. In the case when there is a mapping, find the domain and the range. Graph F and F^{-1}.

3.5   Given are the relations

$$F = \{(x_1,x_2) \mid x_2 = x_1^3\} \quad \text{with } x_1 \in \{-3,-2,-1,0,1,2,3\}$$

and

$$G = \{(x,y) \in \mathbb{R}^2 \mid 9x^2 + 2y^2 = 18\}.$$

Are these relations functions? If so, does the inverse function exist?

3.6   Given are the functions $f : D_f \to \mathbb{R}$ and $g : D_g \to \mathbb{R}$ with

$$f(x) = 2x + 1 \qquad \text{and} \qquad g(x) = x^2 - 2.$$

Find and graph the composite functions $g \circ f$ and $f \circ g$.

3.7   Given are the functions $f : \mathbb{R} \to \mathbb{R}_+$ and $g : \mathbb{R} \to \mathbb{R}$ with

$$f(x) = e^x \qquad \text{and} \qquad g(x) = -x.$$

(a)   Check whether the functions f and g are surjective, injective or bijective. Graph these functions.
(b)   Find f^{-1} and g^{-1} and graph them.
(c)   Find $f \circ g$ and $g \circ f$ and graph them.

3.8 Find $a \in \mathbb{R}$ such that $f : D_f = [a, \infty) \to \mathbb{R}$ with

$$y = f(x) = x^2 + 2x - 3$$

being a bijective function. Find and graph function f^{-1}.

3.9 Find domain, range and the inverse function for function $f : D_f \to \mathbb{R}$ with $y = f(x)$:

(a) $y = \dfrac{\sqrt{x} - 4}{\sqrt{x} + 4}$;   (b) $y = (x - 2)^3$.

3.10 Given are the polynomials $P_5 : \mathbb{R} \to \mathbb{R}$ and $P_2 : \mathbb{R} \to \mathbb{R}$ with

$$P_5(x) = 2x^5 - 6x^4 - 6x^3 + 22x^2 - 12x \qquad \text{and} \qquad P_2(x) = (x - 1)^2.$$

(a) Calculate the quotient P_5/P_2 by polynomial division.
(b) Find all the zeroes of polynomial P_5 and factorize P_5.
(c) Verify Vieta's formulae given in Theorem 3.6.
(d) Draw the graph of the function P_5.

3.11 Check by means of Horner's scheme whether $x_1 = 1, x_2 = -1, x_3 = 2, x_4 = -2$ are zeroes of the polynomial $P_6 : \mathbb{R} \to \mathbb{R}$ with

$$P_6(x) = x^6 + 2x^5 - x^4 - x^3 + 2x^2 - x - 2.$$

Factorize polynomial P_6.

3.12 Find the domain, range and the inverse function for each of the following functions $f_i : D_{f_i} \to \mathbb{R}$ with $y_i = f_i(x)$:

(a) $y_1 = \sin x$, $y_2 = 2 \sin x$, $y_3 = \sin 2x$, $y_4 = \sin x + 2$ and
   $y_5 = \sin(x + 2)$;
(b) $y_1 = e^x$, $y_2 = 2e^x$, $y_3 = e^{2x}$, $y_4 = e^x + 2$ and $y_5 = e^{x+2}$.

Graph the functions given in (a) and (b) and check whether they are odd or even or whether they have none of these properties.

3.13 Given are the following functions $f : D_f \to \mathbb{R}$ with $y = f(x)$:

(a) $y = \ln x^4$;   (b) $y = \ln x^3$;   (c) $y = 3x^2 + 5$;
(d) $y = \sqrt{4 - x^2}$;   (e) $y = 1 + e^{-x}$;   (f) $y = \sqrt{|x|} - x$.

Find the domain and range for each of the above functions and graph these functions. Check where the functions are increasing and whether they are bounded. Which of the functions are odd or even?

4 Differentiation

In economics, there are many problems which require us to take into account how a function value changes with respect to small changes of the independent variable (e.g. input, time, etc.). For example, assume that the price of some product changes slightly. The question is how does this affect the amount of product customers will buy? A useful tool for such investigations is differential calculus, which we treat in this chapter. It is an important field of mathematics with many applications, e.g. graphing functions, determination of extreme points of functions with or without additional constraints. Differential calculus allows us to investigate specific properties of functions such as monotonicity or convexity. For instance in economics, cost, revenue, profit, demand, production or utility functions have to be investigated with respect to their properties. In this chapter, we consider functions $f : D_f \rightarrow \mathbb{R}$ depending on one real variable, i.e. $D_f \subseteq \mathbb{R}$.

4.1 LIMIT AND CONTINUITY

4.1.1 Limit of a function

One of the basic concepts in mathematics is that of a limit (see Definition 2.6 for a sequence). In this section, the limit of a function is introduced. This notion deals with the question of which value does the dependent variable y of a function f with $y = f(x)$ approach as the independent variable x approaches some specific value x_0?

> **Definition 4.1** The real number L is called the *limit* of function $f : D_f \rightarrow \mathbb{R}$ as x tends to x_0 if for any sequence $\{x_n\}$ with $x_n \neq x_0$, $x_n \in D_f$, $n = 1, 2, \ldots$, which converges to x_0, the sequence of the function values $\{f(x_n)\}$ converges to L.

Thus, we say that function f tends to number L as x tends to (but is not equal to) x_0. As an abbreviation we write

$$\lim_{x \to x_0} f(x) = L.$$

Note that limit L must be a (finite) number, otherwise we say that the limit of function f as x tends to x_0 does not exist. If this limit does not exist, we distinguish two cases. If $L = \pm\infty$, we also say that function f is *definitely divergent* as x tends to x_0, otherwise function f is said to be *indefinitely divergent* as x tends to x_0.

In the above definition, the elements of sequence $\{x_n\}$ can be both greater and smaller than x_0. In certain situations, only the limit from one side has to be considered. In the following, we consider such one-sided approaches, where the terms of the sequences are either all greater or all smaller than x_0.

Definition 4.2   The real number L_r (real number L_1) is called the *right-side (left-side) limit* of function $f : D_f \to \mathbb{R}$ as x tends to x_0 from the right side (left side) if for any sequence $\{x_n\}$ with $x_n > x_0$ ($x_n < x_0$), $x_n \in D_f$, $n = 1, 2, \ldots$, the sequence of the function values $\{f(x_n)\}$ converges to L_r (converges to L_1).

We also write

$$\lim_{x \to x_0 + 0} f(x) = L_r \qquad \text{and} \qquad \lim_{x \to x_0 - 0} f(x) = L_1$$

for the right-side and left-side limits, respectively. A relationship between one-sided limits and the limit as introduced in Definition 4.1 is given by the following theorem.

THEOREM 4.1   The limit of a function $f : D_f \to \mathbb{R}$ as x tends to x_0 exists if and only if both the right-side and left-side limits exist and coincide, i.e.

$$\lim_{x \to x_0 + 0} f(x) = \lim_{x \to x_0 - 0} f(x) = \lim_{x \to x_0} f(x).$$

We note that it is not necessary for the existence of a limit of function f as x tends to x_0 that the function value $f(x_0)$ at point x_0 be defined.

Example 4.1   Let function $f : \mathbb{R} \setminus \{0\} \to \mathbb{R}$ with

$$f(x) = \frac{|x|}{x}$$

be given. We want to compute

$$L = \lim_{x \to 0} f(x).$$

However, since

$$\lim_{x \to 0+0} f(x) = \lim_{x \to 0+0} \frac{x}{x} = 1 \qquad \text{and} \qquad \lim_{x \to 0-0} f(x) = \lim_{x \to 0-0} \frac{-x}{x} = -1,$$

the limit of function f as x tends to zero does not exist.

Example 4.2   Let function $f : \mathbb{R} \setminus \{0\} \to \mathbb{R}$ with

$$f(x) = \sin \frac{1}{x}$$

be given. If we want to compute

$$L = \lim_{x \to 0} f(x),$$

we find that even both one-sided limits

$$L_r = \lim_{x \to 0+0} f(x) \qquad \text{and} \qquad L_l = \lim_{x \to 0-0} f(x)$$

do not exist since, as x tends to zero either from the left or from the right, the function values of the sine function oscillate in the interval $[-1, 1]$, i.e. the function values change very quickly between the numbers -1 and 1 (even with increasing frequency as x tends to zero). Thus, the limit of function f as x tends to zero does not exist and function f is indefinitely divergent as x tends to zero.

Next, we give some useful properties of limits.

THEOREM 4.2 Assume that the limits

$$\lim_{x \to x_0} f_1(x) = L_1 \qquad \text{and} \qquad \lim_{x \to x_0} f_2(x) = L_2$$

exist. Then the following limits exist, and we obtain:

(1) $\displaystyle \lim_{x \to x_0} \left[f_1(x) + f_2(x) \right] = \lim_{x \to x_0} f_1(x) + \lim_{x \to x_0} f_2(x) = L_1 + L_2;$

(2) $\displaystyle \lim_{x \to x_0} \left[f_1(x) - f_2(x) \right] = \lim_{x \to x_0} f_1(x) - \lim_{x \to x_0} f_2(x) = L_1 - L_2;$

(3) $\displaystyle \lim_{x \to x_0} \left[f_1(x) \cdot f_2(x) \right] = \lim_{x \to x_0} f_1(x) \cdot \lim_{x \to x_0} f_2(x) = L_1 \cdot L_2;$

(4) $\displaystyle \lim_{x \to x_0} \frac{f_1(x)}{f_2(x)} = \frac{\lim_{x \to x_0} f_1(x)}{\lim_{x \to x_0} f_2(x)} = \frac{L_1}{L_2} \qquad$ provided that $L_2 \neq 0;$

(5) $\displaystyle \lim_{x \to x_0} \sqrt{f_1(x)} = \sqrt{\lim_{x \to x_0} f_1(x)} = \sqrt{L_1} \qquad$ provided that $L_1 \geq 0;$

(6) $\displaystyle \lim_{x \to x_0} [f_1(x)]^n = \left[\lim_{x \to x_0} f_1(x) \right]^n = L_1^n;$

(7) $\displaystyle \lim_{x \to x_0} \left[a^{f_1(x)} \right] = a^{\left[\lim_{x \to x_0} f_1(x) \right]} = a^{L_1}.$

Example 4.3 Given the function $f : D_f \to \mathbb{R}$ with

$$f(x) = \sqrt{\frac{x^2 + 3x - 1}{x}},$$

we compute the limit

$$L = \lim_{x \to 1} f(x).$$

Applying Theorem 4.2, we obtain

$$L = \sqrt{\lim_{x \to 1} \frac{x^2 + 3x - 1}{x}} = \sqrt{\frac{\lim_{x \to 1} x^2 + 3 \lim_{x \to 1} x - 1}{\lim_{x \to 1} x}} = \sqrt{\frac{1 + 3 - 1}{1}} = \sqrt{3}.$$

Example 4.4 Given the function $f : D_f \to \mathbb{R}$ with

$$f(x) = \frac{\sqrt{x} - 2}{x - 4},$$

we compute the limit

$$L = \lim_{x \to 4} f(x).$$

If we apply Theorem 4.2, part (4), and separately determine the limit of the numerator and the denominator, we find that both terms tend to zero, and we cannot find the limit in this way. Therefore, we rationalize the numerator by multiplying the numerator and the denominator by $\sqrt{x} + 2$ and obtain:

$$L = \lim_{x \to 4} \frac{(\sqrt{x} - 2)(\sqrt{x} + 2)}{(x - 4)(\sqrt{x} + 2)} = \lim_{x \to 4} \frac{x - 4}{(x - 4)(\sqrt{x} + 2)} = \lim_{x \to 4} \frac{1}{\sqrt{x} + 2} = \frac{1}{\sqrt{4} + 2} = \frac{1}{4}.$$

4.1.2 Continuity of a function

Definition 4.3 A function $f : D_f \to \mathbb{R}$ is said to be *continuous* at $x_0 \in D_f$ if the limit of function f as x tends to x_0 exists and if this limit coincides with the function value $f(x_0)$, i.e.

$$\lim_{x \to x_0} f(x) = f(x_0).$$

Alternatively, we can give the following equivalent definition of a continuous function at x_0 using the $(\delta - \varepsilon)$ notation.

Definition 4.3* A function $f : D_f \to \mathbb{R}$ is said to be *continuous* at $x_0 \in D_f$ if, for any real number $\varepsilon > 0$, there exists a real number $\delta(\varepsilon)$ such that inequality $|x - x_0| < \delta(\varepsilon)$ implies inequality $|f(x) - f(x_0)| < \varepsilon$.

We illustrate the latter definition in Figure 4.1. Part (a) represents a continuous function at x_0. This means that for any $\varepsilon > 0$ (in particular, for any arbitrarily small positive ε), we can state a number δ depending on ε such that for all x from the open interval $(x_0 - \delta, x_0 + \delta)$ the function values $f(x)$ are within the open interval $(f(x) - \varepsilon, f(x) + \varepsilon)$ (i.e. in the dashed area). In other words, continuity of a function at some point $x_0 \in D_f$ means that small

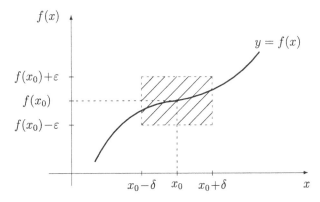

(a) Function f is continuous at x_0

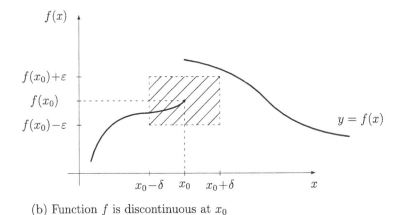

(b) Function f is discontinuous at x_0

Figure 4.1 Continuous and discontinuous functions.

changes in the independent variable x lead to small changes in the dependent variable y. Figure 4.1(b) represents a function which is not continuous at x_0. For some (small) $\varepsilon > 0$, we cannot give a value δ such that for all $x \in (x_0 - \delta, x_0 + \delta)$ the function values are in the dashed area.

If the one-sided limits of a function f as x tends to x_0 are different (or one or both limits do not exist), or if they are identical but the function value $f(x_0)$ is not defined or value $f(x_0)$ is defined but not equal to both one-sided limits, then function f is *discontinuous* at x_0. Next, we classify some types of discontinuities.

If the limit of function f as x tends to x_0 exists but the function value $f(x_0)$ is different or function f is even not defined at point x_0, we have a *removable discontinuity*. In the case when function f is not defined at point x_0, we also say that function f has a *gap* at x_0.

A function f has a *finite jump* at x_0 if both one-sided limits of function f as x tends to $x_0 - 0$ and x tends to $x_0 + 0$ exist and they are different. The following discontinuities characterize situations when at least one of the one-sided limits of function f as x tends to $x_0 \pm 0$ does not exist. If one of the one-sided limits of function f as x tends to $x_0 \pm 0$ exists, but from the other side function f tends to ∞ or $-\infty$, then we say that function f has an *infinite jump* at point x_0. A rational function $f = P/Q$ has a *pole* at point x_0 if $Q(x_0) = 0$ but $P(x_0) \neq 0$. (As a consequence, the function values of f as x tends to $x_0 - 0$ or to $x_0 + 0$ tend either to ∞ or $-\infty$, i.e. function f is definitely divergent as x tends to x_0.) The multiplicity of the zero x_0 of polynomial Q defines the order of the pole. In the case of a pole of even order, the sign of function f does not change 'at' point x_0 while in the case of a pole of odd order the sign of function f changes 'at' point x_0. Finally, function f has an *oscillation point* 'at' x_0 if function f is indefinitely divergent as x tends to x_0 (i.e. neither the limit of function f as x tends to x_0 exist nor function f tends to $\pm\infty$ as x tends to x_0). In the above cases of a (finite or infinite) jump, a pole and an oscillation point, we also say that function f has an *irremovable discontinuity* at point x_0.

For illustration, we consider the following examples.

Example 4.5 Let function $f : D_f \to \mathbb{R}$ with

$$f(x) = \frac{x^2 + x - 2}{x - 1}$$

be given. In this case, we have $x_0 = 1 \notin D_f$, but

$$\lim_{x \to x_0} f(x) = 3$$

(see Figure 4.2). Thus, function f has a gap at point $x_0 = 1$. This is a removable discontinuity, since we can define a function $f^* : \mathbb{R} \to \mathbb{R}$ with

$$f^*(x) = \begin{cases} f(x) & \text{for } x \neq 1 \\ 3 & \text{for } x = 1 \end{cases}$$

which is continuous at point $x_0 = 1$.

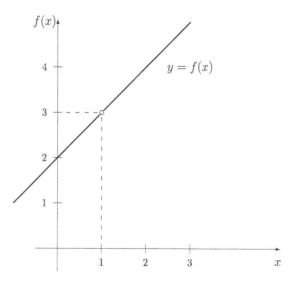

Figure 4.2 Function f with $f(x) = (x^2 + x - 2)/(x - 1)$.

Example 4.6   We consider function $f : D_f \to \mathbb{R}$ with

$$f(x) = \sin \frac{1}{x}.$$

For $x_0 = 0$, we get $x_0 \notin D_f$, and we have already shown in Example 4.2 that function f is indefinitely divergent as x tends to zero. Therefore, point $x_0 = 0$ is an oscillation point.

Analogously to Definition 4.3, we can introduce one-sided continuity. If the limit of function f as x tends to $x_0 + 0$ exists and is equal to $f(x_0)$, then function f is called *right-continuous* at point x_0. In the same way, we can define the left-continuity of a function at a point x_0. Function f is called continuous on the open interval $(a, b) \in D_f$ if it is continuous at all points of (a, b). Analogously, function f is called continuous on the closed interval $[a, b] \subseteq D_f$ if it is continuous on the open interval (a, b), right-continuous at point a and left-continuous at point b. If function f is continuous at all points x of the domain D_f, then we also say that f is continuous.

Properties of continuous functions

First, we give some properties of functions that are continuous at particular points.

THEOREM 4.3   Let functions $f : D_f \to \mathbb{R}$ and $g : D_g \to \mathbb{R}$ be continuous at point $x_0 \in D_f \cap D_g$. Then functions $f + g, f - g, f \cdot g$ and, for $g(x_0) \neq 0$, also function f/g are continuous at x_0.

THEOREM 4.4   Let function $f : D_f \to \mathbb{R}$ be continuous at $x_0 \in D_f$ and function $g : D_g \to \mathbb{R}$ be continuous at point $x_1 = f(x_0) \in D_g$. Then the composite function $g \circ f$ is continuous at point x_0, and we get

$$\lim_{x \to x_0} g(f(x)) = g \left(\lim_{x \to x_0} f(x) \right) = g(f(x_0)).$$

The latter theorem implies that we can 'interchange' the determination of the limit of function f as x tends to x_0 and the calculation of the value of function g. We continue with three properties of functions that are continuous on the closed interval $[a, b]$.

THEOREM 4.5   Let function $f : D_f \to \mathbb{R}$ be continuous on the closed interval $[a, b] \subseteq D_f$. Then function f is bounded on $[a, b]$ and takes its minimal value f_{min} and its maximal value f_{max} at points x_{min} and x_{max}, respectively, belonging to the interval $[a, b]$, i.e.

$$f_{min} = f(x_{min}) \leq f(x) \leq f(x_{max}) = f_{max} \qquad \text{for all } x \in [a, b].$$

Theorem 4.5 does not necessarily hold for an open or half-open interval, e.g. function f with $f(x) = 1/x$ is not bounded on the left-open interval $(0, 1]$.

THEOREM 4.6 (Bolzano's theorem)   Let function $f : D_f \to \mathbb{R}$ be continuous on the closed interval $[a, b] \in D_f$ with $f(a) \cdot f(b) < 0$. Then there exists an $x^* \in (a, b)$ such that $f(x^*) = 0$.

Theorem 4.6 has some importance for finding zeroes of a function numerically (see later in Chapter 4.8). In order to apply a numerical procedure, one needs an interval $[a, b]$ (preferably as small as possible) so that a zero of the function is certainly contained in this interval.

THEOREM 4.7 (intermediate-value theorem) Let function $f : D_f \to \mathbb{R}$ be continuous on the closed interval $[a, b] \subseteq D_f$. Moreover, let f_{min} be the smallest and f_{max} the largest value of function f for $x \in [a, b]$. Then for each $y^* \in [f_{min}, f_{max}]$, there exists an $x^* \in [a, b]$ such that $f(x^*) = y^*$.

The geometrical meaning of Theorem 4.7 is illustrated in Figure 4.3. The graph of any line $y = y^* \in [f_{min}, f_{max}]$ intersects *at least once* the graph of function f. For the particular value y^* chosen in Figure 4.3, there are two such values x_1^* and x_2^* with $f(x_1^*) = f(x_2^*) = y^*$.

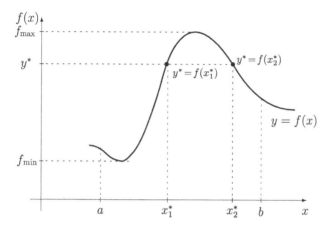

Figure 4.3 Illustration of Theorem 4.7.

4.2 DIFFERENCE QUOTIENT AND THE DERIVATIVE

We now consider changes in the function value $y = f(x)$ in relation to changes in the independent variable x. If we change the value x of the independent variable by some value Δx, the function value may also change by some difference Δy, i.e. we have

$$\Delta y = f(x + \Delta x) - f(x).$$

We now consider the ratio of the changes Δy and Δx and give the following definition.

Definition 4.4 Let $f : D_f \to \mathbb{R}$ and $x_0, x_0 + \Delta x \in (a, b) \subseteq D_f$. The ratio

$$\frac{\Delta y}{\Delta x} = \frac{f(x_0 + \Delta x) - f(x_0)}{\Delta x}$$

is called the *difference quotient* of function f with respect to points $x_0 + \Delta x$ and x_0.

The quotient $\Delta y/\Delta x$ depends on the difference Δx and describes the average change of the value of function f between the points x_0 and $x_0 + \Delta x$. Let us now consider what happens when $\Delta x \to 0$, i.e. the difference in the x values of the two points $x_0 + \Delta x$ and x_0 becomes arbitrarily small.

Definition 4.5 Let $f : D_f \to \mathbb{R}$. Then function f with $y = f(x)$ is said to be *differentiable* at point $x_0 \in (a, b) \subseteq D_f$ if the limit

$$\lim_{\Delta x \to 0} \frac{f(x_0 + \Delta x) - f(x_0)}{\Delta x}$$

exists. The above limit is denoted as

$$\frac{df(x_0)}{dx} = f'(x_0)$$

and is called the *differential quotient* or *derivative* of function f at point x_0.

We only mention that one can define one-sided derivatives in an analogous manner, e.g. the left derivative of function f at some point $x_0 \in D_f$ can be defined by considering only the left-side limit in Definition 4.5. If function f is differentiable at each point x of the open interval $(a, b) \subseteq D_f$, function f is said to be differentiable on the interval (a, b). Analogously, if function f is differentiable at each point $x \in D_f$, function f is said to be differentiable.

If for any $x \in D_f$ the derivative $f'(x)$ exists, we obtain a function f' with $y' = f'(x)$ by assigning to each $x \in D_f$ the value $f'(x)$. We also say that function f' is the first derivative of function f. If function f' is continuous, we say that the original function f is *continuously differentiable*.

In economics, function f' is also referred to as the *marginal* of f or the *marginal function*. It reflects the fact that the derivative characterizes the change in the function value provided that the change in the variable x is sufficiently small, i.e. it can be considered as 'marginal'. This means that the marginal function can be interpreted as the approximate change in function f when variable x increases by one unit from x_0 to $x_0 + 1$, i.e.

$$f'(x_0) \approx f(x_0 + 1) - f(x_0).$$

A *geometric interpretation* of the first derivative is as follows. The value $f'(x_0)$ is the slope of the tangent to the curve $y = f(x)$ at the point $(x_0, f(x_0))$ (see Figure 4.4). Consider the line through the points $(x_0, f(x_0))$ and $(x_0 + \Delta x, f(x_0 + \Delta x))$. For the slope of this line we have

$$\frac{\Delta y}{\Delta x} = \tan \beta.$$

If Δx becomes smaller and tends to zero, the angle of the corresponding line and the x axis tends to α, and we finally obtain

$$f'(x_0) = \lim_{\Delta x \to 0} \frac{\Delta y}{\Delta x} = \tan \alpha.$$

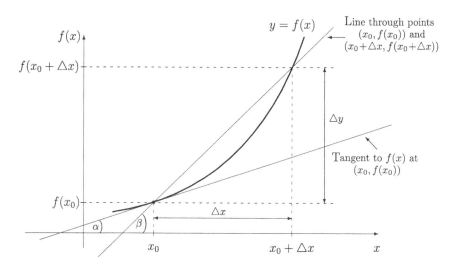

Figure 4.4 Geometrical interpretation of the derivative $f'(x_0)$.

Example 4.7 Let function $f : \mathbb{R} \to \mathbb{R}$ with

$$y = f(x) = \begin{cases} 2x & \text{for} \quad x < 1 \\ 3 - x & \text{for} \quad 1 \leq x \leq 2 \end{cases}$$

be given and let $x_0 = 1$. We obtain

$$\lim_{\Delta x \to 0-0} \frac{f(x_0 + \Delta x) - f(x_0)}{\Delta x} = \lim_{\Delta x \to 0-0} \frac{2(x_0 + \Delta x) - 2x_0}{\Delta x} = \lim_{\Delta x \to 0-0} \frac{2\Delta x}{\Delta x} = 2$$

and

$$\lim_{\Delta x \to 0+0} \frac{f(x_0 + \Delta x) - f(x_0)}{\Delta x} = \lim_{\Delta x \to 0+0} \frac{3 - (x_0 + \Delta x) - (3 - x_0)}{\Delta x}$$

$$= \lim_{\Delta x \to 0+0} \frac{-\Delta x}{\Delta x} = -1.$$

Consequently, the differential quotient of function f at point $x_0 = 1$ does not exist and function f is not differentiable at point $x_0 = 1$. However, since both one-sided limits of function f as x tends to 1 ± 0 exist and are equal to $f(1) = 2$, function f is continuous at point $x_0 = 1$.

The latter example shows that a function being continuous at point x_0 is not necessarily differentiable at this point. However, the converse is true.

THEOREM 4.8 Let function $f : D_f \to \mathbb{R}$ be differentiable at point $x_0 \in D_f$. Then function f is continuous at x_0.

4.3　DERIVATIVES OF ELEMENTARY FUNCTIONS; DIFFERENTIATION RULES

Before giving derivatives of elementary functions, we consider an example for applying Definition 4.5.

Example 4.8　We consider function $f : \mathbb{R} \to \mathbb{R}$ with

$$f(x) = x^3$$

and determine the derivative f' at point x:

$$f'(x) = \lim_{\Delta x \to 0} \frac{f(x + \Delta x) - f(x)}{\Delta x} = \lim_{\Delta x \to 0} \frac{(x + \Delta x)^3 - x^3}{\Delta x}$$

$$= \lim_{\Delta x \to 0} \frac{x^3 + 3x^2 \Delta x + 3x(\Delta x)^2 + (\Delta x)^3 - x^3}{\Delta x}$$

$$= \lim_{\Delta x \to 0} \frac{\Delta x \cdot [3x^2 + 3x\Delta x + (\Delta x)^2]}{\Delta x} = 3x^2.$$

The above formula for the derivative of function f with $f(x) = x^3$ can be generalized to the case of a power function f with $f(x) = x^n$, for which we obtain $f'(x) = nx^{n-1}$.

The determination of the derivative according to Definition 4.5 appears to be unpractical for frequent use in the case of more complicated functions. Therefore, we are interested in an overview on derivatives of elementary functions which we can use when investigating more complicated functions later. Table 4.1 contains the derivatives of some elementary functions of one variable.

Table 4.1 Derivatives of elementary functions

$y = f(x)$	$y' = f'(x)$	D_f	
C	0	$-\infty < x < \infty$,	C is constant
x^n	nx^{n-1}	$-\infty < x < \infty$,	$n \in \mathbb{N}$
x^α	$\alpha x^{\alpha-1}$	$0 < x < \infty$,	$\alpha \in \mathbb{R}$
e^x	e^x	$-\infty < x < \infty$	
a^x	$a^x \ln a$	$-\infty < x < \infty$,	$a > 0$
$\ln x$	$\dfrac{1}{x}$	$0 < x < \infty$	
$\log_a x$	$\dfrac{1}{x \ln a}$	$0 < x < \infty$,	$a > 0$
$\sin x$	$\cos x$	$-\infty < x < \infty$	
$\cos x$	$-\sin x$	$-\infty < x < \infty$	
$\tan x$	$\dfrac{1}{\cos^2 x}$	$x \neq \dfrac{\pi}{2} + k\pi$,	$k \in \mathbb{Z}$
$\cot x$	$-\dfrac{1}{\sin^2 x}$	$x \neq k\pi$,	$k \in \mathbb{Z}$

Sum, product and quotient rules

Next, we present formulas for the derivative of the sum, difference, product and quotient of two functions.

THEOREM 4.9 Let functions $f : D_f \to \mathbb{R}$ and $g : D_g \to \mathbb{R}$ be differentiable at $x \in D_f \cap D_g$. Then the functions $f + g, f - g, f \cdot g$ and, for $g(x) \neq 0$, also function f/g are differentiable at point x, and we have:

(1) $(f + g)'(x) = f'(x) + g'(x)$;
(2) $(f - g)'(x) = f'(x) - g'(x)$;
(3) $(f \cdot g)'(x) = f'(x) \cdot g(x) + f(x) \cdot g'(x)$;
(4) $\left(\dfrac{f}{g}\right)'(x) = \dfrac{f'(x) \cdot g(x) - f(x) \cdot g'(x)}{[g(x)]^2}$.

As a special case of part (3) we obtain: if $f(x) = C$, then $(C \cdot g)'(x) = C \cdot g'(x)$, where C is a constant.

Example 4.9 Let function $f : D_f \to \mathbb{R}$ with

$$f(x) = 4x^4 - 2x + \ln x + \sqrt{x}$$

be given. Using $\sqrt{x} = x^{1/2}$, we obtain

$$f'(x) = 16x^3 - 2 + \frac{1}{x} + \frac{1}{2\sqrt{x}}.$$

Example 4.10 In macroeconomics, it is assumed that for a closed economy

$$Y = C + I,$$

where Y is the national income, C is the consumption and I is the investment. Assume that the consumption linearly depends on the national income, i.e. equation

$$C = a + bY$$

holds, where a and b are parameters. Here $C'(Y) = b$ is called the *marginal propensity to consume*, a parameter which is typically assumed to be between zero and one. If we want to give the national income as a function depending on the investment, we obtain from

$$Y = a + bY + I$$

the function Y as

$$Y = Y(I) = \frac{a + I}{1 - b}.$$

For the derivative $Y'(I)$ we obtain

$$Y'(I) = \frac{dY}{dI} = \frac{1}{1-b}.$$

The latter result can be interpreted such that an increase in I by one unit leads to an increase of Y by $1/(1-b) > 0$ units.

Example 4.11   Let functions $f : \mathbb{R} \to \mathbb{R}$ and $g : \mathbb{R} \to \mathbb{R}$ with

$$f(x) = x^2 \qquad \text{and} \qquad g(x) = e^x$$

be given. Then

$$(f \cdot g)'(x) = 2xe^x + x^2 e^x = xe^x(x+2)$$

$$\left(\frac{f}{g}\right)'(x) = \frac{2xe^x - x^2 e^x}{(e^x)^2} = \frac{xe^x(2-x)}{(e^x)^2} = \frac{x(2-x)}{e^x}.$$

Derivative of composite and inverse functions

Next, we consider composite and inverse functions and give a rule to determine their derivatives.

THEOREM 4.10   Let functions $f : D_f \to \mathbb{R}$ and $g : D_g \to \mathbb{R}$ be continuous. Then:

(1) If function f is differentiable at point $x \in D_f$ and function g is differentiable at point $y = f(x) \in D_g$, then function $g \circ f$ is also differentiable at point x and

$$(g \circ f)'(x) = (g(f))'(x) = g'(f(x)) \cdot f'(x) \qquad \text{(chain rule)}.$$

(2) Let function f be strictly monotone on D_f and differentiable at point $x \in D_f$ with $f'(x) \neq 0$. Then the inverse function f^{-1} with $x = f^{-1}(y)$ is differentiable at point $y = f(x) \in D_{f^{-1}}$ and

$$(f^{-1})'(y) = \frac{1}{f'(x)} = \frac{1}{f'\left(f^{-1}(y)\right)}.$$

An alternative formulation of the chain rule with $h = g(y)$ and $y = f(x)$ is given by

$$\frac{dh}{dx} = \frac{dh}{dy} \cdot \frac{dy}{dx}.$$

The rule given in part (2) of Theorem 4.10 for the derivative of the inverse function f^{-1} can also be written as

$$(f^{-1})' = \frac{1}{f' \circ f^{-1}}.$$

Example 4.12 Suppose that a firm produces only one product. The production cost, denoted by C, depends on the quantity $x \geq 0$ of this product. Let

$$C = f(x) = 4 + \ln(x+1) + \sqrt{3x+1}.$$

Then we obtain

$$C' = f'(x) = \frac{df(x)}{dx} = \frac{1}{x+1} + \frac{3}{2\sqrt{3x+1}}.$$

Assume that the firm produces $x_0 = 133$ units of this product. We get

$$C'(133) = \frac{1}{134} + \frac{3}{2\sqrt{400}} = \frac{1}{134} + \frac{3}{40} \approx 0.08246.$$

So, if the current production is increased by one unit, the production cost increases approximately by 0.08246 units.

Example 4.13 Let function $g : D_g \to \mathbb{R}$ with

$$g(x) = \cos\,[\ln(3x+1)]^2$$

be given which is defined for $x > -1/3$, i.e. $D_g = (-1/3, \infty)$. Setting

$$g = \cos v, \qquad v = h^2, \qquad h = \ln y \qquad \text{and} \qquad y = (3x+1),$$

application of the chain rule yields

$$g'(x) = \frac{dg}{dv} \cdot \frac{dv}{dh} \cdot \frac{dh}{dy} \cdot \frac{dy}{dx},$$

and thus we get

$$g'(x) = -\sin\,[\ln(3x+1)]^2 \cdot 2\ln(3x+1) \cdot \frac{1}{3x+1} \cdot 3$$

$$= -\frac{6}{3x+1} \cdot \sin\,[\ln(3x+1)]^2 \cdot \ln(3x+1).$$

By means of Theorem 4.10, we can also determine the derivatives of the inverse functions of the trigonometric functions. Consider the function f with

$$y = f(x) = \sin x, \qquad x \in \left(-\frac{\pi}{2}, \frac{\pi}{2}\right),$$

and the inverse function

$$x = f^{-1}(y) = \arcsin y.$$

Since $f'(x) = \cos x$, we obtain from Theorem 4.10, part (2):

$$(f^{-1})'(y) = \frac{1}{f'(x)} = \frac{1}{\cos x}.$$

Because $\cos x = +\sqrt{\cos^2 x} = \sqrt{1 - \sin^2 x} = \sqrt{1 - y^2}$ for $x \in (-\pi/2, \pi/2)$, we get

$$(f^{-1})'(y) = \frac{1}{\sqrt{1 - y^2}},$$

and after interchanging variables x and y

$$(f^{-1})'(x) = \frac{1}{\sqrt{1 - x^2}}.$$

Similarly, we can determine the derivatives of the inverse functions of the other trigono-metric functions. We summarize the *derivatives of the inverse functions of the trigonometric functions* in the following overview:

$$f(x) = \arcsin x, \quad f'(x) = \frac{1}{\sqrt{1 - x^2}}, \quad D_f = \{x \in R \mid -1 < x < 1\};$$

$$f(x) = \arccos x, \quad f'(x) = -\frac{1}{\sqrt{1 - x^2}}, \quad D_f = \{x \in R \mid -1 < x < 1\};$$

$$f(x) = \arctan x, \quad f'(x) = \frac{1}{1 + x^2}, \quad D_f = \{x \in R \mid -\infty < x < \infty\};$$

$$f(x) = \text{arccot } x, \quad f'(x) = -\frac{1}{1 + x^2}, \quad D_f = \{x \in R \mid -\infty < x < \infty\}.$$

Logarithmic differentiation

As an application of the chain rule, we consider so-called *logarithmic differentiation*. Let

$$g(x) = \ln f(x) \qquad \text{with} \quad f(x) > 0.$$

We obtain

$$g'(x) = [\ln f(x)]' = \frac{f'(x)}{f(x)}$$

and thus

$$f'(x) = f(x) \cdot g'(x) = f(x) \cdot [\ln f(x)]'.$$

Logarithmic differentiation is particularly useful when considering functions f of type

$$f(x) = u(x)^{v(x)},$$

i.e. both the basis and the exponent are functions depending on variable x.

Example 4.14   Let function $f : D_f \to \mathbb{R}$ with

$$f(x) = x^{\sin x}$$

be given. We set

$$g(x) = \ln f(x) = \ln\left(x^{\sin x}\right) = \sin x \cdot \ln x.$$

Applying the formula of logarithmic differentiation, we get

$$g'(x) = \cos x \cdot \ln x + \sin x \cdot \frac{1}{x}$$

and consequently

$$f'(x) = f(x) \cdot g'(x) = x^{\sin x}\left(\cos x \cdot \ln x + \sin x \cdot \frac{1}{x}\right).$$

Higher-order derivatives

In the following we deal with *higher-order derivatives*. If function f' with $y' = f'(x)$ is again differentiable (see Definition 4.5), function

$$y'' = f''(x) = \frac{df'(x)}{dx} = \frac{d^2 f(x)}{(dx)^2}$$

is called the *second derivative* of function f at point x. We can continue with this procedure and obtain in general:

$$y^{(n)} = f^{(n)}(x) = \frac{df^{n-1}(x)}{dx} = \frac{d^n f(x)}{(dx)^n}, \qquad n \geq 2,$$

which denotes the *nth derivative* of function f at point $x \in D_f$. Notice that we use $f'(x), f''(x), f'''(x)$, and for $n \geq 4$, we use the notation $f^{(n)}(x)$. Higher-order derivatives are used for instance in the next section when investigating specific properties of functions.

Example 4.15   Let function $f : D_f \to \mathbb{R}$ with

$$f(x) = 3x^2 + \frac{1}{x} + e^{2x}$$

be given. We determine all derivatives until fourth order and obtain

$$f'(x) = 6x - \frac{1}{x^2} + 2e^{2x}$$

$$f''(x) = 6 + \frac{2}{x^3} + 4e^{2x}$$

$$f'''(x) = -\frac{6}{x^4} + 8e^{2x}$$

$$f^{(4)}(x) = \frac{24}{x^5} + 16e^{2x}.$$

If the derivatives $f'(x_0), f''(x_0), \ldots, f^{(n)}(x_0)$ all exist, we say that function f is n times differentiable at point $x_0 \in D_f$. If $f^{(n)}$ is continuous at point x_0, then function f is said to be n times continuously differentiable at x_0. Similarly, if function $f^{(n)}$ is continuous, then function f is said to be n times continuously differentiable.

4.4   DIFFERENTIAL; RATE OF CHANGE AND ELASTICITY

In this section, we discuss several possibilities for characterizing the resulting change in the dependent variable y of a function when considering small changes in the independent variable x.

> **Definition 4.6**   Let function $f : D_f \to \mathbb{R}$ be differentiable at point $x_0 \in (a, b) \subseteq D_f$. The *differential* of function f at point x_0 is defined as
>
> $$dy = f'(x_0) \cdot dx.$$

The differential is also denoted as df. Note that dy (or df) is proportional to dx, with $f'(x_0)$ as the factor of proportionality. The differential gives the approximate change in the function value at x_0 when changing the argument by (a small value) dx (i.e. from x_0 to $x_0 + dx$), i.e.

$$\Delta y \approx dy = f'(x_0) \cdot dx.$$

The differential is illustrated in Figure 4.5.

The differential can be used for estimating the maximal absolute error in the function value when the independent variable is only known with some error. Let $|\Delta x| = |dx| \leq \Delta$, then

$$|\Delta y| \approx |dy| = |f'(x_0)| \cdot |dx| \leq |f'(x_0)| \cdot \Delta.$$

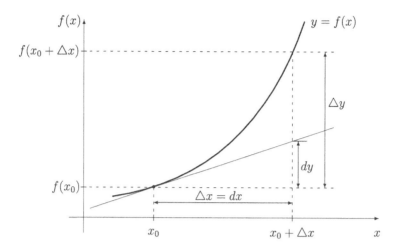

Figure 4.5 The differential dy.

Example 4.16   Let the length of an edge of a cube be determined as $x = 7.5 \pm 0.01$ cm, i.e. the value $x_0 = 7.5$ cm has been measured, and the absolute error is no greater than 0.01, i.e. $|dx| = |\Delta x| \le \Delta = 0.01$. For the surface of the cube we obtain

$$S_0 = S(x_0) = 6x_0^2 = 337.5 \text{ cm}^2.$$

We use the differential to estimate the maximal absolute error of the surface and obtain

$$|\Delta S| \approx |dS| \le |S'(x_0)| \cdot \Delta = |12x_0| \cdot 0.01 = 0.9 \text{ cm}^2,$$

i.e. the maximal absolute error in the surface is estimated by 0.9 cm^2, and we get

$$S_0 \approx 337.5 \pm 0.9 \text{ cm}^3.$$

Moreover, for the estimation of the maximal relative error we obtain

$$\left| \frac{\Delta S}{S_0} \right| \approx \left| \frac{dS}{S_0} \right| \le \frac{0.9}{337.5} \approx 0.00267,$$

i.e. the maximal relative error is estimated by approximately 0.267 per cent.

We have already discussed that the derivative of a function characterizes the change in the function value for a 'very small' change in the variable x. However, in economics often a modified measure is used to describe the change of the value of a function f. The reason for this is that a change in the price of, e.g. bread by 1 EUR would be very big in comparison with the current price, whereas the change in the price of a car by 1 EUR would be very small. Therefore, economists prefer measures for characterizing the change of a function value in relation to this value itself. This leads to the introduction of the proportional rate of change of a function given in the following definition.

Definition 4.7   Let function $f : D_f \to \mathbb{R}$ be differentiable at point $x_0 \in (a, b) \subseteq D_f$ and $f(x_0) \neq 0$. The term

$$\rho_f(x_0) = \frac{f'(x_0)}{f(x_0)}$$

is called the *proportional rate of change* of function $f : D_f \to \mathbb{R}$ at point x_0.

Proportional rates of change are often quoted in percentages or, when time is the independent variable, as percentages per year. This percentage rate of change is obtained by multiplying the value $\rho_f(x_0)$ by 100 per cent.

Example 4.17   Let function $f : D_f \to \mathbb{R}$ with

$$f(x) = x^2 e^{0.1x}$$

be given. The first derivative of function f is given by

$$f'(x) = 2x e^{0.1x} + 0.1 x^2 e^{0.1x} = x e^{0.1x}(2 + 0.1x)$$

and thus, the proportional rate of change at point $x \in D_f$ is calculated as follows:

$$\rho_f(x) = \frac{f'(x)}{f(x)} = \frac{x e^{0.1x}(2 + 0.1x)}{x^2 e^{0.1x}} = \frac{2}{x} + 0.1.$$

We compare the proportional rates of change at points $x_0 = 20$ and $x_1 = 2,000$. For $x_0 = 20$, we get

$$\rho_f(20) = \frac{2}{20} + 0.1 = 0.2$$

which means that the percentage rate of change is 20 per cent. For $x_1 = 2,000$, we get

$$\rho_f(2,000) = \frac{2}{2,000} + 0.1 = 0.101,$$

i.e. the percentage rate of change is 10.1 per cent. Thus, the second percentage rate of change is much smaller than the first one.

Definition 4.8   Let function $f : D_f \to \mathbb{R}$ be differentiable at point $x_0 \in (a, b) \subseteq D_f$ and $f(x_0) \neq 0$. The term

$$\varepsilon_f(x_0) = \frac{x_0 \cdot f'(x_0)}{f(x_0)} = x_0 \cdot \rho_f(x_0)$$

is called the *(point) elasticity* of function $f : D_f \to \mathbb{R}$ at x_0.

In many economic applications, function f represents a demand function and variable x represents price or income. The (point) elasticity of function f corresponds to the ratio of the relative (i.e. percentage) changes of the function values of f and variable x:

$$\varepsilon_f(x_0) = \lim_{\Delta x \to 0} \left(\frac{x_0}{f(x_0)} \cdot \frac{\Delta f(x)}{\Delta x} \right) = \lim_{\Delta x \to 0} \left(\frac{\Delta f(x)}{f(x_0)} : \frac{\Delta x}{x_0} \right).$$

An economic function f is called *elastic* at point $x_0 \in D_f$ if $|\varepsilon_f(x_0)| > 1$. On the other hand, it is called *inelastic* at point x_0 if $|\varepsilon_f(x_0)| < 1$. In the case of $|\varepsilon_f(x_0)| = 1$, function f is of *unit elasticity* at point x_0.

In economics, revenue R is considered as a function of the selling price p by

$$R(p) = p \cdot D(p),$$

where D is the demand function that is usually decreasing, i.e. if the price p rises, the quantity D sold falls. However, the revenue (as the product of price p and quantity D) may rise or fall. For the marginal revenue function, we obtain

$$R'(p) = p \cdot D'(p) + D(p).$$

Assume now that $R'(p) > 0$, i.e. from $p \cdot D'(p) + D(p) > 0$ we get for $D(p) > 0$

$$\varepsilon_D(p) = \frac{p \cdot D'(p)}{D(p)} > -1,$$

where due to $D'(p) \le 0$, we have $|\varepsilon_D(p)| < 1$. Thus, if revenue increases when the price increases, it must be at an inelastic interval of the demand function $D = D(p)$, and in the inelastic case, a small increase in the price will always lead to an increase in revenue. Accordingly, in the elastic case, i.e. if $|\varepsilon_D(p)| > 1$, a small increase in price leads to a decrease in revenue. So an elastic demand function is one for which the quantity demanded is very responsive to price, which can be interpreted as follows. If the price increases by one per cent, the quantity demanded decreases by more than one per cent.

Example 4.18   We consider function $f : (0, \infty) \to \mathbb{R}$ given by

$$f(x) = x^{x+1}$$

and determine the elasticity of function f at point $x \in D_f$. First, we calculate the first derivative f' by applying logarithmic differentiation. We set

$$g(x) = \ln f(x) = (x+1) \ln x.$$

Therefore,

$$g'(x) = [\ln f(x)]' = \frac{f'(x)}{f(x)} = \ln x + \frac{x+1}{x}.$$

Thus, we obtain for the first derivative

$$f'(x) = f(x) \cdot g'(x) = x^{x+1} \cdot \left(\ln x + \frac{x+1}{x} \right).$$

For the elasticity at point $x \in D_f$, we obtain

$$\varepsilon_f(x) = \frac{x \cdot f'(x)}{f(x)} = \frac{x \cdot x^{x+1} \left(\ln x + \frac{x+1}{x} \right)}{x^{x+1}} = x \ln x + x + 1.$$

4.5 GRAPHING FUNCTIONS

To get a quantitative overview on a function $f : D_f \to \mathbb{R}$ resp. on its graph, we determine and investigate:

(1) domain D_f (if not given)
(2) zeroes and discontinuities
(3) monotonicity of the function
(4) extreme points and values
(5) convexity and concavity of the function
(6) inflection points
(7) limits, i.e. how does the function behave when x tends to $\pm\infty$.

Having the detailed information listed above, we can draw the graph of function f. In the following, we discuss the above subproblems in detail.

In connection with functions of one variable, we have already discussed how to determine the domain D_f and we have classified the different types of discontinuities. As far as the determination of zeroes is concerned, we have already considered special cases such as zeroes of a quadratic function. For more complicated functions, where finding the zeroes is difficult or analytically impossible, we give numerical procedures for the approximate determination of zeroes later in this chapter. We start with investigating the monotonicity of a function.

4.5.1 Monotonicity

By means of the first derivative f', we can determine intervals in which a function f is (strictly) increasing or decreasing. In particular, the following theorem can be formulated.

THEOREM 4.11 Let function $f : D_f \to \mathbb{R}$ be differentiable on the open interval (a, b) and let $I = [a, b] \subseteq D_f$. Then:

(1) Function f is increasing on I if and only if $f'(x) \geq 0$ for all $x \in (a, b)$.
(2) Function f is decreasing on I if and only if $f'(x) \leq 0$ for all $x \in (a, b)$.
(3) Function f is constant on I if and only if $f'(x) = 0$ for all $x \in (a, b)$.
(4) If $f'(x) > 0$ for all $x \in (a, b)$, then function f is strictly increasing on I.
(5) If $f'(x) < 0$ for all $x \in (a, b)$, then function f is strictly decreasing on I.

We recall from Chapter 3 that, if a function is (strictly) increasing or (strictly) decreasing on an interval $I \subseteq D_f$, we say that function f is (strictly) monotone on the interval I. Checking a function f for monotonicity requires us to determine the intervals on which function f is monotone and strictly monotone, respectively.

Example 4.19 We investigate function $f : \mathbb{R} \to \mathbb{R}$ with

$$f(x) = e^x(x^2 + 2x + 1)$$

for monotonicity. Differentiating function f, we obtain

$$f'(x) = e^x(x^2 + 2x + 1) + e^x(2x + 2) = e^x(x^2 + 4x + 3).$$

To find the zeroes of function f', we have to solve the quadratic equation

$$x^2 + 4x + 3 = 0$$

(notice that e^x is positive for $x \in \mathbb{R}$) and obtain

$$x_1 = -2 + \sqrt{4 - 3} = -1 \qquad \text{and} \qquad x_2 = -2 - \sqrt{4 - 3} = -3.$$

Since we have distinct zeroes (i.e. the sign of the first derivative changes 'at' each zero) and e.g. $f'(0) = 3 > 0$, we get that $f'(x) > 0$ for $x \in (-\infty, -3) \cup (-1, \infty)$ and $f'(x) < 0$ for $x \in (-3, -1)$. By Theorem 4.11 we get that function f is strictly increasing on the intervals $(-\infty, -3]$ and $[-1, \infty)$ while function f is strictly decreasing on the interval $[-3, -1]$.

4.5.2 Extreme points

First we give the definition of a local and of a global extreme point which can be either a minimum or a maximum.

Definition 4.9 A function $f : D_f \to \mathbb{R}$ has a *local maximum* (*minimum*) at point $x_0 \in D_f$ if there is an interval $(a, b) \subseteq D_f$ containing x_0 such that

$$f(x) \le f(x_0) \qquad (f(x) \ge f(x_0), \text{ respectively}) \tag{4.1}$$

for all points $x \in (a, b)$. Point x_0 is called a *local maximum* (*minimum*) *point*. If inequality (4.1) holds for all points $x \in D_f$, function f has at point x_0 a *global maximum* (*minimum*), and x_0 is called a *global maximum* (*minimum*) *point*.

These notions are illustrated in Figure 4.6. Let $D_f = [a, b]$. In the domain D_f, there are two local minimum points x_2 and x_4 as well as two local maximum points x_1 and x_3. The global maximum point is x_3 and the global minimum point is the left boundary point a. We now look for necessary and sufficient conditions for the existence of a local extreme point in the case of differentiable functions.

THEOREM 4.12 (necessary condition for local optimality) Let function $f : D_f \to \mathbb{R}$ be differentiable on the open interval $(a, b) \subseteq D_f$. If function f has a local maximum or local minimum at point $x_0 \in (a, b)$, then $f'(x_0) = 0$.

A point $x_0 \in (a, b)$ with $f'(x_0) = 0$ is called a *stationary point* (or *critical point*). In searching global maximum and minimum points for a function f in a closed interval $I = [a, b] \subseteq D_f$, we have to search among the following types of points:

(1) points in the open interval (a, b), where $f'(x) = 0$ (stationary points);
(2) end points a and b of I;
(3) points in (a, b), where $f'(x)$ does not exist.

Points in I according to (1) can be found by means of differential calculus. Points according to (2) and (3) have to be checked separately. Returning to Figure 4.6, there are three stationary points x_2, x_3 and x_4. The local maximum point x_1 cannot be found by differential calculus since the function drawn in Figure 4.6 is not differentiable at point x_1.

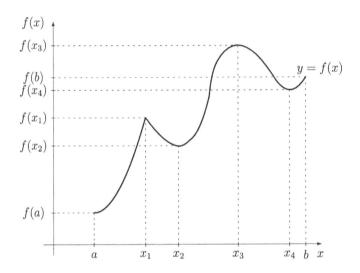

Figure 4.6 Local and global optima of function f on $D_f = [a, b]$.

The following two theorems present sufficient conditions for so-called *isolated* local extreme points, for which in inequality (4.1) in Definition 4.9 the strict inequality holds for all $x \in (a, b)$ different from x_0. First, we give a criterion for deciding whether a stationary point is a local extreme point in the case of a differentiable function which uses only the first derivative of function f.

THEOREM 4.13 (first-derivative test for local extrema)   Let function $f : D_f \to \mathbb{R}$ be differentiable on the open interval $(a, b) \in D_f$ and $x_0 \in (a, b)$ be a stationary point of function f. Then:

(1) If $f'(x) > 0$ for all $x \in (a^*, x_0) \subseteq (a, b)$ and $f'(x) < 0$ for all $x \in (x_0, b^*) \subseteq (a, b)$, then x_0 is a local maximum point of function f.
(2) If $f'(x) < 0$ for all $x \in (a^*, x_0) \subseteq (a, b)$ and $f'(x) > 0$ for all $x \in (x_0, b^*) \subseteq (a, b)$, then x_0 is a local minimum point of function f.
(3) If $f'(x) > 0$ for all $x \in (a^*, x_0) \subseteq (a, b)$ and for all $x \in (x_0, b^*) \subseteq (a, b)$, then x_0 is not a local extreme point of function f. The same conclusion holds if $f'(x) < 0$ on both sides of x_0.

For instance, part (1) of Theorem 4.13 means that, if there exists an interval (a^*, b^*) around x_0 such that function f is strictly increasing to the left of x_0 and strictly decreasing to the right of x_0 in this interval, then x_0 is a local maximum point of function f. The above criterion requires the investigation of monotonicity properties of the marginal function of f. The following theorem presents an alternative by using higher-order derivatives to decide whether a stationary point is a local extreme point or not provided that the required higher-order derivatives exist.

THEOREM 4.14 (higher-order derivative test for local extrema) Let $f : D_f \to \mathbb{R}$ be n times continuously differentiable on the open interval $(a, b) \subseteq D_f$ and $x_0 \in (a, b)$ be a stationary point. If

$$f'(x_0) = f''(x_0) = f'''(x_0) = \cdots = f^{(n-1)}(x_0) = 0 \quad \text{and} \quad f^{(n)}(x_0) \neq 0,$$

where number n is even, then point x_0 is a local extreme point of function f, in particular:

(1) If $f^{(n)}(x_0) < 0$, then function f has at point x_0 a local maximum.
(2) If $f^{(n)}(x_0) > 0$, then function f has at point x_0 a local minimum.

Example 4.20 We determine all local extreme points of function $f : (0, \infty) \to \mathbb{R}$ with

$$f(x) = \frac{1}{x} \cdot \ln^2 3x.$$

To check the necessary condition of Theorem 4.12, we determine

$$f'(x) = -\frac{1}{x^2} \cdot \ln^2 3x + \frac{1}{x} \cdot 2 \ln 3x \cdot \frac{1}{3x} \cdot 3 = \frac{1}{x^2} \cdot \ln 3x (2 - \ln 3x).$$

From equality $f'(x) = 0$, we obtain the following two cases which we need to consider.

Case a $\ln 3x = 0$. Then we obtain $3x = e^0 = 1$ which yields $x_1 = 1/3$.

Case b $2 - \ln 3x = 0$. Then we obtain $\ln 3x = 2$ which yields $x_2 = e^2/3$.

To check the sufficient condition for x_1 and x_2 to be local extreme points according to Theorem 4.14, we determine f'' as follows:

$$f''(x) = \left(-\frac{2}{x^3} \cdot \ln 3x + \frac{1}{x^2} \cdot \frac{1}{3x} \cdot 3\right)(2 - \ln 3x) - \frac{1}{x^2} \cdot \ln 3x \cdot \frac{1}{3x} \cdot 3$$

$$= \left(-\frac{2}{x^3} \cdot \ln 3x + \frac{1}{x^3}\right)(2 - \ln 3x) - \frac{1}{x^3} \cdot \ln 3x$$

$$= \frac{2}{x^3} \cdot (1 - 3 \ln 3x + \ln^2 3x).$$

In particular, we obtain

$$f''\left(\frac{1}{3}\right) = 54 > 0 \quad \text{and} \quad f''\left(\frac{1}{3}e^2\right) = \frac{54}{e^6} \cdot (1 - 3 \cdot 2 + 2^2) = -\frac{54}{e^6} < 0.$$

Hence, $x_1 = 1/3$ is a local minimum point with $f(x_1) = 0$, and $x_2 = e^2/3$ is a local maximum point with $f(x_2) = 12/e^2$.

Example 4.21   A monopolist (i.e. an industry with a single firm) producing a certain product has the demand–price function (also denoted as the inverse demand function)

$$p(x) = -0.04x + 200,$$

which describes the relationship between the (produced and sold) quantity x of the product and the price p at which the product sells. This strictly decreasing function is defined for $0 \le x \le 5,000$ and can be interpreted as follows. If the price tends to 200 units, the number of customers willing to buy the product tends to zero, but if the price tends to zero, the sold quantity of the product tends to 5,000 units. The revenue R of the firm in dependence on the output x is given by

$$R(x) = p(x) \cdot x = (-0.04x + 200) \cdot x = -0.04x^2 + 200x.$$

Moreover, let the cost function C describing the cost of the firm in dependence on the produced quantity x be given by

$$C(x) = 80x + 22,400.$$

This yields the profit function P with

$$P(x) = R(x) - C(x) = -0.04x^2 + 200x - (80x + 22,400) = -0.04x^2 + 120x - 22,400.$$

We determine the production output that maximizes the profit. Looking for stationary points we obtain

$$P'(x) = -0.08x + 120 = 0$$

which yields the point

$$x_P = 1,500.$$

Due to

$$P''(x_P) = -0.08 < 0,$$

the output $x_P = 1,500$ maximizes the profit with $P(1,500) = 67,600$ units. Points with $P(x) = 0$ (i.e. revenue $R(x)$ is equal to cost $C(x)$) are called break-even points. In the example, we obtain from $P(x) = 0$ the equation

$$x^2 - 3,000x + 560,000 = 0,$$

which yields the roots (break-even points)

$$x_1 = 1,500 - \sqrt{1,690,000} = 200 \quad \text{and} \quad x_2 = 1,500 + \sqrt{1,690,000} = 2,800,$$

i.e. an output $x \in (200; \ 2,800)$ leads to a profit for the firm. Finally, we mention that when maximizing revenue, one gets from

$$R'(x) = -0.08x + 200 = 0$$

the stationary point

$$x_R = 2,500,$$

which is, because $R''(2,500) = -0.08 < 0$, indeed the output that maximizes revenue. However, in the latter case, the profit is only $P(2,500) = 27,600$ units.

Example 4.22   The cost function $C : \mathbb{R}_+ \to \mathbb{R}$ of an enterprise producing a quantity x of a product is given by

$$C(x) = 0.1x^2 - 30x + 25,000.$$

Then, the average cost function C_a measuring the cost per unit produced is given by

$$C_a(x) = \frac{C(x)}{x} = 0.1x - 30 + \frac{25,000}{x}.$$

We determine local minimum points of the average cost function C_a. We obtain

$$C_a'(x) = 0.1 - \frac{25,000}{x^2}.$$

The necessary condition for a local extreme point is $C_a'(x) = 0$. This corresponds to

$$0.1x^2 = 25,000$$

which has the two solutions

$$x_1 = 500 \qquad \text{and} \qquad x_2 = -500.$$

Since $x_2 \notin D_C$, the only stationary point is $x_1 = 500$. Checking the sufficient condition, we obtain

$$C_a''(x) = 2 \cdot \frac{25,000}{x^3}$$

and

$$C_a''(500) = \frac{50,000}{125,000,000} > 0,$$

i.e. the produced quantity $x_1 = 500$ minimizes average cost with $C_a(500) = 70$.

Example 4.23   A firm wins an order to design and produce a cylindrical container for transporting a liquid commodity. This cylindrical container should have a given volume V_0. The cost of producing such a cylinder is proportional to its surface. Let R denote the radius and H denote the height of a cylinder. Among all cylinders with given volume V_0, we want to determine that with the smallest surface S (i.e. that with lowest cost). We know that the volume is obtained as

$$V_0 = V(R, H) = \pi R^2 H, \tag{4.2}$$

and the surface can be determined by the formula

$$S = S(R, H) = 2\pi R^2 + 2\pi RH. \tag{4.3}$$

From formula (4.2), we can eliminate variable H which yields

$$H = \frac{V_0}{\pi R^2}. \tag{4.4}$$

Substituting the latter term into formula (4.3), we obtain the surface as a function of one variable, namely in dependence on the radius R, since V_0 is a constant:

$$S = S(R) = 2\pi R^2 + 2\pi R \cdot \frac{V_0}{\pi R^2} = 2\left(\pi R^2 + \frac{V_0}{R}\right).$$

Now we look for the minimal value of $S(R)$ for $0 < R < \infty$. For applying Theorems 4.12 and 4.14, we determine the first and second derivatives of function $S = S(R)$:

$$S'(R) = 2\left(2\pi R - \frac{V_0}{R^2}\right) \quad \text{and} \quad S''(R) = 2\left(2\pi + \frac{2V_0}{R^3}\right).$$

Setting $S'(R) = 0$ yields

$$2\pi R^3 - V_0 = 0 \tag{4.5}$$

and thus the stationary point is obtained as follows:

$$R_1 = \sqrt[3]{\frac{V_0}{2\pi}}.$$

Notice that the other two roots of equation (4.5) are complex and therefore not a candidate point for a minimal surface of the cylinder. Moreover, we obtain

$$S''(R_1) = 2(2\pi + 4\pi) = 12\pi > 0,$$

i.e. the surface becomes minimal for the radius R_1. One can also argue without checking the sufficient condition that the only stationary point with a positive radius must be a minimum point since, both as $R \to 0$ and as $R \to \infty$, the surface of the cylinder tends to infinity so that for R_1, the surface must be at its minimal value. Determining the corresponding value for the height, we obtain from equality (4.4)

$$H_1 = \frac{V_0}{\pi R_1^2} = \frac{V_0}{\pi \left(\sqrt[3]{\frac{V_0}{2\pi}}\right)^2} = \frac{V_0 \cdot (2\pi)^{2/3}}{\pi \cdot V_0^{2/3}} = 2^{2/3} \cdot \sqrt[3]{\frac{V_0}{\pi}} = 2 \cdot \sqrt[3]{\frac{V_0}{2\pi}} = 2R_1.$$

Thus, the surface of a cylinder with given volume is minimal if the height of the cylinder is equal to its diameter, i.e. $H = 2R$.

4.5.3 Convexity and concavity

The definition of a convex and a concave function has already been given in Chapter 3. Now, by means of differential calculus, we can give a criterion to check whether a function is (strictly) convex or concave on a certain interval I.

THEOREM 4.15 Let function $f : D_f \rightarrow \mathbb{R}$ be twice differentiable on the open interval $(a, b) \subseteq D_f$ and let $I = [a, b]$. Then:

(1) Function f is convex on I if and only if $f''(x) \geq 0$ for all $x \in (a, b)$.
(2) Function f is concave on I if and only if $f''(x) \leq 0$ for all $x \in (a, b)$.
(3) If $f''(x) > 0$ for all $x \in (a, b)$, then function f is strictly convex on I.
(4) If $f''(x) < 0$ for all $x \in (a, b)$, then function f is strictly concave on I.

Example 4.24   Let function $f : \mathbb{R} \rightarrow \mathbb{R}$ with

$$f(x) = ae^{bx}, \qquad a, b \in \mathbb{R}_+ \setminus \{0\}$$

be given. We obtain

$$f'(x) = abe^{bx} \qquad \text{and} \qquad f''(x) = ab^2 e^{bx}.$$

For $a, b > 0$, we get $f'(x) > 0$ and $f''(x) > 0$ for all $x \in D_f$, i.e. function f is strictly increasing and strictly convex. In this case, we say that function f has a progressive growth (since it grows faster than a linear function, which has a proportionate growth). Consider now function $g : \mathbb{R}_+ \rightarrow \mathbb{R}$ with

$$g(x) = a \ln(1 + bx), \qquad a, b \in \mathbb{R}_+ \setminus \{0\}.$$

We obtain

$$g'(x) = \frac{ab}{1 + bx} \qquad \text{and} \qquad g''(x) = -\frac{ab^2}{(1 + bx)^2}.$$

For $a, b > 0$, we get $g'(x) > 0$ and $g''(x) < 0$ for all $x \in D_g$, i.e. function g is strictly increasing and strictly concave. In this case, we say that function g has degressive growth.

Example 4.25   We investigate function $f : \mathbb{R} \rightarrow \mathbb{R}$ with

$$f(x) = \frac{2x}{x^2 + 1}$$

for convexity and concavity. We obtain

$$f'(x) = \frac{2(x^2 + 1) - 2x \cdot 2x}{(x^2 + 1)^2} = 2 \cdot \frac{1 - x^2}{(x^2 + 1)^2}$$

$$f''(x) = 2 \cdot \frac{-2x(x^2 + 1)^2 - (1 - x^2) \cdot 2 \cdot (x^2 + 1) \cdot 2x}{(x^2 + 1)^4}$$

$$= 2 \cdot \frac{(x^2 + 1) \cdot \left[-2x(x^2 + 1) - (1 - x^2) \cdot 4x \right]}{(x^2 + 1)^4}$$

$$= 4 \cdot \frac{x^3 - 3x}{(x^2 + 1)^3}.$$

To find the zeroes of the second derivative f'', we have to solve the cubic equation

$$x^3 - 3x = x \cdot (x^2 - 3) = 0$$

which yields the solutions

$$x_1 = 0, \qquad x_2 = \sqrt{3} \qquad \text{and} \qquad x_3 = -\sqrt{3}.$$

Since e.g. $f''(1) = 4 \cdot (1 - 3)/2^3 = -1 < 0$, we have $f''(x) < 0$ for $x \in (x_1, x_2) = (0, \sqrt{3})$, i.e. by Theorem 4.15, function f is strictly concave on the interval $[0, \sqrt{3}]$. Moreover, since we have distinct zeroes, the sign of the second derivative changes 'at' each zero, and by Theorem 4.15 we obtain: function f is strictly convex on $[-\sqrt{3}, 0] \cup [\sqrt{3}, \infty)$ and strictly concave on $(-\infty, -\sqrt{3}] \cup [0, \sqrt{3}]$.

To decide whether a function is convex or concave, the following notion of an inflection point can be helpful.

Definition 4.10 Let function $f : D_f \to \mathbb{R}$ be twice differentiable on the open interval $(a, b) \subseteq D_f$. Point $x_0 \in (a, b)$ is called an *inflection point* of function f when f changes at x_0 from being convex to being concave or vice versa, i.e. if there is an interval $(a^*, b^*) \subseteq (a, b)$ containing x_0 such that either of the following two conditions holds:

(1) $f''(x) \geq 0$ if $a^* < x < x_0$ and $f''(x) \leq 0$ if $x_0 < x < b^*$ or
(2) $f''(x) \leq 0$ if $a^* < x < x_0$ and $f''(x) \geq 0$ if $x_0 < x < b^*$.

Consequently, if the second derivative changes the sign 'at' point x_0, then point x_0 is an inflection point of function f. The notion of an inflection point is illustrated in Figure 4.7. Next, we give a criterion for an inflection point of function f.

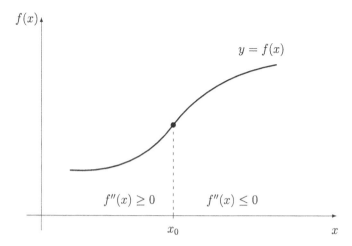

Figure 4.7 Inflection point x_0 of function f.

THEOREM 4.16 Let function $f : D_f \to \mathbb{R}$ be n times continuously differentiable on the open interval $(a, b) \subseteq D_f$. Point $x_0 \in (a, b)$ is an inflection point of function f if and only if

$$f''(x_0) = f'''(x_0) = \cdots = f^{(n-1)}(x_0) = 0 \qquad \text{and} \qquad f^{(n)}(x_0) \neq 0,$$

where n is odd.

Example 4.26 We consider function $f : \mathbb{R} \to \mathbb{R}$ with

$$f(x) = x^4 + 2x^3 - 12x^2 + 4 = 0$$

and determine inflection points. We obtain

$$f'(x) = 4x^3 + 6x^2 - 24x \qquad \text{and} \qquad f''(x) = 12x^2 + 12x - 24.$$

In order to solve $f''(x) = 0$, we have to find the roots of the quadratic equation

$$x^2 + x - 2 = 0,$$

which gives $x_1 = 1$ and $x_2 = -2$ as candidates for an inflection point. Using

$$f'''(x) = 24x + 12,$$

we obtain $f'''(1) = 36 \neq 0$ and $f'''(-2) = -36 \neq 0$, i.e. both points $x_1 = 1$ and $x_2 = -1$ are inflection points of function f.

4.5.4 Limits

We have already discussed some rules for computing limits of sums, differences, products or quotients of functions. However, it was necessary that each of the limits exists, i.e. each of the limits was a finite number. The question we consider now is what happens when we want to determine the limit of a quotient of two functions, and both limits of the function in the numerator and the function in the denominator tend to ∞ as x tends to a specific value x_0. The same question arises when both functions in the numerator and in the denominator tend to zero as x approaches some value x_0.

Definition 4.11 If in a quotient both the numerator and the denominator tend to zero as x tends to x_0, we call such a limit an *indeterminate form* of type "0/0", and we write

$$\lim_{x \to x_0} \frac{f(x)}{g(x)} = \text{``} 0/0 \text{''}.$$

The notion 'indeterminate form' indicates that the limit cannot be found without further examination. There exist six further indeterminate forms as follows:

$$\text{``}\infty/\infty\text{''}, \quad \text{``}0 \cdot \infty\text{''}, \quad \text{``}\infty - \infty\text{''}, \quad \text{``}0^0\text{''}, \quad \text{``}\infty^0\text{''}, \quad \text{``}1^\infty\text{''}.$$

For the two indeterminate forms "0/0" and "∞/∞", the limit can be found by means of the following rule.

THEOREM 4.17 (Bernoulli–l'Hospital's rule) Let functions $f : D_f \to \mathbb{R}$ and $g : D_g \to \mathbb{R}$ both tend to zero as x tends to x_0, or f and g both tend to ∞ as x tends to x_0. Moreover, let f and g be continuously differentiable on the open interval $(a, b) \in D_f \cap D_g$ containing x_0 and $g'(x) \neq 0$ for $x \in (a, b)$. Then

$$\lim_{x \to x_0} \frac{f(x)}{g(x)} = \lim_{x \to x_0} \frac{f'(x)}{g'(x)} = L.$$

Here either the limit exists (i.e. value L is finite) or the limit does not exist.

It may happen that after application of Theorem 4.17, we still have an indeterminate form of type "0/0" or "∞/∞". In that case, we apply Theorem 4.17 repeatedly as long as we have found L. Moreover, we mention that Theorem 4.17 can be applied under appropriate assumptions also to one-sided limits (i.e. $x \to x_0 + 0$ or $x \to x_0 - 0$) as well as to the cases $x \to -\infty$ and $x \to \infty$. To illustrate Theorem 4.17, consider the following example.

Example 4.27 We determine the limit

$$L = \lim_{x \to 0} \frac{e^{\sin 2x} - e^{2x}}{\sin 2x - 2x},$$

which is an indeterminate form of type "0/0". Applying Theorem 4.17 repeatedly three times, we obtain

$$
L = \lim_{x \to 0} \frac{2 \left(e^{\sin 2x} \cos 2x - e^{2x}\right)}{2 \cos 2x - 2} = \lim_{x \to 0} \frac{e^{\sin 2x} \cos 2x - e^{2x}}{\cos 2x - 1} \qquad \text{"0/0"}
$$

$$
= \lim_{x \to 0} \frac{2 \left[e^{\sin 2x} \left(\cos^2 2x - \sin 2x\right) - e^{2x}\right]}{-2 \sin 2x} \qquad \text{"0/0"}
$$

$$
= \lim_{x \to 0} \frac{2 e^{\sin 2x} \cos 2x \left(\cos^2 2x - \sin 2x\right) + e^{\sin 2x} \left(-4 \cos 2x \cdot \sin 2x - 2 \cos 2x\right) - 2 e^{2x}}{-2 \cos 2x}
$$

$$
= 1.
$$

In the above computations, we have to apply Theorem 4.17 three times since after the first and second applications of Theorem 4.17, we still have an indeterminate form of type "0/0".

We note that all other indeterminate forms can be transformed to one of the forms "0/0" or "∞/∞", and they can be treated by Theorem 4.17, too. We discuss these transformations in some more detail.

Let

$$
\lim_{x \to x_0} g(x) = 0 \qquad \text{and} \qquad \lim_{x \to x_0} f(x) = \infty,
$$

which corresponds to an indeterminate form of type "$0 \cdot \infty$". Then we can reduce the latter case to one discussed above by considering the reciprocal expression of one function, i.e. we use either

$$
\lim_{x \to x_0} [f(x) \cdot g(x)] = \lim_{x \to x_0} \frac{g(x)}{\frac{1}{f(x)}}
$$

or

$$
\lim_{x \to 0} [f(x) \cdot g(x)] = \lim_{x \to x_0} \frac{f(x)}{\frac{1}{g(x)}}.
$$

The first limit on the right-hand side is an indeterminate form of type "0/0", and the second one is an indeterminate form of type "∞/∞".

In the case of an indeterminate form of type "$\infty - \infty$", i.e.

$$
\lim_{x \to x_0} [g(x) - f(x)] \qquad \text{with} \qquad \lim_{x \to x_0} g(x) = \infty \quad \text{and} \quad \lim_{x \to x_0} f(x) = \infty,
$$

we can apply the following general transformation:

$$
g(x) - f(x) = \frac{1}{\frac{1}{g(x)}} - \frac{1}{\frac{1}{f(x)}} = \frac{\frac{1}{f(x)} - \frac{1}{g(x)}}{\frac{1}{g(x) f(x)}},
$$

where the right-hand side is an indeterminate form of type "0/0". However, we can often apply an easier reduction to obtain an indeterminate form that we have already discussed.

Example 4.28   We determine

$$L = \lim_{x \to \infty} \left(\sqrt[3]{x^3 - 2x^2} - x \right).$$

Obviously, this is an indeterminate form of type "$\infty - \infty$". We reduce this form to the type "$0/0$" without the above general transformation. Instead of it, we perform for $x \neq 0$ the following algebraic manipulation:

$$\sqrt[3]{x^3 - 2x^2} - x = \left[x^3 \left(1 - \frac{2}{x} \right) \right]^{1/3} - x = x \left(1 - \frac{2}{x} \right)^{1/3} - x \qquad \text{``}\infty - \infty\text{''}$$

$$= x \left[\left(1 - \frac{2}{x} \right)^{1/3} - 1 \right]. \qquad \text{``}\infty \cdot 0\text{''}$$

Thus we obtain

$$\lim_{x \to \infty} \left(\sqrt[3]{x^3 - 2x^2} - x \right) = \lim_{x \to \infty} \frac{\left(1 - \frac{2}{x} \right)^{1/3} - 1}{\frac{1}{x}}$$

which is an indeterminate form of type "$0/0$". By applying Theorem 4.17, we obtain

$$L = \lim_{x \to \infty} \frac{\frac{1}{3} \left(1 - \frac{2}{x} \right)^{-2/3} \cdot \frac{2}{x^2}}{-\frac{1}{x^2}} = \lim_{x \to \infty} \left[-\frac{2}{3} \left(1 - \frac{2}{x} \right)^{-2/3} \right] = -\frac{2}{3}.$$

If we have an indeterminate form of type "0^0", "∞^0" or "1^∞", i.e. the limit to be determined has the structure

$$L = \lim_{x \to x_0} g(x)^{f(x)}$$

with $g(x) > 0$, we take the natural logarithm on both sides of the equation

$$y(x) = g(x)^{f(x)}$$

and obtain

$$\ln y(x) = \ln \left[g(x)^{f(x)} \right] = f(x) \cdot \ln g(x).$$

If now x tends to x_0, we obtain an indeterminate form of type "$0 \cdot \infty$" which we can already treat. Suppose that this limit exists and let

$$L_1 = \lim_{x \to x_0} \left[f(x) \cdot \ln g(x) \right],$$

then we finally get

$$L = e^{L_1}.$$

Notice that the latter is obtained from Theorem 4.4 since the exponential function is continuous.

Example 4.29 We determine the limit

$$L = \lim_{x \to 0} (1+x)^{1/x}$$

which is an indeterminate form of type "1^∞". Setting

$$y(x) = (1+x)^{1/x}$$

and taking the natural logarithm on both sides, we obtain

$$\ln y(x) = \frac{1}{x} \cdot \ln(1+x).$$

Applying Theorem 4.17, we get by Bernoulli–l'Hospital's rule

$$\lim_{x \to 0} \frac{\ln(1+x)}{x} = \lim_{x \to 0} \frac{\frac{1}{1+x}}{1} = \lim_{x \to 0} \frac{1}{1+x} = 1.$$

Thus, we obtain

$$\lim_{x \to 0} \ln y(x) = 1.$$

Therefore,

$$L = \lim_{x \to 0} y(x) = e^1 = e.$$

4.5.5 Further examples

In the following example, we investigate all the properties listed above when graphing functions.

Example 4.30 Let us discuss in detail function $f : D_f \to \mathbb{R}$ with

$$f(x) = \frac{x^2 + 5x + 22}{x - 2}.$$

Function f is defined for all $x \in \mathbb{R}$ with $x \neq 2$, i.e. $D_f = \mathbb{R} \setminus \{2\}$. At point $x_0 = 2$, there is a pole of first order with

$$\lim_{x \to 2-0} f(x) = -\infty \qquad \text{and} \qquad \lim_{x \to 2+0} f(x) = \infty.$$

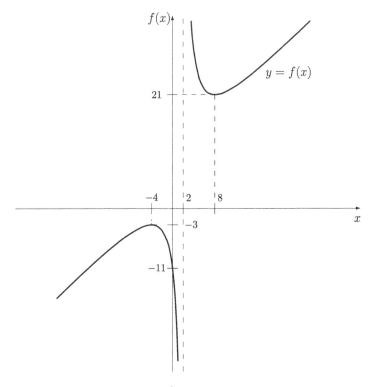

Figure 4.8 Graph of function f with $f(x) = (x^2 + 5x + 22)/(x - 2)$.

From $f(x) = x^2 + 5x + 22 = 0$, we obtain the roots

$$x_1 = -\frac{5}{2} + \sqrt{\frac{25}{4} - 22} \qquad \text{and} \qquad x_2 = -\frac{5}{2} - \sqrt{\frac{25}{4} - 22}.$$

Since

$$\frac{25}{4} - 22 = -\frac{63}{4} < 0,$$

there do not exist real zeroes of function f. We obtain for the first and second derivatives

$$f'(x) = \frac{(2x + 5)(x - 2) - (x^2 + 5x + 22)}{(x - 2)^2} = \frac{x^2 - 4x - 32}{(x - 2)^2}$$

and

$$f''(x) = \frac{(2x - 4)(x - 2)^2 - 2(x - 2)(x^2 - 4x - 32)}{(x - 2)^4} = \frac{72}{(x - 2)^3}.$$

Setting $f'(x) = 0$, we obtain $x^2 - 4x - 32 = 0$ which yields the stationary points

$$x_3 = 2 + \sqrt{4 + 32} = 8 \qquad \text{and} \qquad x_4 = 2 - \sqrt{4 + 32} = -4.$$

From $f''(x_3) > 0$, we get that $x_3 = 8$ is a local minimum point with $f(x_3) = 21$, and from $f''(x_4) < 0$, we get that $x_4 = -4$ is a local maximum point with $f(x_4) = -3$. Since $f'(x) > 0$ for all $x \in (-\infty, -4) \cup (8, \infty)$, function f is strictly increasing on $(-\infty, -4] \cup [8, \infty)$. For $x \in (-4, 2) \cup (2, 8)$, we have $f'(x) < 0$, and thus function f is strictly decreasing on $[-4, 2) \cup (2, 8]$. To determine candidates for an inflection point, we set $f''(x) = 0$. The latter equation has no solution, and thus function f does not have an inflection point. Because $f''(x) > 0$ for $x \in (2, \infty)$, function f is strictly convex on the interval $(2, \infty)$. Since $f''(x) < 0$ for $x \in (-\infty, 2)$, function f is strictly concave on the interval $(-\infty, 2)$. Moreover,

$$\lim_{x \to \infty} f(x) = \infty \qquad \text{and} \qquad \lim_{x \to -\infty} f(x) = -\infty.$$

The graph of function f is given in Figure 4.8.

Example 4.31   Let the relationship between the price p of a good and the resulting demand D for this good be given by the demand function $D = D(p)$, where

$$D = \frac{400 - 10p}{p + 5}, \qquad 0 \le p \le 40. \tag{4.6}$$

First, we determine the extreme points of function $D = D(p)$ and check the function for monotonicity as well as concavity or convexity. We obtain

$$D'(p) = \frac{-10(p + 5) - (400 - 10p) \cdot 1}{(p + 5)^2} = -\frac{450}{(p + 5)^2}.$$

Thus $D'(p) \neq 0$ for all $p \in [0, 40]$, and therefore function $D = D(p)$ cannot have an extreme point since the necessary condition for its existence is not satisfied. Moreover, since $D'(p) < 0$ for all $p \in (0, 40)$, function $D = D(p)$ is strictly decreasing on the closed interval $[0, 40]$. Checking function $D = D(p)$ for convexity and concavity, respectively, we obtain

$$D''(p) = -450(p + 5)^{-3} \cdot (-2) = \frac{900}{(p + 5)^3} > 0$$

for all $p \in (0, 40)$. Therefore, function $D = D(p)$ is strictly convex on the closed interval $[0, 40]$.
Next, we check function $D = D(p)$ for elasticity. We obtain

$$\varepsilon_D(p) = \frac{p \cdot D'(p)}{D(p)} = -\frac{450p(p + 5)}{(p + 5)^2(400 - 10p)} = -\frac{45p}{(p + 5)(40 - p)}.$$

We can determine points where function $D = D(p)$ changes between being elastic and inelastic, i.e. we can check, where equality $|\varepsilon_D(p)| = 1$ holds and obtain

$$\frac{45p}{(p + 5)(40 - p)} = 1.$$

Thus,

$$(p + 5)(40 - p) = 45p$$

which yields the quadratic equation

$$p^2 + 10p - 200 = 0$$

with the two solutions

$$p_1 = -5 + \sqrt{25 + 200} = -5 + 15 = 10 \qquad \text{and} \qquad p_2 = -5 - 15 = -20 < 0.$$

Solution p_2 does not belong to the domain D_D of function $D = D(p)$. Hence, for $p = 10$, function D is of unit elasticity. Since $|\varepsilon_D(p)| < 1$ for $p \in [0, 10)$ and $|\varepsilon_D(p)| > 1$ for $p \in (10, 40]$, we get that function $D = D(p)$ is inelastic for $p \in [0, 10)$ and elastic for $p \in (10, 40]$.

Finally, we look for the function $p = p(D)$ giving the price in dependence on the demand D. Since function $D(p)$ is strictly decreasing on the closed interval $[0, 40]$, the inverse demand function $p = p(D)$ exists. Solving equation (4.6) for p, we obtain

$$D(p + 5) = 400 - 10p$$
$$p(D + 10) = 400 - 5D$$
$$p = \frac{400 - 5D}{D + 10}.$$

Since $D = D(p)$ is strictly monotone, the inverse function $p = p(D)$ is strictly monotone as well. Because $D(0) = 80$ and $D(40) = 0$ (see also the inequality in (4.6)), function $p = p(D)$ is defined for $D \in [0, 80]$.

4.6   MEAN-VALUE THEOREM

Before presenting the mean-value theorem of differential calculus, we start with a special case.

THEOREM 4.18 (Rolle's theorem)   Let function $f : D_f \to \mathbb{R}$ be continuous on the closed interval $[a, b] \in D_f$ with $f(a) = f(b) = 0$ and differentiable on the open interval (a, b). Then there exists a point $x^* \in (a, b)$ with $f'(x^*) = 0$.

Geometrically, Rolle's theorem says that in the case of $f(a) = f(b)$, there exists an interior point x^* of the interval (a, b), where the tangent line is parallel to the x axis. In other words, there is at least one stationary point in the open interval (a, b). The mean-value theorem of differential calculus generalizes Rolle's theorem to the case $f(a) \neq f(b)$.

THEOREM 4.19 (mean-value theorem of differential calculus)   Let function $f : D_f \to \mathbb{R}$ be continuous on the closed interval $[a, b] \in D_f$ and differentiable on the open interval (a, b). Then there exists at least one point $x^* \in (a, b)$ such that

$$f'(x^*) = \frac{f(b) - f(a)}{b - a}.$$

The mean-value theorem of differential calculus is illustrated in Figure 4.9. In this case, we have two points $x_1^* \in (a, b)$ and $x_2^* \in (a, b)$ with the property that the derivative of function f at each of these points is equal to the quotient of $f(b) - f(a)$ and $b - a$.

The mean-value theorem of differential calculus can be used e.g. for the approximate calculation of function values.

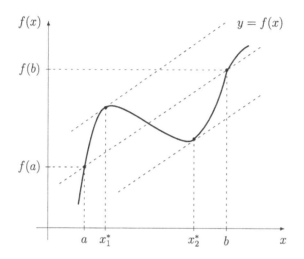

Figure 4.9 Mean-value theorem of differential calculus.

Example 4.32   Using the mean-value theorem, we compute $\sqrt[3]{29}$. Let us consider function $f : \mathbb{R}_+ \to \mathbb{R}$ with

$$f(x) = \sqrt[3]{x}.$$

We get

$$f'(x) = \frac{1}{3} \cdot x^{-2/3}.$$

Applying the mean-value theorem with $a = 27$ and $b = 29$, we get

$$\frac{f(b) - f(a)}{b - a} = \frac{\sqrt[3]{29} - \sqrt[3]{27}}{29 - 27} = \frac{\sqrt[3]{29} - 3}{2} = f'(x^*) = \frac{1}{3} \cdot (x^*)^{-2/3}$$

which can be rewritten as

$$\sqrt[3]{29} = 3 + \frac{1}{3} \cdot (x^*)^{-2/3} \cdot 2.$$

Using $x^* = 27$ (notice that x^* is not from the open interval $(27, 29)$, but it can be chosen since function f is differentiable at $x^* = a = 27$ and moreover, the value of function f' is easily computable at this point), we get

$$\sqrt[3]{29} \approx 3 + \frac{1}{3} \cdot \frac{1}{27^{2/3}} \cdot 2 = 3 + \frac{1}{3} \cdot \frac{2}{(3^3)^{2/3}} = 3 + \frac{2}{27} \approx 3.074.$$

Example 4.33 A vehicle starts at time zero and $s(t)$ gives the distance from the starting point (in metres) in dependence on time t (in seconds). It is known that $s(t) = t^2/2$ and let $a = 0$ and $b = 20$. Then the mean-value theorem reads as

$$\frac{s(20) - s(0)}{20 - 0} = \frac{\frac{1}{2} \cdot 20^2 - 0}{20} = \frac{200}{20} = 10 = s'(t^*) = t^*.$$

The left-hand side fraction gives the average velocity of the vehicle in the interval $[0, 20]$ which is equal to 10 m per second. The mean-value theorem implies that there is at least one value $t^* \in (0, 20)$, where the current velocity of the vehicle is equal to 10 m/s. In the above example, it happens at time $t^* = 10$.

4.7 TAYLOR POLYNOMIALS

Often it is necessary to approximate as closely as possible a 'complicated' function f by a polynomial P_n of some degree n. One way to do this is to require that all n derivatives of function f and of the polynomial P_n, as well as the function values of f and P_n, coincide at some point $x = x_0$ provided that function f is sufficiently often differentiable.

Before considering the general case, we discuss this question for some small values of n. First consider $n = 1$. In this case, we approximate function f around x_0 by a linear function (straight line)

$$P_1(x) = f(x_0) + f'(x_0) \cdot (x - x_0).$$

Obviously, we have $P_1(x_0) = f(x_0)$ and $P_1'(x_0) = f'(x_0)$. For $n = 2$, we approximate function f by a quadratic function (parabola) P_2 with

$$P_2(x) = f(x_0) + f'(x_0) \cdot (x - x_0) + \frac{f''(x_0)}{2!} \cdot (x - x_0)^2.$$

In this case, we also have $P_2(x_0) = f(x_0)$, $P_2'(x_0) = f'(x_0)$ and additionally $P_2''(x_0) = f''(x_0)$. Suppose now that function f is sufficiently often differentiable, and we wish to approximate function f by a polynomial P_n of degree n with the requirements listed above.

THEOREM 4.20 Let function $f : D_f \to \mathbb{R}$ be $n + 1$ times differentiable on the open interval $(a, b) \in D_f$ containing points x_0 and x. Then function f can be written as

$$f(x) = f(x_0) + \frac{f'(x_0)}{1!} \cdot (x - x_0) + \frac{f''(x_0)}{2!} \cdot (x - x_0)^2 + \cdots$$
$$+ \frac{f^{(n)}(x_0)}{n!} \cdot (x - x_0)^n + R_n(x),$$

where

$$R_n(x) = \frac{f^{(n+1)}(x_0 + \lambda(x - x_0))}{(n+1)!} \cdot (x - x_0)^{n+1}, \qquad 0 < \lambda < 1,$$

is Lagrange's form of the remainder.

The above form of representing function f by some polynomial together with some remainder is known as *Taylor's formula*. We still have to explain the meaning of the remainder $R_n(x)$. From equation

$$R_n(x) = f(x) - P_n(x),$$

the remainder gives the difference between the original function f and the approximating polynomial P_n. We are looking for an upper bound for the error when approximating function f by polynomial P_n in some interval $I = (x_0, x)$. To this end, we have to estimate the maximal possible value of the remainder $R_n(x^*)$ for $x^* \in (x_0, x)$. (Note that $x_0 + \lambda(x - x_0)$ with $0 < \lambda < 1$ in $R_n(x)$ above corresponds to some point $x^* \in (x_0, x)$.) If we know that the $(n+1)$th derivative of function f in the interval (x_0, x) is bounded by some constant M, i.e.

$$|f^{(n+1)}(x^*)| \le M \qquad \text{for } x^* \in (x_0, x),$$

then we obtain from Theorem 4.20:

$$|R_n(x)| \le \frac{M}{(n+1)!} \cdot (x - x_0)^{n+1}. \tag{4.7}$$

Inequality (4.7) gives the maximal error when replacing function f with its nth *Taylor polynomial*

$$P_n(x) = \sum_{k=0}^{n} \frac{f^{(k)}(x_0)}{k!} \cdot (x - x_0)^k$$

in the open interval (x_0, x). The above considerations can also be extended to an interval (x, x_0) to the left of point x_0. (Note that we have to replace $(x - x_0)$ in formula (4.7) by $|x - x_0|$.)

Example 4.34 Let function $f : \mathbb{R} \to \mathbb{R}$ with

$$f(x) = e^x$$

be given. We wish to approximate this function by a polynomial of smallest degree such that for $|x| < 0.5$ the maximal error does not exceed 10^{-4}. We choose $x_0 = 0$ and obtain

$$f'(x) = f''(x) = \cdots = f^{(n)}(x) = e^x.$$

Thus, we have $f(0) = f'(0) = \cdots = f^{(n)}(0) = 1$. Consequently,

$$f(x) = 1 + x + \frac{x^2}{2} + \frac{x^3}{6} + \cdots + \frac{x^n}{n!} + R_n(x)$$

with

$$R_n(x) = \frac{f^{(n+1)}(\lambda x)}{(n+1)!} \cdot x^{n+1} = \frac{e^{\lambda x}}{(n+1)!} \cdot x^{n+1},$$

where $0 < \lambda < 1$. For $|x| < 0.5$, we have

$$e^{\lambda x} < e^{1/2} \approx 1.6487 < 2.$$

Thus, we determine n such that

$$R_n(x) < \frac{2 \cdot |x|^{n+1}}{(n+1)!} < \frac{2}{(n+1)! \cdot 2^{n+1}} \leq 10^{-4}.$$

From the latter estimation it follows that we look for the smallest value of n such that

$$(n+1)! \cdot 2^{n+1} \geq 2 \cdot 10^4,$$

or equivalently,

$$(n+1)! \cdot 2^n \geq 10^4.$$

For $n \in \{0, 1, 2, 3, 4\}$, the above inequality is not satisfied. However, for $n = 5$, we get

$$(n+1)! \cdot 2^n = 720 \cdot 32 = 23{,}040 = 2.304 \cdot 10^4 \geq 10^4.$$

We have obtained:

$$R_5(x) = \left| e^x - \sum_{k=0}^{5} \frac{x^k}{k!} \right| \leq 10^{-4} \qquad \text{for } |x| < 0.5,$$

and thus we can approximate function f with $f(x) = e^x$ for $|x| < 0.5$ by the polynomial P_5 with

$$P_5(x) = 1 + x + \frac{x^2}{2} + \frac{x^3}{6} + \frac{x^4}{24} + \frac{x^5}{120},$$

and the error of this approximation does not exceed the value 10^{-4}.

Example 4.35   Let us approximate function $f : [-4, \infty) \to \mathbb{R}$ with

$$f(x) = \sqrt{x+4} = (x+4)^{0.5}$$

by a linear function $P_1(x)$ around $x_0 = 0$. This approximation should be used to give an estimate of $\sqrt{4.02}$.

We get

$$f'(x) = \frac{1}{2} \cdot (x+4)^{-1/2} \qquad \text{and} \qquad f''(x) = \frac{1}{2} \cdot \left(-\frac{1}{2}\right) \cdot (x+4)^{-3/2}.$$

Consequently, for $x_0 = 0$ we get

$$f(0) = 2 \qquad \text{and} \qquad f'(0) = \frac{1}{2} \cdot \frac{1}{2} = \frac{1}{4}.$$

We obtain

$$f(x) = 2 + \frac{1}{4}x - \frac{1}{8}(\lambda x + 4)^{-3/2} x^2, \quad 0 < \lambda < 1. \tag{4.8}$$

In order to estimate $\sqrt{4.02}$, we write $4.02 = 0.02 + 4$ and use representation (4.8). Estimating the maximum error, we get for $x = 0.02$ the inequalities $0 < \lambda x < 0.02$ for $0 < \lambda < 1$, and consequently the inequality $\lambda x + 4 > 4$. Therefore, we have

$$(\lambda x + 4)^{-3/2} < 4^{-3/2} = \frac{1}{8},$$

and we get the following estimate:

$$|R_1(0.02)| = \left| \frac{1}{8}(\lambda \cdot 0.02 + 4)^{-3/2} \cdot \left(\frac{2}{100} \right)^2 \right| < \frac{4}{80,000} \cdot \frac{1}{8} < 10^{-5}.$$

We conclude that

$$\sqrt{4.02} \approx 2 + \frac{1}{4} \cdot 0.02 = 2.005$$

with an error less than 10^{-5}.

4.8 APPROXIMATE DETERMINATION OF ZEROES

In this section, we discuss several algorithms for finding zeroes of a function approximately. As an application of Taylor polynomials we first consider the determination of zeroes by *Newton's method*. We discuss two variants of this method.

The first possibility is to approximate function f about x_0 by its tangent at x_0:

$$f(x) \approx P_1(x) = f(x_0) + f'(x_0) \cdot (x - x_0).$$

Let x_0 be an initial (approximate) value. Then we have $f(x) \approx P_1(x) = 0$, from which we obtain

$$f(x_0) + f'(x_0) \cdot (x - x_0) = 0.$$

We eliminate x and identify it with x_1 which yields:

$$x_1 = x_0 - \frac{f(x_0)}{f'(x_0)}.$$

Now we replace value x_0 on the right-hand side by the new approximate value x_1, and we determine another approximate value x_2. The procedure can be stopped if two successive approximate values x_n and x_{n+1} are sufficiently close to each other. The question that we still have to discuss is whether specific assumptions have to be made on the initial value x_0 in order to ensure that sequence $\{x_n\}$ converges to a zero of function f. The first assumption that we have to make is $x_0 \in [a, b]$ with $f(a) \cdot f(b) < 0$. In the case of a continuous function, this guarantees that the closed interval $[a, b]$ contains at least one zero $\bar{x}$ of function f (see Theorem 4.6). However, the latter condition is still not sufficient. It can be proved that, if function f is differentiable on the open interval $(a^*, b^*) \subseteq D_f$ containing $[a, b]$ and if additionally $f'(x) \neq 0$ and $f''(x) \neq 0$ for $x \in [a, b]$ and $x_0 \in \{a, b\}$ is chosen such that $f(x_0) \cdot f''(x_0) > 0$, then the above procedure converges to the (unique) zero $\bar{x} \in [a, b]$. The procedure is illustrated in Figure 4.10.

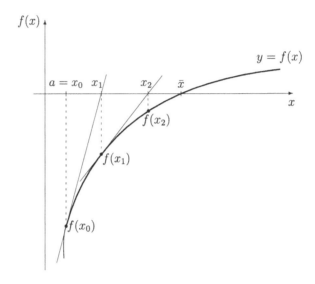

Figure 4.10 Illustration of Newton's method.

A second possibility is that we approximate function f about x_0 by a parabola. This procedure is denoted as *Newton's method of second order*. We get:

$$f(x) \approx P_2(x) = f(x_0) + f'(x_0) \cdot (x - x_0) + \frac{f''(x_0)}{2!} \cdot (x - x_0)^2.$$

Let x_0 be an initial (approximate) value. Then we obtain from $f(x) \approx P_2(x) = 0$

$$f(x_0) + f'(x_0) \cdot (x - x_0) + \frac{f''(x_0)}{2!} \cdot (x - x_0)^2 = 0.$$

If we eliminate x and identify it with x_1 we get:

$$x_{1A,1B} = x_0 - \frac{f'(x_0) \pm \sqrt{[f'(x_0)]^2 - 2f(x_0)f''(x_0)}}{f''(x_0)}.$$

Using now $x_1 = x_{1A}$ or $x_1 = x_{1B}$ (depending on which value is the better approximation of the zero), instead of x_0 on the right-hand side of the above equation, we determine a new approximate value x_2 and so on. Again some assumptions have to be made to ensure convergence of sequence $\{x_n\}$ to a zero. It can be proved that, if function f is differentiable on the open interval $(a^*, b^*) \subseteq D_f$ containing $[a, b]$ and $x_0 \in [a, b]$ with $f(a) \cdot f(b) < 0$ and $f''(x) \neq 0$ for $x \in [a, b]$, then the sequence $\{x_n\}$ determined by Newton's method of second order converges to a zero $\bar{x} \in [a, b]$.

The above two variants of Newton's method use derivatives. Finally, we briefly discuss two derivative-free methods for determining zeroes numerically. The first procedure presents a general iterative algorithm, also known as a *fixed-point procedure*, which transforms the equation $f(x) = 0$ into the form $x = \varphi(x)$. A solution $\bar{x}$ with $\bar{x} = \varphi(\bar{x})$ is denoted as fixed-point. Starting with some point $x_0 \in [a, b]$, we compute iteratively the values

$$x_{n+1} = \varphi(x_n), \qquad n = 0, 1, \ldots.$$

The procedure stops when two successive values x_k and x_{k+1} are 'sufficiently close' to each other. A sufficient condition to ensure convergence of this approach to a fixed point is that there exists a constant $L \in (0, 1)$ with

$$|\varphi'(x)| \leq L \qquad \text{for all } x \in [a, b].$$

The second derivative-free method is known as *regula falsi*. In this case, we approximate function f between $x_0 = a$ and $x_1 = b$ with $f(a) \cdot f(b) < 0$ by a straight line through the points $(a, f(a))$ and $(b, f(b))$. This yields:

$$y - f(a) = \frac{f(b) - f(a)}{b - a} \cdot (x - a).$$

Since we look for a zero $\bar{x}$ of function f, we set $y = 0$ and obtain the approximate value

$$x_2 = a - f(a) \cdot \frac{b - a}{f(b) - f(a)}$$

for the zero. In general, we have $f(x_2) \neq 0$. (In the other case, we have found the exact value of the zero.) Now we check which of both closed intervals $[a, x_2]$ and $[x_2, b]$ contains the zero. If $f(a) \cdot f(x_2) < 0$, then there exists a zero in the interval $[a, x_2]$. We replace b by x_2, and determine a new approximate value x_3. Otherwise, i.e. if $f(x_2) \cdot f(b) < 0$, then interval $[x_2, b]$ contains a zero. In that case, we replace a by x_2 and determine then an approximate value x_3, too. Continuing in this way, we get a sequence $\{x_n\}$ converging to a zero $\bar{x} \in [a, b]$. This procedure is illustrated in Figure 4.11.

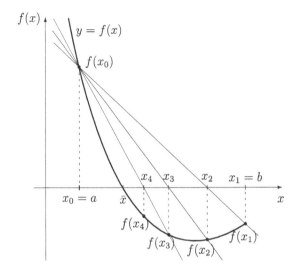

Figure 4.11 Illustration of *regula falsi*.

Example 4.36 Let $f : (0, \infty) \to \mathbb{R}$ with

$$f(x) = x - \lg x - 3$$

be given. We determine the zero contained in the closed interval $[3, 4]$ exactly to three decimal places. First we apply Newton's method. We obtain

$$f'(x) = 1 - \frac{1}{x \ln 10} \qquad \text{and} \qquad f''(x) = \frac{1}{x^2 \ln 10}.$$

Obviously we have $f(3) < 0$, $f(4) > 0$, $f'(x) > 0$ for $x \in [3, 4]$ and $f''(x) > 0$ for $x \in [3, 4]$. Using

$$x_{n+1} = x_n - \frac{f(x_n)}{f'(x_n)} = x_n - \frac{x_n - \lg x_n - 3}{1 - \frac{1}{x_n \ln 10}}, \qquad n = 0, 1, \ldots,$$

we get the results presented in Table 4.2 when starting with $x_0 = 4$. We obtain $f(x_2) < 10^{-5}$, and the value given in the last row in bold face corresponds to the zero rounded to three decimal places.

Table 4.2 Application of Newton's method

n	x_n	$f(x_n)$	$f'(x_n)$
0	4	0.39794	0.89143
1	3.55359	0.00292	0.87779
2	3.55026	0.00000	
3	**3.550**		

Next, we apply the fixed-point procedure. We can rewrite the given function as

$$x = \lg x + 3.$$

Starting with $x_0 = 3$ and using the fixed-point equation

$$x_{n+1} = \varphi(x_n) = \lg x_n + 3, \qquad n = 0, 1, \ldots,$$

we obtain the results given in the second column of Table 4.3. We note that the sufficient convergence condition is satisfied in this case. From

$$\varphi'(x) = \frac{1}{x \ln 10} \qquad \text{and} \qquad \ln 10 > 2,$$

we obtain e.g.

$$|\varphi'(x)| \le \frac{1}{6} < 1 \qquad \text{for } x \in [3, 4].$$

We note that, if one rewrites the given function as $x = 10^{x-3}$, the fixed-point method does not converge to the zero $\bar{x}$. (The reader can easily verify that the sufficient convergence condition is not satisfied in this case.)

Finally, we apply regula falsi. Letting $x_0 = a = 3$ and $x_1 = b = 4$, we get the results presented in the third and fourth columns of Table 4.3.

Table 4.3 Application of derivative-free methods

n	Fixed-point method	Regula falsi	
	x_n	x_n	$f(x_n)$
0	3	3	-0.47712
1	3.47712	4	0.39794
2	3.54122	3.5452	-0.0044
3	3.54915	3.5502	-0.00005
4	3.55012	3.550	
5	3.550		

EXERCISES

4.1 Find the left-side limit and the right-side limit of function $f : D_f \to \mathbb{R}$ as x approaches x_0. Can we conclude from these answers that function f has a limit as x approaches x_0?

(a) $f(x) = \begin{cases} a & \text{for } x \ne x_0 \\ a+1 & \text{for } x = x_0 \end{cases}$, (b) $f(x) = \begin{cases} \sqrt{x} & \text{for } x \le 1 \\ x^2 & \text{for } x > 1 \end{cases}$, $x_0 = 1$;

(c) $f(x) = |x|$, $x_0 = 0$; (d) $f(x) = \begin{cases} x & \text{for } x < 1 \\ x+1 & \text{for } x \ge 1 \end{cases}$, $x_0 = 1$.

4.2   Find the following limits if they exist:

(a) $\lim\limits_{x \to 2} \dfrac{x^3 - 3x^2 + 2x}{x - 2}$;   (b) $\lim\limits_{x \to 2} \dfrac{x^3 - 3x^2}{x - 2}$;   (c) $\lim\limits_{x \to 2} \dfrac{x^3 - 3x^2}{(x - 2)^2}$.

What type of discontinuity is at point $x_0 = 2$?

4.3   Check the continuity of function $f : D_f \to \mathbb{R}$ at point $x = x_0$ and define the type of discontinuity:

(a) $f(x) = \dfrac{\sqrt{x + 7} - 3}{x - 2}$, $x_0 = 2$;

(b) $f(x) = |x - 1|$, $x_0 = 1$;

(c) $f(x) = \begin{cases} e^{x-1} & \text{for } x < 1 \\ 2x & \text{for } x \geq 1 \end{cases}$, $x_0 = 1$;

(d) $f(x) = e^{1/(x-1)}$, $x_0 = 1$.

4.4   Are the following functions $f : D_f \to \mathbb{R}$ differentiable at points $x = x_0$ and $x = x_1$, respectively?

(a) $f(x) = |x - 5| + 6x$, $x_0 = 5, \quad x_1 = 0$;

(b) $f(x) = \begin{cases} \cos x & \text{for } x < 0; \\ 1 + x^2 & \text{for } 0 \leq x \leq 2, \\ 2x + 1 & \text{for } x > 2. \end{cases}$ $x_0 = 0, \quad x_1 = 2$;

4.5   Find the derivative of each of the following functions $f : D_f \to \mathbb{R}$ with:

(a) $y = 2x^3 - 5x - 3\sin x + \sin(\pi/8)$;   (b) $y = (x^4 + 4x)\sin x$;

(c) $y = \dfrac{x^2 - \cos x}{2 + \sin x}$;   (d) $y = (2x^3 - 3x + \ln x)^4$;

(e) $y = \cos(x^3 + 3x^2 - 8)^4$;   (f) $y = \cos^4(x^3 + 3x^2 - 8)$;

(g) $y = \sqrt{\sin(e^x)}$;   (h) $y = \ln(x^2 + 1)$.

4.6   Find and simplify the derivative of each of the following functions $f : D_f \to \mathbb{R}$ with:

(a) $f(x) = (\tan x - 1)\cos x$;   (b) $f(x) = \ln\sqrt{\dfrac{1 + \sin x}{1 - \sin x}}$;

(c) $f(x) = (1 + \sqrt[3]{x})^2$.

4.7   Find the derivatives of the following functions $f : D_f \to \mathbb{R}$ using logarithmic differentiation:

(a) $f(x) = (\tan x)^x$, $D_f = (0, \pi/2)$;   (b) $f(x) = \sin x^{x-1}$, $D_f = (1, \infty)$;

(c) $f(x) = \dfrac{(x + 2)\sqrt{x - 1}}{x^3(x - 2)^2}$.

4.8   Find the third derivatives of the following functions $f : D_f \to \mathbb{R}$ with:

(a) $f(x) = x^2 \sin x$;   (b) $f(x) = \ln(x^2)$;

(c) $f(x) = \dfrac{2x^2}{(x - 2)^3}$;   (d) $f(x) = (x + 1)e^x$.

4.9   Given the total cost function $C : \mathbb{R}_+ \to \mathbb{R}$ with

$$C(x) = 4x^3 - 2x^2 + 4x + 100,$$

where $C(x)$ denotes the total cost in dependence on the output x. Find the marginal cost at $x_0 = 2$. Compute the change in total cost resulting from an increase in the output from 2 to 2.5 units. Give an approximation of the exact value by using the differential.

4.10 Let $D : D_D \to \mathbb{R}_+$ be a demand function, where $D(p)$ is the quantity demand in dependence on the price p. Given $D(p) = 320/p$, calculate approximately the change in quantity demand if p changes from 8 to 10 EUR.

4.11 Given the functions $f : D_f \to \mathbb{R}_+$ with

$$f(x) = 2e^{x/2}$$

and $g : D_g \to \mathbb{R}_+$ with

$$g(x) = 3\sqrt{x}$$

find the proportional rates of change $\rho_f(x)$, $\rho_g(x)$ and the elasticities $\varepsilon_f(x)$ and $\varepsilon_g(x)$ and specify them for $x_0 = 1$ and $x_1 = 100$. For which values $x \in D_f$ are the functions f and g elastic? Give the percentage rate of change of the function value when value x increases by one per cent.

4.12 Given the price–demand function

$$D = D(p) = 1,000e^{-2(p-1)^2}$$

with demand $D > 0$ and price $p > 0$, find the (point) elasticity of demand $\varepsilon_D(p)$. Check at which prices the demand is elastic.

4.13 Find all local extrema, the global maximum and the global minimum of the following functions $f : D_f \to \mathbb{R}$ with $D_f \subseteq [-5, 5]$, where:

(a) $f(x) = x^4 - 3x^3 + x^2 - 5$;   (b) $f(x) = 4 - |x - 3|$;

(c) $f(x) = e^{-x^2/2}$;   (d) $f(x) = \dfrac{x^2}{x - 2}$;

(e) $f(x) = \dfrac{\sqrt{x}}{1 + \sqrt{x}}$.

4.14 Assume that function $f : D_f \to \mathbb{R}$ with

$$f(x) = a \ln x + bx^2 + x$$

has local extrema at point $x_1 = 1$ and at point $x_2 = 2$. What can you conclude about the values of a and b? Check whether they are relative maxima or minima.

4.15 Determine the following limits by Bernoulli–l'Hospital's rule:

(a) $\lim\limits_{x \to 0} \frac{\sin x}{x}$;

(b) $\lim\limits_{x \to 1} \frac{e^{2(x-1)} - x^2}{(x^2 - 1)^2}$;

(c) $\lim\limits_{x \to \infty} (x^2)^{1/x}$;

(d) $\lim\limits_{x \to 0+0} x^x$;

(e) $\lim\limits_{x \to 0} \left(\frac{1}{x \sin x} - \frac{1}{x^2} \right)$;

(f) $\lim\limits_{x \to 0} \frac{1}{x^2} \ln \frac{\sin x}{x}$.

4.16 For the following functions $f : D_f \to \mathbb{R}$, determine and investigate domains, zeroes, discontinuities, monotonicity, extreme points and extreme values, convexity and concavity, inflection points and limits as x tends to $\pm\infty$. Graph the functions f with $f(x)$ given as follows:

(a) $f(x) = \frac{x^2 + 1}{(x - 2)^2}$;

(b) $f(x) = \frac{3x^2 - 4x}{-2x^2 + x}$;

(c) $f(x) = \frac{x^4 + x^3}{x^3 - 2x^2 + x}$;

(d) $f(x) = e^{(x-1)^2/2}$;

(e) $f(x) = \ln \frac{x - 2}{x^2}$;

(f) $f(x) = \sqrt[3]{2x^2 - x^3}$.

4.17 Expand the following functions $f : D_f \to \mathbb{R}$ into Taylor polynomials with corresponding remainder:

(a) $f(x) = \sin \frac{\pi x}{4}$, $x_0 = 2$, $n = 5$;

(b) $f(x) = \ln(x + 1)$, $x_0 = 0$, $n \in \mathbb{N}$;

(c) $f(x) = e^{-x} \sin 2x$, $x_0 = 0$, $n = 4$.

4.18 Calculate $1/\sqrt[5]{e}$ by using Taylor polynomials for function $f : D_f \to \mathbb{R}$ with $f(x) = e^x$. The error should be less than 10^{-6}.

4.19 Determine the zero $\bar{x}$ of function $f : D_f \to \mathbb{R}$ with

$$f(x) = x^3 - 6x + 2 \qquad \text{and} \qquad 0 \le \bar{x} \le 1$$

exactly to four decimal places. Use (a) Newton's method and (b) regula falsi.

4.20 Find the zero $\bar{x}$ of function $f : D_f \to \mathbb{R}$ with

$$f(x) = x - \ln x - 3 \qquad \text{and} \qquad \bar{x} > 1.$$

Determine the value with an error less than 10^{-5} and use Newton's method.

5   Integration

In differential calculus we have determined the derivative f' of a function f. In many applications, a function f is given and we are looking for a function F whose derivative corresponds to function f. For instance, assume that a marginal cost function C' is given, i.e. it is known how cost is changing according to the produced quantity x, and we are looking for the corresponding cost function C. Such a function C can be found by integration, which is the reverse process of differentiation. Another application of integration might be to determine the area enclosed by the graphs of specific functions. In this chapter, we discuss basic integration methods in detail.

5.1   INDEFINITE INTEGRALS

We start with the definition of an antiderivative of a function f.

Definition 5.1   A function $F : D_F \to \mathbb{R}$ differentiable on an interval $I \subseteq D_F$ is called an *antiderivative* of the function $f : D_f = D_F \to \mathbb{R}$ if

$$F'(x) = f(x) \qquad \text{for all } x \in D_F.$$

Obviously, the antiderivative is not uniquely determined since we can add any constant the derivative of which is always equal to zero. In particular, we get the following theorem.

THEOREM 5.1   If function $F : D_F \to \mathbb{R}$ is any antiderivative of function $f : D_f \to \mathbb{R}$, then all the antiderivatives F^* of function f are of the form

$$F^*(x) = F(x) + C,$$

where $C \in \mathbb{R}$ is any constant.

By means of the antiderivative, we can now introduce the notion of the indefinite integral, which is as follows.

Definition 5.2   Let function $F : D_F \to \mathbb{R}$ be an antiderivative of function f. The *indefinite integral* of function f, denoted by $\int f(x)\, dx$, is defined as

$$\int f(x)\, dx = F(x) + C,$$

where $C \in \mathbb{R}$ is any constant.

Function f is also called the *integrand*, and as we see from the above definition, the indefinite integral of function f gives the infinitely many antiderivatives of the integrand f. The notation dx in the integral indicates that x is the *variable of integration*, and C denotes the *integration constant*.

The relationship between differentiation and integration can be seen from the following two formulas. We have

$$\frac{d}{dx} \int f(x)\, dx = \frac{d}{dx} [F(x) + C] = f(x),$$

and

$$\int F'(x)\, dx = \int f(x)\, dx = F(x) + C.$$

The first formula says that, if we differentiate the obtained antiderivative, we again obtain the integrand of the indefinite integral. In this way, one can easily check whether the indefinite integral has been found correctly. Conversely, if we differentiate function F and find then the corresponding indefinite integral with the integrand F', the result differs from function F only by some constant C.

5.2   INTEGRATION FORMULAS AND METHODS

5.2.1   Basic indefinite integrals and rules

From the considerations about differential calculus in Chapter 4, we are already able to present some antiderivatives. Their validity can be easily checked by differentiating the right-hand side, where we must obtain the integrand of the corresponding left-hand side integral. Unfortunately, it is not possible to find analytically an antiderivative for any given function. Determining an antiderivative for a given function is often much harder than differentiating the function.

Some indefinite integrals

$$\int x^n\, dx \;=\; \frac{x^{n+1}}{n+1} + C \qquad (n \in \mathbb{Z},\ n \neq -1)$$

$$\int x^r\, dx \;=\; \frac{x^{r+1}}{r+1} + C \qquad (r \in \mathbb{R},\ r \neq -1,\ x > 0)$$

$$\int \frac{1}{x}\, dx \;=\; \ln |x| + C \qquad (x \neq 0)$$

$$\int e^x \, dx \quad = \quad e^x + C$$

$$\int a^x \, dx \quad = \quad \frac{a^x}{\ln a} + C \qquad (a > 0, \, a \neq 1)$$

$$\int \sin x \, dx \quad = \quad -\cos x + C$$

$$\int \cos x \, dx \quad = \quad \sin x + C$$

$$\int \frac{dx}{\cos^2 x} \quad = \quad \tan x + C \qquad \left(x \neq \frac{\pi}{2} + k\pi, \, k \in \mathbb{Z} \right)$$

$$\int \frac{dx}{\sin^2 x} \quad = \quad -\cot x + C \qquad (x \neq k\pi, \, k \in \mathbb{Z})$$

$$\int \frac{dx}{\sqrt{1 - x^2}} \quad = \quad \arcsin x + C$$

$$\int \frac{dx}{1 + x^2} \quad = \quad \arctan x + C$$

Next, we give two basic rules for indefinite integrals:

(1) $\displaystyle\int k \cdot f(x) \, dx = k \cdot \int f(x) \, dx$ \qquad\qquad (constant-factor rule);

(2) $\displaystyle\int [f(x) \pm g(x)] \, dx = \int f(x) \, dx \pm \int g(x) \, dx$ \qquad (sum–difference rule).

Rule (1) says that we can write a constant factor in front of the integral, and rule (2) says that, if the integrand is the sum (or difference) of two functions, we can determine the indefinite integral as the sum (or difference) of the corresponding two integrals. Using the given list of definite integrals and the two rules above, we are now able to find indefinite integrals for some simple functions.

Example 5.1 Given is the integral

$$I = \int \left(2x^3 + 3^x - 2\sin x \right) dx.$$

Applying the rules for indefinite integrals, we can split the integral into several integrals and solve each of them by using the list of indefinite integrals given above. We obtain:

$$I = 2 \int x^3 \, dx + \int 3^x \, dx - 2 \int \sin x \, dx$$

$$= 2 \cdot \frac{x^4}{4} + \frac{3^x}{\ln 3} - 2(-\cos x) + C = \frac{1}{2}x^4 + \frac{3^x}{\ln 3} + 2\cos x + C.$$

Example 5.2 We wish to find

$$I = \int \left(\frac{\sqrt{x} - 2x^2}{\sqrt[3]{x}} \right) dx.$$

Using power rules and the indefinite integral for a power function as integrand, we can transform the given integral as follows:

$$I = \int \left(\frac{\sqrt{x}}{\sqrt[3]{x}} - 2 \cdot \frac{x^2}{\sqrt[3]{x}} \right) dx = \int \left(\frac{x^{1/2}}{x^{1/3}} - 2 \cdot \frac{x^2}{x^{1/3}} \right) dx$$

$$= \int \left(x^{1/6} - 2x^{5/3} \right) dx = \frac{x^{7/6}}{\frac{7}{6}} - 2 \cdot \frac{x^{8/3}}{\frac{8}{3}} + C = \frac{6}{7} \cdot \sqrt[6]{x^7} - \frac{3}{4} \cdot \sqrt[3]{x^8} + C.$$

The antiderivatives and rules presented so far are sufficient only for a few specific indefinite integrals. Hence, it is necessary to consider some more general integration methods which allow us to find broader classes of indefinite integrals. One of these methods is integration by substitution.

5.2.2 Integration by substitution

The aim of this method is to transform a given integral in such a way that the resulting integral can be easily found. This is done by introducing a new variable t by means of an appropriate substitution $t = g(x)$ or $x = g^{-1}(t)$. The integration method by substitution results from the chain rule of differential calculus.

THEOREM 5.2 Suppose that function $f : D_f \to \mathbb{R}$ has an antiderivative F and function $g : D_g \to \mathbb{R}$ with $R_g \subseteq D_f$ is continuously differentiable on an open interval $(a, b) \in D_g$. Then function $z = f \circ g$ exists with $z = (f \circ g)(x) = f(g(x))$ and setting $t = g(x)$, we obtain

$$\int f(g(x)) \cdot g'(x) \, dx = \int f(t) \, dt = F(t) + C = F(g(x)) + C.$$

The symbol $\circ$ stands for the composition of functions as introduced in Chapter 3. Theorem 5.2 states that, if the integrand is the product of a composite function $f \circ g$ and the derivative g' of the inside function, then the antiderivative is given by the composite function $F \circ g$, where function F is an antiderivative of function f. The validity of Theorem 5.2 can be easily proved by differentiating the composite function $F \circ g$ (using the chain rule).

Before considering several examples, we give some special cases of Theorem 5.2 with specific inside functions g:

(1) Let $g(x) = ax + b$. In this case, we get $g'(x) = a$, and Theorem 5.2 reads as

$$\int f(ax + b) \, dx = \frac{1}{a} \int f(ax + b) \cdot a \, dx = \frac{1}{a} \int f(t) \, dt = \frac{1}{a} \cdot F(t) + C$$

$$= \frac{1}{a} \cdot F(ax + b) + C.$$

Function g describes a linear substitution.

(2) Let $f(g(x)) = [g(x)]^n$. Then Theorem 5.2 turns into

$$\int [g(x)]^n \cdot g'(x) \, dx = \frac{1}{n+1} \cdot [g(x)]^{n+1} + C.$$

(3) Let $f(g(x)) = 1/[g(x)]$. Then Theorem 5.2 reads as

$$\int \frac{g'(x)}{g(x)}\, dx = \ln|g(x)| + C.$$

(4) Let $f(g(x)) = e^{g(x)}$. Then Theorem 5.2 corresponds to the equality

$$\int e^{g(x)} \cdot g'(x)\, dx = e^{g(x)} + C.$$

We illustrate the above method by the following examples.

Example 5.3 Let us find

$$\int (3x + 2)^4 \, dx.$$

This integral is of type (1) above, i.e. we have $g(x) = 3x + 2$ and $f(t) = t^4$. Using $g'(x) = 3$, we obtain

$$\int (3x + 2)^4 \, dx = \frac{1}{3} \int (3x + 2)^4 \cdot 3 \, dx = \frac{1}{3} \int t^4 \, dt = \frac{1}{3} \cdot \frac{t^5}{5} + C$$

$$= \frac{1}{15} \cdot (3x + 2)^5 + C.$$

Example 5.4 Consider the integral

$$\int 5x^2 e^{x^3} \, dx.$$

Setting $t = g(x) = x^3$, we obtain

$$\frac{dg}{dx} = g'(x) = 3x^2.$$

The application of Theorem 5.2 yields (see also special case (4) above with $g(x) = x^3$)

$$\int 5x^2 e^{x^3} \, dx = \frac{5}{3} \int 3x^2 e^{x^3} \, dx = \frac{5}{3} \int e^t \, dt = \frac{5}{3} e^t + C = \frac{5}{3} e^{x^3} + C.$$

Example 5.5 We want to find the integral

$$\int \frac{5x + 8}{3x^2 + 1} \, dx.$$

The first step is to split the above integral into two integrals:

$$\int \frac{5x+8}{3x^2+1}\, dx = \int \frac{5x}{3x^2+1}\, dx + \int \frac{8}{3x^2+1}\, dx.$$

We next discuss how we can find both integrals by substitution. The first integral is very similar to case (3) of Theorem 5.2. If we had $6x$ in the numerator, we could immediately apply this case. So, we multiply both the numerator and the denominator by 6 and obtain

$$\int \frac{5x}{3x^2+1}\, dx = \frac{5}{6}\int \frac{6x}{3x^2+1} = \frac{5}{6}\ln(3x^2+1)+C_1.$$

The second integral is similar to one of the basic integrals given in Chapter 5.2.1. If the integrand is $1/(1+x^2)$, then we obtain from this list of indefinite integrals the antiderivative $\arctan x$. Thus we transform the second integral as follows:

$$\int \frac{8}{3x^2+1}\, dx = 8\int \frac{1}{\left[(\sqrt{3}\cdot x)^2+1\right]}\, dx = \frac{8}{\sqrt{3}}\int \frac{1}{t^2+1}\, dt.$$

In the latter integral, we can apply the linear substitution $t = \sqrt{3}\cdot x$ (or equivalently, we apply immediately the formula given for integrals of type (1) after Theorem 5.2). Now the indefinite integral can be found immediately from the list of given integrals:

$$\frac{8}{\sqrt{3}}\cdot \int \frac{1}{1+t^2}\, dt = \frac{8}{\sqrt{3}}\arctan t + C_2 = \frac{8}{\sqrt{3}}\arctan\left(\sqrt{3}\cdot x\right) + C_2.$$

Combining both results and rationalizing all denominators, we obtain

$$\int \frac{5x+8}{3x^2+1}\, dx = \frac{5}{6}\ln(3x^2+1)+\frac{8}{3}\sqrt{3}\arctan\left(\sqrt{3}\cdot x\right) + C.$$

Sometimes we do not see immediately that the integrand is of the type $f(g(x))\cdot g'(x)$ (or can be easily transformed into this form). If we try to apply some substitution $t = g(x)$ and if, by using the differential $dt = g'(x)\, dx$, it is possible to replace all terms in the original integrand by some terms depending on the new variable t, then we can successfully apply integration by substitution. We illustrate this by the following examples.

Example 5.6   We determine

$$\int \sqrt{\frac{e^{2x}}{1-e^{2x}}}\, dx$$

and apply the substitution $t = e^x$. By differentiation, we get

$$\frac{dt}{dx} = e^x$$

which can be rewritten as

$$dx = \frac{dt}{e^x} = \frac{dt}{t}.$$

Using the substitution and the latter equality, we can replace all terms depending on x in the original integral by some term depending on t, and we obtain

$$\int \sqrt{\frac{e^{2x}}{1 - e^{2x}}}\, dx = \int \sqrt{\frac{t^2}{1 - t^2}} \cdot \frac{dt}{t} = \int \frac{t}{\sqrt{1 - t^2}} \cdot \frac{dt}{t} = \int \frac{dt}{\sqrt{1 - t^2}} = \arcsin t + C.$$

The latter indefinite integral has been taken from the list in Chapter 5.2.1. Substituting back again, we get

$$\int \sqrt{\frac{e^{2x}}{1 - e^{2x}}}\, dx = \arcsin e^x + C.$$

Example 5.7   Consider the integral

$$\int \frac{dx}{x\sqrt{x^2 - 9}}.$$

We apply the substitution $t = \sqrt{x^2 - 9}$ and obtain by differentiation

$$\frac{dt}{dx} = \frac{x}{\sqrt{x^2 - 9}},$$

which yields

$$\frac{dt}{x} = \frac{dx}{\sqrt{x^2 - 9}}.$$

Replacing $dx/\sqrt{x^2 - 9}$ in the integral, we would still have x^2 in the denominator. In order to apply integration by substitution, we must be able also to replace this term by some term in the variable t. Using the above substitution again, we can replace x^2 in the denominator by solving equation $t = \sqrt{x^2 - 9}$ for x^2 which yields $x^2 = t^2 + 9$. Hence we obtain

$$\int \frac{dx}{x\sqrt{x^2 - 9}} = \int \frac{dt}{t^2 + 9} = \frac{1}{9} \int \frac{dt}{\left(\frac{t}{3}\right)^2 + 1} = \frac{1}{3} \arctan \frac{t}{3} + C$$

$$= \frac{1}{3} \arctan \frac{\sqrt{x^2 - 9}}{3} + C.$$

Notice that in the last step, when determining the indefinite integral, we have again applied a substitution, namely $z = t/3$ (or equivalently, type (1) of the integrals after Theorem 5.2 has been used).

Example 5.8   Let us consider the integral

$$\int \frac{dx}{\sin x}.$$

In this case, we can apply the substitution

$$\tan \frac{x}{2} = t$$

which can always be applied when the integrand is a rational function of the trigonometric functions $\sin x$ and $\cos x$. Solving the above substitution for x yields

$$x = 2 \arctan t.$$

By differentiation, we obtain

$$\frac{dx}{dt} = \frac{2}{1 + t^2}.$$

Now, we still have to replace $\sin x$ by some function depending only on variable t. This can be done by using the addition theorems for the sine function (see property (1) of trigonometric functions in Chapter 3.3.3 and its special form given by formulas (3.8)). We have

$$\sin x = \frac{2 \cdot \sin \frac{x}{2} \cdot \cos \frac{x}{2}}{1} = \frac{2 \sin \frac{x}{2} \cdot \cos \frac{x}{2}}{\sin^2 \frac{x}{2} + \cos^2 \frac{x}{2}} = \frac{2 \tan \frac{x}{2}}{1 + \tan^2 \frac{x}{2}} = \frac{2t}{1 + t^2}.$$

To get the second to last fraction above, we have divided both the numerator and the denominator by $\cos^2(x/2)$. Then, we get for the given integral

$$\int \frac{dx}{\sin x} = \int \frac{1 + t^2}{2t} \cdot \frac{2dt}{1 + t^2} = \int \frac{dt}{t} = \ln |t| + C = \ln \left| \tan \frac{x}{2} \right| + C.$$

5.2.3   Integration by parts

Another general integration method is integration by parts. The formula for this method is obtained from the formula for the differentiation of a product of two functions u and v:

$$[u(x) \cdot v(x)]' = u'(x) \cdot v(x) + u(x) \cdot v'(x).$$

Integrating now both sides of the above equation leads to the following theorem which gives us the formula for integration by parts.

THEOREM 5.3   Let $u : D_u \to \mathbb{R}$ and $v : D_v \to \mathbb{R}$ be two functions differentiable on some open interval $I = (a, b) \subseteq D_u \cap D_v$. Then:

$$\int u(x) \cdot v'(x)\, dx = u(x) \cdot v(x) - \int u'(x) \cdot v(x)\, dx.$$

The application of integration by parts requires that we can find an antiderivative of function v' and an antiderivative of function $u' \cdot v$. If we are looking for an antiderivative of a product of two functions, the successful use of integration by parts depends on an appropriate choice of functions u and v'. Integration by parts can, for instance, be applied to the following types of integrals:

(1) $\displaystyle\int P_n(x) \cdot \ln x \, dx,$

(2) $\displaystyle\int P_n(x) \cdot \sin ax \, dx,$

(3) $\displaystyle\int P_n(x) \cdot \cos ax \, dx,$

(4) $\displaystyle\int P_n(x) \cdot e^{ax} \, dx,$

where $P_n(x) = a_n x^n + a_{n-1} x^{n-1} + \ldots + a_1 x + a_0$ is a polynomial of degree n.

In most cases above, polynomial P_n is taken as function u which has to be differentiated within the application of Theorem 5.3. (As a consequence, the derivative u' is a polynomial of smaller degree.) However, in case (1) it is usually preferable to take P_n as function v' which has to be integrated within the application of integration by parts (so that the logarithmic function is differentiated). We illustrate integration by parts by some examples.

Example 5.9   Let us find

$$\int (x^2 + 2) \sin x \, dx.$$

This is an integral of type (2) above, and we set

$$u(x) = x^2 + 2 \qquad \text{and} \qquad v'(x) = \sin x.$$

Now we obtain

$$u'(x) = 2x \qquad \text{and} \qquad v(x) = -\cos x.$$

Hence,

$$\int (x^2 + 2) \sin x \, dx = -(x^2 + 2) \cos x - \int (-2x \cos x) \, dx$$

$$= -(x^2 + 2) \cos x + 2 \int x \cos x \, dx.$$

If we now apply integration by parts again to the latter integral with

$$u(x) = x \qquad \text{and} \qquad v'(x) = \cos x,$$

we get

$$u'(x) = 1 \qquad \text{and} \qquad v(x) = \sin x.$$

This yields

$$\int (x^2 + 2) \sin x \, dx = -(x^2 + 2) \cos x + 2 \left(x \sin x - \int \sin x \, dx \right)$$

$$= -(x^2 + 2) \cos x + 2x \sin x + 2 \cos x + C$$

$$= 2x \sin x - x^2 \cos x + C.$$

Notice that integration constant C has to be written as soon as no further integral appears on the right-hand side.

Example 5.10   Let us determine

$$\int \ln x \, dx.$$

Although the integrand $f(x) = \ln x$ is here not written as a product of two functions, we can nevertheless apply integration by parts by introducing factor one, i.e. we set

$$u(x) = \ln x \qquad \text{and} \qquad v'(x) = 1.$$

Then we obtain

$$u'(x) = \frac{1}{x} \qquad \text{and} \qquad v(x) = x$$

which leads to

$$\int \ln x \, dx = x \ln x - \int dx = x(\ln x - 1) + C.$$

Example 5.11   We determine

$$\int \sin^2 x \, dx.$$

In this case, we set

$$u(x) = \sin x \qquad \text{and} \qquad v'(x) = \sin x,$$

and we obtain with

$$u'(x) = \cos x \qquad \text{and} \qquad v(x) = - \cos x$$

by applying integration by parts

$$\int \sin^2 x \, dx = - \sin x \cos x + \int \cos^2 x \, dx.$$

Now one could again apply integration by parts to the integral on the right-hand side. However, doing this we only obtain the identity $\int \sin^2 x \, dx = \int \sin^2 x \, dx$ which does not yield a solution to the problem. Instead we use the equality

$$\cos^2 x = 1 - \sin^2 x$$

(see property (8) of trigonometric functions in Chapter 3.3.3), and the above integral can be rewritten as follows:

$$\int \sin^2 x \, dx = -\sin x \cos x + \int (1 - \sin^2 x) \, dx$$

$$= -\sin x \cos x + x - \int \sin^2 x \, dx.$$

Now we can add $\int \sin^2 x \, dx$ to both sides, divide the resulting equation by two, introduce the integration constant C and obtain

$$\int \sin^2 x \, dx = \frac{1}{2} (x - \sin x \cos x) + C.$$

Often, one has to combine both discussed integration methods. Consider the following two examples.

Example 5.12   We determine

$$\int \frac{x}{\sin^2 x} \, dx.$$

Although the integrand does not have one of the special forms (1) to (4) given after Theorem 5.3, the application of integration by parts is worthwhile. Setting

$$u(x) = x \qquad \text{and} \qquad v'(x) = \frac{1}{\sin^2 x},$$

we get

$$u'(x) = 1 \qquad \text{and} \qquad v(x) = -\cot x.$$

Function v is obtained from the list of indefinite integrals given in Chapter 5.2.1. This leads to

$$\int \frac{x}{\sin^2 x} \, dx = -x \cot x + \int \cot x \, dx.$$

It remains to find the integral on the right-hand side. Using

$$\cot x = \frac{\cos x}{\sin x},$$

we can apply integration by substitution. Setting $g(x) = \sin x$ the integrand takes the form $g'(x)/g(x)$ (see case (3) after Theorem 5.2), and so we obtain

$$\int \frac{x}{\sin^2 x}\, dx = -x \cot x + \ln|\sin x| + C.$$

Example 5.13   We find

$$\int \sin \sqrt{x}\, dx.$$

First, we apply integration by substitution and set $t = \sqrt{x}$. This gives

$$\frac{dt}{dx} = \frac{1}{2\sqrt{x}} = \frac{1}{2t}$$

which can be rewritten as

$$2t\, dt = dx.$$

Replacing $\sqrt{x}$ and dx, we get

$$\int \sin \sqrt{x}\, dx = 2 \int t \sin t\, dt.$$

The substitution has been successfully applied since in the right-hand side integral, only terms depending on variable t and dt occur. This is an integral of type (2) given after Theorem 5.3, and we apply integration by parts. This yields

$$u(t) = t \qquad \text{and} \qquad v'(t) = \sin t$$

and

$$u'(t) = 1 \qquad \text{and} \qquad v(t) = -\cos t.$$

Thus,

$$2\int t \sin t\, dt = 2 \cdot \left(-t \cos t + \int \cos t\, dt\right)$$
$$= 2 \cdot (-t \cos t + \sin t) + C.$$

After substituting back, we have the final result

$$\int \sin \sqrt{x}\, dx = 2 \cdot \left(-\sqrt{x} \cdot \cos \sqrt{x} + \sin \sqrt{x}\right) + C.$$

5.3 THE DEFINITE INTEGRAL

In this section, we start with consideration of the following problem. Given is a function f with $y = f(x) \geq 0$ for all $x \in [a, b] \subseteq D_f$. How can we compute the area A under the graph of function f from a to b assuming that function f is continuous on the closed interval $[a, b]$? To derive a formula for answering this question, we subdivide $[a, b]$ into n subintervals of equal length by choosing points

$$a = x_0 < x_1 < x_2 < \ldots < x_{n-1} < x_n = b.$$

Let l_i (resp. u_i) be the point of the closed interval $[x_{i-1}, x_i]$, where function f takes the minimum (maximum) value, i.e.

$$f(l_i) = \min\{f(x) \mid x \in [x_{i-1}, x_i]\},$$
$$f(v_i) = \max\{f(x) \mid x \in [x_{i-1}, x_i]\},$$

and let $\Delta x_i = x_i - x_{i-1}$. Note that for a continuous function, the existence of the function values $f(l_i)$ and $f(u_i)$ in the interval $[x_{i-1}, x_i]$ follows from Theorem 4.5.

Then we can give a lower bound $A^n_{\min}$ and an upper bound $A^n_{\max}$ in dependence on number n for the area A as follows (see Figure 5.1):

$$A^n_{\min} = \sum_{i=1}^{n} f(l_i) \cdot \Delta x_i,$$

$$A^n_{\max} = \sum_{i=1}^{n} f(u_i) \cdot \Delta x_i.$$

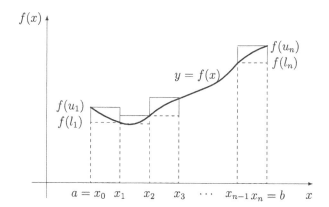

Figure 5.1 The definition of the definite integral.

Since for each $x \in [x_{i-1}, x_i]$, we have $f(l_i) \leq f(x) \leq f(u_i)$, $i \in \{1, 2, \ldots, n\}$, inequalities $A^n_{\min} \leq A \leq A^n_{\max}$ hold. We observe that, if n increases (i.e. the lengths of the intervals become smaller), the lower and upper bounds for the area improve. Therefore, we consider the limits of both bounds as the number of intervals tends to ∞ (or equivalently, the lengths of the intervals tend to zero). If both limits of the sequences $\{A^n_{\min}\}$ and $\{A^n_{\max}\}$ as n tends to

∞ exist and are equal, we say that the definite integral of function f over the interval $[a, b]$ exists. Formally we can summarize this in the following definition.

Definition 5.3 Let function $f : D_f \to \mathbb{R}$ be continuous on the closed interval $[a, b] \subseteq D_f$. If the limits of the sequences $\{A^n_{min}\}$ and $\{A^n_{max}\}$ as n tends to ∞ exist and coincide, i.e.

$$\lim_{n\to\infty} A^n_{min} = \lim_{n\to\infty} A^n_{max} = I,$$

then I is called the *definite (Riemann) integral* of function f over the closed interval $[a, b] \subseteq D_f$.

We write

$$I = \int_a^b f(x) \, dx$$

for the definite integral of function f over the interval $[a, b]$. The numbers a and b, respectively, are denoted as *lower and upper limits of integration*. The Riemann integral is named after the German mathematician Riemann. We can state the following property.

THEOREM 5.4 Let function $f : D_f \to \mathbb{R}$ be continuous on the closed interval $[a, b] \subseteq D_f$. Then the definite integral of function f over the interval $[a, b]$ exists.

We only mention that there exist further classes of functions for which the definite integral can be appropriately defined, e.g. for functions that are bounded on the closed interval $[a, b]$ having at most a finite number of discontinuities in $[a, b]$.

Note that the evaluation of a definite integral according to Definition 5.3 may be rather complicated or even impossible. There is a much easier way by means of an antiderivative of function f which is presented in the following theorem.

THEOREM 5.5 (Newton–Leibniz's formula) Let function $f : D_f \to \mathbb{R}$ be continuous on the closed interval $[a, b] \subseteq D_f$ and function F be an antiderivative of f. Then the definite integral of function f over $[a, b]$ is given by the change in the antiderivative between $x = a$ and $x = b$:

$$\int_a^b f(x)dx = F(x) \Big|_a^b = F(b) - F(a).$$

From Theorem 5.5 we see that also for a definite integral, the main difficulty is to find an antiderivative of the integrand f. Therefore, we again have to apply one of the methods presented in Chapter 5.2 for finding an antiderivative. In the following, we give some properties of the definite integral which can immediately be derived from the definition of the definite integral.

Properties of the definite integral

(1) $\displaystyle\int_a^a f(x)\,dx = 0;$

(2) $\displaystyle\int_a^b f(x)\,dx = -\int_b^a f(x)\,dx;$

(3) $\displaystyle\int_a^b k \cdot f(x)\,dx = k \cdot \int_a^b f(x)\,dx \qquad (k \in \mathbb{R});$

(4) $\displaystyle\int_a^b f(x)\,dx = \int_a^c f(x)\,dx + \int_c^b f(x)\,dx \qquad (a \le c \le b);$

(5) $\displaystyle\left| \int_a^b f(x)\,dx \right| \le \int_a^b |f(x)|\,dx.$

Moreover, the following property holds:

$$\int_a^t f(x)\,dx = F(x)\Big|_a^t = F(t) - F(a) = G(t).$$

Using the latter formulation, we obtain

$$\frac{d}{dt}\int_a^t f(x)\,dx = G'(t) = \frac{d}{dt}[F(t) + C] = f(t).$$

The first property expresses that the definite integral with a variable upper limit can be considered as a function G depending on this limit. The latter property states that the derivative of the definite integral with respect to the upper limit of integration is equal to the integrand as a function evaluated at that limit.

THEOREM 5.6   Let functions $f : D_f \to \mathbb{R}$ and $g : D_g \to \mathbb{R}$ be continuous on the closed interval $[a, b] \subseteq D_f \cap D_g$ with $f(x) \le g(x)$ for $x \in [a, b]$. Then

$$\int_a^b f(x)\,dx \le \int_a^b g(x)\,dx.$$

Theorem 5.6 is also known as the *monotonicity* property of the definite integral. In the derivation of the definite integral we have considered the case when function f is non-negative in the closed interval $[a, b] \subseteq D_f$. We have seen that the definite integral gives the area enclosed by the function f and the x axis between $x = a$ and $x = b$. In the case when function f is negative in the closed interval $[a, b]$, it follows from Theorem 5.6 that the definite integral has a negative value (using function g identically equal to zero, the result follows from the monotonicity property). In this case, the definite integral gives the area enclosed by the function f and the x axis between $x = a$ and $x = b$ with negative sign. As a consequence, if function f has zeroes in the interval $[a, b]$, one has to split this interval into subintervals in order to get the area enclosed by the function and the x axis between $x = a$ and $x = b$.

We continue with some examples for evaluating definite integrals. The first example illustrates the above comment.

Example 5.14   We wish to determine the area A enclosed by function f with $f(x) = \cos x$ with the x axis between $a = 0$ and $b = \pi$. Function f is non-negative in the closed interval $[0, \pi/2]$ and non-positive in the closed interval $[\pi/2, \pi]$. Therefore, by using property (4) and the above comment, the area A is obtained as follows:

$$A = \int_0^{\frac{\pi}{2}} \cos x \; dx - \int_{\frac{\pi}{2}}^{\pi} \cos x \; dx$$

$$= \sin x \Big|_0^{\frac{\pi}{2}} - \sin x \Big|_{\frac{\pi}{2}}^{\pi}$$

$$= (1 - 0) - (0 - 1) = 2.$$

It is worth noting that we would get value zero when evaluating the definite integral of the cosine function between $a = 0$ and $b = \pi$ (since the area between the cosine function and the x axis in the first subinterval is equal to the area between the x axis and the cosine function in the second subinterval).

Example 5.15   We evaluate the definite integral

$$\int_1^2 \frac{dx}{x(1 + \ln x)} \; .$$

Applying integration by substitution, we set

$$t = 1 + \ln x$$

and obtain by differentiation

$$\frac{dt}{dx} = \frac{1}{x} \; .$$

Inserting both terms into the integral, we get

$$\int_1^2 \frac{dx}{x(1 + \ln x)} = \int_{t(1)}^{t(2)} \frac{dt}{t} \; .$$

Thus we obtain

$$\int_{t(1)}^{t(2)} \frac{dt}{t} = \ln|t| \Big|_{t(1)}^{t(2)} = \ln|1 + \ln x| \Big|_1^2 = \ln(1 + \ln 2) - \ln(1 + \ln 1) = \ln(1 + \ln 2).$$

In the above computations, we did not transform the limits of the definite integral into the corresponding t values but we have used the above substitution again after having found an antiderivative. Of course, we can also transform the limits of integration (in this case, we get $t(1) = 1 + \ln 1 = 1$ and $t(2) = 1 + \ln 2$) and insert the obtained values directly into the obtained antiderivative $\ln|t|$.

Example 5.16 The marginal cost of a firm manufacturing a single product is given by

$$C'(x) = 6 - \frac{60}{x+1}, \quad 0 \le x \le 1000,$$

where x is the quantity produced, and the marginal cost is given in EUR. If the quantity produced changes from 300 to 400, the change in cost is obtained by Newton–Leibniz's formula as follows:

$$C(400) - C(300) = \int_{300}^{400} C'(x) \, dx$$

$$= \int_{300}^{400} \left(6 - \frac{60}{x+1} \right) dx = (6x - 60 \ln |x+1|) \Big|_{300}^{400}$$

$$\approx (2,400 - 359.64) - (1,800 - 342.43) = 582.79 \text{ EUR}.$$

Thus, the cost increases by 582.79 EUR when production is increased from 300 units to 400 units.

Example 5.17 We want to compute the area enclosed by the graphs of the two functions $f_1 : \mathbb{R} \to \mathbb{R}$ and $f_2 : \mathbb{R} \to \mathbb{R}$ given by

$$f_1(x) = x^2 - 4 \quad \text{and} \quad f_2(x) = 2x - x^2.$$

We first determine the points of intercept of both functions and obtain from

$$x^2 - 4 = 2x - x^2$$

the quadratic equation

$$x^2 - x - 2 = 0$$

which has the two real solutions

$$x_1 = -1 \quad \text{and} \quad x_2 = 2.$$

The graphs of both functions are parabolas which intersect only in these two points $(-1, -3)$ and $(2, 0)$. To compute the enclosed area A, we therefore have to evaluate the definite integral

$$A = \int_{-1}^{2} [f_2(x) - f_1(x)] \, dx$$

which yields

$$A = \int_{-1}^{2} \left[(2x - x^2) - (x^2 - 4) \right] dx = \int_{-1}^{2} (-2x^2 + 2x + 4) \, dx$$

$$= 2 \int_{-1}^{2} (-x^2 + x + 2) \, dx = 2 \left(-\frac{x^3}{3} + \frac{x^2}{2} + 2x \right) \Big|_{-1}^{2}$$

$$= 2 \left[\left(-\frac{8}{3} + 2 + 4 \right) - \left(\frac{1}{3} + \frac{1}{2} - 2 \right) \right] = 9.$$

Thus, the area enclosed by the graphs of the given functions is equal to nine squared units.

THEOREM 5.7 (mean-value theorem for integrals)   Let function $f : D_f \to \mathbb{R}$ be continuous on the closed interval $[a, b] \subseteq D_f$. Then there exists a real number $x^* \in [a, b]$ such that

$$M = f(x^*) = \frac{1}{b - a} \int_{a}^{b} f(x) \, dx.$$

This theorem is graphically illustrated for the case $f(x) \geq 0$ in Figure 5.2. That is, there is at least one value $x^* \in [a, b]$ such that the dashed area

$$I = \int_{a}^{b} f(x) \, dx$$

is equal to the area of the rectangle with the lengths $b - a$ and function value $f(x^*)$ (where value x^* must be suitably determined).

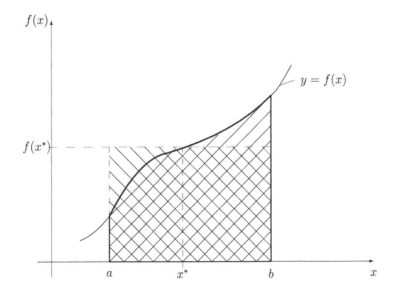

Figure 5.2 The mean-value theorem for integrals.

5.4   APPROXIMATION OF DEFINITE INTEGRALS

There are several reasons why it may not be possible to evaluate a definite integral. For some functions, there does not exist an antiderivative that can be determined analytically. As an example, we can mention here function f with $f(x) = e^{-x^2}$, which is often applied in probability theory and statistics, or function g with $g(x) = (\sin x)/x$. Sometimes it may be too time-consuming to determine an antiderivative, or function f may be given only as a set of points (x, y) experimentally determined.

In such cases, we want to determine the definite integral approximately by applying numerical methods. In the following, we present a few methods and give some comments on the precision of the approximate value obtained. Approximate methods divide the closed interval $[a, b]$ into n subintervals of equal length $h = (b - a)/n$, and so we get the subintervals $[x_0, x_1], [x_1, x_2], \ldots, [x_{n-1}, x_n]$, where $a = x_0$ and $b = x_n$. Within each interval, we now replace function f by some other function which is 'close to the original one' and for which the integration can easily be performed. In all the methods below, we replace function f by a polynomial of small degree.

Approximation by rectangles

In this case, we approximate function f by a step function (in each interval a constant function is used), i.e. we approximate the definite integral by the sum of the areas of rectangles. We obtain

$$\int_a^b f(x)\, dx \approx \frac{b - a}{n} \cdot [f(x_0) + f(x_1) + \cdots + f(x_{n-1})] = I_R.$$

The above approximation is illustrated in Figure 5.3. This formula gives a value between the lower and upper Riemann sum for a specific value of n. When applying approximation by rectangles, the error ΔI_R of the approximate value of the definite integral can be estimated as follows:

$$\Delta I_R = \left| \int_a^b f(x)\, dx - I_R \right| \le c \cdot \frac{(b - a)^2}{n} \cdot \left| \max_{a \le x \le b} f'(x) \right|,$$

where $c \in [0, 1)$.

Approximation by trapeziums

An alternative approximation is obtained when in each closed interval $[x_{i-1}, x_i]$ function f is replaced by a line segment (i.e. a linear function) through the points $(x_{i-1}, f(x_{i-1}))$ and $(x_i, f(x_i))$. This situation is illustrated in Figure 5.4. In this way, we get the following approximation formula for the definite integral:

$$\int_a^b f(x)\, dx \approx \frac{b - a}{n} \cdot \left[\frac{f(a) + f(b)}{2} + f(x_1) + f(x_2) + \cdots + f(x_{n-1}) \right] = I_{TR}.$$

By the above formula, we approximate the definite integral by the sum I_{TR} of the areas of n trapeziums. Assuming an equal number n of subintervals, often the approximation by

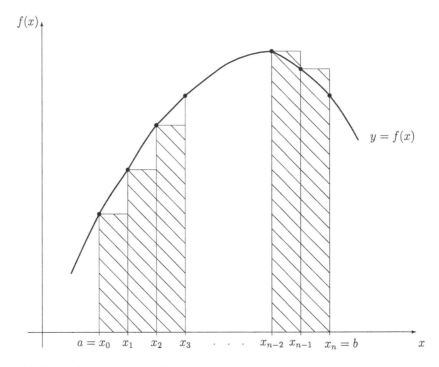

Figure 5.3 Approximation by rectangles.

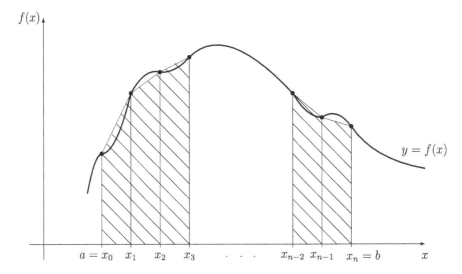

Figure 5.4 Approximation by trapeziums.

trapeziums gives better approximate values than the approximation by rectangles. This is particularly true when the absolute value of the first derivative in the closed interval $[a, b]$ can become large and thus we may have big differences in the function values in some of the subintervals.

When determining the definite integral by the above approximation by trapeziums, we get the following estimation for the maximum error ΔI_{TR} of the approximate value I_{TR}:

$$\Delta I_{TR} = \left| \int_a^b f(x)dx - I_{TR} \right| \leq \frac{(b-a)^3}{12n^2} \cdot \left| \max_{a \leq x \leq b} f''(x) \right|.$$

From the above formula we see that the maximum error can become large when the absolute value of the second derivative of function f is large for some value in the closed interval $[a, b]$. Moreover, the smaller the interval length h is, the better the approximation.

Kepler's formula

Here we consider the special case when the closed interval $[a, b]$ is divided only into two subintervals $[a, (a + b)/2]$ and $[(a + b)/2, b]$ of equal length. But now we approximate function f by a quadratic function (parabola) which is uniquely defined by three points, and here we use the points $(a, f(a))$, $((a + b)/2, f((a + b)/2))$ and $(b, f(b))$ (see Figure 5.5).

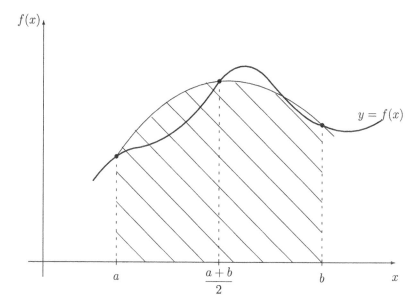

Figure 5.5 Kepler's formula.

This leads to Kepler's formula:

$$\int_a^b f(x)\, dx \approx \frac{b-a}{6} \cdot \left[f(a) + 4f\left(\frac{a+b}{2}\right) + f(b) \right] = I_K.$$

The error ΔI_K of the approximate value I_K of the definite integral can be estimated as follows:

$$\Delta I_K = \left| \int_a^b f(x)dx - I_K \right| \leq \frac{(b-a)^5}{2,880} \cdot \left| \max_{a \leq x \leq b} f^{(4)}(x) \right|.$$

The formula shows that a large absolute value of the fourth derivative for some value $x \in [a, b]$ can lead to a large error of the approximate value. If function f is a polynomial of degree no greater than three, then the value I_K is even the exact value of the definite integral since for each values $x \in [a, b]$, the fourth derivative is equal to zero in this case.

Simpson's formula

This formula is a generalization of Kepler's formula. The only difference is that we do not divide the closed interval $[a, b]$ into only two subintervals but into a larger even number $n = 2m$ of subintervals. Now we apply Kepler's formula to any two successive intervals. This leads to Simpson's formula:

$$\int_a^b f(x)\, dx \approx \frac{b-a}{6m} \cdot [f(a) + 4(f(x_1) + f(x_3) + \cdots + f(x_{2m-1}))$$

$$+ 2(f(x_2) + f(x_4) + \cdots + f(x_{2m-2})) + f(b)] = I_S.$$

We can give the following estimate for the maximum error ΔI_S of the approximate value I_S:

$$\Delta I_S = \left| \int_a^b f(x)dx - I_S \right| \leq \frac{(b-a)^5}{180} \cdot n \cdot \left| \max_{a \leq x \leq b} f^{(4)}(x) \right|.$$

Example 5.18　We illustrate approximation by rectangles, trapeziums and by Simpson's formula by looking at the example

$$\int_2^6 f(x)dx = \int_2^6 \frac{e^x}{x}\, dx,$$

where we use $n = 8$ subintervals for each method. We get the function values given in Table 5.1.

Table 5.1 Function values for Example 5.18

i	x_i	$f(x_i)$
0	2	3.6945
1	2.5	4.8730
2	3	6.6952
3	3.5	9.4616
4	4	13.6495
5	4.5	20.0038
6	5	29.6826
7	5.5	44.4894
8	6	67.2381

If we use approximation by rectangles, we get

$$\int_2^6 f(x)dx \approx \frac{b-a}{n} \cdot [f(x_0) + f(x_1) + \cdots + f(x_7)]$$

$$\approx \frac{4}{8} \cdot 132.5497 = 66.2749.$$

Applying approximation by trapeziums, we get

$$\int_2^6 f(x)\, dx \approx \frac{b-a}{n} \cdot \left[\frac{f(x_0) + f(x_8)}{2} + f(x_1) + f(x_2) + \cdots + f(x_7) \right]$$

$$\approx \frac{4}{8} \cdot 164.3215 = 82.1608.$$

If we apply Simpson's formula, we get

$$\int_2^6 \frac{e^x}{x}\, dx \approx \frac{b-a}{6m} \cdot [f(x_0) + 4f(x_1) + 2f(x_2) + 4f(x_3) + 2f(x_4) + 4f(x_5) + 2f(x_6)$$

$$+ 4f(x_7) + f(x_8)] \approx \frac{4}{24} \cdot 486.2984 \approx 81.0497.$$

5.5 IMPROPER INTEGRALS

So far, we have made the following two basic assumptions in defining the definite integral of a function f over the interval $[a, b]$.

(1) The limits of integration a and b are both *finite*.
(2) The function is continuous and therefore *bounded* on the closed interval $[a, b]$.

If either of the conditions (1) and (2) is not satisfied, the definite integral is called an *improper integral*. We now consider both cases separately.

5.5.1 Infinite limits of integration

In this case, we define the improper integral appropriately as the limit of certain definite integrals with finite limits of integration.

Definition 5.4   Let one or both limits of integration be infinite. Then we define the *improper integrals* as follows:

$$\int_a^\infty f(x)\,dx = \lim_{b\to\infty} \int_a^b f(x)\,dx$$

$$\int_{-\infty}^b f(x)\,dx = \lim_{a\to-\infty} \int_a^b f(x)\,dx$$

$$\int_{-\infty}^\infty f(x)\,dx = \lim_{\substack{b\to\infty\\a\to-\infty}} \int_a^b f(x)\,dx$$

provided that these limits exist. If a limit does not exist, the corresponding improper integral has no value and is said to be divergent.

By using Newton–Leibniz's formula for definite integrals, we can evaluate the improper integrals given above as follows:

Evaluation of improper integrals:

$$\int_a^\infty f(x)\,dx = \lim_{b\to\infty} F(x)\Big|_a^b = \lim_{b\to\infty} F(b) - F(a)$$

$$\int_{-\infty}^b f(x)\,dx = \lim_{a\to-\infty} F(x)\Big|_a^b = F(b) - \lim_{a\to-\infty} F(a)$$

$$\int_{-\infty}^\infty f(x)\,dx = \lim_{\substack{b\to\infty\\a\to-\infty}} F(x)\Big|_a^b = \lim_{b\to\infty} F(b) - \lim_{a\to-\infty} F(a)$$

Example 5.19   We evaluate the improper integral

$$\int_0^\infty \frac{1}{3} e^{-2x}\,dx.$$

We obtain

$$\int_0^\infty \frac{1}{3} e^{-2x}\,dx = \frac{1}{3} \cdot \lim_{b\to\infty} \int_0^b e^{-2x}\,dx = \frac{1}{3} \cdot \left[\frac{1}{2} \cdot \lim_{b\to\infty} (-e^{-2x}) \right] \Big|_0^b$$

$$= \frac{1}{6} \cdot \left[\lim_{b\to\infty} (-e^{-2b}) - (-e^0) \right] = \frac{1}{6} \cdot (0 + 1) = \frac{1}{6}.$$

5.5.2   Unbounded integrands

Similarly to case (1), we define the improper integral again as a limit of certain definite integrals.

Definition 5.5 Let function $f : D_f \to \mathbb{R}$ be continuous on the right-open interval $[a, b) \subseteq D_f$ but definitely divergent as x tends to $b - 0$. Then we define the *improper integral* $\int_a^b f(x)\, dx$ as

$$\int_a^b f(x)\, dx = \lim_{t \to b-0} \int_a^t f(x)\, dx$$

provided that this limit exists. Otherwise the improper integral has no value and is said to be divergent.

Let function f be continuous on the left-open interval $(a, b] \subseteq D_f$ but definitely divergent as x tends to $a + 0$. Then we define the *improper integral* $\int_a^b f(x)\, dx$ as

$$\int_a^b f(x)\, dx = \lim_{t \to a+0} \int_t^b f(x)\, dx$$

provided that this limit exists. Otherwise the improper integral has no value and is said to be divergent.

Example 5.20 We evaluate the improper integral

$$\int_0^1 \frac{dx}{\sqrt[3]{x}}.$$

Since function $1/\sqrt[3]{x}$ is definitely divergent as x tends to $0 + 0$, we apply Definition 5.5 and obtain

$$\int_0^1 \frac{dx}{\sqrt[3]{x}} = \lim_{t \to 0+0} \int_t^1 x^{-1/3}\, dx = \lim_{t \to 0+0} \frac{x^{2/3}}{\frac{2}{3}} \Big|_t^1 = \lim_{t \to 0+0} \left(\frac{3}{2} \cdot x^{2/3}\right) \Big|_t^1$$

$$= \left(\frac{3}{2} - \frac{3}{2} \lim_{t \to 0+0} \sqrt[3]{t^2}\right) = \frac{3}{2}.$$

Thus the above improper integral has the value $3/2$.

Finally, we consider an example where a point x_0 exists in the interior of the closed interval $[a, b]$, where function f is definitely divergent as x tends to x_0. This case can be reduced to the consideration of two integrals according to Definition 5.5.

Example 5.21 Consider the improper integral

$$\int_{-1}^1 \frac{dx}{x^2}.$$

This integrand has a discontinuity (pole of second order) at $x = 0$. Therefore we must partition the integral into two integrals, each of them with an unbounded integrand according to one of the cases presented in Definition 5.5, and we obtain

$$
\int_{-1}^{1} \frac{dx}{x^2} = \lim_{t_1 \to 0-0} \int_{-1}^{t_1} \frac{dx}{x^2} + \lim_{t_2 \to 0+0} \int_{t_2}^{1} \frac{dx}{x^2}
$$

$$
= \lim_{t_1 \to 0-0} \left(-\frac{1}{x} \right) \Big|_{-1}^{t_1} + \lim_{t_2 \to 0+0} \left(-\frac{1}{x} \right) \Big|_{t_2}^{1}
$$

$$
= \lim_{t_1 \to 0-0} \left(-\frac{1}{t_1} + 1 \right) + \lim_{t_2 \to 0+0} \left(-1 + \frac{1}{t_2} \right) = \infty.
$$

Thus, this improper integral has no finite value. Notice that, if we did not take into consideration that the integrand is definitely divergent as x tends to zero and applied Newton–Leibniz's formula for evaluating a definite integral, we would obtain the value -2 for the above integral (which is obviously false since the integrand is a non-negative function in the interval so that the area cannot be negative).

5.6 SOME APPLICATIONS OF INTEGRATION

In this section, we discuss some economic applications of integration.

5.6.1 Present value of a continuous future income flow

Assume that in some time interval $[0, T]$ an income is received continuously at a rate of $f(t)$ EUR per year. Contrary to the problems considered in Chapter 2, we now assume that interest is compounded *continuously* at a rate of interest i. Moreover, we denote by $P(t)$ the *present value* of all payments made over the time interval $[0, t]$. In other words, the value $P(t)$ gives the amount of money one would have to deposit at time zero in order to have at time T the amount which would result from depositing continuously the income flow $f(t)$ over the time interval $[0, T]$.

In Chapter 2, we discussed the case of a discrete compounding of some amount. Let A be the amount due for payment after t years with a rate of interest i per year. Then the present value of this amount A is equal to

$$
P = A \cdot (1 + i)^{-t}
$$

(see Chapter 2). In the case of m payment periods per year, the present value would be

$$
P = A \cdot \left(1 + \frac{i}{m} \right)^{-mt}.
$$

If we assume continuous compounding, we have to consider the question of what happens if $m \to \infty$. We set $n = m/i$. Notice that from $m \to \infty$ it follows that $n \to \infty$ as well. Now the present value P of the amount A is given by

$$
P = A \left[\left(1 + \frac{1}{n} \right)^{n} \right]^{-it}.
$$

Using the limit

$$\lim_{n\to\infty} \left(1 + \frac{1}{n}\right)^n = e,$$

we find that, in the case of a continuous compounding, the present value P of amount A is given by

$$P = A \cdot e^{-it}.$$

Now returning to our original problem, we can say that $f(t) \cdot \Delta t$ is approximately equal to the income received in the closed interval $[t, t + \Delta t]$. The present value of this amount at time zero is therefore equal to $f(t) \cdot \Delta t \cdot e^{-it}$. Taking into account that this present value is equal to the difference $P(t + \Delta t) - P(t)$, we obtain

$$\frac{P(t + \Delta t) - P(t)}{\Delta t} \approx f(t) \cdot e^{-it}.$$

The left-hand side is the difference quotient of function P (considering points t and $t + \Delta t$). Taking the limit of this difference quotient as Δt tends to zero, we obtain

$$P'(t) = f(t) \cdot e^{-it}.$$

Evaluating now the definite integral of this integrand from 0 to T, we get

$$P(T) - P(0) = \int_0^T P'(t) \, dt = \int_0^T f(t) \cdot e^{-it} \, dt.$$

Since $P(0) = 0$ by definition, we finally obtain

$$P(T) = \int_0^T f(t) \cdot e^{-it} \, dt.$$

Example 5.22   Let function $f : D_f \to \mathbb{R}$ with $f(t) = 30t + 100$ (in EUR) describe the annual rate of an income flow at time t continuously received over the years from time $t = 0$ to time $t = 5$. Moreover, assume that the rate of interest is $i = 0.06$ (or, which is the same, $p = i \cdot 100$ per cent $= 6$ per cent) compounded continuously. Applying the formula derived above, the present value at time zero is obtained by

$$P(5) = \int_0^5 (30t + 100) \cdot e^{-0.06t} \, dt.$$

Applying integration by parts with

$$u(t) = 30t + 100 \qquad \text{and} \qquad v'(t) = e^{-0.06t},$$

we obtain

$$u'(t) = 30 \qquad \text{and} \qquad v(t) = -\frac{1}{0.06} \cdot e^{-0.06t},$$

and thus

$$P(5) = -\frac{1}{0.06} \cdot (30t + 100) \cdot e^{-0.06t}\Big|_0^5 + \frac{30}{0.06} \int_0^5 e^{-0.06t}\, dt$$

$$= -\frac{1}{0.06} \cdot (30t + 100) \cdot e^{-0.06t}\Big|_0^5 - \frac{30}{(0.06)^2} \cdot e^{-0.06t}\Big|_0^5$$

$$= -\frac{1}{0.06} \cdot e^{-0.06t} \cdot \left(30t + 100 + \frac{30}{0.06}\right)\Big|_0^5$$

$$= -\frac{1}{0.06} \cdot e^{-0.06t} \cdot (30t + 600)\Big|_0^5$$

$$= -\frac{1}{0.06} \cdot e^{-0.3} \cdot (150 + 600) + \left(\frac{1}{0.06} \cdot e^0 \cdot 600\right)$$

$$= -\frac{750}{0.06} \cdot e^{-0.3} + \frac{600}{0.06} = 739.77 \text{ EUR},$$

i.e. the present value at time zero is equal to 739.77 EUR.

5.6.2   Lorenz curves

These are curves that characterize the income distribution among the population. The Lorenz curve L is defined for $0 \le x \le 1$, where x represents the percentage of the population (as a decimal), and $L(x)$ gives the income share of $100x$ per cent of the poorest part of the population. Therefore, for each Lorenz curve, we have $L(0) = 0$, $L(1) = 1$, $0 \le L(x) \le x$, and L is an increasing and convex function on the closed interval $[0, 1]$.

Then the Gini coefficient G is defined as follows:

$$G = 2 \int_0^1 [x - L(x)]\, dx.$$

Factor two is used as a scaling factor to guarantee that the Gini coefficient is always between zero and one. The smaller the Gini coefficient is, the fairer is the income distributed among the population. The Lorenz curve and the Gini coefficient are illustrated in Figure 5.6. They can also be used e.g. for measuring the concentration in certain industrial sectors.

Example 5.23   Assume that for Nowhereland the Lorenz curve describing the income distribution among the population is given by

$$L(x) = \frac{3}{5}x^2 + \frac{2}{5}x, \quad D_L = [0, 1].$$

First we note that this is a Lorenz curve since $L(0) = 0, L(1) = 1$ and the difference

$$x - L(x) = x - \left(\frac{3}{5}x^2 + \frac{2}{5}x\right) = \frac{3}{5}x - \frac{3}{5}x^2 = \frac{3}{5}x(1 - x)$$

is non-negative for $0 \leq x \leq 1$ since all factors are non-negative in the latter product representation. (Notice also that function L is strictly increasing and strictly convex on the closed interval $[0, 1]$.) We obtain

$$G = 2 \int_0^1 [x - L(x)] \, dx = 2 \int_0^1 \left(\frac{3}{5}x - \frac{3}{5}x^2 \right) dx$$

$$= 2 \left(\frac{3}{10}x^2 - \frac{3}{15}x^3 \right) \Big|_0^1 = \frac{1}{5}.$$

Since the Gini coefficient is rather small, the income is 'rather equally' distributed among the population (e.g. $L(0.25) = 0.1375$ means that 25 per cent of the poorest population still have an income share of 13.75 per cent and $L(0.5) = 0.35$ means that 50 per cent of the poorest population still have an income share of 35 per cent).

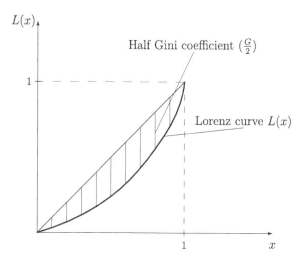

Figure 5.6 Lorenz curve and Gini coefficient.

5.6.3 Consumer and producer surplus

Assume that for a certain product there is a demand function D and a supply function S, both depending on the price p of the product. The demand function is decreasing while the supply function is increasing. For some price p^*, there is an equilibrium called the *market price*, i.e. demand is equal to supply: $D(p^*) = S(p^*)$. Some consumers are willing to pay a higher price than p^* until a certain maximum price p_{max} is reached. On the other hand, some producers could be willing to sell the product at a lower price than p^*, which means that the supply $S(p)$ increases from the minimum price p_{min}. Assuming that the price p can be considered as a continuous variable, the customer surplus CS is obtained as

$$CS = \int_{p^*}^{p_{max}} D(p) \, dp,$$

while the producer surplus *PS* is obtained as

$$PS = \int_{p_{min}}^{p^*} S(p) \, dp.$$

The consumer surplus *CS* can be interpreted as the total sum that customers save when buying the product at the market price instead of the price they would be willing to pay. Analogously, the producer surplus is the total sum that producers earn when all products are sold at the market price instead of the price they would be willing to accept. The producer surplus *PS* and the consumer surplus *CS* are illustrated in Figure 5.7.

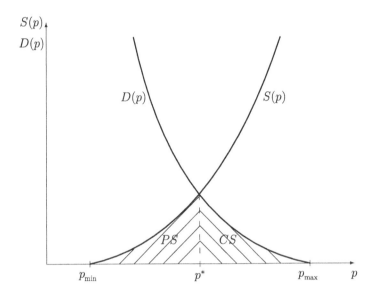

Figure 5.7 Producer and consumer surplus.

Example 5.24   Assume that the demand function *D* is given by

$$D(p) = 5p^2 - 190p + 1,805$$

and the supply function *S* is given by

$$S(p) = 20p^2 - 160p + 320.$$

First, we mention that we get the minimum price $p_{min} = 4$ from $S(p) = 0$ and the maximum price $p_{max} = 19$ from $D(p) = 0$. Moreover, we note that the demand function is in fact strictly decreasing due to $D'(p) = 10p - 190 < 0$ for $p < 19$, and the supply function is strictly increasing since $S'(p) = 40p - 160 > 0$ for $p > 4$. The market price p^* is obtained from $D(p^*) = S(p^*)$:

$$5(p^*)^2 - 190p^* + 1,805 = 20(p^*)^2 - 160p^* + 320$$

which yields

$$15(p^*)^2 + 30p^* - 1,485 = 0,$$

or, after dividing by 15, equivalently,

$$(p^*)^2 + 2p^* - 99 = 0.$$

From the latter quadratic equation, we get the zeroes

$$p_1^* = -1 + \sqrt{1 + 99} = 9 \qquad \text{and} \qquad p_2^* = -1 - \sqrt{1 + 99} = -11.$$

Since the second root is negative, the market price is $p^* = p_1^* = 9$. Hence, the consumer surplus is obtained as

$$CS = \int_9^{19} D(p)\,dp = \int_9^{19} \left(5p^2 - 190p + 1,805\right) dp$$

$$= \left(5 \cdot \frac{p^3}{3} - 190 \cdot \frac{p^2}{2} + 1,805p\right) \Big|_9^{19} = \left(\frac{5}{3} \cdot p^3 - 95p^2 + 1,805p\right) \Big|_9^{19}$$

$$= \left(\frac{34,295}{3} - 34,295 + 34,295\right) - (1,215 - 7,695 + 16,245) = 1,666.67.$$

The producer surplus is obtained as

$$PS = \int_4^9 S(p)\,dp = \int_4^9 (20p^2 - 160p + 320)\,dp$$

$$= \left(20 \cdot \frac{p^3}{3} - 160 \cdot \frac{p^2}{2} + 320p\right) \Big|_4^9 = \left(\frac{20}{3} \cdot p^3 - 80p^2 + 320p\right) \Big|_4^9$$

$$= 4,860 - 6,480 + 2,880 - \left(\frac{1,280}{3} - 1,280 + 1,280\right) = 833.33.$$

EXERCISES

5.1   Use the substitution rule to find the following indefinite integrals:

(a) $\displaystyle\int e^{\sin x} \cos x\,dx;$ (b) $\displaystyle\int \frac{\ln x}{x}\,dx;$ (c) $\displaystyle\int \frac{5}{1 - 4x}\,dx;$

(d) $\displaystyle\int \frac{dx}{e^{3-2x}};$ (e) $\displaystyle\int \frac{x\,dx}{\sqrt{x^2 + 1}};$ (f) $\displaystyle\int \frac{x^3\,dx}{\sqrt{1 + x^2}};$

(g) $\displaystyle\int \frac{e^{2x} - 2e^x}{e^{2x} + 1}\,dx;$ (h) $\displaystyle\int \frac{dx}{\sqrt{2 - 9x^2}};$ (i) $\displaystyle\int \frac{\cos^3 x}{\sin^2 x}\,dx;$

(j) $\displaystyle\int \frac{dx}{1 - \cos x};$ (k) $\displaystyle\int \frac{dx}{2\sin x + \sin 2x}.$

5.2   Use integration by parts to find the following indefinite integrals:

(a) $\int x^2 e^x \, dx;$ (b) $\int e^x \cos x \, dx;$ (c) $\int \dfrac{x}{\cos^2 x} \, dx;$

(d) $\int \cos^2 x \, dx;$ (e) $\int x^2 \ln x \, dx;$ (f) $\int \ln(x^2 + 1) \, dx.$

5.3   Evaluate the following definite integrals:

(a) $\int_{-1}^{2} x^2 \, dx;$ (b) $\int_{0}^{4} \dfrac{dx}{1 + \sqrt{x}};$ (c) $\int_{0}^{\frac{\pi}{2}} \sin^3 x \, dx;$

(d) $\int_{0}^{4} \dfrac{x \, dx}{\sqrt{1 + 2x}};$ (e) $\int_{0}^{t} \dfrac{dx}{2x - 1}, \quad t < \dfrac{1}{2};$ (f) $\int_{0}^{\frac{\pi}{2}} \sin x \cos^2 x \, dx;$

(g) $\int_{-1}^{0} \dfrac{dx}{x^2 + 2x + 2}.$

5.4   A firm intends to pre-calculate the development of cost, sales and profit of a new product for the first five years after launching it. The calculations are based on the following assumptions:

- t denotes the time (in years) from the introduction of the product beginning with $t = 0$;
- $C(t) = 1,000 \cdot \left[4 - (2e^t)/(e^t + 1)\right]$ is the cost as a function of $t \in \mathbb{R}, t \geq 0$;
- $S(t) = 10,000 \cdot t^2 \cdot e^{-t}$ are the sales as a function of $t \in \mathbb{R}, t \geq 0$.

(a) Calculate total cost, total sales and total profit for a period of four years.
(b) Find average sales per year and average cost per year for this period.
(c) Find the total profit as a function of the time t.

5.5   (a) Find

$$\int_{0}^{2\pi} \sin x \, dx$$

and compute the area enclosed by function $f : \mathbb{R} \to \mathbb{R}$ with $f(x) = \sin x$ and the x axis.

(b) Compute the area enclosed by the two functions $f_1 : \mathbb{R} \to \mathbb{R}$ and $f_2 : \mathbb{R} \to \mathbb{R}$ given by

$$f_1(x) = x^3 - 4x \quad \text{and} \quad f_2(x) = 3x + 6.$$

5.6   The throughput $q = q(t)$ (output per time unit) of a continuously working production plant is given by a function depending on time t:

$$q(t) = q_0 \cdot \left[1 - \left(\dfrac{t}{10}\right)^2\right].$$

The throughput decreases for $t = 0$ up to $t = 10$ from q_0 to 0. One overhaul during the time interval $[0, T]$ with $T < 10$ means that the throughput goes up to q_0. After that it decreases as before.

(a) Graph the function q with regard to the overhaul.

(b) Let $t_0 = 4$ be the time of overhaul. Find the total output for the time interval $[0, T]$ with $T > 4$.

(c) Determine the time t_0 of overhaul which maximizes the total output in the interval $[0, T]$.

5.7 Determine the following definite integral numerically:

$$\int_0^1 \frac{dx}{1 + x^2}.$$

(a) Use approximation by trapeziums with $n = 10$.

(b) Use Kepler's formula.

(c) Use Simpson's formula with $n = 10$.

Compare the results of (a), (b) and (c) with the exact value.

5.8 Evaluate the following improper integrals:

(a) $\int_{-\infty}^0 e^x \, dx;$ (b) $\int_1^\infty \frac{dx}{x^2 + 2x + 1};$ (c) $\int_0^\infty \lambda e^{-\lambda x} \, dx;$

(d) $\int_0^\infty \lambda x^2 e^{-\lambda x} \, dx;$ (e) $\int_0^4 \frac{dx}{\sqrt{x}};$ (f) $\int_0^6 \frac{2x - 1}{(x + 1)(x - 2)} \, dx.$

5.9 Let function f with

$$f(t) = 20t + 200$$

(in EUR) describe the annual rate of an income flow at time t continuously received over the years from time $t = 0$ to time $t = 6$. Interest is compounded continuously at a rate of 4 per cent p.a. Evaluate the present value at time zero.

5.10 Given are a demand function D and a supply function S depending on price p as follows:

$$D(p) = 12 - 2p \qquad \text{and} \qquad S(p) = \frac{8}{7}p - \frac{4}{7}.$$

Find the equilibrium price p^* and evaluate customer surplus CS and producer surplus PS. Illustrate the result graphically.

6 Vectors

In economics, an ordered n-tuple often describes a bundle of commodities such that the ith value represents the quantity of the ith commodity. This leads to the concept of a vector which we will introduce next.

6.1 PRELIMINARIES

Definition 6.1 A *vector* **a** is an ordered n-tuple of real numbers $a_1, a_2, \ldots, a_n$. The numbers $a_1, a_2, \ldots, a_n$ are called the *components* (or *coordinates*) of vector **a**.

We write

$$\mathbf{a} = \begin{pmatrix} a_1 \\ a_2 \\ \vdots \\ a_n \end{pmatrix} = (a_1, a_2, \ldots, a_n)^{\mathrm{T}}$$

for a *column* vector. For the so-called *transposed* vector $\mathbf{a}^{\mathrm{T}}$ which is a *row* vector obtained by taking the column as a row, we write:

$$\mathbf{a}^{\mathrm{T}} = (a_1, a_2, \ldots, a_n) = \begin{pmatrix} a_1 \\ a_2 \\ \vdots \\ a_n \end{pmatrix}^{\mathrm{T}}.$$

We use letters in bold face to denote vectors. We have defined above a vector **a** always as a column vector, and if we write this vector as a row vector, this is indicated by an upper-case T which stands for 'transpose'. The convenience of using the superscript T will become clear when matrices are discussed in the next chapter. If vector **a** has n components, we say that vector **a** has *dimension n* or that **a** is an n-dimensional vector or n-vector. If not noted differently, we index components of a vector by subscripts while different vectors are indexed by superscripts.

Definition 6.2 The *n-dimensional (Euclidean) space* $\mathbb{R}^n$ is defined as the set of all real *n*-tuples:

$$\mathbb{R}^n = \left\{ \left. \begin{pmatrix} a_1 \\ a_2 \\ \vdots \\ a_n \end{pmatrix} \right| a_i \in \mathbb{R}, \ i = 1, 2, \ldots, n \right\}$$

Similarly, $\mathbb{R}^n_+$ stands for the set of all non-negative real *n*-tuples. We can graph a vector as an arrow in the *n*-dimensional space which can be interpreted as a *displacement* of a starting point P resulting in a terminal point Q. For instance, if point P has the coordinates $(p_1, p_2, \ldots, p_q)$ and point Q has the coordinates $(q_1, q_2, \ldots, q_n)$, then

$$\mathbf{a} = \overline{PQ} = \begin{pmatrix} q_1 - p_1 \\ q_2 - p_2 \\ \cdots \\ q_n - p_n \end{pmatrix}.$$

It is often assumed that the starting point P is the origin of the coordinate system. In this case, the components of vector $\mathbf{a}$ are simply the coordinates of point Q and, therefore, a row vector $\mathbf{a}^\mathrm{T}$ can be interpreted as a *point* (i.e. a *location*) in the *n*-dimensional Euclidean space. In the case of $n = 2$, we can illustrate vectors in the plane, e.g. the vectors

$$\mathbf{a} = \begin{pmatrix} 4 \\ 3 \end{pmatrix} \quad \text{and} \quad \mathbf{b} = \begin{pmatrix} -1 \\ 2 \end{pmatrix}$$

are illustrated in Figure 6.1. Finding the terminal point of vector $\mathbf{a}$ means that we are going four units to the right and three units up from the origin. Similarly, to find the terminal point of $\mathbf{b}$, we are going one unit to the left and two units up from the origin.

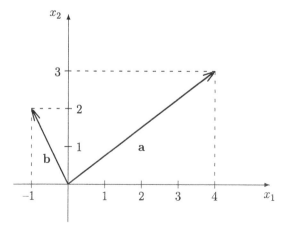

Figure 6.1 Representation of two-dimensional vectors $\mathbf{a}$ and $\mathbf{b}$.

Next, we introduce some relations on vectors of the *same* dimension.

Definition 6.3 Let $\mathbf{a}, \mathbf{b} \in \mathbb{R}^n$ with $\mathbf{a} = (a_1, a_2, \ldots, a_n)^{\mathrm{T}}$ and $\mathbf{b} = (b_1, b_2, \ldots, b_n)^{\mathrm{T}}$.
The vectors $\mathbf{a}$ and $\mathbf{b}$ are said to be *equal* if all their corresponding components are equal, i.e. $a_i = b_i$ for all $i = 1, 2, \ldots, n$.
We write $\mathbf{a} \leq \mathbf{b}$ if $a_i \leq b_i$ and $\mathbf{a} \geq \mathbf{b}$ if $a_i \geq b_i$ for all $i = 1, 2, \ldots, n$. Analogously, we write $\mathbf{a} < \mathbf{b}$ if $a_i < b_i$ and $\mathbf{a} > \mathbf{b}$ if $a_i > b_i$ for all $i = 1, 2, \ldots, n$.

Remark Note that not every pair of n-dimensional vectors may be compared by the relations $\leq$ and $\geq$, respectively. For instance, for the vectors

$$\mathbf{a} = \begin{pmatrix} 3 \\ -1 \\ 2 \end{pmatrix} \quad \text{and} \quad \mathbf{b} = \begin{pmatrix} 2 \\ 0 \\ 2 \end{pmatrix},$$

we have neither $\mathbf{a} \leq \mathbf{b}$ nor $\mathbf{a} \geq \mathbf{b}$.

Example 6.1 Consider the vectors

$$\mathbf{a} = \begin{pmatrix} 2 \\ 4 \\ 3 \end{pmatrix}, \quad \mathbf{b} = \begin{pmatrix} 1 \\ -2 \\ 3 \end{pmatrix} \quad \text{and} \quad \mathbf{c} = \begin{pmatrix} 0 \\ 3 \\ 2 \end{pmatrix}.$$

Then we have $\mathbf{a} \geq \mathbf{b}$, $\mathbf{a} > \mathbf{c}$, but vectors $\mathbf{b}$ and $\mathbf{c}$ cannot be compared, i.e. we have neither $\mathbf{b} \leq \mathbf{c}$ nor $\mathbf{c} \leq \mathbf{b}$.

Special vectors

Finally, we introduce some special vectors. By $\mathbf{e}^i$ we denote the *ith unit vector*, where the ith component is equal to one and all other components are equal to zero, i.e.

$$\mathbf{e}^i = \begin{pmatrix} 0 \\ \vdots \\ 0 \\ 1 \\ 0 \\ \vdots \\ 0 \end{pmatrix} \quad \leftarrow i\text{th component}$$

The n-dimensional *zero vector* is a vector containing only zeroes as the n components:

$$\mathbf{0} = \begin{pmatrix} 0 \\ 0 \\ \vdots \\ 0 \end{pmatrix}.$$

6.2 OPERATIONS ON VECTORS

We start with the operations of adding two vectors and multiplying a vector by a real number (scalar).

Definition 6.4 Let $\mathbf{a}, \mathbf{b} \in \mathbb{R}^n$ with $\mathbf{a} = (a_1, a_2, \ldots, a_n)^T$ and $\mathbf{b} = (b_1, b_2, \ldots, b_n)^T$. The **sum** of the two vectors $\mathbf{a}$ and $\mathbf{b}$ is the n-dimensional vector $\mathbf{a} + \mathbf{b}$ obtained by adding each component of $\mathbf{a}$ to the corresponding component of $\mathbf{b}$:

$$\mathbf{a} + \mathbf{b} = \begin{pmatrix} a_1 \\ a_2 \\ \vdots \\ a_n \end{pmatrix} + \begin{pmatrix} b_1 \\ b_2 \\ \vdots \\ b_n \end{pmatrix} = \begin{pmatrix} a_1 + b_1 \\ a_2 + b_2 \\ \vdots \\ a_n + b_n \end{pmatrix}$$

Definition 6.5 Let $\mathbf{a} \in \mathbb{R}^n$ with $\mathbf{a} = (a_1, a_2, \ldots, a_n)^T$ and $\lambda \in \mathbb{R}$. The product of number (or scalar) λ and vector $\mathbf{a}$ is the n-dimensional vector $\lambda \mathbf{a}$ whose components are λ times the corresponding components of $\mathbf{a}$:

$$\lambda \, \mathbf{a} = \lambda \cdot \begin{pmatrix} a_1 \\ a_2 \\ \vdots \\ a_n \end{pmatrix} = \begin{pmatrix} \lambda a_1 \\ \lambda a_2 \\ \vdots \\ \lambda a_n \end{pmatrix}.$$

The operation of multiplying a vector by a scalar is known as *scalar multiplication*.

Using Definitions 6.4 and 6.5, we can now define the difference of two vectors.

Definition 6.6 Let $\mathbf{a}, \mathbf{b} \in \mathbb{R}^n$ be n-dimensional vectors. Then the *difference* between the vectors $\mathbf{a}$ and $\mathbf{b}$ is defined by

$$\mathbf{a} - \mathbf{b} = \mathbf{a} + (-1)\mathbf{b}.$$

According to Definition 6.6, the difference vector is obtained by subtracting the components of vector $\mathbf{b}$ from the *corresponding* components of vector $\mathbf{a}$. Notice that the sum and the difference of two vectors are only defined when both vectors $\mathbf{a}$ and $\mathbf{b}$ have the *same dimension*.

Example 6.2 Let

$$\mathbf{a} = \begin{pmatrix} 4 \\ 1 \end{pmatrix} \quad \text{and} \quad \mathbf{b} = \begin{pmatrix} 1 \\ 3 \end{pmatrix}.$$

Then we obtain

$$\mathbf{a} + \mathbf{b} = \begin{pmatrix} 4 \\ 1 \end{pmatrix} + \begin{pmatrix} 1 \\ 3 \end{pmatrix} = \begin{pmatrix} 5 \\ 4 \end{pmatrix} \quad \text{and} \quad \mathbf{a} - \mathbf{b} = \begin{pmatrix} 4 \\ 1 \end{pmatrix} - \begin{pmatrix} 1 \\ 3 \end{pmatrix} = \begin{pmatrix} 3 \\ -2 \end{pmatrix}.$$

Applying Definition 6.5, we obtain

$$3\,\mathbf{a} = 3 \cdot \begin{pmatrix} 4 \\ 1 \end{pmatrix} = \begin{pmatrix} 12 \\ 3 \end{pmatrix} \quad \text{and} \quad (-2)\,\mathbf{b} = (-2) \cdot \begin{pmatrix} 1 \\ 3 \end{pmatrix} = \begin{pmatrix} -2 \\ -6 \end{pmatrix}.$$

The sum and difference of the two vectors, as well as the scalar multiplication, are geometrically illustrated in Figure 6.2. The sum of the two vectors $\mathbf{a}$ and $\mathbf{b}$ is the vector obtained when adding vector $\mathbf{b}$ to the terminal point of vector $\mathbf{a}$. The resulting vector from the origin to the terminal point of vector $\mathbf{b}$ gives the sum $\mathbf{a} + \mathbf{b}$.

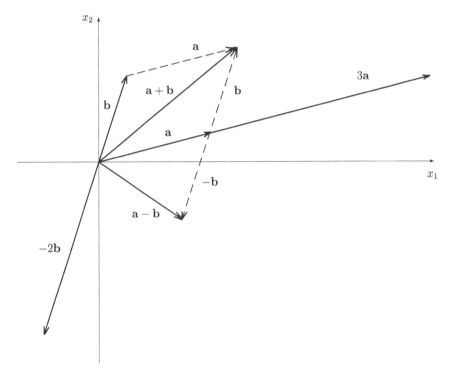

Figure 6.2 Vector operations.

We see that multiplication of a vector by a positive scalar λ does not change the orientation of the vector, while multiplication by a negative scalar reverses the orientation of the vector. The difference $\mathbf{a} - \mathbf{b}$ of two vectors means that we add vector $\mathbf{b}$ with opposite orientation to the terminal point of vector $\mathbf{a}$, i.e. we add vector $-\mathbf{b}$ to vector $\mathbf{a}$.

Next, we summarize some rules for the vector operations introduced above. Let $\mathbf{a}, \mathbf{b}, \mathbf{c} \in \mathbb{R}^n$ and $\lambda, \mu \in \mathbb{R}$.

Rules for vector addition and scalar multiplication

(1) $\mathbf{a} + \mathbf{b} = \mathbf{b} + \mathbf{a},$ $\lambda\mathbf{a} = \mathbf{a}\lambda;$
     (commutative laws)
(2) $(\mathbf{a} + \mathbf{b}) + \mathbf{c} = \mathbf{a} + (\mathbf{b} + \mathbf{c}),$ $(\lambda\mu)\mathbf{a} = \lambda(\mu\mathbf{a});$
     (associative laws)
(3) $\lambda(\mathbf{a} + \mathbf{b}) = \lambda\mathbf{a} + \lambda\mathbf{b},$ $(\lambda + \mu)\mathbf{a} = \lambda\mathbf{a} + \mu\mathbf{a};$
     (distributive laws)
(4) $\mathbf{a} + \mathbf{0} = \mathbf{a},$ $\mathbf{a} + (-\mathbf{a}) = \mathbf{0};$
     $(\mathbf{0} \in \mathbb{R}^n)$
(5) $1\,\mathbf{a} = \mathbf{a}.$

The validity of the above rules follows immediately from Definitions 6.4 and 6.5 and the validity of the commutative, associative and distributive laws for the set of real numbers.

Definition 6.7   The *scalar product* of two n-dimensional vectors $\mathbf{a} = (a_1, a_2, \ldots, a_n)^\mathrm{T}$ and $\mathbf{b} = (b_1, b_2, \ldots, b_n)^\mathrm{T}$ is defined as follows:

$$\mathbf{a}^\mathrm{T} \cdot \mathbf{b} = (a_1, a_2, \ldots, a_n) \cdot \begin{pmatrix} b_1 \\ b_2 \\ \vdots \\ b_n \end{pmatrix} = a_1 b_1 + a_2 b_2 + \ldots + a_n b_n = \sum_{i=1}^{n} a_i b_i.$$

The scalar product is also known as the *inner product*. Note also that the scalar product of two vectors is not a vector, but a *number* (i.e. a *scalar*) and that $\mathbf{a}^\mathrm{T} \cdot \mathbf{b}$ is defined only if $\mathbf{a}$ and $\mathbf{b}$ are both of the *same dimension*. In order to guarantee consistence with operations in later chapters, we define the scalar product in such a way that the first vector is written as a row vector $\mathbf{a}^\mathrm{T}$ and the second vector is written as a column vector $\mathbf{b}$. The commutative and distributive laws are valid for the scalar product, i.e.

$$\mathbf{a}^\mathrm{T} \cdot \mathbf{b} = \mathbf{b}^\mathrm{T} \cdot \mathbf{a} \qquad \text{and} \qquad \mathbf{a}^\mathrm{T} \cdot (\mathbf{b} + \mathbf{c}) = \mathbf{a}^\mathrm{T} \cdot \mathbf{b} + \mathbf{a}^\mathrm{T} \cdot \mathbf{c}$$

for $\mathbf{a}, \mathbf{b}, \mathbf{c} \in \mathbb{R}^n$. It is worth noting that the associative law does not necessarily hold for the scalar product, i.e. in general we have

$$\mathbf{a} \cdot (\mathbf{b}^\mathrm{T} \cdot \mathbf{c}) \neq (\mathbf{a}^\mathrm{T} \cdot \mathbf{b}) \cdot \mathbf{c}.$$

Example 6.3   Assume that a firm produces three products with the quantities $x_1 = 30$, $x_2 = 40$ and $x_3 = 10$, where x_i denotes the quantity of product i. Moreover, the cost of production is 20 EUR per unit of product 1, 15 EUR per unit of product 2 and 40 EUR per unit of product 3. Let $\mathbf{c} = (c_1, c_2, c_3)^\mathrm{T}$ be the cost vector, where the ith component describes the cost per unit of product i and $\mathbf{x} = (x_1, x_2, x_3)^\mathrm{T}$. The total cost of production is obtained as the scalar product of vectors **c** and **x**, i.e. with

$$\mathbf{c} = \begin{pmatrix} 20 \\ 15 \\ 40 \end{pmatrix} \qquad \text{and} \qquad \mathbf{x} = \begin{pmatrix} 30 \\ 40 \\ 10 \end{pmatrix}$$

we obtain

$$\mathbf{c}^{\mathrm{T}} \cdot \mathbf{x} = (20, 15, 40) \cdot \begin{pmatrix} 30 \\ 40 \\ 10 \end{pmatrix} = 20 \cdot 30 + 15 \cdot 40 + 40 \cdot 10$$

$$= 600 + 600 + 400 = 1,600.$$

We have found that the total cost of production for the three products is 1,600 EUR.

Definition 6.8   Let **a** $\in \mathbb{R}$ with $\mathbf{a} = (a_1, a_2, \dots, a_n)^{\mathrm{T}}$. The *(Euclidean) length* (or *norm*) of vector **a**, denoted by $|\mathbf{a}|$, is defined as

$$|\mathbf{a}| = \sqrt{a_1^2 + a_2^2 + \dots + a_n^2}.$$

A vector with length one is called a *unit vector* (remember that we have already introduced the specific unit vectors $\mathbf{e}^1, \mathbf{e}^2, \dots, \mathbf{e}^n$ which obviously have length one). Each non-zero n-dimensional vector **a** can be written as the product of its length $|\mathbf{a}|$ and an n-dimensional unit vector $\mathbf{e}^{(a)}$ pointing in the same direction as the vector **a** itself, i.e.

$$\mathbf{a} = |\mathbf{a}| \cdot \mathbf{e}^{(a)}.$$

Example 6.4   Let vector

$$\mathbf{a} = \begin{pmatrix} -2 \\ 3 \\ 6 \end{pmatrix}$$

be given. We are looking for a unit vector pointing in the same direction as vector **a**. Using

$$|\mathbf{a}| = \sqrt{(-2)^2 + 3^2 + 6^2} = \sqrt{49} = 7,$$

we find the corresponding unit vector

$$\mathbf{e}^{(a)} = \frac{1}{|\mathbf{a}|} \cdot \mathbf{a} = \frac{1}{7} \cdot \begin{pmatrix} -2 \\ 3 \\ 6 \end{pmatrix} = \begin{pmatrix} -2/7 \\ 3/7 \\ 6/7 \end{pmatrix}.$$

Using Definition 6.8, we can define the *(Euclidean) distance* between the n-vectors $\mathbf{a} = (a_1, a_2, \dots, a_n)^{\mathrm{T}}$ and $\mathbf{b} = (b_1, b_2, \dots, b_n)^{\mathrm{T}}$ as follows:

$$|\mathbf{a} - \mathbf{b}| = \sqrt{(a_1 - b_1)^2 + (a_2 - b_2)^2 + \dots + (a_n - b_n)^2}.$$

The distance between the two-dimensional vectors **a** and **b** is illustrated in Figure 6.3. It corresponds to the length of the vector connecting the terminal points of vectors **a** and **b**.

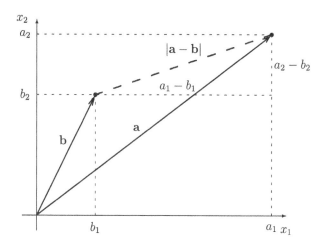

Figure 6.3 Distance between the vectors **a** and **b**.

Example 6.5 Let vectors

$$\mathbf{a} = \begin{pmatrix} 3 \\ 2 \\ -3 \end{pmatrix} \quad \text{and} \quad \mathbf{b} = \begin{pmatrix} -1 \\ 1 \\ 5 \end{pmatrix}$$

be given. The distance between both vectors is given by

$$|\mathbf{a} - \mathbf{b}| = \sqrt{[3 - (-1)]^2 + (2 - 1)^2 + (-3 - 5)^2} = \sqrt{81} = 9.$$

Next, we present some further rules for the scalar product of two vectors and the length of a vector. Let $\mathbf{a}, \mathbf{b} \in \mathbb{R}^n$ and $\lambda \in \mathbb{R}$.

Further rules for the scalar product and the length

(1) $|\mathbf{a}| = \sqrt{\mathbf{a}^T \cdot \mathbf{a}} \geq 0;$
(2) $|\mathbf{a}| = 0 \iff \mathbf{a} = \mathbf{0};$
(3) $|\lambda \mathbf{a}| = |\lambda| \cdot |\mathbf{a}|;$
(4) $|\mathbf{a} + \mathbf{b}| \leq |\mathbf{a}| + |\mathbf{b}|;$
(5) $\mathbf{a}^T \cdot \mathbf{b} = |\mathbf{a}| \cdot |\mathbf{b}| \cdot \cos(\mathbf{a}, \mathbf{b});$
(6) $|\mathbf{a}^T \cdot \mathbf{b}| \leq |\mathbf{a}| \cdot |\mathbf{b}|$ (Cauchy–Schwarz inequality).

In rule (5), $\cos(\mathbf{a}, \mathbf{b})$ denotes the cosine value of the angle between vectors $\mathbf{a}$ and $\mathbf{b}$. We illustrate the Cauchy–Schwarz inequality by the following example.

Example 6.6   Let

$$\mathbf{a} = \begin{pmatrix} 2 \\ -1 \\ 3 \end{pmatrix} \qquad \text{and} \qquad \mathbf{b} = \begin{pmatrix} 5 \\ -4 \\ -1 \end{pmatrix}.$$

For the lengths of vectors $\mathbf{a}$ and $\mathbf{b}$, we obtain

$$|\mathbf{a}| = \sqrt{2^2 + (-1)^2 + 3^2} = \sqrt{14} \qquad \text{and} \qquad |\mathbf{b}| = \sqrt{5^2 + (-4)^2 + (-1)^2} = \sqrt{42}.$$

The scalar product of vectors $\mathbf{a}$ and $\mathbf{b}$ is obtained as

$$\mathbf{a}^T \cdot \mathbf{b} = 2 \cdot 5 + (-1) \cdot (-4) + 3 \cdot (-1) = 10 + 4 - 3 = 11.$$

Cauchy–Schwarz's inequality says that the absolute value of the scalar product of two vectors $\mathbf{a}$ and $\mathbf{b}$ is never greater than the product of the lengths of both vectors. For the example, this inequality turns into

$$|\mathbf{a}^T \cdot \mathbf{b}| = 11 \leq |\mathbf{a}| \cdot |\mathbf{b}| = \sqrt{14} \cdot \sqrt{42} \approx 24.2487.$$

Example 6.7   Using rule (5) above, which can also be considered as an alternative equivalent definition of the scalar product, and the definition of the scalar product, according to Definition 6.7, one can easily determine the angle between two vectors $\mathbf{a}$ and $\mathbf{b}$ of the same dimension. Let

$$\mathbf{a} = \begin{pmatrix} 3 \\ -1 \\ 2 \end{pmatrix} \qquad \text{and} \qquad \mathbf{b} = \begin{pmatrix} 2 \\ 1 \\ 2 \end{pmatrix}.$$

Then we obtain

$$\begin{aligned}
\cos(\mathbf{a}, \mathbf{b}) &= \frac{\mathbf{a}^T \cdot \mathbf{b}}{|\mathbf{a}| \cdot |\mathbf{b}|} \\
&= \frac{3 \cdot 2 + (-1) \cdot 1 + 2 \cdot 2}{\sqrt{3^2 + (-1)^2 + 2^2} \cdot \sqrt{2^2 + 1^2 + 2^2}} = \frac{9}{\sqrt{14} \cdot \sqrt{9}} = \frac{3}{\sqrt{14}} \approx 0.80178.
\end{aligned}$$

We have to find the smallest positive argument of the cosine function which gives the value 0.80178. Therefore, the angle between vectors $\mathbf{a}$ and $\mathbf{b}$ is approximately equal to $36.7°$.

Next, we consider *orthogonal* vectors. Consider the triangle given in Figure 6.4 formed by the three two-dimensional vectors $\mathbf{a}, \mathbf{b}$ and $\mathbf{a} - \mathbf{b}$. Denote the angle between vectors $\mathbf{a}$ and $\mathbf{b}$ by γ. From the Pythagorean theorem we know that angle γ is equal to $90°$ if and only if the sum of the squared lengths of vectors **a** and **b** is equal to the squared length of vector **a** − **b**. Thus, we have:

$$\gamma = 90° \iff |\mathbf{a}|^2 + |\mathbf{b}|^2 = |\mathbf{a} - \mathbf{b}|^2$$
$$\iff \mathbf{a}^T \cdot \mathbf{a} + \mathbf{b}^T \cdot \mathbf{b} = (\mathbf{a} - \mathbf{b})^T \cdot (\mathbf{a} - \mathbf{b})$$
$$\iff \mathbf{a}^T \cdot \mathbf{a} + \mathbf{b}^T \cdot \mathbf{b} = \mathbf{a}^T \cdot \mathbf{a} - \mathbf{a}^T \cdot \mathbf{b} - \mathbf{b}^T \cdot \mathbf{a} + \mathbf{b}^T \cdot \mathbf{b}$$
$$\iff \mathbf{a}^T \cdot \mathbf{b} = 0.$$

The latter equality has been obtained since $\mathbf{a}^T \cdot \mathbf{b} = \mathbf{b}^T \cdot \mathbf{a}$. For two-dimensional vectors we have seen that the angle between them is equal to $90°$ if and only if the scalar product is equal to zero. We say in this case that vectors **a** and **b** are *orthogonal* (or *perpendicular*) and write $\mathbf{a} \perp \mathbf{b}$. The above considerations can be generalized to the n-dimensional case and we define orthogonality accordingly:

$$\mathbf{a} \perp \mathbf{b} \iff \mathbf{a}^T \cdot \mathbf{b} = 0,$$

where $\mathbf{a}, \mathbf{b} \in \mathbb{R}^n$.

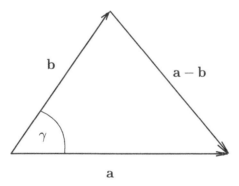

Figure 6.4 Triangle formed by vectors **a**, **b** and **a** − **b**.

Example 6.8 The three-dimensional vectors

$$\mathbf{a} = \begin{pmatrix} 3 \\ -1 \\ 2 \end{pmatrix} \quad \text{and} \quad \mathbf{b} = \begin{pmatrix} 4 \\ 6 \\ -3 \end{pmatrix}$$

are orthogonal since

$$\mathbf{a}^T \cdot \mathbf{b} = 3 \cdot 4 + (-1) \cdot 6 + 2 \cdot (-3) = 0.$$

6.3   LINEAR DEPENDENCE AND INDEPENDENCE

In this section, we discuss one of the most important concepts in linear algebra. Before introducing linearly dependent and linearly independent vectors, we give the following definition.

Definition 6.9 Let $\mathbf{a}^i$, $i = 1, 2, \ldots, m$, be n-dimensional vectors, and λ_i, $i = 1, 2, \ldots, m$, be real numbers. Then the n-vector $\mathbf{a}$ given by

$$\mathbf{a} = \lambda_1 \mathbf{a}^1 + \lambda_2 \mathbf{a}^2 + \ldots + \lambda_m \mathbf{a}^m \quad = \quad \sum_{i=1}^{m} \lambda_i \mathbf{a}^i \tag{6.1}$$

is called a *linear combination* of the vectors $\mathbf{a}^1, \mathbf{a}^2, \ldots, \mathbf{a}^m$. If

$$\lambda_i \geq 0, \ i = 1, 2, \ldots, m, \qquad \text{and} \qquad \sum_{i=1}^{m} \lambda_i = 1$$

in representation (6.1), then $\mathbf{a}$ is called a **convex combination** of the vectors $\mathbf{a}^1, \mathbf{a}^2, \ldots, \mathbf{a}^m$.

The set of all convex combinations of the two vectors $\mathbf{a}^1$ and $\mathbf{a}^2$ is illustrated in Figure 6.5. It is the set of all vectors whose terminal points are on the line connecting the terminal points of vectors $\mathbf{a}^1$ and $\mathbf{a}^2$. Therefore, both vectors $\mathbf{c}^1$ and $\mathbf{c}^2$ can be written as a convex combination of the vectors $\mathbf{a}^1$ and $\mathbf{a}^2$. Notice that for $\lambda_1 = 1$ and $\lambda_2 = 1 - \lambda_1 = 0$ we obtain vector $\mathbf{a}^1$, whereas for $\lambda_1 = 0$ and $\lambda_2 = 1 - \lambda_1 = 1$ we obtain vector $\mathbf{a}^2$. Note that a convex combination of some vectors is also a special linear combination of these vectors.

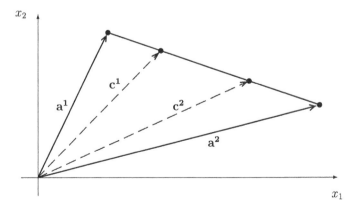

Figure 6.5 Set of convex combinations of vectors $\mathbf{a}^1$ and $\mathbf{a}^2$.

Definition 6.10 The m n-dimensional vectors $\mathbf{a}^1, \mathbf{a}^2, \ldots, \mathbf{a}^m \in \mathbb{R}^n$ are **linearly dependent** if there exist numbers λ_i, $i = 1, 2, \ldots, m$, not all equal to zero, such that

$$\sum_{i=1}^{m} \lambda_i \mathbf{a}^i = \lambda_1 \mathbf{a}^1 + \lambda_2 \mathbf{a}^2 + \cdots + \lambda_m \mathbf{a}^m = \mathbf{0}. \tag{6.2}$$

If equation (6.2) only holds when $\lambda_1 = \lambda_2 = \ldots = \lambda_m = 0$, then the vectors $\mathbf{a}^1, \mathbf{a}^2, \ldots, \mathbf{a}^m$ are said to be **linearly independent**.

Since two vectors are equal if they coincide in all components, the above equation (6.2) represents n linear equations with the variables $\lambda_1, \lambda_2, \ldots, \lambda_m$. In Chapter 8, we deal with the solution of such systems in detail.

Remark From Definition 6.10 we obtain the following equivalent characterization of linearly dependent and independent vectors.

(1) A set of m vectors $\mathbf{a}^1, \mathbf{a}^2, \ldots, \mathbf{a}^m \in \mathbb{R}^n$ is linearly dependent if and only if at least one of the vectors can be written as a linear combination of the others.
(2) A set of m vectors $\mathbf{a}^1, \mathbf{a}^2, \ldots, \mathbf{a}^m \in \mathbb{R}^n$ is linearly independent if and only if none of the vectors can be written as a linear combination of the others.

Example 6.9 Let

$$\mathbf{a}^1 = \begin{pmatrix} 3 \\ 1 \end{pmatrix} \qquad \text{and} \qquad \mathbf{a}^2 = \begin{pmatrix} -9 \\ -3 \end{pmatrix}.$$

In this case, we have $\mathbf{a}^2 = -3\mathbf{a}^1$ which can be written as

$$3\mathbf{a}^1 + 1\mathbf{a}^2 = \mathbf{0}.$$

We can conclude that equation (6.2) holds with $\lambda_1 = 3$ and $\lambda_2 = 1$, and thus vectors $\mathbf{a}^1$ and $\mathbf{a}^2$ are linearly dependent (see Figure 6.6).

Example 6.10 Let

$$\mathbf{a}^1 = \begin{pmatrix} 3 \\ 1 \end{pmatrix} \qquad \text{and} \qquad \mathbf{b}^2 = \begin{pmatrix} -1 \\ 2 \end{pmatrix}.$$

In this case, equation $\lambda_1 \mathbf{a}^1 + \lambda_2 \mathbf{b}^2 = \mathbf{0}$ reduces to

$$3\lambda_1 - \lambda_2 = 0$$
$$\lambda_1 + 2\lambda_2 = 0.$$

This is a system of two linear equations with two variables λ_1 and λ_2 which can easily be solved. Multiplying the first equation by two and adding it to the second equation, we obtain $\lambda_1 = 0$ and then $\lambda_2 = 0$ as the only solution of this system. Therefore, both vectors $\mathbf{a}^1$ and $\mathbf{b}^2$ are linearly independent (see Figure 6.6).

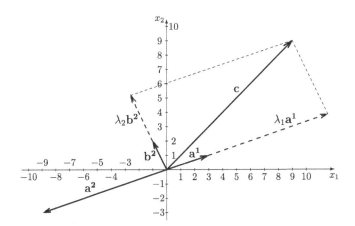

Figure 6.6 Linearly dependent and independent vectors.

The above examples illustrate that in the case of two vectors of the two-dimensional Euclidean space $\mathbb{R}^2$, we can easily decide whether they are linearly dependent or independent. Two two-dimensional vectors are linearly dependent if and only if one vector can be written as a multiple of the other vector, i.e.

$$\mathbf{a}^2 = -\frac{\lambda_1}{\lambda_2} \cdot \mathbf{a}^1, \quad \lambda_2 \neq 0$$

(see Figure 6.6). On the other hand, in the two-dimensional space every three vectors are linearly dependent. This is also illustrated in Figure 6.6. Vector $\mathbf{c}$ can be written as a linear combination of the linearly independent vectors $\mathbf{a}^1$ and $\mathbf{b}^2$, i.e.

$$\mathbf{c} = \lambda_1 \mathbf{a}^1 + \lambda_2 \mathbf{b}^2,$$

from which we obtain

$$1\mathbf{c} - \lambda_1 \mathbf{a}^1 - \lambda_2 \mathbf{b}^2 = \mathbf{0}. \tag{6.3}$$

By Definition 6.10, these vectors are linearly dependent since e.g. the scalar of vector $\mathbf{c}$ in representation (6.3) is different from zero.

Considering 3-vectors, three vectors are linearly dependent if one of them can be written as a linear combination of the other two vectors which means that the third vector belongs to the plane spanned by the other two vectors. If the three vectors do not belong to the same plane, these vectors are linearly independent. Four vectors in the three-dimensional Euclidean space are always linearly dependent. In general, we can say that in the n-dimensional Euclidean space $\mathbb{R}^n$, there are no more than n linearly independent vectors.

Example 6.11   Let us consider the three vectors

$$
\mathbf{a}^1 = \begin{pmatrix} 2 \\ 0 \\ 0 \end{pmatrix}, \qquad \mathbf{a}^2 = \begin{pmatrix} 1 \\ 1 \\ 0 \end{pmatrix} \qquad \text{and} \qquad \mathbf{a}^3 = \begin{pmatrix} 3 \\ 2 \\ 1 \end{pmatrix},
$$

and investigate whether they are linearly dependent or independent. Using Definition 6.10, we obtain

$$
\lambda_1 \begin{pmatrix} 2 \\ 0 \\ 0 \end{pmatrix} + \lambda_2 \begin{pmatrix} 1 \\ 1 \\ 0 \end{pmatrix} + \lambda_3 \begin{pmatrix} 3 \\ 2 \\ 1 \end{pmatrix} = \begin{pmatrix} 2\lambda_1 + \lambda_2 + 3\lambda_3 \\ \lambda_2 + 2\lambda_3 \\ \lambda_3 \end{pmatrix} = \begin{pmatrix} 0 \\ 0 \\ 0 \end{pmatrix}.
$$

Considering the third component of the above vectors, we obtain $\lambda_3 = 0$. Substituting $\lambda_3 = 0$ into the second component, we get from $\lambda_2 + 2\lambda_3 = 0$ the only solution $\lambda_2 = 0$ and considering finally the first component, we obtain from $2\lambda_1 + \lambda_2 + 3\lambda_3 = 0$ the only solution $\lambda_1 = 0$. Since vector $\lambda^T = (\lambda_1, \lambda_2, \lambda_3) = (0, 0, 0)$ is the only solution, the above three vectors are linearly independent.

Example 6.12   The set $\{\mathbf{e}^1, \mathbf{e}^2, \ldots, \mathbf{e}^n\}$ of n-dimensional unit vectors in the space $\mathbb{R}^n$ obviously constitutes a set of linearly independent vectors, and any n-dimensional vector $\mathbf{a}^T = (a_1, a_2, \ldots, a_n)$ can be immediately written as a linear combination of these unit vectors:

$$
\mathbf{a} = a_1 \begin{pmatrix} 1 \\ 0 \\ \vdots \\ 0 \end{pmatrix} + a_2 \begin{pmatrix} 0 \\ 1 \\ \vdots \\ 0 \end{pmatrix} + \cdots + a_n \begin{pmatrix} 0 \\ 0 \\ \vdots \\ 1 \end{pmatrix} = \sum_{i=1}^{n} a_i \, \mathbf{e}^i.
$$

In this case, the scalars of the linear combination of the unit vectors are simply the components of vector $\mathbf{a}$.

6.4   VECTOR SPACES

We have discussed several properties of vector operations so far. In this section, we introduce the notion of a vector space. This is a set of elements (not necessarily only vectors of real numbers) which satisfy certain rules listed in the following definition.

Definition 6.11   Given a set $V = \{a, b, c, \ldots\}$ of vectors (or other mathematical objects), for which an addition and a scalar multiplication are defined, suppose that the following properties hold ($\lambda, \mu \in \mathbb{R}$):

(1) $a + b = b + a$;
(2) $(a + b) + c = a + (b + c)$;
(3) there exists a vector $0 \in V$ such that for all $a \in V$ the equation
   $a + 0 = a$ holds (0 is the zero or neutral element with respect to addition);
(4) for each $a \in V$ there exists a uniquely determined element $x \in V$ such that
   $a + x = 0$ ($x = -a$ is the inverse element of a with respect to addition);
(5) $(\lambda\mu)a = \lambda(\mu a)$;
(6) $1 \cdot a = a$;
(7) $\lambda(a + b) = \lambda a + \lambda b$;
(8) $(\lambda + \mu)a = \lambda a + \mu a$.

If for any $a, b \in V$, inclusion $a + b \in V$ and for any $\lambda \in \mathbb{R}$ inclusion $\lambda a \in V$ hold, then V is called a *linear space* or *vector space*.

As mentioned before, the elements of a vector space do not necessarily need to be vectors since other 'mathematical objects' may also obey the above rules (1) to (8). Next, we give some examples of vector spaces satisfying the rules listed in Definition 6.11, where in each case an addition and a scalar multiplication is defined in the usual way.

Examples of vector spaces

(1) the n-dimensional space $\mathbb{R}^n$;
(2) the set of all n-vectors $a \in \mathbb{R}^n$ that are orthogonal to some fixed n-vector $b \in \mathbb{R}^n$;
(3) the set of all sequences $\{a_n\}$;
(4) the set $C[a, b]$ of all continuous functions on the closed interval $[a, b]$;
(5) the set of all polynomials

$$P_n(x) = a_n x^n + a_{n-1} x^{n-1} + \cdots + a_1 x + a_0$$

of a degree of at most n.

To prove this, one has to verify the validity of the rules (1) to (8) given in Definition 6.11 and to show that the sum of any two elements as well as multiplication by a scalar again gives an element of this space. For instance, consider the set of all polynomials of degree n. The sum of two such polynomials

$$P_n^1(x) = a_n x^n + a_{n-1} x^{n-1} + \cdots + a_1 x + a_0$$
$$P_n^2(x) = b_n x^n + b_{n-1} x^{n-1} + \cdots + b_1 x + b_0$$

gives

$$P_n^1(x) + P_n^2(x) = (a_n + b_n)x^n + (a_{n-1} + b_{n-1})x^{n-1} + \cdots + (a_1 + b_1)x + (a_0 + b_0),$$

i.e. the sum of these polynomials is again a polynomial of degree n. By multiplying a polynomial P_n^1 of degree n by a real number λ, we obtain again a polynomial of degree n:

$$\lambda P_n^1 = (\lambda a_n)x^n + (\lambda a_{n-1})x^{n-1} + \cdots + (\lambda a_1)x + \lambda a_0.$$

Basis of a vector space; change of basis

Next, we introduce the notion of the basis of a vector space.

Definition 6.12 A set $B = \{\mathbf{b}^1, \mathbf{b}^2, \ldots, \mathbf{b}^n\}$ of linearly independent vectors of a vector space V is called a *basis* of V if any vector $\mathbf{a} \in V$ can be written as a linear combination

$$\mathbf{a} = \lambda_1 \mathbf{b}^1 + \lambda_2 \mathbf{b}^2 + \cdots + \lambda_n \mathbf{b}^n$$

of the basis vectors $\mathbf{b}^1, \mathbf{b}^2, \ldots, \mathbf{b}^n$. The number $n = |B|$ of vectors contained in the basis gives the *dimension* of vector space V.

For an n-dimensional vector space V, we also write: dim $V = n$. Obviously, the set $B_c = \{\mathbf{e}^1, \mathbf{e}^2, \ldots, \mathbf{e}^n\}$ of unit vectors constitutes a basis of the n-dimensional Euclidean space $\mathbb{R}^n$ (see also Example 6.12). This basis B_c is also called the *canonical basis*. The notion of a basis of a vector space is a fundamental concept in linear algebra which we will need again later when discussing various algorithms.

Remark (1) An equivalent definition of a basis is that a *maximal set of linearly independent vectors* of a vector space V constitutes a basis and therefore, the dimension of a vector space is given by the maximal number of linearly independent vectors.

(2) We say that V is *spanned* by the vectors $\mathbf{b}^1, \mathbf{b}^2, \ldots, \mathbf{b}^n$ of B since any vector of the vector space V can be 'generated' by means of the basis vectors.

(3) The dimension of a vector space is *not* necessarily equal to the number of components of its vectors. If a set of n-dimensional vectors is given, they can contain *at most* n linearly independent vectors. However, *any* n linearly independent vectors of an n-dimensional vector space constitute a basis.

Next, we establish whether an arbitrary vector of a vector space can be written as a linear combination of the basis vectors in a *unique* way.

THEOREM 6.1   Let $B = \{b^1, b^2, \ldots, b^n\}$ be a basis of an n-dimensional vector space V. Then any vector $c \in V$ can be uniquely written as a linear combination of the basis vectors from set B.

PROOF   We prove the theorem indirectly. Assume there exist two different linear combinations of the given basis vectors from B which are equal to vector c:

$$c = \lambda_1 b^1 + \lambda_2 b^2 + \cdots + \lambda_n b^n \tag{6.4}$$

and

$$c = \mu_1 b^1 + \mu_2 b^2 + \cdots + \mu_n b^n, \tag{6.5}$$

i.e. there exists an index i with $1 \le i \le n$ such that $\lambda_i \ne \mu_i$. By subtracting equation (6.5) from equation (6.4), we get

$$0 = (\lambda_1 - \mu_1) b^1 + (\lambda_2 - \mu_2) b^2 + \cdots + (\lambda_n - \mu_n) b^n$$

Since the basis vectors $b^1, b^2, \ldots, b^n$ are linearly independent by Definition 6.12, we must have

$$\lambda_1 - \mu_1 = 0, \quad \lambda_2 - \mu_2 = 0, \quad \ldots, \quad \lambda_n - \mu_n = 0,$$

which is equivalent to

$$\lambda_1 = \mu_1, \quad \lambda_2 = \mu_2, \quad \ldots, \quad \lambda_n = \mu_n,$$

i.e. we have obtained a contradiction. Thus, any vector of V can be uniquely written as a linear combination of the given basis vectors.   ∎

While the *dimension of a vector space is uniquely determined, the basis is not uniquely determined*. This leads to the question of whether we can replace a particular vector in the basis by some other vector not contained in the basis such that again a basis is obtained. The following theorem shows that there is an easy way to answer this question, and from the proof of the following theorem we derive an algorithm for the replacement of a vector in the basis by some other vector (provided this is possible). The resulting procedure is a basic part of some algorithms for the solution of systems of linear equations or linear inequalities which we discuss in Chapter 8.

THEOREM 6.2 (Steinitz's theorem)   Let set $B = \{b^1, b^2, \ldots, b^n\}$ be a basis of an n-dimensional vector space V and let vector a^k be given by

$$a^k = \lambda_1 b^1 + \lambda_2 b^2 + \cdots + \lambda_k b^k + \cdots + \lambda_n b^n$$

with $\lambda_k \ne 0$. Then the set $B^* = \{b^1, b^2, \ldots, b^{k-1}, a^k, b^{k+1}, \ldots, b^n\}$ is also a basis of V, i.e. vector b^k contained in basis B can be replaced by the vector a^k to obtain another basis.

PROOF   Let us consider the following linear combination of the zero vector:

$$\mu_1 b^1 + \mu_2 b^2 + \cdots + \mu_{k-1} b^{k-1} + \mu_k a^k + \mu_{k+1} b^{k+1} + \cdots + \mu_n b^n = 0. \tag{6.6}$$

By substituting the linear combination

$$\mathbf{a}^k = \lambda_1 \mathbf{b}^1 + \lambda_2 \mathbf{b}^2 + \cdots + \lambda_n \mathbf{b}^n$$

into equation (6.6), we obtain:

$$(\mu_1 + \mu_k \lambda_1)\mathbf{b}^1 + (\mu_2 + \mu_k \lambda_2)\mathbf{b}^2 + \cdots + (\mu_{k-1} + \mu_k \lambda_{k-1})\mathbf{b}^{k-1}$$
$$+ (\mu_k \lambda_k)\mathbf{b}^k + (\mu_{k+1} + \mu_k \lambda_{k+1})\mathbf{b}^{k+1} + \cdots + (\mu_n + \mu_k \lambda_n)\mathbf{b}^n = \mathbf{0}.$$

Since the vectors of the set $B = \{\mathbf{b}^1, \mathbf{b}^2, \ldots, \mathbf{b}^n\}$ constitute a basis, they are linearly independent and all the scalars in the above linear combination must be equal to zero, i.e. we get

$$(\mu_i + \mu_k \lambda_i) = 0 \quad \text{for} \quad i = 1, 2, \ldots, n, \ i \neq k \quad \text{and} \quad \mu_k \lambda_k = 0.$$

Since by assumption $\lambda_k \neq 0$, we get first $\mu_k = 0$ and then, using the latter result, $\mu_i = 0$ for all i with $1 \leq i \leq n$, $i \neq k$, i.e. all the scalars in the linear combination (6.6) must be equal to zero. Hence, the vectors of set $B^* = \{\mathbf{b}^1, \mathbf{b}^2, \ldots, \mathbf{b}^{k-1}, \mathbf{a}^k, \mathbf{b}^{k+1}, \ldots, \mathbf{b}^n\}$ are linearly independent by Definition 6.10 and they constitute a basis. ∎

We know that any basis of a certain vector space consists of the same number of vectors. So, if we want to remove one vector from the current basis, we have to add exactly one other vector to the remaining ones such that the resulting set of vectors is again linearly independent. We now look for a procedure for performing such an interchange of two vectors described in the proof of Theorem 6.2. To this end, assume that $B = \{\mathbf{b}^1, \ldots, \mathbf{b}^{k-1}, \mathbf{b}^k, \mathbf{b}^{k+1}, \ldots, \mathbf{b}^n\}$ is a basis and $B^* = \{\mathbf{b}^1, \ldots, \mathbf{b}^{k-1}, \mathbf{a}^k, \mathbf{b}^{k+1}, \ldots, \mathbf{b}^n\}$ is another basis, where vector $\mathbf{b}^k$ has been replaced by vector $\mathbf{a}^k$. According to Theorem 6.2, we must have $\lambda_k \neq 0$ in the linear combination of vector

$$\mathbf{a}^k = \lambda_1 \mathbf{b}^1 + \cdots + \lambda_k \mathbf{b}^k + \cdots + \lambda_n \mathbf{b}^n \tag{6.7}$$

of the vectors of basis B since otherwise a replacement of vector $\mathbf{b}^k$ by vector $\mathbf{a}^k$ is not possible. Let us consider an arbitrary vector $\mathbf{c}$ and its linear combinations of the basis vectors of bases B and B^*, respectively:

$$\mathbf{c} = \alpha_1 \mathbf{b}^1 + \cdots + \alpha_{k-1} \mathbf{b}^{k-1} + \alpha_k \mathbf{b}^k + \alpha_{k+1} \mathbf{b}^{k+1} + \cdots + \alpha_n \mathbf{b}^n \tag{6.8}$$

and

$$\mathbf{c} = \beta_1 \mathbf{b}^1 + \cdots + \beta_{k-1} \mathbf{b}^{k-1} + \beta_k \mathbf{a}^k + \beta_{k+1} \mathbf{b}^{k+1} + \cdots + \beta_n \mathbf{b}^n. \tag{6.9}$$

By substituting representation (6.7) of vector $\mathbf{a}^k$ into representation (6.9), we obtain

$$\mathbf{c} = \beta_1 \mathbf{b}^1 + \cdots + \beta_{k-1} \mathbf{b}^{k-1} + \beta_k (\lambda_1 \mathbf{b}^1 + \cdots + \lambda_k \mathbf{b}^k + \cdots + \lambda_n \mathbf{b}^n)$$
$$+ \beta_{k+1} \mathbf{b}^{k+1} + \cdots + \beta_n \mathbf{b}^n$$
$$= (\beta_1 + \beta_k \lambda_1) \, \mathbf{b}^1 + \cdots + (\beta_{k-1} + \beta_k \lambda_{k-1}) \, \mathbf{b}^{k-1} + \beta_k \lambda_k \, \mathbf{b}^k$$
$$+ (\beta_{k+1} + \beta_k \lambda_{k+1}) \, \mathbf{b}^{k+1} + \cdots + (\beta_n + \beta_k \lambda_n) \, \mathbf{b}^n.$$

Comparing now the scalars of the vectors from the basis $B = \{b^1, b^2, \ldots, b^n\}$ in both representations of vector c (i.e in representation (6.8) and the last equality above) first for k and then for all remaining $i \neq k$, we first obtain

$$\beta_k = \frac{\alpha_k}{\lambda_k} \tag{6.10}$$

and then from

$$\alpha_i = \beta_i + \beta_k \lambda_i, \quad i = 1, 2, \ldots, n, \ i \neq k,$$

it follows by means of equality (6.10)

$$\beta_i = \alpha_i - \frac{\lambda_i}{\lambda_k} \alpha_k, \quad i = 1, 2, \ldots, n, \ i \neq k.$$

In order to transform the linear combination of vector c of the basis vectors of B into the linear combination of the basis vectors of B^*, we can therefore use the scheme given in Table 6.1. The last column describes the operation that has to be performed in order to get the elements of the current row: e.g. in row $n+2$, the notation 'row $2 - (\lambda_2/\lambda_k)$ row k' means that we have to take the corresponding element of row 2 (i.e. α_2 in the c column), and then we have to subtract λ_2/λ_k times the corresponding element of row k (i.e. α_k in the c column) which gives the new element β_2 in row $n+2$ and the c column. The transformation formula above for all rows different from k is also called the *rectangle formula* since exactly four elements forming a rectangle are required to determine the corresponding new element. This is illustrated in the scheme above for the determination of element β_2, where the corresponding four elements in row 2 and row k are underlined.

Table 6.1 Tableau for Steinitz's procedure

Row	Basis vectors	a^k	c	Operation
1	b^1	λ_1	α_1	
2	b^2	$\underline{\lambda_2}$	$\underline{\alpha_2}$	
$\vdots$	$\vdots$	$\vdots$	$\vdots$	
k	b^k	$\underline{\lambda_k}$	$\underline{\alpha_k}$	
$\vdots$	$\vdots$	$\vdots$	$\vdots$	
n	b^n	λ_n	α_n	
$n+1$	b^1	0	$\beta_1 = \alpha_1 - \frac{\lambda_1}{\lambda_k}\alpha_k$	row $1 - \frac{\lambda_1}{\lambda_k}$ row k
$n+2$	b^2	0	$\beta_2 = \alpha_2 - \frac{\lambda_2}{\lambda_k}\alpha_k$	row $2 - \frac{\lambda_2}{\lambda_k}$ row k
$\vdots$	$\vdots$	$\vdots$	$\vdots$	$\vdots$
$n+k$	a^k	1	$\beta_k = \frac{\alpha_k}{\lambda_k}$	$\frac{1}{\lambda_k}$ row k
$\vdots$	$\vdots$	$\vdots$	$\vdots$	$\vdots$
$2n$	b^n	0	$\beta_n = \alpha_n - \frac{\lambda_n}{\lambda_k}\alpha_k$	row $n - \frac{\lambda_n}{\lambda_k}$ row k

In columns 3 and 4, we have the corresponding scalars of vectors $\mathbf{a}^k$ and $\mathbf{c}$ in the linear combinations of the basis vectors of B (row 1 up to row n) and of the basis vectors of B^* (row $n+1$ up to row $2n$). In particular, from the first n rows we get

$$\mathbf{a}^k = \lambda_1 \mathbf{b}^1 + \cdots + \lambda_k \mathbf{b}^k + \cdots + \lambda_n \mathbf{b}^n$$

and

$$\mathbf{c} = \alpha_1 \mathbf{b}^1 + \cdots + \alpha_k \mathbf{b}^k + \cdots + \alpha_n \mathbf{b}^n.$$

From the last n rows, we get the linear combinations

$$\mathbf{a}^k = 0\mathbf{b}^1 + \cdots + 0\mathbf{b}^{k-1} + 1\mathbf{a}^k + 0\mathbf{b}^{k+1} + \cdots + 0\mathbf{b}^n$$

and

$$\mathbf{c} = \left(\alpha_1 - \frac{\lambda_1}{\lambda_k}\alpha_k\right)\mathbf{b}^1 + \cdots + \left(\frac{\alpha_k}{\lambda_k}\right)\mathbf{a}^k + \cdots + \left(\alpha_n - \frac{\lambda_n}{\lambda_k}\alpha_k\right)\mathbf{b}^n.$$

If we have representations of several vectors with respect to basis B and we look for their corresponding representations in the new basis B^*, we simply add some more columns (for each vector one column) in the above scheme, and by performing the same operations as indicated in the last column, we get all the representations with respect to the new basis B^*. The above procedure of replacing a vector in the current basis is illustrated by the following example.

Example 6.13 Let a basis $B = \{\mathbf{b}^1, \mathbf{b}^2, \mathbf{b}^3, \mathbf{b}^4\}$ of the four-dimensional space $\mathbb{R}^4$ with

$$\mathbf{b}^1 = \begin{pmatrix} 1 \\ -1 \\ 1 \\ -2 \end{pmatrix}, \quad \mathbf{b}^2 = \begin{pmatrix} 3 \\ -5 \\ 2 \\ -1 \end{pmatrix}, \quad \mathbf{b}^3 = \begin{pmatrix} -2 \\ 1 \\ -2 \\ 2 \end{pmatrix} \quad \text{and} \quad \mathbf{b}^4 = \begin{pmatrix} -1 \\ 0 \\ -1 \\ 1 \end{pmatrix}$$

be given. We do not prove here that these four vectors indeed constitute a basis in $\mathbb{R}^4$. According to Definition 6.10, we have to solve a system of four linear equations with four variables which we treat later in Chapter 8 in detail. Moreover, let

$$\mathbf{a}^3 = 4\mathbf{b}^1 - 2\mathbf{b}^2 + 2\mathbf{b}^3 - 6\mathbf{b}^4$$

be the vector which enters the basis instead of vector $\mathbf{b}^3$. In addition, let vector $\mathbf{c}$ be given as

$$\mathbf{c} = \mathbf{b}^1 + 2\mathbf{b}^2 + 4\mathbf{b}^3 + 3\mathbf{b}^4.$$

Notice that

$$\mathbf{c} = \begin{pmatrix} -4 \\ -7 \\ -6 \\ 7 \end{pmatrix}, \tag{6.11}$$

i.e. the representation of this vector by means of the unit vectors e^1, e^2, e^3, e^4 as basis vectors is

$$c = -4e^1 - 7e^2 - 6e^3 + 7e^4.$$

Applying the tableau given in Table 6.1 to find the linear combination of vector c of the new basis vectors, we obtain the results given in Table 6.2. From row 5 to row 8 of the c column, we get the representation of vector c by means of the basis $B^* = \{b^1, b^2, a^3, b^4\}$:

$$c = -7b^1 + 6b^2 + 2a^3 + 15b^4.$$

Table 6.2 The change of the basis in Example 6.13

Row	Basis vectors	b^3	c	Operation
1	b^1	4	1	
2	b^2	-2	2	
3	b^3	2	4	
4	b^4	-6	3	
5	b^1	0	-7	row 1 $-$ 2 row 3
6	b^2	0	6	row 2 $+$ row 3
7	a^3	1	2	$\frac{1}{2}$ row 3
8	b^4	0	15	row 4 $+$ 3 row 3

We can easily check that our computations are correct:

$$c = -7\begin{pmatrix} 1 \\ -1 \\ 1 \\ -2 \end{pmatrix} + 6\begin{pmatrix} 3 \\ -5 \\ 2 \\ -1 \end{pmatrix} + 2\begin{pmatrix} 0 \\ 8 \\ 2 \\ -8 \end{pmatrix} + 15\begin{pmatrix} -1 \\ 0 \\ -1 \\ 1 \end{pmatrix} = \begin{pmatrix} -4 \\ -7 \\ -6 \\ 7 \end{pmatrix},$$

i.e. for vector c we get the same representation with respect to the basis vectors e^1, e^2, e^3, e^4 as before (see equality (6.11)).

If a basis $B = \{b^1, b^2, \ldots, b^n\}$ should be replaced by a basis $B^* = \{a^1, a^2, \ldots, a^n\}$, then we can apply consecutively the above procedure by replacing in each step a vector $b^i, 1 \leq i \leq n$, by a vector $a^j, 1 \leq j \leq n$, provided that the assumption of Theorem 6.2 is satisfied.

===

EXERCISES

6.1   Given are the vectors

$$a = \begin{pmatrix} 2 \\ 1 \\ -1 \end{pmatrix}, \quad b = \begin{pmatrix} 1 \\ -4 \\ -2 \end{pmatrix} \quad \text{and} \quad c = \begin{pmatrix} 2 \\ 2 \\ 6 \end{pmatrix}.$$

(a) Find vectors $\mathbf{a} + \mathbf{b} - \mathbf{c}$, $\mathbf{a} + 3\mathbf{b}$, $\mathbf{b} - 4\mathbf{a} + 2\mathbf{c}$, $\mathbf{a} + 3(\mathbf{b} - 2\mathbf{c})$.

(b) For which of the vectors $\mathbf{a}$, $\mathbf{b}$ and $\mathbf{c}$ do the relations $>$ or $\geq$ hold?

(c) Find the scalar products $\mathbf{a}^T \cdot \mathbf{b}$, $\mathbf{a}^T \cdot \mathbf{c}$, $\mathbf{b}^T \cdot \mathbf{c}$. Which of the vectors $\mathbf{a}, \mathbf{b}$ and $\mathbf{c}$ are orthogonal? What is the angle between the vectors $\mathbf{b}$ and $\mathbf{c}$?

(d) Compute vectors $(\mathbf{a}^T \cdot \mathbf{b}) \cdot \mathbf{c}$ and $\mathbf{a} \cdot (\mathbf{b}^T \cdot \mathbf{c})$.

(e) Compare number $|\mathbf{b} + \mathbf{c}|$ with number $|\mathbf{b}| + |\mathbf{c}|$ and number $|\mathbf{b}^T \cdot \mathbf{c}|$ with number $|\mathbf{b}| \cdot |\mathbf{c}|$.

6.2 Find α and β so that vectors

$$\mathbf{a} = \begin{pmatrix} 2 \\ -1 \\ \alpha \end{pmatrix} \quad \text{and} \quad \mathbf{b} = \begin{pmatrix} \beta \\ 4 \\ -2 \end{pmatrix}$$

are orthogonal.

6.3 (a) What is the distance between the following points: $(1, 2, 3)$ and $(4, -1, 2)$ in the three-dimensional Euclidean space $\mathbb{R}^3$?

(b) Illustrate the following sets of points in $\mathbb{R}^2$: $\mathbf{a} \geq \mathbf{b}$ and $|\mathbf{a}| \geq |\mathbf{b}|$.

6.4 Given are the vectors

$$\mathbf{a}^1 = \begin{pmatrix} 1 \\ 0 \end{pmatrix} \quad \text{and} \quad \mathbf{a}^2 = \begin{pmatrix} -1 \\ 1 \end{pmatrix}.$$

Find out which of the vectors

$$\begin{pmatrix} -2 \\ 1 \end{pmatrix}, \quad \begin{pmatrix} 2 \\ 3 \end{pmatrix} \quad \text{and} \quad \begin{pmatrix} 0 \\ 0.5 \end{pmatrix}$$

are linear combinations of $\mathbf{a}^1$ and $\mathbf{a}^2$. Is one of the above vectors a convex combination of vectors $\mathbf{a}^1$ and $\mathbf{a}^2$? Graph all these vectors.

6.5 Given are the vectors

$$\mathbf{a}^1 = \begin{pmatrix} 4 \\ 2 \end{pmatrix}, \quad \mathbf{a}^2 = \begin{pmatrix} 1 \\ 4 \end{pmatrix}, \quad \mathbf{a}^3 = \begin{pmatrix} 3 \\ 0 \end{pmatrix} \quad \text{and} \quad \mathbf{a}^4 = \begin{pmatrix} 3 \\ 2 \end{pmatrix}.$$

Show that vector $\mathbf{a}^4$ can be expressed as a convex linear combination of vectors $\mathbf{a}^1$, $\mathbf{a}^2$ and $\mathbf{a}^3$. Find the convex combinations of vectors $\mathbf{a}^1$, $\mathbf{a}^2$ and $\mathbf{a}^3$ graphically.

6.6 Are the vectors

$$\mathbf{a}^1 = \begin{pmatrix} 1 \\ 0 \\ 0 \end{pmatrix}, \quad \mathbf{a}^2 = \begin{pmatrix} 1 \\ -2 \\ 1 \end{pmatrix} \quad \text{and} \quad \mathbf{a}^3 = \begin{pmatrix} 5 \\ 4 \\ -2 \end{pmatrix}$$

linearly independent?

6.7 Do the two vectors

$$\mathbf{a}^1 = \begin{pmatrix} 2 \\ -1 \end{pmatrix} \quad \text{and} \quad \mathbf{a}^2 = \begin{pmatrix} -4 \\ 2 \end{pmatrix}$$

span the two-dimensional space? Do they constitute a basis? Graph the vectors and illustrate their linear combinations.

6.8   Do the vectors

$$\begin{pmatrix} 1 \\ 0 \\ 0 \\ 1 \end{pmatrix}, \quad \begin{pmatrix} 0 \\ 0 \\ 1 \\ 0 \end{pmatrix}, \quad \begin{pmatrix} 0 \\ 1 \\ 0 \\ 1 \end{pmatrix} \quad \text{and} \quad \begin{pmatrix} 1 \\ 0 \\ 1 \\ 0 \end{pmatrix}$$

constitute a basis in $\mathbb{R}^4$?

6.9   Let vectors

$$\mathbf{a}^1 = \begin{pmatrix} 1 \\ 0 \\ 3 \end{pmatrix}, \quad \mathbf{a}^2 = \begin{pmatrix} 0 \\ 1 \\ 0 \end{pmatrix} \quad \text{and} \quad \mathbf{a}^3 = \begin{pmatrix} 1 \\ 0 \\ -1 \end{pmatrix}$$

constitute a basis in $\mathbb{R}^3$.

(a) Express vector

$$\mathbf{a} = \begin{pmatrix} 3 \\ 3 \\ -3 \end{pmatrix}$$

as a linear combination of the three vectors $\mathbf{a}^1$, $\mathbf{a}^2$ and $\mathbf{a}^3$ above.

(b) Find all other bases for the three-dimensional space which include vector $\mathbf{a}$ and vectors from the set $\{\mathbf{a}^1, \mathbf{a}^2, \mathbf{a}^3\}$.

(c) Express vector

$$\mathbf{b} = 2\mathbf{a}^1 + 2\mathbf{a}^2 + 3\mathbf{a}^3 = \begin{pmatrix} 5 \\ 2 \\ 3 \end{pmatrix}$$

by the basis vectors $\mathbf{a}^1$, $\mathbf{a}^2$ and $\mathbf{a}$.

7 Matrices and determinants

7.1 MATRICES

We start with an introductory example.

Example 7.1   Assume that a firm uses three raw materials denoted by R_1, R_2 and R_3 to produce four intermediate products S_1, S_2, S_3 and S_4. These intermediate products are partially also used to produce two final products F_1 and F_2. The numbers of required units of the intermediate products are independent of the use of intermediate products as input for the two final products. Table 7.1 gives the number of units of each raw material which are required for the production of one unit of each of the intermediate products. Table 7.2 gives the number of units of each intermediate product necessary to produce one unit of each of the final products.

The firm intends to produce 80 units of S_1, 60 units of S_2, 100 units of S_3 and 50 units of S_4 as well as 70 units of F_1 and 120 units of F_2. The question is: how many units of the raw materials are necessary to produce the required numbers of the intermediate and final products?

Table 7.1 Raw material requirements for the intermediate products

Raw material	S_1	S_2	S_3	S_4
R_1	2	1	4	0
R_2	3	2	1	3
R_3	1	4	0	5

Table 7.2 Intermediate product requirements for the final products

Raw material	F_1	F_2
S_1	2	3
S_2	4	0
S_3	1	4
S_4	3	1

To produce the required units of the intermediate products, we need

$$2 \cdot 80 + 1 \cdot 60 + 4 \cdot 100 + 0 \cdot 50 = 620$$

units of raw material R_1. Similarly, we need

$$3 \cdot 80 + 2 \cdot 60 + 1 \cdot 100 + 3 \cdot 50 = 610$$

units of raw material R_2 and

$$1 \cdot 80 + 4 \cdot 60 + 0 \cdot 100 + 5 \cdot 50 = 570$$

units of raw material R_3. Summarizing the above considerations, the vector $\mathbf{y}^S$ of the required units of raw materials for the production of the intermediate products is given by

$$\mathbf{y}^S = \begin{pmatrix} 620 \\ 610 \\ 570 \end{pmatrix},$$

where the kth component gives the number of units of R_k required for the production of the intermediate products.

Next, we calculate how many units of each raw material are required for the production of the final products. Since the intermediate products are used for the production of the final products (see Table 7.2), we find that for the production of one unit of final product F_1 the required amount of raw material R_1 is

$$2 \cdot 2 + 1 \cdot 4 + 4 \cdot 1 + 0 \cdot 3 = 12.$$

Similarly, to produce one unit of final product F_2 requires

$$2 \cdot 3 + 1 \cdot 0 + 4 \cdot 4 + 0 \cdot 1 = 22$$

units of R_1. To produce one unit of final product F_1 requires

$$3 \cdot 2 + 2 \cdot 4 + 1 \cdot 1 + 3 \cdot 3 = 24$$

units of R_2. Continuing in this way, we get Table 7.3, describing how many units of each raw material are required for the production of each of the final products.

Table 7.3 Raw material requirements for the final products

Raw material	F_1	F_2
R_1	12	22
R_2	24	16
R_3	33	8

Therefore, for the production of the final products, there are

$$12 \cdot 70 + 22 \cdot 120 = 3,480$$

units of raw material R_1,

$$24 \cdot 70 + 16 \cdot 120 = 3,600$$

units of raw material R_2 and finally

$$33 \cdot 70 + 8 \cdot 120 = 3,270$$

units of raw material R_3 required. The vector $\mathbf{y}^F$ containing as components the units of each raw material required for the production of the final products is then given by

$$\mathbf{y}^F = \begin{pmatrix} 3,480 \\ 3,600 \\ 3,270 \end{pmatrix}.$$

So the amount of the individual raw materials required for the total production of the intermediate and final products is obtained as the sum of the vectors $\mathbf{y}^S$ and $\mathbf{y}^F$. Denoting this sum vector by $\mathbf{y}$, we obtain

$$\mathbf{y} = \begin{pmatrix} 620 \\ 610 \\ 575 \end{pmatrix} + \begin{pmatrix} 3,480 \\ 3,600 \\ 3,270 \end{pmatrix} = \begin{pmatrix} 4,100 \\ 4,210 \\ 3,845 \end{pmatrix}.$$

The question is whether we can simplify the above computations by introducing some formal apparatus. In the following, we use matrices and define operations such as addition or multiplication in an appropriate way.

Definition 7.1 A *matrix* A is a rectangular array of elements (numbers or other mathematical objects, e.g. functions) a_{ij} of the form

$$A = (a_{ij}) = \begin{pmatrix} a_{11} & a_{12} & \cdots & a_{1n} \\ a_{21} & a_{22} & \cdots & a_{2n} \\ \vdots & \vdots & & \vdots \\ a_{m1} & a_{m2} & \cdots & a_{mn} \end{pmatrix}.$$

Any element (or entry) a_{ij} has two indices, a row index i and a column index j. The matrix A is said to have the *order* or *dimension* $m \times n$ (read: m by n).
If $m = n$, matrix A is called a *square matrix*.

For a matrix A of order $m \times n$, we also write $A = A_{(m,n)}$, or $A = (a_{ij})_{(m,n)}$, or simply $A = (a_{ij})$.

Definition 7.2 Let a matrix A of order $m \times n$ be given. The *transpose* A^T of matrix A is obtained by interchanging the rows and columns of A, i.e. the first column becomes the first row, the first row becomes the first column and so on. Thus:

$$A = (a_{ij}) \quad \Longrightarrow \quad A^T = (a_{ij}^*) \quad \text{with} \quad a_{ji}^* = a_{ij} \quad \text{for } 1 \le j \le n \text{ and } 1 \le i \le m.$$

Obviously, matrix A^T in Definition 7.2 is of order $n \times m$.

Example 7.2 Let

$$A = \begin{pmatrix} 2 & 3 & 4 & 1 \\ 7 & -1 & 0 & 4 \end{pmatrix}.$$

Since matrix A is of order 2×4, matrix A^T is of order 4×2, and we get

$$A = \begin{pmatrix} 2 & 7 \\ 3 & -1 \\ 4 & 0 \\ 1 & 4 \end{pmatrix}.$$

Remark A vector

$$\mathbf{a} = \begin{pmatrix} a_1 \\ a_2 \\ \vdots \\ a_m \end{pmatrix}$$

is a special matrix with one column (i.e. a matrix of order $m \times 1$). Analogously, a transposed vector $\mathbf{a}^T = (a_1, a_2, \dots, a_m)$ is a special matrix consisting only of one row.

Definition 7.3 Two matrices A and B of the same order $m \times n$ are *equal* if corresponding elements are equal, i.e.

$$a_{ij} = b_{ij} \qquad \text{for } 1 \le i \le m \quad \text{and} \quad 1 \le j \le n.$$

So only for matrices of the same order $m \times n$ can we decide whether both matrices are equal.

Definition 7.4 A matrix A of order $n \times n$ is called *symmetric* if $A = A^T$, i.e. equality $a_{ij} = a_{ji}$ holds for $1 \le i, j \le n$. Matrix A is called *antisymmetric* if $A = -A^T$, i.e. $a_{ij} = -a_{ji}$ for $1 \le i, j \le n$.

As a consequence from Definition 7.4, we obtain: if A is antisymmetric, then we must have $a_{ii} = 0$ for $i = 1, 2, \ldots, n$.

Special matrices

We finish this section with some matrices of special structure.

Definition 7.5 A matrix $D = (d_{ij})$ of order $n \times n$ with

$$D = \begin{pmatrix} d_1 & 0 & \cdots & 0 \\ 0 & d_2 & \cdots & 0 \\ \vdots & \vdots & & \vdots \\ 0 & 0 & \cdots & d_n \end{pmatrix}, \quad \text{i.e. } d_{ij} = \begin{cases} d_i & \text{for} \quad 1 \leq i,j \leq n \text{ and } i = j, \\ 0 & \text{for} \quad 1 \leq i,j \leq n \text{ and } i \neq j, \end{cases}$$

is called a *diagonal matrix*. A diagonal matrix $I = (i_{ij})$ of order $n \times n$ with

$$I = \begin{pmatrix} 1 & 0 & \cdots & 0 \\ 0 & 1 & \cdots & 0 \\ \vdots & \vdots & & \vdots \\ 0 & 0 & \cdots & 1 \end{pmatrix}, \quad \text{i.e. } i_{ij} = \begin{cases} 1 & \text{for} \quad 1 \leq i,j \leq n \text{ and } i = j \\ 0 & \text{for} \quad 1 \leq i,j \leq n \text{ and } i \neq j \end{cases}$$

is called an *identity matrix*.

Definition 7.6 A matrix $U = (u_{ij})$ of order $n \times n$ with

$$U = \begin{pmatrix} u_{11} & u_{12} & \cdots & u_{1n} \\ 0 & u_{22} & \cdots & u_{2n} \\ \vdots & \vdots & & \vdots \\ 0 & 0 & \cdots & u_{nn} \end{pmatrix}, \quad \text{i.e. } u_{ij} = 0 \text{ for } 1 \leq i,j \leq n \text{ and } i > j$$

is called an *upper triangular matrix*. A matrix $L = (l_{ij})$ of order $n \times n$ with

$$L = \begin{pmatrix} l_{11} & 0 & \cdots & 0 \\ l_{21} & l_{22} & \cdots & 0 \\ \vdots & \vdots & & \vdots \\ l_{n1} & l_{n2} & \cdots & l_{nn} \end{pmatrix}, \quad \text{i.e. } l_{ij} = 0 \text{ for } 1 \leq i,j \leq n \text{ and } i < j$$

is called a *lower triangular matrix*.

Notice that the matrices given in Definitions 7.5 and 7.6 are defined only in the case of a *square* matrix.

Definition 7.7 A matrix $O = (o_{ij})$ of order $m \times n$ with

$$O = \begin{pmatrix} 0 & 0 & \cdots & 0 \\ 0 & 0 & \cdots & 0 \\ \vdots & \vdots & & \vdots \\ 0 & 0 & \cdots & 0 \end{pmatrix}, \quad \text{i.e. } o_{ij} = 0 \quad \text{for } 1 \le i \le m \text{ and } 1 \le j \le n$$

is called a *zero matrix*.

7.2 MATRIX OPERATIONS

In the following, we discuss matrix operations such as addition and multiplication and their properties.

Definition 7.8 Let $A = (a_{ij})$ and $B = (b_{ij})$ be two matrices of order $m \times n$. The *sum* $A + B$ is defined as the $m \times n$ matrix $(a_{ij} + b_{ij})$, i.e.

$$A + B = (a_{ij})_{(m,n)} + (b_{ij})_{(m,n)} = (a_{ij} + b_{ij})_{(m,n)}.$$

Thus, the sum of two matrices of the same order is obtained when corresponding elements at the same position in both matrices are added. The zero matrix O is the *neutral element* with respect to matrix addition, i.e. we have

$$A + O = O + A = A,$$

where matrix O has the same order as matrix A.

Definition 7.9 Let $A = (a_{ij})$ be an $m \times n$ matrix and $\lambda \in \mathbb{R}$. The product of the scalar λ and the matrix A is the $m \times n$ matrix $\lambda A = (\lambda a_{ij})$, i.e. any element of matrix A is multiplied by the scalar λ. The operation of multiplying a matrix by a scalar is called *scalar multiplication*.

Using Definitions 7.8 and 7.9, we can define the difference of two matrices as follows.

Definition 7.10 Let $A = (a_{ij})$ and $B = (b_{ij})$ be matrices of order $m \times n$. Then the *difference* of matrices A and B is defined as

$$A - B = A + (-1)B.$$

Consequently, matrix $A - B$ is given by the $m \times n$ matrix $(a_{ij} - b_{ij})$, i.e.

$$A - B = (a_{ij})_{(m,n)} - (b_{ij})_{(m,n)} = (a_{ij} - b_{ij})_{(m,n)}.$$

Example 7.3   Let

$$A = \begin{pmatrix} 3 & 1 & 2 \\ 0 & -2 & 3 \\ 1 & 4 & 5 \end{pmatrix}, \quad B = \begin{pmatrix} 1 & 2 & 0 \\ 3 & 1 & -1 \\ 4 & 2 & -2 \end{pmatrix} \quad \text{and} \quad C = \begin{pmatrix} 1 & 2 & 5 \\ 2 & 0 & 3 \\ 2 & -3 & 1 \end{pmatrix}.$$

We compute $2A + 3B - C$ and obtain

$$2A + 3B - C = \begin{pmatrix} 2 \cdot 3 & 2 \cdot 1 & 2 \cdot 2 \\ 2 \cdot 0 & 2 \cdot (-2) & 2 \cdot 3 \\ 2 \cdot 1 & 2 \cdot 4 & 2 \cdot 5 \end{pmatrix} + \begin{pmatrix} 3 \cdot 1 & 3 \cdot 2 & 3 \cdot 0 \\ 3 \cdot 3 & 3 \cdot 1 & 3 \cdot (-1) \\ 3 \cdot 4 & 3 \cdot 2 & 3 \cdot (-2) \end{pmatrix}$$

$$- \begin{pmatrix} 1 & 2 & 5 \\ 2 & 0 & 3 \\ 2 & -3 & 1 \end{pmatrix}$$

$$= \begin{pmatrix} 6 & 2 & 4 \\ 0 & -4 & 6 \\ 2 & 8 & 10 \end{pmatrix} + \begin{pmatrix} 3 & 6 & 0 \\ 9 & 3 & -3 \\ 12 & 6 & -6 \end{pmatrix} - \begin{pmatrix} 1 & 2 & 5 \\ 2 & 0 & 3 \\ 2 & -3 & 1 \end{pmatrix}$$

$$= \begin{pmatrix} 8 & 6 & -1 \\ 7 & -1 & 0 \\ 12 & 17 & 3 \end{pmatrix}.$$

Next, we give some rules for adding two matrices A and B of the same order and for multiplying a matrix by some real number (scalar multiplication). Let A, B, C be matrices of order $m \times n$ and $\lambda, \mu \in \mathbb{R}$.

Rules for matrix addition and scalar multiplication

(1) $A + B = B + A$;
   (commutative law)
(2) $(A + B) + C = A + (B + C)$;
   (associative law)
(3) $\lambda(A + B) = \lambda A + \lambda B$; $(\lambda + \mu)A = \lambda A + \mu A$.
   (distributive laws)

We have already introduced the notion of a vector space in Chapter 6. Using matrix addition and scalar multiplication as introduced in Definitions 7.8 and 7.9, we can extend the rules presented above and get the following result.

THEOREM 7.1   The set of all matrices of order $m \times n$ constitutes a vector space.

Next, we introduce the multiplication of two matrices of specific orders.

Definition 7.11 Let $A = (a_{ij})$ be a matrix of order $m \times p$ and $B = (b_{ij})$ be a matrix of order $p \times n$. The *product* AB is a matrix of order $m \times n$ which is defined by

$$
AB = \begin{pmatrix}
a_{11}b_{11} + \cdots + a_{1p}b_{p1} & a_{11}b_{12} + \cdots + a_{1p}b_{p2} & \cdots & a_{11}b_{1n} + \cdots + a_{1p}b_{pn} \\
a_{21}b_{11} + \cdots + a_{2p}b_{p1} & a_{21}b_{12} + \cdots + a_{2p}b_{p2} & \cdots & a_{21}b_{1n} + \cdots + a_{2p}b_{pn} \\
\vdots & \vdots & \vdots & \vdots \\
a_{m1}b_{11} + \cdots + a_{mp}b_{p1} & a_{m1}b_{12} + \cdots + a_{mp}b_{p2} & \cdots & a_{m1}b_{1n} + \cdots + a_{mp}b_{pn}
\end{pmatrix}.
$$

Notice that the product AB is defined only when the number of columns of matrix A is equal to the number of rows of matrix B. For calculating the product $A_{(m,p)}B_{(p,n)}$, we can use *Falk's scheme* which is as follows:

				b_{11}	b_{12}	$\cdots$	b_{1n}
				b_{21}	b_{22}	$\cdots$	b_{2n}
	AB			$\vdots$	$\vdots$		$\vdots$
				b_{p1}	b_{p2}	$\cdots$	b_{pn}
a_{11}	a_{12}	$\cdots$	a_{1p}	c_{11}	c_{12}	$\cdots$	c_{1n}
a_{21}	a_{22}	$\cdots$	a_{2p}	c_{21}	c_{22}	$\cdots$	c_{2n}
$\vdots$	$\vdots$		$\vdots$	$\vdots$	$\vdots$		$\vdots$
a_{m1}	a_{m2}	$\cdots$	a_{mp}	c_{m1}	c_{m2}	$\cdots$	c_{mn}

with $\quad c_{ij} = \displaystyle\sum_{k=1}^{p} a_{ik}b_{kj}\quad$ for $i = 1, 2, \ldots, m$ and $j = 1, 2, \ldots, n$.

From the above scheme we again see that element c_{ij} is obtained as the scalar product of the ith row vector of matrix A and the jth column vector of matrix B.

If we have to perform more than one matrix multiplication, we can successively apply Falk's scheme. Assuming that the corresponding products of n matrices are defined, we can (due to the validity of the associative law) perform the multiplications either starting from the left or from the right. In the former case, we obtain $C = A_1A_2A_3 \cdots A_n$ according to $C = [(A_1A_2)A_3] \cdots A_n$, i.e. by using Falk's scheme repeatedly we obtain

	A_2	A_3	$\cdots$	A_n
A_1	A_1A_2	$A_1A_2A_3$	$\cdots$	C

or in the latter case $C = A_1[A_2 \ldots (A_{n-1}A_n)]$, i.e. we obtain by using Falk's scheme repeatedly:

	A_n
A_{n-1}	$A_{n-1}A_n$
$\vdots$	$\vdots$
A_2	$A_2A_3 \ldots A_n$
A_1	C

Next, we discuss some properties of matrix multiplication.

(1) Matrix multiplication is not commutative, i.e. in general we have $AB \neq BA$. It may even happen that only one of the possible products of two matrices is defined but not the other. For instance, let A be a matrix of order 2×4 and B be a matrix of order 4×3. Then the product AB is defined and gives a product matrix of order 2×3. However, the product BA is not defined since matrix B has three columns but matrix A has only two rows.

(2) For matrices $A_{(m,p)}, B_{(p,r)}$ and $C_{(r,n)}$, we have $A(BC) = (AB)C$, i.e. matrix multiplication is associative provided that the corresponding products are defined.

(3) For matrices $A_{(m,p)}, B_{(p,n)}$ and $C_{(p,n)}$, we have $A(B+C) = AB+AC$, i.e. the distributive law holds provided that B and C have the same order and the product of matrices A and $B + C$ is defined.

(4) The identity matrix I of order $n \times n$ is the *neutral element* of matrix multiplication of square matrices of order $n \times n$, i.e.

$$AI = IA = A.$$

Let A be a square matrix. Then we write $AA = A^2$, and in general $A^n = AA \ldots A$, where factor A occurs n times, is known as the *nth power* of matrix A.

Example 7.4 Let matrices

$$A = \begin{pmatrix} 3 & -2 & 6 \\ 4 & 1 & 3 \\ 5 & 4 & 0 \end{pmatrix} \qquad \text{and} \qquad B = \begin{pmatrix} 2 & 3 \\ 4 & 1 \\ 1 & 5 \end{pmatrix}$$

be given. The product BA is not defined since matrix B has two columns but matrix A has three rows. The product AB is defined according to Definition 7.11, and the resulting product matrix $C = AB$ is of the order 3×2. Applying Falk's scheme, we obtain

$$
\begin{array}{ccc|cc}
 & & & 2 & 3 \\
 & AB & & 4 & 1 \\
 & & & 1 & 5 \\
\hline
3 & -2 & 6 & 4 & 37 \\
4 & 1 & 3 & 15 & 28 \\
5 & 4 & 0 & 26 & 19
\end{array}
$$

i.e. we have obtained

$$C = AB = \begin{pmatrix} 4 & 37 \\ 15 & 28 \\ 26 & 19 \end{pmatrix}.$$

Example 7.5 Three firms 1, 2 and 3 share a market for a certain product. Currently, firm 1 has 25 per cent of the market, firm 2 has 55 per cent and firm 3 has 20 per cent of the market.

We can summarize this in a so-called market share vector **s**, where component s_i is a real number between zero and one giving the current percentage of firm i as a decimal so that the sum of all components is equal to one. In this example, the corresponding market share vector $\mathbf{s} = (s_1, s_2, s_3)^{\mathrm{T}}$ is given by

$$\mathbf{s} = \begin{pmatrix} 0.25 \\ 0.55 \\ 0.20 \end{pmatrix}.$$

In the course of one year, the following changes occur.

(1) Firm 1 keeps 80 per cent of its customers, while losing 5 per cent to firm 2 and 15 per cent to firm 3.
(2) Firm 2 keeps 65 per cent of its customers, while losing 15 per cent to firm 1 and 20 per cent to firm 3.
(3) Firm 3 keeps 75 per cent of its customers, while losing 15 per cent to firm 1 and 10 per cent to firm 2.

We compute the market share vector $\mathbf{s}^*$ after the above changes. To do this, we introduce a matrix $T = (t_{ij})$, where t_{ij} is the percentage (as a decimal) of customers of firm j who become a customer of firm i within the next year. Matrix T is called a transition matrix. In this example, matrix T is as follows:

$$\begin{pmatrix} 0.80 & 0.15 & 0.15 \\ 0.05 & 0.65 & 0.10 \\ 0.15 & 0.20 & 0.75 \end{pmatrix}.$$

To get the percentage of customers of firm 1 after the course of the year, we have to compute

$$s_1^* = 0.80s_1 + 0.15s_2 + 0.15s_3.$$

Similarly, we can compute the values s_2^* and s_3^*, and we find that vector $\mathbf{s}^*$ is obtained as the product of matrix T and vector **s**:

$$\mathbf{s}^* = T\mathbf{s} = \begin{pmatrix} 0.80 & 0.15 & 0.15 \\ 0.05 & 0.65 & 0.10 \\ 0.15 & 0.20 & 0.75 \end{pmatrix} \begin{pmatrix} 0.25 \\ 0.55 \\ 0.20 \end{pmatrix} = \begin{pmatrix} 0.3125 \\ 0.3900 \\ 0.2975 \end{pmatrix}.$$

Hence, after one year, firm 1 has 31.25 per cent of the customers, firm 2 has 39 per cent and firm 3 has 29.75 per cent.

Example 7.6 Consider again the data given in Example 7.1. Introducing matrix $R_{(3,4)}^S$ as the matrix giving the raw material requirements for the intermediate products as in Table 7.1 and matrix $S_{(4,2)}^F$ as the matrix of the intermediate product requirements for the final products as in Table 7.2, we get the raw material requirements for the final products described by matrix $R_{(3,2)}^F$ by matrix multiplication:

$$R_{(3,2)}^F = R_{(3,4)}^S \cdot S_{(4,2)}^F.$$

Let vectors $x^S_{(4,1)}$ and $x^F_{(2,1)}$ give the number of units of each of the intermediate and final products, respectively, where the ith component refers to the ith product. Then we obtain the vector y of the total raw material requirements as follows:

$$
\begin{aligned}
y_{(3,1)} &= y^S_{(3,1)} + y^F_{(3,1)} \\
&= R^S_{(3,4)} \cdot x^S_{(4,1)} + R^F_{(3,2)} \cdot x^F_{(2,1)} \\
&= R^S_{(3,4)} \cdot x^S_{(4,1)} + R^S_{(3,4)} \cdot S^F_{(4,2)} \cdot x^F_{(2,1)}.
\end{aligned}
$$

The indicated orders of the matrices confirm that all the products and sums are defined.

We now return to transposes of matrices and summarize the following rules, where A and B are $m \times n$ matrices, C is an $n \times p$ matrix and $\lambda \in \mathbb{R}$.

Rules for transposes of matrices

(1) $(A^T)^T = A$;
(2) $(A + B)^T = A^T + B^T$, $\quad (A - B)^T = A^T - B^T$;
(3) $(\lambda A)^T = \lambda A^T$;
(4) $(AC)^T = C^T A^T$.

Definition 7.12 A matrix A of order $n \times n$ is said to be *orthogonal* if $A^T A = I$.

As a consequence of Definition 7.12, we find that in an orthogonal matrix A, the scalar product of the ith row vector and the jth column vector with $i \neq j$ is equal to zero, i.e. these vectors are orthogonal (cf. Chapte 6.2).

Example 7.7 Matrix

$$
A = \begin{pmatrix} \frac{1}{2} & -\frac{1}{2}\sqrt{3} \\ \frac{1}{2}\sqrt{3} & \frac{1}{2} \end{pmatrix}
$$

is orthogonal since

$$
A^T A = \begin{pmatrix} \frac{1}{2} & \frac{1}{2}\sqrt{3} \\ -\frac{1}{2}\sqrt{3} & \frac{1}{2} \end{pmatrix} \begin{pmatrix} \frac{1}{2} & -\frac{1}{2}\sqrt{3} \\ \frac{1}{2}\sqrt{3} & \frac{1}{2} \end{pmatrix} = \begin{pmatrix} 1 & 0 \\ 0 & 1 \end{pmatrix} = I.
$$

7.3 DETERMINANTS

Determinants can be used to answer the question of whether the inverse of a matrix exists and to find such an inverse matrix. They can be used e.g. as a tool for solving systems of linear equations (this topic is discussed in detail in Chapter 8) or for finding eigenvalues (see Chapter 10).

Let

$$A = \begin{pmatrix} a_{11} & a_{12} & \cdots & a_{1n} \\ a_{21} & a_{22} & \cdots & a_{2n} \\ \vdots & \vdots & & \vdots \\ a_{n1} & a_{n2} & \cdots & a_{nn} \end{pmatrix}$$

be a square matrix and A_{ij} denote the submatrix obtained from A by deleting the ith row and jth column. It is clear that A_{ij} is a square matrix of order $(n-1) \times (n-1)$.

Definition 7.13 The *determinant* of a matrix A of order $n \times n$ with numbers as elements is a number, assigned to matrix A by the following rule:

$$\det A = |A| = \sum_{j=1}^{n} (-1)^{j+1} \cdot a_{1j} \cdot |A_{1j}|.$$

For $n = 1$, we define $|A| = a_{11}$.

Whereas a matrix of order $m \times n$ is a rectangular array of $m \cdot n$ elements, determinants are defined only for square matrices and in contrast to matrices, a determinant is a number provided that the elements of the matrix are numbers as well. According to Definition 7.13, a determinant of a matrix of order $n \times n$ can be found by means of n determinants of matrices of order $(n-1) \times (n-1)$. The rule given in Definition 7.13 can also be applied when the elements of the matrices are e.g. functions or mathematical terms.

For $n = 2$ and matrix

$$A = \begin{pmatrix} a_{11} & a_{12} \\ a_{21} & a_{22} \end{pmatrix},$$

we get

$$|A| = a_{11}a_{22} - a_{12}a_{21}.$$

For $n = 3$ and matrix

$$A = \begin{pmatrix} a_{11} & a_{12} & a_{13} \\ a_{21} & a_{22} & a_{23} \\ a_{31} & a_{32} & a_{33} \end{pmatrix},$$

we get

$$|A| = a_{11} \cdot |A_{11}| - a_{12} \cdot |A_{12}| + a_{13} \cdot |A_{13}|$$

$$= a_{11} \cdot \begin{vmatrix} a_{22} & a_{23} \\ a_{32} & a_{33} \end{vmatrix} - a_{12} \cdot \begin{vmatrix} a_{21} & a_{23} \\ a_{31} & a_{33} \end{vmatrix} + a_{13} \cdot \begin{vmatrix} a_{21} & a_{22} \\ a_{31} & a_{32} \end{vmatrix}$$

$$= a_{11}(a_{22}a_{33} - a_{32}a_{23}) - a_{12}(a_{21}a_{33} - a_{31}a_{23}) + a_{13}(a_{21}a_{32} - a_{31}a_{22})$$

$$= a_{11}a_{22}a_{33} + a_{12}a_{23}a_{31} + a_{13}a_{21}a_{32} - a_{11}a_{23}a_{32} - a_{12}a_{21}a_{33} - a_{13}a_{22}a_{31}.$$

The latter computations for $n = 3$ can be done as follows. We add the first two columns at the end as fourth and fifth columns. Then we compute the products of the three diagonals from the left top to the right bottom and add them, and from this value we subtract the sum of the products of the three diagonals from the left bottom to the right top. This procedure is known as *Sarrus's rule* (see Figure 7.1) and works *only* for the case $n = 3$.

Determinants of square submatrices are called *minors*. The *order* of a minor is determined by its number of rows (or columns). A minor $|A_{ij}|$ multiplied by $(-1)^{i+j}$ is called a *cofactor*. The following theorem gives, in addition to Definition 7.13, an alternative way of finding the determinant of a matrix of order $n \times n$.

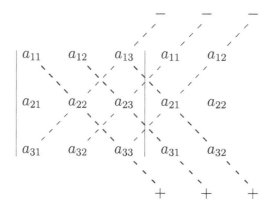

Figure 7.1 Sarrus's rule.

THEOREM 7.2 (Laplace's theorem, cofactor expansion of a determinant)   Let A be a matrix of order $n \times n$. Then the determinant of matrix A is equal to the sum of the products of the elements of one row or column with the corresponding cofactors, i.e.

$$|A| = \sum_{j=1}^{n}(-1)^{i+j} \cdot a_{ij} \cdot |A_{ij}| \qquad \text{(expansion of a determinant by row } i)$$

and

$$|A| = \sum_{i=1}^{n}(-1)^{i+j} \cdot a_{ij} \cdot |A_{ij}| \qquad \text{(expansion of a determinant by column } j).$$

Theorem 7.2 contains Definition 7.13 as a special case. While Definition 7.13 requires cofactor expansion by the first row to evaluate the determinant of matrix A, Theorem 7.2 indicates that we can choose *one arbitrary row or column* of matrix A to which we apply cofactor expansion. Therefore, computations are simplified if we choose one row or column with many zeroes in matrix A.

Example 7.8 We evaluate the determinant of matrix

$$A = \begin{pmatrix} 2 & 3 & 5 \\ 1 & 0 & 2 \\ -1 & -4 & 2 \end{pmatrix}$$

by applying Theorem 7.2 and performing cofactor expansion by the second column. We get

$$|A| = (-1)^3 \cdot 3 \cdot \begin{vmatrix} 1 & 2 \\ -1 & 2 \end{vmatrix} + (-1)^4 \cdot 0 \cdot \begin{vmatrix} 2 & 5 \\ -1 & 2 \end{vmatrix} + (-1)^5 \cdot (-4) \cdot \begin{vmatrix} 2 & 5 \\ 1 & 2 \end{vmatrix}$$

$$= (-3) \cdot [2 - (-2)] + 0 + 4 \cdot (4 - 5) = -12 + 0 - 4 = -16.$$

According to Theorem 7.2, the first determinant of order 2×2 on the right-hand side of the first row above is the minor $|A_{12}|$ obtained by crossing out in matrix A the first row and the second column, i.e.

$$|A_{12}| = \begin{vmatrix} a_{21} & a_{23} \\ a_{31} & a_{33} \end{vmatrix} = \begin{vmatrix} 1 & 2 \\ -1 & 2 \end{vmatrix}.$$

Accordingly, the other minors A_{22} and A_{32} are obtained by crossing out the second column as well as the second and third rows, respectively.

We now give some *properties of determinants*.

THEOREM 7.3 Let A be an $n \times n$ matrix. Then $|A| = |A^T|$.

This is a consequence of Theorem 7.2 since we can apply cofactor expansion by the elements of either one row or one column. Therefore, the determinant of matrix A and the determinant of the transpose A^T are always equal. In the case of a triangular matrix, we can easily evaluate the determinant, as the following theorem shows.

THEOREM 7.4 Let A be an $n \times n$ (lower or upper) triangular matrix. Then

$$|A| = a_{11} \cdot a_{22} \cdot \ldots \cdot a_{nn} = \prod_{i=1}^{n} a_{ii}.$$

As a corollary of Theorem 7.4, we find that the determinant of an identity matrix I is equal to one, i.e. $|I| = 1$.

If we evaluate a determinant using Theorem 7.2, it is desirable that the determinant has an appropriate structure, e.g. computations are simplified if many elements of one row or of one column are equal to zero. For this reason, we are looking for some rules that allow us to evaluate a determinant in an easier form.

THEOREM 7.5 Let A be an $n \times n$ matrix. Then:

(1) If we interchange in A two rows (or two columns), then we get for the resulting matrix A^*: $|A^*| = -|A|$.

(2) If we multiply all elements of a row (or all elements of a column) by $\lambda \in \mathbb{R}$, then we get for the resulting matrix A^*: $|A^*| = \lambda \cdot |A|$.

(3) If we add to all elements of a row (or to all elements of a column) λ times the corresponding elements of another row (column), then we get for the resulting matrix A^*: $|A^*| = |A|$.

COROLLARY 7.1 For the $n \times n$ matrix $B = \lambda A$, we obtain: $|B| = |\lambda A| = \lambda^n \cdot |A|$.

The latter corollary is obtained by a repeated application of part (2) of Theorem 7.5.

THEOREM 7.6 Let A and B be matrices of order $n \times n$. Then $|AB| = |A| \cdot |B|$.

It is worth noting that in general $|A + B| \neq |A| + |B|$. Next, we consider two examples of evaluating determinants.

Example 7.9 We evaluate the determinant of matrix

$$A = \begin{pmatrix} 1 & -2 & 4 & 3 \\ -2 & 5 & -6 & 4 \\ 3 & -4 & 16 & 5 \\ 4 & -11 & 20 & 10 \end{pmatrix}.$$

We apply Theorem 7.5 to generate a determinant having the same value, in which all elements are equal to zero below the diagonal:

$$|A| = \begin{vmatrix} 1 & -2 & 4 & 3 \\ -2 & 5 & -6 & 4 \\ 3 & -4 & 16 & 5 \\ 4 & -11 & 20 & 10 \end{vmatrix} = \begin{vmatrix} 1 & -2 & 4 & 3 \\ 0 & 1 & 2 & 10 \\ 0 & 2 & 4 & -4 \\ 0 & -3 & 4 & -2 \end{vmatrix} = \begin{vmatrix} 1 & -2 & 4 & 3 \\ 0 & 1 & 2 & 10 \\ 0 & 0 & 0 & -24 \\ 0 & 0 & 10 & 28 \end{vmatrix}$$

$$= -\begin{vmatrix} 1 & -2 & 4 & 3 \\ 0 & 1 & 2 & 10 \\ 0 & 0 & 10 & 28 \\ 0 & 0 & 0 & -24 \end{vmatrix} = 240.$$

In the first transformation step, we have generated zeroes in rows 2 to 4 of the first column. To this end, we have multiplied the first row by 2 and added it to the second row, yielding the new second row. Analogously, we have multiplied the first row by -3 and added it to the third row, and we have multiplied the first row by -4 and added it to the fourth row. In the next transformation step, we have generated zeroes in rows 3 and 4 of the second column. This means we have multiplied the second row (of the second determinant) by -2 and added it to the third row, and we have multiplied the second row by 3 and added it to the fourth row (application of part (3) of Theorem 7.5). Additionally, we have interchanged rows 3 and 4, which changes the sign of the determinant. Finally, we applied Theorem 7.4.

Example 7.10   We want to determine for which values of t the determinant

$$|A| = \begin{vmatrix} 3 & 1 & 2 \\ 2+2t & 0 & 4 \\ 1 & 2-t & 0 \end{vmatrix}$$

is equal to zero. We first apply expansion by column 3 according to Theorem 7.2 and obtain

$$|A| = 2 \cdot \begin{vmatrix} 2+2t & 0 \\ 1 & 2-t \end{vmatrix} - 4 \cdot \begin{vmatrix} 3 & 1 \\ 1 & 2-t \end{vmatrix} + 0 \cdot \begin{vmatrix} 3 & 1 \\ 2+2t & 0 \end{vmatrix}$$

$$= 2 \cdot [(2+2t) \cdot (2-t)] - 4 \cdot (6 - 3t - 1) = -4t^2 + 16t - 12.$$

From $|A| = 0$, we obtain

$$-4t^2 + 16t - 12 = 0$$

which corresponds to

$$t^2 - 4t + 3 = 0.$$

This quadratic equation has the two real roots $t_1 = 1$ and $t_2 = 3$. Thus, for $t_1 = 1$ and $t_2 = 3$, we get $|A| = 0$. To find the value of $|A|$, we did not apply Theorem 7.5. Using Theorem 7.5, we can transform the determinant such that we have many zeroes in one row or column (which simplifies our remaining computations when applying Theorem 7.2). Multiplying each element of row 1 in the initial determinant by -2 and adding each element to the corresponding element of row 2, we obtain

$$|A| = \begin{vmatrix} 3 & 1 & 2 \\ -4+2t & -2 & 0 \\ 1 & 2-t & 0 \end{vmatrix}.$$

In this case, when expanding by column 3, we have to determine the value of only one subdeterminant. We get

$$|A| = 2 \begin{vmatrix} -4+2t & -2 \\ 1 & 2-t \end{vmatrix}$$

which is equal to the already obtained value $-4t^2 + 16t - 12$.

The following theorem presents some cases when the determinant of a matrix A is equal to zero.

THEOREM 7.7   Let A be a matrix of order $n \times n$. Assume that one of the following propositions holds:

(1) Two rows (columns) are equal.
(2) All elements of a row (column) of A are equal to zero.
(3) A row (column) is the sum of multiples of other rows (columns).

Then $|A| = 0$.

We next introduce the notions of a singular and a regular matrix.

Definition 7.14 A square matrix A is said to be *singular* if $|A| = 0$ and *regular* (or *non-singular*) if $|A| \neq 0$.

We now consider a first possibility to solve special systems of linear equations. This approach is named after the German mathematician Cramer and uses determinants to find the values of the variables.

Cramer's rule

Let

$$A_{(n,n)} \cdot \mathbf{x}_{(n,1)} = \mathbf{b}_{(n,1)}$$

be a system of linear equations, i.e.

$$a_{11}x_1 + a_{12}x_2 + \cdots + a_{1n}x_n = b_1$$
$$a_{21}x_1 + a_{22}x_2 + \cdots + a_{2n}x_n = b_2$$
$$\vdots \qquad\qquad \vdots \;\; \vdots$$
$$a_{n1}x_1 + a_{n2}x_2 + \cdots + a_{nn}x_n = b_n,$$

and we assume that A is regular (i.e. $|A| \neq 0$). Moreover, let $A_j(\mathbf{b})$ denote the matrix which is obtained if the jth column of A is replaced by vector $\mathbf{b}$, i.e.

$$|A_j(\mathbf{b})| = \begin{vmatrix} a_{11} & a_{12} & \cdots & a_{1,j-1} & b_1 & a_{1,j+1} & \cdots & a_{1n} \\ a_{21} & a_{22} & \cdots & a_{2,j-1} & b_2 & a_{2,j+1} & \cdots & a_{2n} \\ \vdots & \vdots & & \vdots & \vdots & \vdots & & \vdots \\ a_{n1} & a_{n2} & \cdots & a_{n,j-1} & b_n & a_{n,j+1} & \cdots & a_{nn} \end{vmatrix}.$$

Then

$$x_j = \frac{|A_j(\mathbf{b})|}{|A|} \qquad \text{for } j = 1, 2, \dots, n$$

is the *unique* solution of the system $A\mathbf{x} = \mathbf{b}$ of linear equations.

Cramer's rule makes it possible to solve special systems of linear equations. However, this rule is appropriate only when the determinant of matrix A is different from zero (and thus a unique solution of the system of linear equations exists). It is also a disadvantage of this method that, if we obtain $|A| = 0$, we must stop our computations and we have to apply some more general method for solving systems of linear equations, as we will discuss in Chapter 8. Moreover, from a practical point of view, Cramer's rule is applicable only in the case of a rather small number of n.

Example 7.11   Consider the system of linear equations

$$
\begin{aligned}
3x_1 &+ 4x_2 &+ 2x_3 &= 1 \\
x_1 &- x_2 &- 3x_3 &= 7 \\
2x_1 & &+ x_3 &= 4
\end{aligned}
$$

which we solve by applying Cramer's rule. We first evaluate the determinant of matrix

$$
A = \begin{pmatrix} 3 & 4 & 2 \\ 1 & -1 & -3 \\ 2 & 0 & 1 \end{pmatrix}.
$$

Using expansion by row 3 according to Theorem 7.2, we obtain

$$
|A| = \begin{vmatrix} 3 & 4 & 2 \\ 1 & -1 & -3 \\ 2 & 0 & 1 \end{vmatrix} = 2 \cdot \begin{vmatrix} 4 & 2 \\ -1 & -3 \end{vmatrix} - 0 \cdot \begin{vmatrix} 3 & 2 \\ 1 & -3 \end{vmatrix} + 1 \cdot \begin{vmatrix} 3 & 4 \\ 1 & -1 \end{vmatrix}
$$

$$
= 2 \cdot (-12 + 2) - 0 + 1 \cdot (-3 - 4) = -27.
$$

In the above computations, we have decided to expand by row 3 since there is already one zero contained in this row and therefore we have to evaluate only two minors of order two. For this reason, one could also choose expansion by column 2. Since $|A| \neq 0$, we know now that the given system has a unique solution which can be found by Cramer's rule. Continuing, we get

$$
|A_1(\mathbf{b})| = \begin{vmatrix} 1 & 4 & 2 \\ 7 & -1 & -3 \\ 4 & 0 & 1 \end{vmatrix} = 4 \cdot \begin{vmatrix} 4 & 2 \\ -1 & -3 \end{vmatrix} + 1 \cdot \begin{vmatrix} 1 & 4 \\ 7 & -1 \end{vmatrix}
$$

$$
= 4 \cdot (-12 + 2) + 1 \cdot (-1 - 28) = -69;
$$

$$
|A_2(\mathbf{b})| = \begin{vmatrix} 3 & 1 & 2 \\ 1 & 7 & -3 \\ 2 & 4 & 1 \end{vmatrix} = \begin{vmatrix} 0 & -20 & 11 \\ 1 & 7 & -3 \\ 0 & -10 & 7 \end{vmatrix} = -1 \cdot \begin{vmatrix} -20 & 11 \\ -10 & 7 \end{vmatrix}
$$

$$
= -1 \cdot (-140 + 110) = 30
$$

and

$$
|A_3(\mathbf{b})| = \begin{vmatrix} 3 & 4 & 1 \\ 1 & -1 & 7 \\ 2 & 0 & 4 \end{vmatrix} = -4 \cdot \begin{vmatrix} 1 & 7 \\ 2 & 4 \end{vmatrix} + (-1) \cdot \begin{vmatrix} 3 & 1 \\ 2 & 4 \end{vmatrix}
$$

$$
= -4 \cdot (4 - 14) - 1 \cdot (12 - 2) = 30.
$$

For finding $|A_1(\mathbf{b})|$ and $|A_3(\mathbf{b})|$, we have used Theorem 7.2. In the former case, we have again applied expansion by row 3, and in the latter case, we have applied expansion by column 2. For finding $|A_2(\mathbf{b})|$, we have first used Theorem 7.5, part (3). Since there are no zero elements, we have transformed the determinant such that in one column or row (in our case column 1) all but one elements are equal to zero so that the application of Theorem 7.2

is reduced to finding the value of one minor of order two. By Cramer's rule, we get

$$x_1 = \frac{|A_1(\mathbf{b})|}{|A|} = \frac{69}{27} = \frac{23}{9};$$

$$x_2 = \frac{|A_2(\mathbf{b})|}{|A|} = -\frac{30}{27} = -\frac{10}{9};$$

$$x_3 = \frac{|A_3(\mathbf{b})|}{|A|} = -\frac{30}{27} = -\frac{10}{9}.$$

7.4 LINEAR MAPPINGS

Definition 7.15 A mapping $A : \mathbb{R}^n \to \mathbb{R}^m$ is called *linear* if

$$A(\mathbf{x}^1 + \mathbf{x}^2) = A(\mathbf{x}^1) + A(\mathbf{x}^2) \qquad \text{for all } \mathbf{x}^1, \mathbf{x}^2 \in \mathbb{R}^n$$

and

$$A(\lambda \mathbf{x}) = \lambda A(\mathbf{x}) \qquad \text{for all } \lambda \in \mathbb{R} \text{ and } \mathbf{x} \in \mathbb{R}^n.$$

A linear mapping is therefore defined in such a way that the image of the sum of two vectors is equal to the (vector) sum of the two images, and the image of the multiple of a vector is equal to the multiple of the image of the vector.

A linear mapping $A : \mathbb{R}^n \to \mathbb{R}^m$ can be described by means of a matrix $A = (a_{ij})$ of order $m \times n$ such that

$$\mathbf{x} = \begin{pmatrix} x_1 \\ x_2 \\ \vdots \\ x_n \end{pmatrix} \in \mathbb{R}^n \longmapsto \mathbf{y} = \begin{pmatrix} y_1 \\ y_2 \\ \vdots \\ y_m \end{pmatrix} = \begin{pmatrix} a_{11} & a_{12} & \cdots & a_{1n} \\ a_{21} & a_{22} & \cdots & a_{2n} \\ \vdots & \vdots & & \vdots \\ a_{m1} & a_{m2} & \cdots & a_{mn} \end{pmatrix} \cdot \begin{pmatrix} x_1 \\ x_2 \\ \vdots \\ x_n \end{pmatrix} \in \mathbb{R}^m$$

Definition 7.16 The set of all n-dimensional vectors $\mathbf{x}$ which are mapped by $A :$ $\mathbb{R}^n \to \mathbb{R}^m$ into the m-dimensional zero vector $\mathbf{0}$ is called the *kernel* of the mapping, abbreviated ker A, i.e.

$$\ker A = \{\mathbf{x}_{(n,1)} \in \mathbb{R}^n \mid A_{(m,n)} \cdot \mathbf{x}_{(n,1)} = \mathbf{0}_{(m,1)}\}.$$

The kernel of a linear mapping is also called *null space*. Determining the kernel of a linear mapping requires the solution of a system of linear equations with the components of vector $\mathbf{x}$ as unknowns, which we will treat in detail in Chapter 8. The following theorem shows how a composition of two linear mappings can be described by a matrix.

THEOREM 7.8   Let $B : \mathbb{R}^n \to \mathbb{R}^s$ and $A : \mathbb{R}^s \to \mathbb{R}^m$ be linear mappings. Then the composite mapping $A \circ B : \mathbb{R}^n \to \mathbb{R}^m$ is a linear mapping described by matrix

$$C_{(m,n)} = A_{(m,s)} \cdot B_{(s,n)}.$$

Example 7.12   Assume that a firm produces by means of q raw materials $R_1, R_2, \ldots, R_q$ the m intermediate products $S_1, S_2, \ldots, S_m$, and with these intermediate products and with the q raw materials the n final products $F_1, F_2, \ldots, F_n$. Denote by

r_{ij}^S the number of units of raw material R_i which are necessary for the production of one unit of intermediate product S_j,

s_{jk}^F the number of units of intermediate product S_j which are necessary for the production of one unit of final product F_k,

r_{ik}^F the number of units of raw material R_i which are additionally necessary for the production of one unit of final product F_k.

We introduce the matrices $R^S = (r_{ij}^S)$ of order $q \times m$, $S^F = (s_{ij}^F)$ of order $m \times n$ and $R^F = (r_{ij}^F)$ of order $q \times n$ and denote by $\mathbf{x}^F = (x_1^F, x_2^F, \ldots, x_n^F)^{\mathrm{T}}$ the production vector of the final products and by $\mathbf{x}^S = (x_1^S, x_2^S, \ldots, x_m^S)$ the production vector of the intermediate products. We want to determine the required vector $\mathbf{y}$ of raw materials. First, raw materials according to the matrix equation

$$\mathbf{y}_{(q,1)}^1 = R_{(q,n)}^F \cdot \mathbf{x}_{(n,1)}^F$$

are required for the final products. Moreover, we get for vector $\mathbf{x}^S$ the following matrix equation:

$$\mathbf{x}_{(m,1)}^S = S_{(m,n)}^F \cdot \mathbf{x}_{(n,1)}^F,$$

and for the production of intermediate products given by vector $\mathbf{x}^S$, the required vector $\mathbf{y}^2$ of raw materials is given by

$$\mathbf{y}_{(q,1)}^2 = R_{(q,m)}^S \cdot \mathbf{x}_{(m,1)}^S = R_{(q,m)}^S \cdot S_{(m,n)}^F \cdot \mathbf{x}_{(n,1)}^F.$$

Thus, we get the following relationship between the q-vector $\mathbf{y}$ of required raw materials and the n-dimensional vector $\mathbf{x}^F$:

$$\begin{aligned} \mathbf{y}_{(q,1)} &= \mathbf{y}_{(q,1)}^1 + \mathbf{y}_{(q,1)}^2 \\ &= (R^F + R^S \cdot S^F)_{(q,n)} \cdot \mathbf{x}_{(n,1)}^F, \end{aligned}$$

i.e. $R^F + R^S \circ S^F$ represents a linear mapping from the n-space $\mathbb{R}_+^n$ into the q-space $\mathbb{R}_+^q$. This linear mapping can be described in the following way:

$$\mathbf{x}^F \in \mathbb{R}_+^n \longmapsto (R^F + R^S \circ S^F)(\mathbf{x}^F) = R^F \cdot \mathbf{x}^F + R^S \cdot S^F \cdot \mathbf{x}^F = \mathbf{y} \in \mathbb{R}_+^q,$$

i.e. by this linear mapping a feasible n-dimensional production vector of the final products is mapped into a q-dimensional vector of required raw materials.

Next, we introduce the inverse mapping of a linear mapping.

THEOREM 7.9   Let $A : \mathbb{R}^n \to \mathbb{R}^n$ be a linear mapping described by a matrix A of order $n \times n$, i.e.

$$
\mathbf{x} = \begin{pmatrix} x_1 \\ x_2 \\ \vdots \\ x_n \end{pmatrix} \in \mathbb{R}^n \longmapsto A\mathbf{x} = \mathbf{y} = \begin{pmatrix} y_1 \\ y_2 \\ \vdots \\ y_n \end{pmatrix} \in \mathbb{R}^n,
$$

and let matrix A be regular. Then there exists a unique inverse mapping A^{-1} such that

$$
\mathbf{y} = \begin{pmatrix} y_1 \\ y_2 \\ \vdots \\ y_n \end{pmatrix} \in \mathbb{R}^n \longmapsto A^{-1}\mathbf{y} = \mathbf{x} = \begin{pmatrix} x_1 \\ x_2 \\ \vdots \\ x_n \end{pmatrix} \in \mathbb{R}^n.
$$

Obviously, the composite mapping $A \circ A^{-1} = A^{-1} \circ A$ is the identical mapping I.

7.5   THE INVERSE MATRIX

Definition 7.17   Given is a square matrix A. If there exists a matrix A^{-1} such that

$$
AA^{-1} = A^{-1}A = I,
$$

then we say that A^{-1} is an *inverse* or *inverse matrix* of A.

We note that the inverse A^{-1} characterizes the *inverse mapping* of a linear mapping described by matrix A. The following theorem answers the question: under which condition does the inverse of a matrix A exist?

THEOREM 7.10   Let A be a matrix of order $n \times n$. Then:

(1) If matrix A is regular, then there exists a unique inverse matrix A^{-1}.
(2) If matrix A is singular, then A does not have an inverse.

If the inverse A^{-1} of matrix A exists, we also say that matrix A is *invertible*. According to Theorem 7.10, a square matrix A is invertible if and only if $|A| \neq 0$.

Solving equations by matrix inversion

Consider the matrix equations $AX = B$ and $YA = C$, where matrix B of order $n \times m$ and matrix C of order $m \times n$ are given. The matrices X and Y are assumed to be unknown. From the above equations, it follows that matrix X has the order $n \times m$ and matrix Y has the order $m \times n$. For $|A| \neq 0$, the inverse of matrix A exists and we get

$$AX = B \quad \Longleftrightarrow \quad X = A^{-1}B;$$
$$YA = C \quad \Longleftrightarrow \quad Y = CA^{-1}.$$

The equations on the right-hand side are obtained by multiplying in the former case, equation $AX = B$ from the *left* by A^{-1} and in the latter case, equation $YA = C$ from the *right* by A^{-1}. Remember that matrix multiplication is not commutative.

Example 7.13   Let the matrix equation

$$4X = X(2B - A) + 3(A + X)$$

be given, where A, B and X are $n \times n$ matrices. We solve the above equation for matrix X and obtain:

$$4X = 2XB - XA + 3A + 3X$$
$$X - 2XB + XA = 3A$$
$$X(I - 2B + A) = 3A$$
$$X = 3A(I - 2B + A)^{-1}.$$

In the second to last step, we have factored out matrix X from the left and we have used $X = XI$. Thus, if the inverse of matrix $I - 2B + A$ exists, matrix X is uniquely determined.

The following theorem presents a first possibility of computing the inverse of a matrix.

THEOREM 7.11   Let A be a regular matrix of order $n \times n$. Then the inverse matrix A^{-1} is given by

$$A^{-1} = \frac{1}{|A|}\left((-1)^{i+j} \cdot |A_{ij}|\right)^{\mathrm{T}} = \frac{1}{|A|}\begin{pmatrix} +|A_{11}| & -|A_{21}| & +\cdots & \pm|A_{n1}| \\ -|A_{12}| & +|A_{22}| & -\cdots & \mp|A_{n2}| \\ \vdots & \vdots & & \vdots \\ \pm|A_{1n}| & \mp|A_{2n}| & \pm\cdots & +|A_{nn}| \end{pmatrix}.$$

The matrix $\left((-1)^{i+j} \cdot |A_{ij}|\right)^{\mathrm{T}}$ is the transpose of the matrix of the cofactors, which is called the *adjoint* of matrix A and denoted by $adj(A)$.

To determine the inverse of a matrix A of order $n \times n$, the evaluation of a determinant of order $n \times n$ and of n^2 minors of order $n - 1$ is required. Thus, with increasing order of n, the application of Theorem 7.11 becomes rather time-consuming.

Example 7.14 We consider the matrix

$$A = \begin{pmatrix} 1 & 2 & -1 \\ 2 & 1 & 0 \\ -1 & 0 & 1 \end{pmatrix},$$

and we want to determine the inverse A^{-1} of matrix A. In order to apply Theorem 7.2, we first evaluate the determinant of A:

$$|A| = \begin{vmatrix} 1 & 2 & -1 \\ 2 & 1 & 0 \\ -1 & 0 & 1 \end{vmatrix} = \begin{vmatrix} 1 & 2 & -1 \\ 2 & 1 & 0 \\ 0 & 2 & 0 \end{vmatrix} = -1 \begin{vmatrix} 2 & 1 \\ 0 & 2 \end{vmatrix} = -4.$$

In the above computations, we have first added rows 1 and 3 and then applied cofactor expansion by column 3. Calculating the minors, we obtain

$$|A_{11}| = \begin{vmatrix} 1 & 0 \\ 0 & 1 \end{vmatrix} = 1; \quad |A_{12}| = \begin{vmatrix} 2 & 0 \\ -1 & 1 \end{vmatrix} = 2; \quad |A_{13}| = \begin{vmatrix} 2 & 1 \\ -1 & 0 \end{vmatrix} = 1;$$

$$|A_{21}| = \begin{vmatrix} 2 & -1 \\ 0 & 1 \end{vmatrix} = 2; \quad |A_{22}| = \begin{vmatrix} 1 & -1 \\ -1 & 1 \end{vmatrix} = 0; \quad |A_{23}| = \begin{vmatrix} 1 & 2 \\ -1 & 0 \end{vmatrix} = 2;$$

$$|A_{31}| = \begin{vmatrix} 2 & -1 \\ 1 & 0 \end{vmatrix} = 1; \quad |A_{32}| = \begin{vmatrix} 1 & -1 \\ 2 & 0 \end{vmatrix} = 2; \quad |A_{33}| = \begin{vmatrix} 1 & 2 \\ 2 & 1 \end{vmatrix} = -3.$$

With the above computations, we get the inverse matrix

$$A^{-1} = \frac{1}{|A|} \begin{pmatrix} |A_{11}| & -|A_{21}| & |A_{31}| \\ -|A_{12}| & |A_{22}| & -|A_{32}| \\ |A_{13}| & -|A_{23}| & |A_{33}| \end{pmatrix} = \frac{1}{-4} \begin{pmatrix} 1 & -2 & 1 \\ -2 & 0 & -2 \\ 1 & -2 & -3 \end{pmatrix}$$

$$= \begin{pmatrix} -\dfrac{1}{4} & \dfrac{1}{2} & -\dfrac{1}{4} \\ \dfrac{1}{2} & 0 & \dfrac{1}{2} \\ -\dfrac{1}{4} & \dfrac{1}{2} & \dfrac{3}{4} \end{pmatrix}.$$

Example 7.15 Consider the matrix equation

$$AX = X - B$$

with

$$A = \begin{pmatrix} 3 & a \\ 10 & 6 \end{pmatrix} \quad \text{and} \quad B = \begin{pmatrix} 1 & 0 \\ -2 & 1 \end{pmatrix},$$

where $a \in \mathbb{R}$. We want to determine X provided that this matrix is uniquely determined. Replacing X by IX, where I is the identity matrix, and solving the above matrix equation for X, we obtain first

$$(I - A)X = B$$

and then

$$X = (I - A)^{-1}B$$

with

$$I - A = \begin{pmatrix} -2 & -a \\ -10 & -5 \end{pmatrix}.$$

To check whether the inverse of matrix $I - A$ exists, we determine

$$|I - A| = \begin{vmatrix} -2 & -a \\ -10 & -5 \end{vmatrix} = 10 - 10a = 10(1 - a).$$

For $a \neq 1$, we have $|I - A| \neq 0$, and thus the inverse of $I - A$ exists and the given matrix X is uniquely determined. On the contrary, for $a = 1$, the inverse of $I - A$ does not exist, and thus matrix X is not uniquely determined. We continue and obtain for $a \neq 1$

$$(I - A)^{-1} = \frac{1}{10(1 - a)} \begin{pmatrix} -5 & a \\ 10 & -2 \end{pmatrix}.$$

By multiplying matrices $(I - A)^{-1}$ and B, we finally obtain

$$X = \frac{1}{10(1 - a)} \begin{pmatrix} -5 - 2a & a \\ 14 & -2 \end{pmatrix} = \begin{pmatrix} -\dfrac{5 + 2a}{10(1 - a)} & \dfrac{a}{10(1 - a)} \\ \dfrac{7}{5(1 - a)} & -\dfrac{1}{5(1 - a)} \end{pmatrix}.$$

We now summarize some rules for operating with the inverses of matrices, assuming that the matrices A and B are of order $n \times n$ and that the inverses A^{-1} and B^{-1} exist.

Rules for calculations with inverses

(1) $(A^{-1})^{-1} = A$;
(2) $(A^{T})^{-1} = (A^{-1})^{T}$;
(3) $(AB)^{-1} = B^{-1}A^{-1}$;
(4) $(\lambda A)^{-1} = \dfrac{1}{\lambda} \cdot A^{-1} \qquad (\lambda \in \mathbb{R} \setminus \{0\})$;

(5) $|A^{-1}| = \dfrac{1}{|A|}$.

We prove the validity of rule (5). Using $|I| = 1$ and Theorem 7.6, we obtain

$$1 = |I| = |AA^{-1}| = |A| \cdot |A^{-1}|$$

from which rule (5) follows. As a generalization of rule (3), we obtain

$$(A^{n})^{-1} = (A^{-1})^{n} \qquad \text{for all } n \in \mathbb{N}.$$

We have seen that, if the order of the matrix is large, the determination of the inverse of the matrix can be rather time-consuming. In some cases, it is possible to apply an easier approach. The following theorem treats such a case, which occurs (as we discuss later) in several economic applications.

THEOREM 7.12 Let C be a triangular matrix of order $n \times n$ with $c_{ii} = 0$ for $i = 1, 2, \ldots, n$ and $A = I - C$, where I is the identity matrix of order $n \times n$. Then the inverse A^{-1} is given by

$$A^{-1} = (I - C)^{-1} = I + C + C^2 + \cdots + C^{n-1}.$$

The advantage of the formula presented in Theorem 7.12 over the formula given in Theorem 7.11 is that the determination of the inverse is done without using determinants. The latter formula uses only matrix multiplication and addition, and the matrices to be considered contain a lot of zeroes as elements.

7.6 AN ECONOMIC APPLICATION: INPUT–OUTPUT MODEL

We finish this chapter with an important application of matrices in economics. Assume that we have a set of n firms each of them producing one good only. Production of each good j requires an input of a_{ij} units of good i per unit of good j produced. (The coefficients a_{ij} are also known as *input–output* coefficients.) Production takes place with fixed techniques (i.e. the values a_{ij} do not change). Let $x = (x_1, x_2, \ldots, x_n)^T$ be the vector giving the total amount of goods produced, let matrix $A = (a_{ij})$ of order $n \times n$ be the so-called *technology* or *input–output* matrix and let $y = (y_1, y_2, \ldots, y_n)^T$ be the demand vector for the use of the n goods. Considering the ith good, there are $a_{ij}x_j$ units required as input for the production of x_j units of good j, and y_i units are required as final customer demand. Therefore, the amount x_i of good i has to satisfy the equation

$$x_i = a_{i1}x_1 + a_{i2}x_2 + \cdots + a_{in}x_n + y_i, \quad i = 1, 2, \ldots, n.$$

Expressing the latter n equations in matrix notation, we get the equation

$$x = Ax + y$$

which can be rewritten as

$$Ix - Ax = (I - A)x = y,$$

where I is the identity matrix of order $n \times n$. The above model expresses that vector x, giving the total output of goods produced, is equal to the sum of vector Ax describing the internal consumption of the goods and vector y representing the customer demand. The model is referred to as an *input–output* or *Leontief* model. The customer demand vector y is in general different from the zero vector, and in this case we have an *open* Leontief model. The equation

$$(I - A)x = y$$

represents a linear mapping $\mathbb{R}^n \to \mathbb{R}^n$ described by matrix $I - A$. If the total possible output x is known, we are interested in getting the possible amount y of goods left for the customer. Conversely, a customer demand vector y can be given and we ask for the total output vector x

required to satisfy the customer demand. In the latter case, we have to consider the inverse mapping

$$\mathbf{x} = (I - A)^{-1}\mathbf{y}.$$

So we have to determine the inverse of matrix $I - A$ to answer this question in the latter case.

Example 7.16   Consider a numerical example for an open input–output model. Let matrix A and vector $\mathbf{y}$ be given as follows:

$$A = \begin{pmatrix} \dfrac{1}{5} & \dfrac{1}{5} & 0 \\[6pt] \dfrac{2}{5} & \dfrac{3}{5} & \dfrac{1}{5} \\[6pt] \dfrac{3}{5} & \dfrac{2}{5} & \dfrac{1}{5} \end{pmatrix} \qquad \text{and} \qquad \mathbf{y} = \begin{pmatrix} 2 \\ 0 \\ 1 \end{pmatrix}.$$

Then we get

$$I - A = \begin{pmatrix} \dfrac{4}{5} & -\dfrac{1}{5} & 0 \\[6pt] -\dfrac{2}{5} & \dfrac{2}{5} & -\dfrac{1}{5} \\[6pt] -\dfrac{3}{5} & -\dfrac{2}{5} & \dfrac{4}{5} \end{pmatrix} = \dfrac{1}{5} \begin{pmatrix} 4 & -1 & 0 \\ -2 & 2 & -1 \\ -3 & -2 & 4 \end{pmatrix}.$$

Setting

$$B = \begin{pmatrix} 4 & -1 & 0 \\ -2 & 2 & -1 \\ -3 & -2 & 4 \end{pmatrix},$$

we have $I - A = B/5$. Instead of inverting matrix $I - A$, we invert matrix B, which has only integers, and finally take into account that matrix equation $(I - A)^{-1} = 5 \cdot B^{-1}$ holds. First, we obtain $|B| = 13$ and thus the inverse of matrix B exists. Applying Theorem 7.11, we get

$$B^{-1} = \dfrac{1}{13} \begin{pmatrix} 6 & 4 & 1 \\ 11 & 16 & 4 \\ 10 & 11 & 6 \end{pmatrix}.$$

Then we get

$$(I - A)^{-1} = 5B^{-1} = \dfrac{5}{13} \begin{pmatrix} 6 & 4 & 1 \\ 11 & 16 & 4 \\ 10 & 11 & 6 \end{pmatrix}.$$

Finally, we obtain vector $\mathbf{x}$ as follows:

$$\mathbf{x} = \dfrac{5}{13} \begin{pmatrix} 6 & 4 & 1 \\ 11 & 16 & 4 \\ 10 & 11 & 6 \end{pmatrix} \begin{pmatrix} 2 \\ 0 \\ 1 \end{pmatrix} = \dfrac{5}{13} \begin{pmatrix} 13 \\ 26 \\ 26 \end{pmatrix} = \begin{pmatrix} 5 \\ 10 \\ 10 \end{pmatrix}.$$

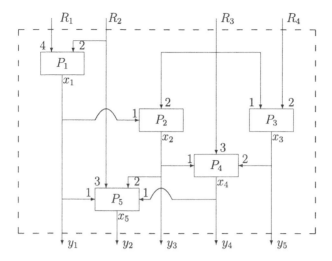

Figure 7.2 Relationships between raw materials and products in Example 7.17.

Next, we consider a second example of this type of problem.

Example 7.17 A firm produces by means of four raw materials R_1, R_2, R_3 and R_4 five products P_1, P_2, P_3, P_4 and P_5, where some of these products are also used as intermediate products. The relationships are given in the graph presented in Figure 7.2. The numbers beside the arrows describe how many units of raw material R_k and product P_i, respectively, are necessary for one unit of $P_j, j = 1, 2, \ldots, 5$. Vector $\mathbf{x} = (x_1, x_2, x_3, x_4, x_5)^\mathrm{T}$ describes the produced units (total output) of product P_i and $\mathbf{y} = (y_1, y_2, y_3, y_4, y_5)^\mathrm{T}$ denotes the final demand (export) for the output of the products P_i.

(1) We first determine a relationship between vectors $\mathbf{x}$ and $\mathbf{y}$. Let the technology matrix

$$A = \begin{pmatrix} 0 & 1 & 0 & 0 & 1 \\ 0 & 0 & 0 & 1 & 2 \\ 0 & 0 & 0 & 2 & 0 \\ 0 & 0 & 0 & 0 & 1 \\ 0 & 0 & 0 & 0 & 0 \end{pmatrix}$$

be given. Then

$$\mathbf{x} = A\mathbf{x} + \mathbf{y}$$

or, correspondingly,

$$(I - A)\mathbf{x} = \mathbf{y}.$$

In detail, we have

$$
\begin{pmatrix}
1 & -1 & 0 & 0 & -1 \\
0 & 1 & 0 & -1 & -2 \\
0 & 0 & 1 & -2 & 0 \\
0 & 0 & 0 & 1 & -1 \\
0 & 0 & 0 & 0 & 1
\end{pmatrix}
\begin{pmatrix}
x_1 \\ x_2 \\ x_3 \\ x_4 \\ x_5
\end{pmatrix}
=
\begin{pmatrix}
y_1 \\ y_2 \\ y_3 \\ y_4 \\ y_5
\end{pmatrix}.
$$

(2) We now calculate the final demand when the total output is given by $\mathbf{x} = (220, 110, 120, 40, 20)^{\mathrm{T}}$. We obtain

$$
\mathbf{y} = (I - A)\mathbf{x} =
\begin{pmatrix}
1 & -1 & 0 & 0 & -1 \\
0 & 1 & 0 & -1 & -2 \\
0 & 0 & 1 & -2 & 0 \\
0 & 0 & 0 & 1 & -1 \\
0 & 0 & 0 & 0 & 1
\end{pmatrix}
\begin{pmatrix}
220 \\ 110 \\ 120 \\ 40 \\ 20
\end{pmatrix}
=
\begin{pmatrix}
90 \\ 30 \\ 40 \\ 20 \\ 20
\end{pmatrix}.
$$

(3) Let $\mathbf{y} = (60, 30, 40, 10, 20)^{\mathrm{T}}$ be the given final demand vector. We determine the production vector $\mathbf{x}$ and the required units of raw materials for this case. To this end, we need the inverse of matrix $I - A$. Since $I - A$ is an upper triangular matrix with all diagonal elements equal to zero, we can apply Theorem 7.12 for determining $(I - A)^{-1}$, and we obtain

$$
(I - A)^{-1} = I + A + A^2 + A^3 + A^4.
$$

Using

$$
A^2 =
\begin{pmatrix}
0 & 0 & 0 & 1 & 2 \\
0 & 0 & 0 & 0 & 1 \\
0 & 0 & 0 & 0 & 2 \\
0 & 0 & 0 & 0 & 0 \\
0 & 0 & 0 & 0 & 0
\end{pmatrix},
\quad
A^3 =
\begin{pmatrix}
0 & 0 & 0 & 0 & 1 \\
0 & 0 & 0 & 0 & 0 \\
0 & 0 & 0 & 0 & 0 \\
0 & 0 & 0 & 0 & 0 \\
0 & 0 & 0 & 0 & 0
\end{pmatrix}
$$

we get that A^4 is the zero matrix and therefore

$$
(I - A)^{-1} =
\begin{pmatrix}
1 & 1 & 0 & 1 & 4 \\
0 & 1 & 0 & 1 & 3 \\
0 & 0 & 1 & 2 & 2 \\
0 & 0 & 0 & 1 & 1 \\
0 & 0 & 0 & 0 & 1
\end{pmatrix}.
$$

Then we get

$$
\mathbf{x} = (I - A)^{-1}\mathbf{y} =
\begin{pmatrix}
1 & 1 & 0 & 1 & 4 \\
0 & 1 & 0 & 1 & 3 \\
0 & 0 & 1 & 2 & 2 \\
0 & 0 & 0 & 1 & 1 \\
0 & 0 & 0 & 0 & 1
\end{pmatrix}
\begin{pmatrix}
60 \\ 30 \\ 40 \\ 10 \\ 20
\end{pmatrix}
=
\begin{pmatrix}
180 \\ 100 \\ 100 \\ 30 \\ 20
\end{pmatrix}.
$$

For the required vector of raw materials $\mathbf{r} = (r_1, r_2, r_3, r_4)^T$, where r_i denotes the quantity of required units of raw material R_i, we obtain from Figure 7.2:

$$
\begin{array}{rcl}
r_1 & = & 4x_1 \\
r_2 & = & 2x_1 \quad\quad\quad\quad\quad\quad\quad + \quad 3x_5 \\
r_3 & = & \quad\quad 2x_2 \;+\; x_3 \;+\; 3x_4 \\
r_4 & = & \quad\quad 2x_3
\end{array}
$$

i.e.

$$
\mathbf{r} = \begin{pmatrix} r_1 \\ r_2 \\ r_3 \\ r_4 \end{pmatrix} = \begin{pmatrix} 4 & 0 & 0 & 0 & 0 \\ 2 & 0 & 0 & 0 & 3 \\ 0 & 2 & 1 & 3 & 0 \\ 0 & 0 & 2 & 0 & 0 \end{pmatrix} \begin{pmatrix} x_1 \\ x_2 \\ x_3 \\ x_4 \\ x_5 \end{pmatrix}.
$$

For the determined vector $\mathbf{x} = (180, 100, 100, 30, 20)^T$, we get the following vector $\mathbf{r}$:

$$
\mathbf{r} = \begin{pmatrix} r_1 \\ r_2 \\ r_3 \\ r_4 \end{pmatrix} = \begin{pmatrix} 720 \\ 420 \\ 390 \\ 200 \end{pmatrix}.
$$

EXERCISES

7.1 Given are the matrices

$$
A = \begin{pmatrix} 3 & 4 & 1 \\ 1 & 2 & -2 \end{pmatrix}; \quad B = \begin{pmatrix} 2 & 1 \\ 1 & 2 \\ -1 & 0 \end{pmatrix};
$$

$$
C = \begin{pmatrix} 1 & 0 & -1 \\ -1 & 1 & 1 \end{pmatrix} \quad \text{and} \quad D = \begin{pmatrix} 3 & 1 & 0 \\ 4 & 2 & 2 \end{pmatrix}.
$$

(a) Find the transposes. Check whether some matrices are equal.
(b) Calculate $A + D$, $A - D$, $A^T - B$ and $C - D$.
(c) Find $A + 3(B^T - 2D)$.

7.2 Find a symmetric and an antisymmetric matrix so that their sum is equal to

$$
A = \begin{pmatrix} 2 & -1 & 0 \\ 1 & 1 & -1 \\ 3 & 0 & -2 \end{pmatrix}.
$$

7.3 Calculate all defined products of matrices A and B:

(a) $A = \begin{pmatrix} 4 & 2 \\ 3 & 2 \end{pmatrix}; \quad B = \begin{pmatrix} 3 & 5 & -1 \\ 7 & 6 & 2 \end{pmatrix};$

(b) $A = \begin{pmatrix} 4 & 3 & 5 & 3 \\ 2 & 5 & 0 & 1 \end{pmatrix}; \quad B = \begin{pmatrix} 1 & 2 & 6 \\ 4 & 2 & 3 \\ 4 & 5 & 2 \\ 3 & 4 & 5 \end{pmatrix};$

(c) $A = \begin{pmatrix} 2 & 3 & 4 & 5 \end{pmatrix}$; $B = \begin{pmatrix} -3 \\ 6 \\ -3 \\ 2 \end{pmatrix}$;

(d) $A = \begin{pmatrix} 2 & -1 & 3 \\ -4 & 2 & -6 \end{pmatrix}$; $B = \begin{pmatrix} 2 & 3 \\ 1 & 0 \\ -1 & -2 \end{pmatrix}$;

(e) $A = \begin{pmatrix} 2 & 3 & 1 \\ 1 & 5 & 2 \end{pmatrix}$; $B = \begin{pmatrix} x \\ y \\ z \end{pmatrix}$.

7.4   Use the matrices given in Exercise 7.3 (d) and verify the equalities

$$A^\mathrm{T} B^\mathrm{T} = (BA)^\mathrm{T} \qquad \text{and} \qquad B^\mathrm{T} A^\mathrm{T} = (AB)^\mathrm{T}.$$

7.5   Given are the matrices

$$A = \begin{pmatrix} 1 & 0 \\ 7 & -4 \\ 5 & 3 \end{pmatrix}; \qquad B = \begin{pmatrix} -1 \\ 2 \\ 0 \\ 7 \end{pmatrix} \qquad \text{and} \qquad C = \begin{pmatrix} -2 & 0 & 1 & 0 \\ 0 & 3 & 0 & 6 \end{pmatrix}.$$

(a)   Find the dimension of a product of all three matrices if possible.
(b)   Test the associative law of multiplication with the given matrices.

7.6   Calculate all powers of the following matrices:

(a) $A = \begin{pmatrix} 0 & 2 & 8 & 1 \\ 0 & 0 & 7 & 3 \\ 0 & 0 & 0 & 5 \\ 0 & 0 & 0 & 0 \end{pmatrix}$;   (b) $B = \begin{pmatrix} \cos\alpha & \sin\alpha \\ \sin\alpha & -\cos\alpha \end{pmatrix}$.

7.7   A firm produces by means of two raw materials R_1 and R_2 three intermediate products S_1, S_2 and S_3, and with these intermediate products two final products F_1 and F_2. The numbers of units of R_1 and R_2 necessary for the production of 1 unit of $S_j, j \in \{1, 2, 3\}$, and the numbers of units of S_1, S_2 and S_3 necessary to produce 1 unit of F_1 and F_2, are given in the following tables:

	S_1	S_2	S_3
R_1	2	3	5
R_2	5	4	1

	F_1	F_2
S_1	6	0
S_2	1	4
S_3	3	2

Solve the following problems by means of matrix operations.

(a)   How many raw materials are required when 1,000 units of F_1 and 2,000 units of F_2 have to be produced?
(b)   The costs for one unit of raw material are 3 EUR for R_1 and 5 EUR for R_2. Calculate the costs for intermediate and final products.

7.8   Given is the matrix

$$A = \begin{pmatrix} 2 & 3 & 0 \\ -1 & 2 & 4 \\ 0 & 5 & 1 \end{pmatrix}.$$

(a) Find the submatrices A_{12} and A_{22}.
(b) Calculate the minors $|A_{12}|$ and $|A_{22}|$.
(c) Calculate the cofactors of the elements a_{11}, a_{21}, a_{31} of matrix A.
(d) Evaluate the determinant of matrix A.

7.9   Evaluate the following determinants:

(a) $\begin{vmatrix} 2 & 1 & 6 \\ -1 & 0 & 3 \\ 3 & 2 & 9 \end{vmatrix}$;

(b) $\begin{vmatrix} 1 & 2 & 0 & 0 \\ 2 & 1 & 2 & 0 \\ 0 & 2 & 1 & 2 \\ 0 & 0 & 2 & 1 \end{vmatrix}$;

(c) $\begin{vmatrix} 1 & 0 & 1 & 2 \\ 1 & 2 & 3 & 4 \\ 4 & 3 & 2 & 1 \\ 2 & 1 & 4 & 0 \end{vmatrix}$;

(d) $\begin{vmatrix} 2 & 7 & 4 & 1 \\ 3 & 1 & 4 & 0 \\ 5 & 1 & 0 & 0 \\ 2 & 0 & 0 & 0 \end{vmatrix}$;

(e) $\begin{vmatrix} -1 & 2 & 4 & 3 \\ 2 & -4 & -8 & -6 \\ 7 & 1 & 5 & 0 \\ 1 & 5 & 0 & 1 \end{vmatrix}$;

(f) $\begin{vmatrix} 3 & 3 & 3 & \cdots & 3 & 3 \\ 3 & 0 & 3 & \cdots & 3 & 3 \\ 3 & 3 & 0 & \cdots & 3 & 3 \\ \vdots & & & & & \\ 3 & 3 & 3 & \cdots & 3 & 0 \end{vmatrix}_{(n,n)}$.

7.10   Find the solutions x of the following equations:

(a) $\begin{vmatrix} -1 & x & x \\ 2 & -1 & 2 \\ 2 & 2 & -1 \end{vmatrix} = 27$;

(b) $\begin{vmatrix} x & 1 & 2 \\ 3 & x & -1 \\ 4 & x & -2 \end{vmatrix} = 2$.

7.11   Find the solution of the following system of equations by Cramer's rule:

$$\begin{array}{rrrr} 2x_1 & +4x_2 & +3x_3 & = & 1 \\ 3x_1 & -6x_2 & -2x_3 & = & -2. \\ -5x_1 & +8x_2 & +2x_3 & = & 4 \end{array}$$

7.12   Let $A : \mathbb{R}^3 \to \mathbb{R}^3$ be a linear mapping described by matrix

$$A = \begin{pmatrix} 3 & 1 & 0 \\ -1 & 2 & 4 \\ 4 & 1 & 5 \end{pmatrix}.$$

Find the kernel of this mapping.

7.13   Given are three linear mappings described by the following systems of equations:

$$\begin{array}{rrrrr} u_1 & = & v_1 & -v_2 & +v_3 \\ u_2 & = & 2v_1 & -v_2 & -v_3 \\ u_3 & = & -v_1 & +v_2 & +2v_3, \end{array}$$

$$
\begin{aligned}
v_1 &= -w_1 & &+w_3 \\
v_2 &= w_1 & +2w_2 & -w_3 \\
v_3 &= & w_2 & -2w_3,
\end{aligned}
$$

$$
\begin{aligned}
w_1 &= x_1 & -x_2 & -x_3 \\
w_2 &= -x_1 & -2x_2 & +3x_3 \\
w_3 &= 2x_1 & & +x_3.
\end{aligned}
$$

Find the composite mapping $\mathbf{x} \in \mathbb{R}^3 \mapsto \mathbf{u} \in \mathbb{R}^3$.

7.14   Find the inverse of each of the following matrices:

(a) $A = \begin{pmatrix} 1 & 0 & 3 \\ 4 & 1 & 2 \\ 0 & 1 & 1 \end{pmatrix}$;

(b) $B = \begin{pmatrix} 2 & -3 & 1 \\ 3 & 4 & -2 \\ 5 & 1 & -1 \end{pmatrix}$;

(c) $C = \begin{pmatrix} 1 & 3 & -2 \\ 0 & 2 & 4 \\ 0 & 0 & -1 \end{pmatrix}$;

(d) $D = \begin{pmatrix} 1 & 0 & -1 & 2 \\ 2 & -1 & -2 & 3 \\ -1 & 2 & 2 & -4 \\ 0 & 1 & 2 & -5 \end{pmatrix}$.

7.15   Let

$$
A = \begin{pmatrix} 1 & 2 & -1 & 0 & 4 \\ 0 & 1 & 2 & 3 & 0 \\ 0 & 0 & 1 & -3 & -1 \\ 0 & 0 & 0 & 1 & 2 \\ 0 & 0 & 0 & 0 & 1 \end{pmatrix}.
$$

Find the inverse matrix by means of equality $A = I - C$.

7.16   Given are the matrices

$$
A = \begin{pmatrix} -2 & 5 \\ 1 & -3 \end{pmatrix} \qquad \text{and} \qquad B = \begin{pmatrix} 1 & 4 \\ -2 & 9 \end{pmatrix}.
$$

Find $(AB)^{-1}$ and $B^{-1}A^{-1}$.

7.17   Given are the following matrix equations:

(a) $(XA)^{\mathrm{T}} = B$;   (b) $XA = B - 2X$;   (c) $AXB = C$;
(d) $A(XB)^{-1} = C$;   (e) $C^{\mathrm{T}}XA + (X^{\mathrm{T}}C)^{\mathrm{T}} = I - 3C^{\mathrm{T}}X$.

Find matrix X.

7.18 Given is an open input–output model (Leontief model) with

$$A = \begin{pmatrix} 0 & 0.2 & 0.1 & 0.3 \\ 0 & 0 & 0.2 & 0.5 \\ 0 & 0 & 0 & 0 \\ 0 & 0.4 & 0 & 0 \end{pmatrix}.$$

Let $\mathbf{x}$ be the total vector of goods produced and $\mathbf{y}$ the final demand vector.

(a) Explain the economic meaning of the elements of A.
(b) Find the linear mapping which maps all the vectors $\mathbf{x}$ into the set of all final demand vectors $\mathbf{y}$.
(c) Is a vector

$$\mathbf{x} = \begin{pmatrix} 100 \\ 200 \\ 200 \\ 400 \end{pmatrix}$$

of goods produced possible for some final demand vector $\mathbf{y}$?
(d) Find the inverse mapping of that obtained in (b) and interpret it economically.

7.19 A firm produces by means of three factors R_1, R_2 and R_3 five products $P_1, P_2, \ldots, P_5$, where some of these products are also used as intermediate products. The relationships are given in the graph presented in Figure 7.3. The numbers beside the arrows describe how many units of R_i respectively P_i are necessary for one unit of P_j. Let p_i denote

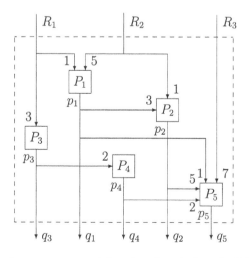

Figure 7.3 Relationships between raw materials and products in Exercise 7.19.

the produced units (output) of P_i and q_i denote the final demand for the output of $P_i, i \in \{1, 2, \ldots, 5\}$.

(a) Find a linear mapping $\mathbf{p} \in \mathbb{R}_+^5 \mapsto \mathbf{q} \in \mathbb{R}_+^5$.

(b) Find the inverse mapping.

(c) Let $\mathbf{r}^T = (r_1, r_2, r_3)$ be the vector which contains the required units of the factors R_1, R_2, R_3. Find a linear mapping $\mathbf{q} \mapsto \mathbf{r}$. Calculate $\mathbf{r}$ when $\mathbf{q} = (50, 40, 30, 20, 10)^T$.

8 Linear equations and inequalities

Many problems in economics can be modelled as a system of linear equations or a system of linear inequalities. In this chapter, we consider some basic properties of such systems and discuss general solution procedures.

8.1 SYSTEMS OF LINEAR EQUATIONS

8.1.1 Preliminaries

At several points in previous chapters we have been confronted with systems of linear equations. For instance, deciding whether a set of given vectors is linearly dependent or linearly independent can be answered via the solution of such a system. The following example of determining feasible production programmes also leads to a system of linear equations.

Example 8.1 Assume that a firm uses three raw materials R_1, R_2 and R_3 for the production of four goods G_1, G_2, G_3 and G_4. There are the following amounts of raw materials available: 120 units of R_1, 150 units of R_2 and 180 units of R_3. Table 8.1 gives the raw material requirements per unit of each good. Denote by x_i the quantity of good G_i, $i \in \{1, 2, 3, 4\}$. We are interested in all possible production programmes which fully use the available amounts of the raw materials.

Considering raw material R_1 we get the following equation:

$$1x_1 + 2x_2 + 1x_3 + 3x_4 = 120.$$

Table 8.1 Raw material requirements for the goods G_i, $i \in \{1, 2, 3, 4\}$

Raw material	Goods			
	G_1	G_2	G_3	G_4
R_1	1	2	1	3
R_2	2	0	3	1
R_3	1	4	2	4

Here $1x_1$ is the amount of raw material R_1 necessary for the production of good G_1, $2x_2$ the amount of R_1 for good G_2, $1x_3$ the amount of R_1 for good G_3 and $3x_4$ is the amount of R_1 required for the production of good G_4. Since all 120 units of raw material R_1 should be used, we have an equation. Similarly we obtain an equation for the consumption of the other two raw materials. Thus, we get the following system of three linear equations with four variables:

$$
\begin{array}{rcrcrcrcr}
x_1 & + & 2x_2 & + & x_3 & + & 3x_4 & = & 120 \\
2x_1 & & & + & 3x_3 & + & x_4 & = & 150 \\
x_1 & + & 4x_2 & + & 2x_3 & + & 4x_4 & = & 180
\end{array}
$$

Moreover, we are interested only in solutions for which all values $x_i, i \in \{1,2,3,4\}$, are non-negative. Considering e.g. the production programme $x_1 = 40$, $x_2 = 15$, $x_3 = 20$ and $x_4 = 10$, we can easily check that all equations are satisfied, i.e. this production programme is feasible, and there exists at least one solution of this system of linear equations. In order to describe all feasible production programmes, we have to find all solutions satisfying the above equations such that all values x_i, $i \in \{1,2,3,4\}$, are non-negative.

Often, it is not desired that all raw materials should necessarily be fully used. In the latter case it is required only that for each raw material R_i the available amount is not exceeded. Then all equality signs in the above equations have to be replaced by an inequality sign of the form $\leq$, and we obtain a system of linear inequalities which is discussed in Chapter 8.2.

In the following, we discuss general methods for solving systems of linear equations. We answer the question of whether a system has a solution, whether an existing solution is uniquely determined and how the set of all solutions can be determined in the general case.

Definition 8.1   The system

$$
\begin{array}{l}
a_{11}x_1 + a_{12}x_2 + \cdots + a_{1n}x_n = b_1 \\
a_{21}x_1 + a_{22}x_2 + \cdots + a_{2n}x_n = b_2 \\
\quad\vdots \qquad\qquad\qquad \vdots \quad \vdots \\
a_{m1}x_1 + a_{m2}x_2 + \cdots + a_{mn}x_n = b_m
\end{array}
\qquad (8.1)
$$

is called a *system of linear equations*, where $x_1, x_2, \ldots, x_n$ are the *unknowns* or *variables*, $a_{11}, a_{12}, \ldots, a_{mn}$ are the *coefficients* and $b_1, b_2, \ldots, b_m$ are called the *right-hand sides*.

As an abbreviation, system (8.1) can also be written in *matrix representation*:

$$A\mathbf{x} = \mathbf{b},$$

where the left-hand side corresponds to the multiplication of a matrix of order $m \times n$ by a vector (matrix of order $n \times 1$) resulting in the m-vector $\mathbf{b}$, i.e.

$$\begin{pmatrix} a_{11} & a_{12} & \cdots & a_{1n} \\ a_{21} & a_{22} & \cdots & a_{2n} \\ \vdots & \vdots & & \vdots \\ a_{m1} & a_{m2} & \cdots & a_{mn} \end{pmatrix} \begin{pmatrix} x_1 \\ x_2 \\ \vdots \\ x_n \end{pmatrix} = \begin{pmatrix} b_1 \\ b_2 \\ \vdots \\ b_m \end{pmatrix}.$$

Analogously, we can write a system of linear equations in *vector representation* as follows:

$$\sum_{j=1}^{n} x_j \, \mathbf{a}^j = \mathbf{b},$$

where

$$\mathbf{a}^j = \begin{pmatrix} a_{1j} \\ a_{2j} \\ \vdots \\ a_{mj} \end{pmatrix}, \quad j = 1, 2, \ldots, n,$$

are the column vectors of matrix A. The above left-hand side represents a linear combination of the column vectors of matrix A with the values of the variables as scalars, i.e.:

$$x_1 \begin{pmatrix} a_{11} \\ a_{21} \\ \vdots \\ a_{m1} \end{pmatrix} + x_2 \begin{pmatrix} a_{12} \\ a_{22} \\ \vdots \\ a_{m2} \end{pmatrix} + \cdots + x_n \begin{pmatrix} a_{1n} \\ a_{2n} \\ \vdots \\ a_{mn} \end{pmatrix} = \begin{pmatrix} b_1 \\ b_2 \\ \vdots \\ b_m \end{pmatrix}.$$

Next, we introduce some basic notions.

Definition 8.2   If we have $b_i = 0, i = 1, 2, \ldots, m$, in system (8.1), then this system is called *homogeneous*. If we have $b_k \neq 0$ for at least one $k \in \{1, 2, \ldots, m\}$, then system (8.1) is called *non-homogeneous*.

Definition 8.3   A vector $\mathbf{x} = (x_1, x_2, \ldots, x_n)^{\mathrm{T}}$ which satisfies $A\mathbf{x} = \mathbf{b}$ is called a *solution* of system (8.1). The set

$$S = \{\mathbf{x} \in \mathbb{R}^n \mid A\mathbf{x} = \mathbf{b}\}$$

is called the *set of solutions* or the *general solution* of system (8.1).

8.1.2 Existence and uniqueness of a solution

Next, we investigate in which cases system (8.1) has a solution. To this end, we introduce the notion of the rank of a matrix. For this purpose, the following property is useful.

THEOREM 8.1 Let A be a matrix of order $m \times n$. Then the maximum number of linearly independent column vectors of A coincides with the maximum number of linearly independent row vectors of A.

Definition 8.4 Let A be a matrix of order $m \times n$. The *rank* of matrix A, written $r(A)$, is the maximum number p of linearly independent column (or according to Theorem 8.1 equivalently, row) vectors in A. If A is any zero matrix, we set $r(A) = 0$.

As an obvious consequence of Definition 8.4, we obtain $r(A) = p \le \min\{m, n\}$. The following theorem gives a first criterion to determine the rank of a matrix A.

THEOREM 8.2 The rank $r(A)$ of matrix A is equal to the order of the largest minor of A that is different from zero.

We recall that a minor of a matrix A was defined as a determinant of a square submatrix of A. The above criterion can be used when transforming the determinant of a matrix A in such a way that the order of the largest minor can be easily obtained. Otherwise, it can be rather time-consuming to determine the rank of a matrix by applying Theorem 8.2. Consider the following two examples.

Example 8.2 Let

$$A = \begin{pmatrix} 1 & 2 & 0 \\ 4 & 6 & 2 \\ 3 & 2 & 4 \end{pmatrix}.$$

We obtain $|A| = 0$ which means that matrix A cannot have rank three. However, e.g. for the minor obtained from matrix A by deleting the last row and column, we get

$$|A_{33}| = \begin{vmatrix} 1 & 2 \\ 4 & 6 \end{vmatrix} = -2,$$

i.e. there is a minor of order two which is different from zero, and thus matrix A has rank two.

Example 8.3 Let us consider the matrix

$$A = \begin{pmatrix} 1 - x & -1 & 1 \\ 1 & 1 - x & 3 \\ 1 & 0 & 1 \end{pmatrix}$$

with $x \in \mathbb{R}$. We determine the rank of matrix A in dependence on the value of $x \in \mathbb{R}$. Expanding $|A|$ by the third row, we get

$$
|A| = 1 \cdot \begin{vmatrix} -1 & 1 \\ 1-x & 3 \end{vmatrix} + 1 \cdot \begin{vmatrix} 1-x & -1 \\ 1 & 1-x \end{vmatrix}
$$

$$
= -3 - (1-x) + \left[(1-x)(1-x) + 1 \right]
$$

$$
= (-4 + x) + (x^2 - 2x + 2)
$$

$$
= x^2 - x - 2.
$$

We determine the roots of the equation $x^2 - x - 2 = 0$ and obtain $x_1 = -1$ and $x_2 = 2$. Thus, we get $|A| \neq 0$ for the case $x \neq -1$ and $x \neq 2$, i.e. due to Theorem 8.2, we have $r(A) = 3$ for $x \in \mathbb{R} \setminus \{-1, 2\}$. For $x \in \{-1, 2\}$, we obtain for the minor formed by rows 1 and 3 as well as columns 2 and 3

$$
|A_{21}| = \begin{vmatrix} -1 & 1 \\ 0 & 1 \end{vmatrix} = -1 \neq 0.
$$

Thus, we obtain $r(A) = 2$ for the case $x = -1$ and for the case $x = 2$ because the order of the largest minor with a value different from zero is two.

Definition 8.5 We define the $m \times (n+1)$ *augmented matrix*

$$
A_b = (A \mid b) = \begin{pmatrix} a_{11} & a_{12} & \cdots & a_{1n} & b_1 \\ a_{21} & a_{22} & \cdots & a_{2n} & b_2 \\ \vdots & \vdots & & \vdots & \vdots \\ a_{m1} & a_{m2} & \cdots & a_{mn} & b_m \end{pmatrix}
$$

as the coefficient matrix A expanded by an additional column containing vector b of the right-hand side.

Obviously, we have $r(A) \leq r(A_b)$ since matrix A_b contains an additional column vector in comparison with matrix A. Moreover, since the augmented matrix differs from the coefficient matrix A by exactly one column, there are only two cases possible: either $r(A_b) = r(A)$ or $r(A_b) = r(A) + 1$.

Definition 8.6 If system (8.1) has at least one solution, it is said to be *consistent*. If this system has no solution, it is said to be *inconsistent*.

Next, we present a necessary and sufficient condition for the case where a system of linear equations has at least one solution.

THEOREM 8.3 System (8.1) is consistent if and only if the rank of the coefficient matrix A is equal to the rank of the augmented matrix $A_b = (A \mid b)$, i.e.

$$\text{system } A\mathbf{x} = \mathbf{b} \text{ is consistent} \quad \Longleftrightarrow \quad r(A) = r(A_b).$$

Since for a homogeneous system of equations, the augmented matrix contains matrix A plus an additional zero (column) vector, the number of linearly independent (column or row) vectors of the augmented matrix is always equal to the number of linearly independent vectors of matrix A. This leads to the following corollary.

COROLLARY 8.1 A homogeneous system $A\mathbf{x} = \mathbf{0}$ is always consistent.

Indeed, for system $A\mathbf{x} = \mathbf{0}$ we have $r(A) = r(A_b)$. We can note that a homogeneous system has at least the so-called trivial solution $\mathbf{x}^{\mathrm{T}} = (x_1, x_2, \ldots, x_n) = (0, 0, \ldots, 0)$.

Next, we deal with the following question. If system (8.1) is consistent, when is the solution *uniquely* determined? An answer is given by the following theorem.

THEOREM 8.4 Consider the system $A\mathbf{x} = \mathbf{b}$ of linear equations, where A is a matrix of order $m \times n$, and let this system be consistent. Then:

(1) If $r(A) = r(A_b) = n$, then solution $\mathbf{x} = (x_1, x_2, \ldots, x_n)^{\mathrm{T}}$ is uniquely determined.
(2) If $r(A) = r(A_b) = p < n$, then there exist infinitely many solutions. In this case, the set of solutions forms an $(n - p)$-dimensional vector space.

In case (2) of Theorem 8.4 we say that the set of solutions has dimension $n - p$. Let us consider part (2) of Theorem 8.4 in a bit more detail. In this case, we can select $n - p$ variables that can be chosen freely. Having their values fixed, the remaining variables are uniquely determined. We denote the $n - p$ arbitrarily chosen variables as *free* variables, and we say that the system of linear equations has $n - p$ *degrees of freedom*.

8.1.3 Elementary transformation; solution procedures

Solution procedures for systems of linear equations transform the given system into a 'system with easier structure'. The following theorem characterizes some transformations of a given system of linear equations such that the set of solutions does not change.

THEOREM 8.5 The set of solutions of system (8.1) does not change if one of the following transformations is applied:

(1) An equation is multiplied by a number $\lambda \neq 0$ or it is divided by a number $\lambda \neq 0$.
(2) Two equations are interchanged.
(3) A multiple of one equation is added to another equation.

Operations (1) to (3) are called *elementary* or *equivalent* transformations. By such elementary transformations, the rank $r(A)$ of a matrix A does not change either (see also rules for evaluating determinants given in Chapter 7.3). Finally, we introduce a special form of a system of linear equations and a solution as follows.

Definition 8.7 A system $A\mathbf{x} = \mathbf{b}$ of $p = r(A)$ linear equations, where in each equation one variable occurs only in this equation and it has the coefficient $+1$, is called a system of linear equations in *canonical form*. These eliminated variables are called *basic variables* (bv), while the remaining variables are called *non-basic variables* (nbv).

Hence the number of basic variables of a system of linear equations in canonical form is equal to the rank of matrix A. As a consequence of Definition 8.7, if a system of linear equations $A\mathbf{x} = \mathbf{b}$ is given in canonical form, the coefficient matrix A always contains an identity matrix. If $r(A) = p = n$, the identity matrix I is of order $n \times n$, i.e. the system has the form

$$I\mathbf{x}^{\mathrm{B}} = \mathbf{b},$$

where $\mathbf{x}^{\mathrm{B}}$ is the vector of the basic variables. (Note that columns might have been interchanged in matrix A to get the identity matrix, which means that the order of the variables in vector $\mathbf{x}^{\mathrm{B}}$ is different from that in vector $\mathbf{x}$.) If $r(A) = p < n$, the order of the identity submatrix is $p \times p$. In the latter case, the system can be written as

$$I\mathbf{x}^{\mathrm{B}} + A_N \mathbf{x}^{\mathrm{N}} = \mathbf{b},$$

where $\mathbf{x}^{\mathrm{B}}$ is the p-vector of the basic variables, $\mathbf{x}^{\mathrm{N}}$ is the $(n - p)$-vector of the non-basic variables and A_N is the submatrix of A formed by the column vectors belonging to the non-basic variables. (Again column interchanges in matrix A might have been applied.) This canonical form, from which the general solution can easily be derived, is used in one of the solution procedures described in this subsection.

Definition 8.8 A solution $\mathbf{x}$ of a system of equations $A\mathbf{x} = \mathbf{b}$ in canonical form, where each non-basic variable has the value zero, is called a *basic solution*.

Thus, if matrix A is of order $p \times n$ with $r(A) = p < n$, then at least $n - p$ variables are equal to zero in a basic solution of the system $A\mathbf{x} = \mathbf{b}$. The number of possible basic solutions of a given system of linear equations is determined by the number of different possibilities of choosing p basic variables. That is, one has to find among the column vectors of matrix A all possibilities of p linearly independent vectors belonging to the p basic variables. There exist at most $\binom{n}{p}$ basic solutions (see Chapter 1.3 on Combinatorics).

One method of solving systems of linear equations has already been discussed in Chapter 7.3, but remember that Cramer's rule is applicable only in special cases. The usual methods of solving systems of linear equations apply elementary transformations mentioned in Theorem 8.5 to transform the given system into a form from which the solution can be easily obtained. The methods typically used transform the original system into either

(1) a canonical form according to Definition 8.7 (*pivoting procedure* or *Gauss–Jordan elimination*) or into
(2) a 'triangular' or echelon form (*Gaussian elimination*).

It is worth noting that the notation for both procedures in the literature is not used in a standard way; in particular Gaussian elimination is often also denoted as pivoting. The reason is that both procedures are variants of the same strategy: simplify the given system of linear equations in such a way that the solution can be easily obtained from the final form of the system. We now discuss both methods in detail.

Pivoting

First, we discuss the *pivoting procedure*. The transformation of the original system into a canonical form (possibly including less than m equations) is based on the following theorem and the remark given below.

THEOREM 8.6   Let $Ax = b$ be a given system of m linear equations with n variables and $r(A) = p < \min\{m, n\}$. Then the augmented matrix $A_b = (A \mid b)$ can be transformed by applying Theorem 8.5 and column interchanges into the form

$$(A^* \mid b^*) = \begin{pmatrix} 1 & 0 & \cdots & 0 & a^*_{1,p+1} & \cdots & a^*_{1n} & b^*_1 \\ 0 & 1 & \cdots & 0 & a^*_{2,p+1} & \cdots & a^*_{2n} & b^*_2 \\ \vdots & \vdots & & \vdots & \vdots & & \vdots & \vdots \\ 0 & 0 & \cdots & 1 & a^*_{p,p+1} & \cdots & a^*_{pn} & b^*_p \\ 0 & 0 & \cdots & 0 & 0 & \cdots & 0 & b^*_{p+1} \\ 0 & 0 & \cdots & 0 & 0 & \cdots & 0 & 0 \\ \vdots & \vdots & & \vdots & \vdots & & \vdots & \vdots \\ 0 & 0 & \cdots & 0 & 0 & \cdots & 0 & 0 \end{pmatrix} \qquad (8.2)$$

with $b^*_{p+1} = 0$ or $b^*_{p+1} \neq 0$.

It is easy to see that the matrix A^* given in (8.2) (and therefore also the original coefficient matrix A) have rank p. In terms of Theorem 8.2, this means that matrix A^* has a minor of order p whose value is different from zero. This can easily be seen by taking the identity submatrix obtained in the left upper part (printed in bold face) whose determinant is equal to one. However, there is no minor of a larger order than p whose value is different from zero. (If we add one row and one column, the value of the determinant is equal to zero since there is one row containing only zero entries.) Notice also that the first p rows in representation (8.2) describe a system of linear equations in canonical form.

Remark   In the case when $r(A) = p$ is not smaller than $\min\{m, n\}$, we can transform matrix A into one of the three following forms $(A^* \mid b^*)$:

(1) If $m < n$, then

$$(A^* \mid b^*) = \begin{pmatrix} 1 & 0 & \cdots & 0 & a^*_{1,m+1} & \cdots & a^*_{1n} & b^*_1 \\ 0 & 1 & \cdots & 0 & a^*_{2,m+1} & \cdots & a^*_{2n} & b^*_2 \\ \vdots & \vdots & & \vdots & \vdots & & \vdots & \vdots \\ 0 & 0 & \cdots & 1 & a^*_{m,m+1} & \cdots & a^*_{mn} & b^*_m \end{pmatrix}. \qquad (8.3)$$

(2) If $m > n$, then

$$
(A^* \mid b^*) = \begin{pmatrix}
1 & 0 & \cdots & 0 & b_1^* \\
0 & 1 & \cdots & 0 & b_2^* \\
\vdots & \vdots & & \vdots & \vdots \\
0 & 0 & \cdots & 1 & b_n^* \\
0 & 0 & \cdots & 0 & b_{n+1}^* \\
0 & 0 & \cdots & 0 & 0 \\
\vdots & \vdots & & \vdots & \vdots \\
0 & 0 & \cdots & 0 & 0
\end{pmatrix}
\tag{8.4}
$$

with $b_{n+1}^* = 0$ or $b_{n+1}^* \neq 0$.

(3) If $m = n$, then

$$
(A^* \mid b^*) = \begin{pmatrix}
1 & 0 & \cdots & 0 & b_1^* \\
0 & 1 & \cdots & 0 & b_2^* \\
\vdots & \vdots & & \vdots & \vdots \\
0 & 0 & \cdots & 1 & b_m^*
\end{pmatrix}.
\tag{8.5}
$$

Each of the matrices in (8.2) to (8.5) contains an identity submatrix. The order of this identity matrix is the rank of the coefficient matrix A originally given. In the transformed matrix A^*, each column corresponds to some variable. Since column interchanges were allowed, we simply denote the variable belonging to the first column as the first basic variable x_{B1}, the variable belonging to the second column as the second basic variable x_{B2} and so on. (If no column interchanges were applied, we have the natural numbering $x_1, x_2, \ldots, x_p$.) Accordingly, we denote the variables belonging to the columns which do not form the identity submatrix as the non-basic variables $x_{N1}, x_{N2}, \ldots, x_{N,n-p}$.

We now discuss how the solutions can be found from the system in canonical form given by the first rows of matrices (8.2) to (8.5) (including the identity matrix).

In the cases defined by matrices (8.3) and (8.5), there always exists a solution ($r(A) = r(A_b) = m$). In particular, in the case of matrix (8.5) we have the unique solution

$$
x_{B1} = b_1^*, \qquad x_{B2} = b_2^*, \qquad \ldots, \qquad x_{Bn} = b_n^*.
$$

(Note that variables might have been interchanged.) In the case of matrix (8.3), we have $n - m$ degrees of freedom, i.e. $n - m$ variables belonging to columns $m + 1, m + 2, \ldots, n$ can be chosen arbitrarily.

In the case of matrix (8.4), there exists a unique solution

$$
x_{B1} = b_1^*, \qquad x_{B2} = b_2^*, \qquad \ldots, \qquad x_{Bn} = b_n^*,
$$

provided that $b_{n+1}^* = 0$, otherwise the system has no solution. In the case of $p < \min\{m, n\}$ considered in Theorem 8.6, the system is consistent if and only if $b_{p+1}^* = 0$, and the system of the first p equations obtained by the transformation of matrix $(A \mid b)$ into $(A^* \mid b^*)$ is a system in canonical form which can be written as

$$
\begin{aligned}
x_{B1} &&+\ a^*_{1,p+1}x_{N1} &+\ \cdots &+\ a^*_{1n}x_{N,n-p} &= b^*_1 \\
&x_{B2} &+\ a^*_{2,p+1}x_{N1} &+\ \cdots &+\ a^*_{2n}x_{N,n-p} &= b^*_2 \\
&\ddots &\vdots & &\vdots &\vdots \\
&\quad x_{Bp} +\ a^*_{p,p+1}x_{N1} &+\ \cdots &+\ a^*_{pn}x_{N,n-p} &= b^*_p.
\end{aligned}
$$

Since $r(A) = r(A^*) = p$, there are $n - p$ degrees of freedom, i.e. $n - p$ variables can be chosen arbitrarily. We can rewrite the system in canonical form in terms of the basic variables as follows:

$$
\begin{aligned}
x_{B1} &= b^*_1 - a^*_{1,p+1}x_{N1} - \cdots - a^*_{1n}x_{N,n-p} \\
x_{B2} &= b^*_2 - a^*_{2,p+1}x_{N1} - \cdots - a^*_{2n}x_{N,n-p} \\
\vdots\ &\ \ \vdots \qquad\quad \vdots \qquad\qquad\quad \vdots \\
x_{Bp} &= b^*_p - a^*_{p,p+1}x_{N1} - \cdots - a^*_{pn}x_{N,n-p}
\end{aligned}
$$

x_{Nj} arbitrary for $j = 1, 2, \ldots, n - p$.

We emphasize that, if $x_{B1}, x_{B2}, \ldots, x_{Bp}$ are the basic variables, this means that the column vectors of matrix A belonging to these variables are linearly independent and form a basis of the vector space spanned by the corresponding p column vectors of matrix A. If we choose $x_{Nj} = 0, j = 1, 2, \ldots, n - p$, we obtain the *basic solution*

$$
x_{B1} = b^*_1, \qquad x_{B2} = b^*_2, \qquad \ldots, \qquad x_{Bp} = b^*_p.
$$

Next, we discuss how we can transform the augmented matrix $A_b = (A \mid b)$ into matrix $(A^* \mid b^*)$ by elementary transformations in a systematic way from which the canonical form can be established.

These formulas correspond to those which have been presented in Chapter 6.4 when considering the replacement of one vector in a basis by another vector of a vector space. Assume that variable x_l should become the basic variable in the kth equation of the system. This can be done only if $a_{kl} \neq 0$. The element a_{kl} is called the *pivot* or *pivot element*. Accordingly, row k is called the *pivot row*, and column l is called the *pivot column*. The transformation formulas distinguish between the pivot row and all remaining rows. They are as follows.

Transformation formulas

(1) Pivot row k:

$$
\bar{a}_{kj} = \frac{a_{kj}}{a_{kl}}, \quad j = 1, 2, \ldots, n
$$

$$
\bar{b}_k = \frac{b_k}{a_{kl}}
$$

(2) Remaining rows $i = 1, 2, \ldots, m, \ i \neq k$:

$$
\bar{a}_{ij} = a_{ij} - \frac{a_{il}}{a_{kl}} \cdot a_{kj} = a_{ij} - \bar{a}_{kj} \cdot a_{il}, \quad j = 1, 2, \ldots, n
$$

$$
\bar{b}_i = b_i - \frac{a_{il}}{a_{kl}} \cdot b_k = b_i - \bar{b}_k \cdot a_{il}
$$

These transformations can be performed by using the following tableaus, where rows 1 to m in the first tableau give the initial system, and rows $m+1$ up to $2m$ in the second tableau give the system of equations after the transformation. The column bv indicates the basic variables in the corresponding rows (i.e. after the pivoting transformation, x_l is basic variable in the kth equation of the system, while we do not in general have basic variables in the other equations yet). The last column describes the operation which has to be done in order to get the corresponding row.

Row	bv	x_1	x_2	$\cdots$	x_l	$\cdots$	x_n	b	Operation
1	$-$	a_{11}	a_{12}	$\cdots$	a_{1l}	$\cdots$	a_{1n}	b_1	
2	$-$	a_{21}	$\mathbf{a_{22}}$	$\cdots$	$\mathbf{a_{2l}}$	$\cdots$	a_{2n}	b_2	
$\vdots$		$\vdots$	$\vdots$		$\vdots$		$\vdots$	$\vdots$	
k	$-$	a_{k1}	$\mathbf{a_{k2}}$	$\cdots$	$\mathbf{a_{kl}}$	$\cdots$	a_{kn}	b_k	
$\vdots$		$\vdots$	$\vdots$		$\vdots$		$\vdots$	$\vdots$	
m	$-$	a_{m1}	a_{m2}	$\cdots$	a_{ml}	$\cdots$	a_{mn}	b_m	
$m+1$	$-$	$\bar{a}_{11}$	$\bar{a}_{12}$	$\cdots$	0	$\cdots$	$\bar{a}_{1n}$	$\bar{b}_1$	row $1 - \frac{a_{1l}}{a_{kl}}$ row k
$m+2$	$-$	$\bar{a}_{21}$	$\bar{a}_{22}$	$\cdots$	0	$\cdots$	$\bar{a}_{2n}$	$\bar{b}_2$	row $2 - \frac{a_{2l}}{a_{kl}}$ row k
$\vdots$		$\vdots$	$\vdots$		$\vdots$		$\vdots$	$\vdots$	
$m+k$	x_l	$\bar{a}_{k1}$	$\bar{a}_{k2}$	$\cdots$	1	$\cdots$	$\bar{a}_{kn}$	$\bar{b}_k$	$\frac{1}{a_{kl}}$ row k
$\vdots$		$\vdots$	$\vdots$		$\vdots$		$\vdots$	$\vdots$	
$2m$	$-$	$\bar{a}_{m1}$	$\bar{a}_{m2}$	$\cdots$	0	$\cdots$	$\bar{a}_{mn}$	$\bar{b}_m$	row $m - \frac{a_{ml}}{a_{kl}}$ row k

Here all elements in the pivot column (except the pivot element) are equal to zero in the new tableau. After the transformation, we have the basic variable x_l in row $m+k$ which occurs with coefficient $+1$ now only in equation k of the system and the transformed equation can be found now in row $m+k$ of the second tableau. To illustrate the computations in all rows except the pivot row, consider the determination of element $\bar{a}_{22}$. We have to consider four elements: the number a_{22} on the original position, the element a_{2l} of the second row standing in the pivot column, the element a_{k2} of the pivot row standing in the second column and the pivot element a_{kl}. (The corresponding elements are printed in bold face in the tableau.) These elements form a rectangle, and therefore the rule for computing the values in all rows except the pivot row in the new scheme is also known as the *rectangle rule*.

Example 8.4   Let us consider the following system of linear equations:

$$
\begin{array}{rcrcrcrcl}
x_1 & & & + & 3x_3 & - & 2x_4 & = & 5 \\
3x_1 & + & x_2 & + & 4x_3 & + & x_4 & = & 5 \\
4x_1 & + & x_2 & + & 7x_3 & + & x_4 & = & 10 \\
2x_1 & + & x_2 & + & x_3 & + & 3x_4 & = & 0.
\end{array}
$$

Applying the pivoting procedure, we get the following sequence of tableaus. (Hereafter, the pivot element is always printed in bold face, and the last column is added in order to describe the operation that yields the corresponding row):

Row	bv	x_1	x_2	x_3	x_4	b	Operation
1	–	1	0	3	−2	5	
2	–	3	1	4	1	5	
3	–	4	1	7	1	10	
4	–	2	1	1	3	0	
5	x_1	1	0	3	−2	5	row 1
6	–	0	1	−5	7	−10	row 2 − 3 row 1
7	–	0	1	−5	9	−10	row 3 − 4 row 1
8	–	0	1	−5	7	−10	row 4 − 2 row 1
9	x_1	1	0	3	−2	5	row 5
10	x_2	0	1	−5	7	−10	row 6
11	–	0	0	0	2	0	row 7 − row 6
12	–	0	0	0	0	0	row 8 − row 6
13	x_1	1	0	3	0	5	row 9 + row 11
14	x_2	0	1	−5	0	−10	row 10 − $\frac{7}{2}$ row 11
15	x_4	0	0	0	1	0	$\frac{1}{2}$ row 11
16	–	0	0	0	0	0	row 12

We avoid the interchange of the column of x_3 with the column of x_4 to get formally the structure given in Theorem 8.6. In the first two pivoting steps, we have always chosen element one as the pivot element, which ensures that all above tableaus contain only integers. In the second tableau (i.e. rows 5 up to 8), we could already delete one row since rows 6 and 8 are identical (therefore, these row vectors are linearly dependent). From the last tableau we also see that the rank of the coefficient matrix is equal to three (since we have a 3×3 identity submatrix, which means that the largest minor with a value different from zero is of order three). Since x_1, x_2 and x_4 are basic variables, we have found that the column vectors belonging to these variables (i.e. the first, second and fourth columns of A) are linearly independent and constitute a basis of the space generated by the column vectors of matrix A. From rows 13 to 16, we can rewrite the system in terms of the basic variables:

$$x_1 = 5 - 3x_3$$

$$x_2 = -10 + 5x_3$$

$$x_4 = 0$$

(x_3 arbitrary).

Setting now $x_3 = t$ with $t \in \mathbb{R}$, we get the following set of solutions of the considered system:

$$x_1 = 5 - 3t; \qquad x_2 = -10 + 5t; \qquad x_3 = t; \qquad x_4 = 0; \qquad t \in \mathbb{R}.$$

Since we know that one variable can be chosen arbitrarily, we have selected the non-basic variable x_3 as the free variable. From the last tableau we see that we could not choose x_4 as the free variable since it must have value zero in any solution of the given system of linear equations. However, we can easily see from the transformed system of linear equations, whether one or several variables are uniquely determined so that they cannot be taken as free variables.

In the latter example, there was (due to $n - p = 1$) one variable that can be chosen arbitrarily. In this case, we also say that we have a *one-parametric* set of solutions.

Gaussian elimination

Next, we discuss *Gaussian elimination*. This procedure is based on the following theorem.

THEOREM 8.7   Let $Ax = b$ be a given system of m linear equations with n variables and $r(A) = p < \min\{m, n\}$. Then the augmented matrix $A_b = (A \mid b)$ can be transformed by applying Theorem 8.5 and column interchanges into the form

$$(A^* \mid b^*) = \begin{pmatrix} a_{11}^* & a_{12}^* & \cdots & a_{1p}^* & a_{1,p+1}^* & \cdots & a_{1n}^* & b_1^* \\ 0 & a_{22}^* & \cdots & a_{2p}^* & a_{2,p+1}^* & \cdots & a_{2n}^* & b_2^* \\ \vdots & \vdots & & \vdots & \vdots & & \vdots & \vdots \\ 0 & 0 & \cdots & a_{pp}^* & a_{p,p+1}^* & \cdots & a_{pn}^* & b_p^* \\ 0 & 0 & \cdots & 0 & 0 & \cdots & 0 & b_{p+1}^* \\ 0 & 0 & \cdots & 0 & 0 & \cdots & 0 & 0 \\ \vdots & \vdots & & \vdots & \vdots & & \vdots & \vdots \\ 0 & 0 & \cdots & 0 & 0 & \cdots & 0 & 0 \end{pmatrix} \tag{8.6}$$

with $a_{11}^* \cdot a_{22}^* \cdot \ldots \cdot a_{pp}^* \neq 0$ and $b_{p+1}^* = 0$ or $b_{p+1}^* \neq 0$.

In terms of Theorem 8.2, the transformed matrix A^* in Theorem 8.7 (and the original matrix A too) possesses a minor of order p whose value is different from zero. This is the minor formed by the first p rows and columns in representation (8.6). Since all diagonal elements are different from zero, the value of the determinant is equal to the product of these diagonal elements (see Theorem 7.4). However, matrix A^* (and matrix A too) does not have a minor of order $p + 1$ which is different from zero (in each minor of matrix A^*, there would be one row containing only zero entries, and by Theorem 7.7 in Chapter 7.3, this determinant is equal to zero).

If p is not smaller than the minimum of m and n, we can transform the augmented matrix A_b similarly to Theorem 8.7, as described in the following remark.

Remark   In the case when rank $r(A) = p$ is not smaller than $\min\{m, n\}$, we can transform the augmented matrix $A_b = (A \mid b)$ by elementary transformations and column interchanges to get one of the following special cases:

(1) If $m < n$, then

$$(A^* \mid b^*) = \begin{pmatrix} a_{11}^* & a_{12}^* & \cdots & a_{1m}^* & a_{1,m+1}^* & \cdots & a_{1n}^* & b_1^* \\ 0 & a_{22}^* & \cdots & a_{2m}^* & a_{2,m+1}^* & \cdots & a_{2n}^* & b_2^* \\ \vdots & \vdots & & \vdots & \vdots & & \vdots & \vdots \\ 0 & 0 & \cdots & a_{mm}^* & a_{m,m+1}^* & \cdots & a_{mn}^* & b_m^* \end{pmatrix} \tag{8.7}$$

with $a_{11}^* \cdot a_{22}^* \cdot \ldots \cdot a_{mm}^* \neq 0$.

(2) If $m > n$, then

$$(A^* \mid b^*) = \begin{pmatrix} a_{11}^* & a_{12}^* & \cdots & a_{1n}^* & b_1^* \\ 0 & a_{22}^* & \cdots & a_{2n}^* & b_2^* \\ \vdots & \vdots & & \vdots & \vdots \\ 0 & 0 & \cdots & a_{nn}^* & b_n^* \\ 0 & 0 & \cdots & 0 & b_{n+1}^* \\ 0 & 0 & \cdots & 0 & 0 \\ \vdots & \vdots & & \vdots & \vdots \\ 0 & 0 & \cdots & 0 & 0 \end{pmatrix} \tag{8.8}$$

with $a_{11}^* \cdot a_{22}^* \cdot \ldots \cdot a_{nn}^* \neq 0$ and $b_{n+1}^* = 0$ or $b_{n+1}^* \neq 0$.

(3) If $m = n$, then

$$(A^* \mid b^*) = \begin{pmatrix} a_{11}^* & a_{12}^* & \cdots & a_{1n}^* & b_1^* \\ 0 & a_{22}^* & \cdots & a_{2n}^* & b_2^* \\ \vdots & \vdots & & \vdots & \vdots \\ 0 & 0 & \cdots & a_{nn}^* & b_n^* \end{pmatrix} \tag{8.9}$$

with $a_{11}^* \cdot a_{22}^* \cdot \ldots \cdot a_{nn}^* \neq 0$.

Considering the largest order of a minor different from zero, we conclude that in case (1) above, matrix A has rank m, in case (2) matrix A has rank n and in case (3), matrix A has rank $m = n$. The discussion of the consistency of the system of linear equations and the selection of the free variables is the same as for the pivoting procedure.

In all matrices (8.6) to (8.9), an upper triangular submatrix with non-zero entries on the diagonal is contained. We next describe the systematic generation of the triangular form given in Theorem 8.7 and the above remark. Assume that $a_{11} \neq 0$ (otherwise we interchange two rows or columns). Then we transform all equations except the first one, where the new kth equation, $k \in \{2, 3, \ldots, m\}$, is obtained by multiplying the first equation by $-a_{k1}/a_{11}$ and adding the resulting equation to the original kth equation. Element a_{11} is also denoted as the *pivot* or *pivot element*. This leads to the following system:

$$\begin{array}{lclclcl}
a_{11}x_1 & + & a_{12}x_2 & + & \cdots & + & a_{1n}x_n & = b_1 \\
 & & \bar{a}_{22}x_2 & + & \cdots & + & \bar{a}_{2n}x_n & = b_2 \\
 & & \vdots & & & & \vdots & \vdots \\
 & & \bar{a}_{m2}x_2 & + & \cdots & + & \bar{a}_{mn}x_n & = b_m,
\end{array}$$

where

$$\bar{a}_{kj} = a_{kj} - \frac{a_{k1}}{a_{11}} \cdot a_{1j}, \quad k = 2, 3, \ldots, m, \ j = 2, 3, \ldots, n.$$

By the above transformation we have obtained a system where variable x_1 occurs only in the first equation, i.e. all elements below the pivot element are now equal to zero. Now we apply this procedure to equations $2, 3, \ldots, m$ provided that $\bar{a}_{22} \neq 0$ (otherwise we interchange two rows or columns) and so on until a triangular form according to (8.6) to (8.9) has been obtained.

From the triangular system, we determine the values of the variables by 'back substitution' (i.e. we determine the value of one variable from the last equation, then the value of a second variable from the second to last equation and so on). It is worth noting that only the first equations belonging to the triangular submatrix are necessary to find the general solution, while the remaining equations are superfluous and can be skipped. We illustrate Gaussian elimination by the following examples.

Example 8.5 Consider the following system of linear equations:

$$\begin{aligned}
x_1 + x_2 + x_3 &= 3 \\
x_1 - x_2 + 2x_3 &= 2 \\
4x_1 + 6x_2 - x_3 &= 9.
\end{aligned}$$

Applying Gaussian elimination we obtain the following tableaus.

Row	x_1	x_2	x_3	b	Operation
1	1	1	1	3	
2	1	−1	2	2	
3	4	6	−1	9	
4	1	1	1	3	row 1
5	0	−2	1	−1	row 2 − row 1
6	0	2	−5	− 3	row 3 − 4 row 1
7	1	1	1	3	row 4
8	0	−2	1	−1	row 5
9	0	0	−4	−4	row 6 + row 5

From rows 7 to 9, we see that both the coefficient matrix A and the augmented matrix A_b have rank three. Therefore, the given system is consistent and has a unique solution. Moreover, we get the following triangular system:

$$\begin{aligned}
x_1 + x_2 + x_3 &= 3 \\
- 2x_2 + x_3 &= -1 \\
- 4x_3 &= -4.
\end{aligned}$$

Applying back substitution, we get from the last equation $x_3 = 1$. Then we obtain from the second equation

$$-2x_2 = -1 - x_3 = -2$$

which yields $x_2 = 1$, and finally from the first equation

$$x_1 = 3 - x_2 - x_3 = 3 - 1 - 1$$

which gives $x_1 = 1$.

Example 8.6   We solve the following system of linear equations:

$$
\begin{aligned}
x_1 &+ x_2 &+ x_3 &= 2 \\
3x_1 &+ 2x_2 &+ x_3 &= 2 \\
2x_1 &+ 3x_2 &+ 4x_3 &= 1.
\end{aligned}
$$

Applying Gaussian elimination, we obtain the following tableaus.

Row	x_1	x_2	x_3	b	*Operation*
1	1	1	1	2	
2	3	2	1	2	
3	2	3	4	1	
4	1	1	1	2	row 1
5	0	−1	−2	−4	row 2 − 3 row 1
6	0	1	2	−3	row 3 − 2 row 1
7	1	1	1	2	row 4
8	0	−1	−2	−4	row 5
9	0	0	0	−7	row 6 + row 5

From row 9 we see that the considered system has no solution since this equation

$$0x_1 + 0x_2 + 0x_3 = -7$$

leads to a contradiction. The rank of the coefficient matrix A is equal to two but the rank of the augmented matrix A_b is equal to three.

8.1.4   General solution

We now investigate how the set of solutions of a system of linear equations can be written in terms of vectors. The following theorem describes the general solution of a *homogeneous system* of linear equations.

THEOREM 8.8   Let $A\mathbf{x} = \mathbf{0}$ be a homogeneous system of m linear equations with n variables and $r(A) = p < n$. Then there exist besides the zero vector $n - p$ further linearly independent solutions $\mathbf{x}^1, \mathbf{x}^2, \ldots, \mathbf{x}^{n-p} \in \mathbb{R}^n$ of the system $A\mathbf{x} = \mathbf{0}$, and the set of solutions S_H can be written as

$$
S_H = \left\{ \mathbf{x}^H \in \mathbb{R}^n \mid \mathbf{x}^H = \lambda_1 \mathbf{x}^1 + \lambda_2 \mathbf{x}^2 + \cdots + \lambda_{n-p} \mathbf{x}^{n-p}, \quad \lambda_1, \lambda_2, \ldots, \lambda_{n-p} \in \mathbb{R} \right\}.
$$

According to Definition 7.15 the set S_H corresponds to the kernel of a linear mapping described by matrix A. In order to present the general solution, we need $n - p$ *linearly independent* solution vectors. To illustrate how these vectors can be found, we consider the following example of a homogeneous system of linear equations and determine the general solution of this system.

Example 8.7   Let $Ax = 0$ with

$$A = \begin{pmatrix} 1 & 2 & 0 & 1 & 1 \\ -1 & -3 & 1 & -2 & 1 \\ 0 & -1 & 1 & -1 & 2 \\ 1 & 1 & 1 & 0 & 3 \end{pmatrix}.$$

Applying Gaussian elimination, we obtain the following tableaus:

Row	x_1	x_2	x_3	x_4	x_5	b	Operation
1	1	2	0	1	1	0	
2	-1	-3	1	-2	1	0	
3	0	-1	1	-1	2	0	
4	1	1	1	0	3	0	
5	1	2	0	1	1	0	row 1
6	0	-1	1	-1	2	0	row 1 + row 2
7	0	-1	1	-1	2	0	row 3
8	0	-1	1	-1	2	0	-row 1 + row 4

We can stop our computations here because rows 7 and 8 are identical to row 6 and thus we can drop rows 7 and 8. (If we continue applying Gaussian elimination, we obtain two rows containing only zeroes.) Hence we use a triangular system with the following two equations obtained from rows 5 and 6:

$$\begin{aligned} x_1 + 2x_2 \quad\quad + x_4 + x_5 &= 0 \\ - x_2 + x_3 - x_4 + 2x_5 &= 0 \end{aligned} \tag{8.10}$$

Thus, we have $r(A) = 2$. Because $n = 5$, there are three linearly independent solutions x^1, x^2 and x^3. They can be obtained for instance by setting exactly one of three variables that can be chosen arbitrarily to be equal to one and the other two of them to be equal to zero, i.e.

$$\bar{x}^1 : x_3^1 = 1, \quad x_4^1 = 0, \quad x_5^1 = 0;$$
$$\bar{x}^2 : x_3^2 = 0, \quad x_4^2 = 1, \quad x_5^2 = 0;$$
$$\bar{x}^3 : x_3^3 = 0, \quad x_4^3 = 0, \quad x_5^3 = 1.$$

Now the remaining components of each of the three vectors are uniquely determined. It is clear that the resulting vectors are linearly independent since the matrix formed by the three vectors $\bar{x}^1, \bar{x}^2$ and $\bar{x}^3$ is a 3×3 identity matrix. Determining the remaining components from system (8.10), we obtain

$$x^1 = \begin{pmatrix} -2 \\ 1 \\ 1 \\ 0 \\ 0 \end{pmatrix}, \quad x^2 = \begin{pmatrix} 1 \\ -1 \\ 0 \\ 1 \\ 0 \end{pmatrix} \quad \text{and} \quad x^3 = \begin{pmatrix} -5 \\ 2 \\ 0 \\ 0 \\ 1 \end{pmatrix}.$$

Thus, according to Theorem 8.8 we get the following set S_H of solutions:

$$S_H = \left\{ \mathbf{x}^H \in \mathbb{R}^5 \,\middle|\, \mathbf{x}^H = \lambda_1 \begin{pmatrix} -2 \\ 1 \\ 1 \\ 0 \\ 0 \end{pmatrix} + \lambda_2 \begin{pmatrix} 1 \\ -1 \\ 0 \\ 1 \\ 0 \end{pmatrix} + \lambda_3 \begin{pmatrix} -5 \\ 2 \\ 0 \\ 0 \\ 1 \end{pmatrix}; \ \lambda_1, \lambda_2, \lambda_3 \in \mathbb{R} \right\}.$$

The following theorem characterizes the general solution of a *non-homogeneous system* of linear equations.

THEOREM 8.9   Let $A\mathbf{x} = \mathbf{b}$ be a system of m linear equations with n variables and $r(A) = p < n$. Moreover, let

$$S_H = \left\{ \mathbf{x}^H \in \mathbb{R}^n \mid \mathbf{x}^H = \lambda_1 \mathbf{x}^1 + \lambda_2 \mathbf{x}^2 + \cdots + \lambda_{n-p} \mathbf{x}^{n-p}, \quad \lambda_1, \lambda_2, \ldots, \lambda_{n-p} \in \mathbb{R} \right\}$$

denote the general solution of the homogeneous system $A\mathbf{x} = \mathbf{0}$ and $\mathbf{x}^N$ be a solution of the non-homogeneous system $A\mathbf{x} = \mathbf{b}$. Then the set of solutions S of system $A\mathbf{x} = \mathbf{b}$ can be written as

$$S = \left\{ \mathbf{x} \in \mathbb{R}^n \mid \mathbf{x} = \mathbf{x}^H + \mathbf{x}^N, \quad \mathbf{x}^H \in S_H \right\}$$
$$= \left\{ \mathbf{x} \in \mathbb{R}^n \mid \mathbf{x} = \lambda_1 \mathbf{x}^1 + \lambda_2 \mathbf{x}^2 + \cdots + \lambda_{n-p} \mathbf{x}^{n-p} + \mathbf{x}^N, \quad \lambda_1, \lambda_2, \ldots, \lambda_{n-p} \in \mathbb{R} \right\}.$$

Theorem 8.9 says that the general solution of a non-homogeneous system of linear equations is obtained as the sum of the general solution of the corresponding homogeneous system (with vector $\mathbf{b}$ replaced by the zero vector $\mathbf{0}$) and a particular solution of the non-homogeneous system.

As we see in this and the next chapters, the pivoting procedure is also the base for solving linear systems of inequalities and linear programming problems. The Gaussian elimination procedure is sometimes advantageous when the system contains parameters. Finally, we give a more complicated example of a system of linear equations containing a parameter a, and we discuss the solution in dependence on the value of this parameter.

Example 8.8   We determine all solutions of the following system of linear equations:

$$
\begin{array}{rcrcrcl}
a x_1 & + & x_2 & + & x_3 & = & 1 \\
x_1 & + & a x_2 & + & x_3 & = & a \\
x_1 & + & x_2 & + & a x_3 & = & a^2,
\end{array}
$$

where $a \in \mathbb{R}$. We interchange the first and third equations. Applying now Gaussian elimination, this gives the first pivot 1 (in the case of taking the original first equation, we would have pivot a, and this means that we have to exclude the case $a = 0$ in order to guarantee that the pivot is different from zero). We obtain the following tableaus.

Row	x_1	x_2	x_3	**b**	Operation
1	1	a	1	a	
2	1	1	a	a^2	
3	a	1	1	1	
4	1	a	1	a	row 1
5	0	$1-a$	$a-1$	a^2-a	row 2 $-$ row 1
6	0	$-a^2+1$	$-a+1$	$-a^2+1$	row 3 $- a$ row 1
7	1	a	1	a	row 4
8	0	$1-a$	$a-1$	a^2-a	row 5
9	0	$-a^2-a+2$	0	$1-a$	row 6 $+$ row 5

In row 5, we have taken $a - 1$ in the x_3 column as pivot and this generates a zero below this element. (Notice that this implies $a - 1 \neq 0$.) In the third tableau (rows 7 to 9), we can still interchange the columns belonging to variables x_2 and x_3 (to generate formally a triangular matrix). We first consider the case when the coefficient of variable x_2 in row 9 of the above scheme is equal to zero, i.e. $-a^2 - a + 2 = 0$. For this quadratic equation we get two real solutions $a_1 = 1$ and $a_2 = -2$. If $a = a_1 = 1$, then the right-hand side $1 - a$ is also equal to zero. Moreover, in this case also all elements of row 8 are equal to zero, but row 7 contains non-zero elements. Hence both the coefficient matrix and the augmented coefficient matrix have rank one: $r(A) = r(A \mid b) = 1$, and there exist infinitely many solutions (two variables can be chosen arbitrarily). Choosing variables $x_2 = s \in \mathbb{R}$ and $x_3 = t \in \mathbb{R}$ arbitrarily, we get $x_1 = 1 - s - t$.

This solution can be alternatively written using Theorem 8.9. To find $\mathbf{x}^H$, we determine x_1^1 using $x_2^1 = s = 1, x_3^1 = t = 0$ which yields $x_1^1 = -1$ (notice that $x_1 = -s - t$ in the homogeneous system) and x_1^2 using $x_2^2 = s = 0, x_3^2 = t = 1$ which yields $x_1^2 = -1$. To get a particular solution $\mathbf{x}^N$ of the non-homogeneous system, we set $x_2^N = x_3^N = 0$ which yields $x_1^N = 1$. Therefore, we get the general solution S as follows:

$$S = \left\{ \mathbf{x} \in \mathbb{R}^3 \mid \mathbf{x} = \mathbf{x}^H + \mathbf{x}^N \right\}$$

$$= \left\{ \mathbf{x} \in \mathbb{R}^3 \, \middle| \, \mathbf{x} = \lambda_1 \begin{pmatrix} -1 \\ 1 \\ 0 \end{pmatrix} + \lambda_2 \begin{pmatrix} -1 \\ 0 \\ 1 \end{pmatrix} + \begin{pmatrix} 1 \\ 0 \\ 0 \end{pmatrix} ; \quad \lambda_1, \lambda_2 \in \mathbb{R} \right\}.$$

If $a = a_2 = -2$, then the right-hand side $1 - a = 3$ is different from zero. In this case, we have $r(A) = 2$ but $r(A_b) = 3$. Therefore, the system of linear equations is inconsistent, and there is not a solution of the system.

Consider now the remaining cases with $a \neq 1$ and $a \neq -2$. In all these cases we have $r(A) = r(A_b) = 3$, and thus there exists a uniquely determined solution. Using rows 7 to 9 in the above tableau, we get

$$x_2 = \frac{1-a}{-a^2-a+2} = -\frac{1-a}{(a-1)(a+2)} = \frac{1}{a+2}$$

$$x_3 = \frac{a^2-a-(1-a)x_2}{a-1} = \frac{a(a-1)-(1-a)x_2}{a-1} = a+x_2$$

$$= a+\frac{1}{a+2} = \frac{a^2+2a+1}{a+2} = \frac{(a+1)^2}{a+2}$$

$$x_1 = a-x_3-ax_2 = a-\frac{(a+1)^2}{a+2}-\frac{a}{a+2}$$

$$= \frac{a^2+2a-a^2-2a-1-a}{a+2} = -\frac{a+1}{a+2}.$$

We have already discussed a method for determining the inverse of a regular matrix A (see Chapter 7.5). The method presented there requires us to evaluate a lot of determinants so that this is only efficient for matrices of order $n \times n$ with a small value of n. Of course, we can also determine matrix A^{-1} via the solution of systems of linear equations with different right-hand side vectors, as in the following section.

8.1.5 Matrix inversion

The inverse $X = A^{-1}$ of a matrix A of order $n \times n$ satisfies the matrix equation

$$AX = I,$$

assuming that $|A| \neq 0$ (notice that this corresponds to the condition $r(A) = n$). Let the column vectors of matrix X be:

$$\mathbf{x}^1 = \begin{pmatrix} x_{11} \\ x_{21} \\ \vdots \\ x_{n1} \end{pmatrix}, \quad \mathbf{x}^2 = \begin{pmatrix} x_{12} \\ x_{22} \\ \vdots \\ x_{n2} \end{pmatrix}, \dots, \quad \mathbf{x}^n = \begin{pmatrix} x_{1n} \\ x_{2n} \\ \vdots \\ x_{nn} \end{pmatrix}.$$

Then the matrix equation $AX = I$ can be written as

$$A(\mathbf{x}^1\,\mathbf{x}^2\,\dots\,\mathbf{x}^n) = I = (\mathbf{e}^1\,\mathbf{e}^2\,\dots\,\mathbf{e}^n)$$

which is equivalent to the following n systems of linear equations:

$$A\mathbf{x}^1 = \mathbf{e}^1, \qquad A\mathbf{x}^2 = \mathbf{e}^2, \qquad \dots, \qquad A\mathbf{x}^n = \mathbf{e}^n.$$

These n systems of n linear equations differ only in the right-hand side vector and all the transformations of the coefficient matrix A within the application of a solution method are the same. Thus, applying for instance the pivoting procedure, we can use the following scheme for solving simultaneously these systems of n linear equations, i.e. if $r(A) = n$, we obtain after n pivoting steps the inverse $A^{-1} = X = (x_{ij})$. In the following tableaus we have assumed that no rows and columns have been interchanged (so the variables occur according to their natural numbering), i.e. the elements on the diagonal have successively been chosen as pivots. Of course, we can analogously also apply the Gaussian elimination procedure to solve the above systems with different right-hand side vectors.

Row	bv	A				I			
1	—	a_{11}	a_{12}	$\cdots$	a_{1n}	1	0	$\cdots$	0
2	—	a_{21}	a_{22}	$\cdots$	a_{2n}	0	1	$\cdots$	0
$\vdots$		$\vdots$	$\vdots$		$\vdots$	$\vdots$	$\vdots$		$\vdots$
n	—	a_{n1}	a_{n2}	$\cdots$	a_{nn}	0	0	$\cdots$	1
$\vdots$	$\vdots$	$\vdots$				$\vdots$			
$n(n-1)+1$	x_1	1	0	$\cdots$	0	x_{11}	x_{12}	$\cdots$	x_{1n}
$n(n-1)+2$	x_2	0	1	$\cdots$	0	x_{21}	x_{22}	$\cdots$	x_{2n}
$\vdots$	$\vdots$	$\vdots$	$\vdots$	$\vdots$		$\vdots$	$\vdots$		$\vdots$
n^2	x_n	0	0	$\cdots$	1	x_{n1}	x_{n2}	$\cdots$	x_{nn}
				I			A^{-1}		

Example 8.9 We consider matrix

$$A = \begin{pmatrix} 1 & 3 & 1 \\ 2 & -1 & 1 \\ 1 & 1 & 1 \end{pmatrix}$$

and determine the inverse A^{-1} by means of the pivoting procedure. The computations are shown in the following scheme.

Row	bv	A			I			Operation
1	—	1	3	1	1	0	0	
2	—	2	−1	1	0	1	0	
3	—	1	1	1	0	0	1	
4	x_1	1	3	1	1	0	0	row 1
5	—	0	−7	−1	−2	1	0	row 2 − 2 row 1
6	—	0	−2	0	−1	0	1	row 3 − row 1
7	x_1	1	0	$\frac{4}{7}$	$\frac{1}{7}$	$\frac{3}{7}$	0	row 4 + $\frac{3}{7}$ row 5
8	x_2	0	1	$\frac{1}{7}$	$\frac{2}{7}$	$-\frac{1}{7}$	0	$-\frac{1}{7}$ row 5
9	—	0	0	$\frac{2}{7}$	$-\frac{3}{7}$	$-\frac{2}{7}$	1	row 6 − $\frac{2}{7}$ row 5
10	x_1	1	0	0	1	1	−2	row 7 − 2 row 9
11	x_2	0	1	0	$\frac{1}{2}$	0	$-\frac{1}{2}$	row 8 − $\frac{1}{2}$ row 9
12	x_3	0	0	1	$-\frac{3}{2}$	−1	$\frac{7}{2}$	$\frac{7}{2}$ row 9

Since it is possible to generate an identity matrix on the left-hand side (see rows 10 to 12), the rank of matrix A is equal to three and the inverse A^{-1} exists. From rows 10 to 12, we obtain

$$A^{-1} = \begin{pmatrix} 1 & 1 & -2 \\ \frac{1}{2} & 0 & -\frac{1}{2} \\ -\frac{3}{2} & -1 & \frac{7}{2} \end{pmatrix}.$$

Computing the product $A^{-1}A$ gives the identity matrix and confirms the correctness of our computations.

8.2   SYSTEMS OF LINEAR INEQUALITIES

8.2.1   Preliminaries

We start this section with the following introductory example.

Example 8.10   A drink consisting of orange juice and champagne is to be mixed for a party. The ratio of orange juice to champagne has to be at least 1 : 2. The total quantity (volume) of the drink must not be more than 30 l, and at least 4 l more orange juice than champagne are to be used.

We denote by x_1 the quantity of orange juice in litres and by x_2 the quantity of champagne in litres. Then we get the following constraints:

$$
\begin{array}{rcrcr}
x_1 & : & x_2 & \geq & 1 \; : \; 2 \\
x_1 & - & x_2 & \leq & 4 \\
x_1 & + & x_2 & \leq & 30 \\
& & x_1, x_2 & \geq & 0.
\end{array}
$$

Hereafter, the notation $x_1, x_2 \geq 0$ means that both variables are non-negative: $x_1 \geq 0, x_2 \geq 0$. The first inequality considers the requirement that the ratio of orange juice and champagne (i.e. $x_1 : x_2$) should be at least 1 : 2. The second constraint takes into account that at most 4 l orange juice more than champagne are to be used (i.e. $x_1 \leq x_2 + 4$), and the third constraint ensures that the quantity of the drink is no more than 30 l. Of course, both quantities of orange juice and champagne have to be non-negative.

The first inequality can be rewritten by multiplying both sides by $-2x_2$ (note that the inequality sign changes) and putting both variables on the left-hand side so that we obtain the following inequalities:

$$
\begin{array}{rcrcr}
-2x_1 & + & 1x_2 & \leq & 0 \\
x_1 & - & x_2 & \leq & 4 \\
x_1 & + & x_2 & \leq & 30 \\
& & x_1, x_2 & \geq & 0.
\end{array}
$$

In this section, we deal with the solution of such systems of linear inequalities with non-negative variables.

In general, we define a system of linear inequalities as follows.

Definition 8.9   The system

$$
\begin{array}{cccccccc}
a_{11}x_1 & + & a_{12}x_2 & + & \cdots & + & a_{1n}x_n & R_1 & b_1 \\
a_{21}x_1 & + & a_{22}x_2 & + & \cdots & + & a_{2n}x_n & R_2 & b_2 \\
& & \vdots & & & & & \vdots & \vdots \\
a_{m1}x_1 & + & a_{m2}x_2 & + & \cdots & + & a_{mn}x_n & R_m & b_m
\end{array}
\tag{8.11}
$$

is called a *system of linear inequalities* with the coefficients a_{ij}, the right-hand sides b_i and the variables x_i. Here $R_i \in \{\leq, =, \geq\}$, $i = 1, 2, \ldots, m$, means that one of the three relations $\leq, =,$ or $\geq$ should hold, and we assume that at least one inequality occurs in the given system.
The inequalities

$$
x_j \geq 0, \qquad j \in J \subseteq \{1, 2, \ldots, n\}
\tag{8.12}
$$

are called *non-negativity constraints*.

The constraints (8.11) and (8.12) are called a *system of linear inequalities with* $|J|$ *non-negativity constraints*.

In the following, we consider a system of linear inequalities with n non-negativity constraints (i.e. $J = \{1, 2, \ldots, n\}$) which can be formulated in matrix form as follows:

$$
Ax \; R \; b, \quad x \geq 0,
\tag{8.13}
$$

where $\mathbf{R} = (R_1, R_2, \ldots, R_m)^T$ denotes the vector of the relation symbols with $R_i \in \{\leq, =, \geq\}$, $i = 1, 2, \ldots, m$.

Definition 8.10   A vector $\mathbf{x} = (x_1, x_2, \ldots, x_n)^T \in \mathbb{R}^n$ which satisfies the system $Ax \; R \; b$ is called a *solution*. If a solution $\mathbf{x}$ also satisfies the non-negativity constraints $\mathbf{x} \geq \mathbf{0}$, it is called a *feasible solution*. The set

$$
M = \left\{ \mathbf{x} \in \mathbb{R}^n \mid Ax \; R \; b, \; \mathbf{x} \geq \mathbf{0} \right\}
$$

is called the *set of feasible solutions* or the *feasible region* of a system (8.13).

8.2.2   Properties of feasible solutions

First, we introduce the notion of a convex set.

Definition 8.11   A set M is called *convex* if for any two vectors $\mathbf{x}^1, \mathbf{x}^2 \in M$, any convex combination $\lambda \mathbf{x}^1 + (1 - \lambda)\mathbf{x}^2$ with $0 \leq \lambda \leq 1$ also belongs to set M.

The definition of a convex set is illustrated in Figure 8.1. In Figure 8.1(a), set M is convex since every point of the connecting straight line between the terminal points of vectors $\mathbf{x}^1$ and $\mathbf{x}^2$ also belongs to set M (for arbitrary vectors $\mathbf{x}^1$ and $\mathbf{x}^2$ ending in M). However, set M^* in Figure 8.1(b) is not convex, since for the chosen vectors $\mathbf{x}^1$ and $\mathbf{x}^2$ not every point on the connecting straight line between the terminal points of both vectors belongs to set M^*.

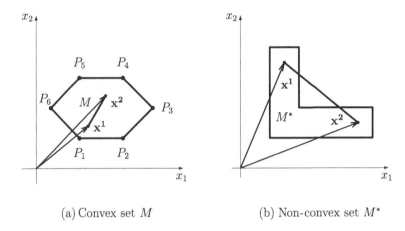

(a) Convex set M (b) Non-convex set M^*

Figure 8.1 A convex set and a non-convex set.

Definition 8.12   A vector (point) $\mathbf{x} \in M$ is called the *extreme point* (or *corner point* or *vertex*) of the convex set M if $\mathbf{x}$ cannot be written as a proper convex combination of two other vectors of M, i.e. $\mathbf{x}$ cannot be written as

$$\mathbf{x} = \lambda \mathbf{x}^1 + (1 - \lambda)\mathbf{x}^2 \qquad \text{with } \mathbf{x}^1, \mathbf{x}^2 \in M \text{ and } 0 < \lambda < 1.$$

Returning to Figure 8.1, set M in part (a) has six extreme points $\mathbf{x}^{(1)}, \mathbf{x}^{(2)}, \ldots, \mathbf{x}^{(6)}$ or equivalently the terminal points $P_1, P_2, \ldots, P_6$ of the corresponding vectors (here and in the following chapter, we always give the corresponding points P_i in the figures).

In the case of two variables, we can give the following *geometric interpretation* of a system of linear inequalities. Assume that the constraints are given as inequalities. The constraints $a_{i1}x_1 + a_{i2}x_2 \, R_i \, b_i$, $R_i \in \{\leq, \geq\}$, $i = 1, 2, \ldots, m$, are half-planes which are bounded by the lines $a_{i1}x_1 + a_{i2}x_2 = b_i$. The ith constraint can also be written in the form

$$\frac{x_1}{s_{i1}} + \frac{x_2}{s_{i2}} = 1,$$

where $s_{i1} = b_i/a_{i1}$ and $s_{i2} = b_i/a_{i2}$ are the intercepts of the line with the x_1 axis and the x_2 axis, respectively (see Figure 8.2).

The non-negativity constraints $x_1 \geq 0$ and $x_2 \geq 0$ represent the non-negative quadrant in the two-dimensional space. Thus, when considering a system of m inequalities with two non-negative variables, the feasible region is described by the intersection of m half-planes

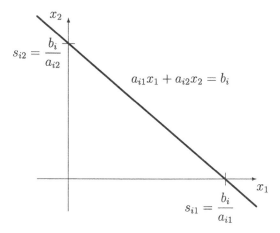

Figure 8.2 Representation of a line by x_1 and x_2 intercepts.

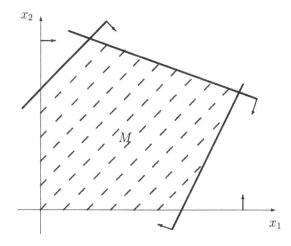

Figure 8.3 The set of solutions as the intersection of half-planes.

with the non-negative quadrant. This is illustrated for the case $m = 3$ in Figure 8.3, where the feasible region M is dashed. In Figure 8.3 and figures which follow we use arrows to indicate which of the resulting half-planes of each constraint satisfies the corresponding inequality constraint. The arrows at the coordinate axes indicate that both variables have to be non-negative.

In general, we can formulate the following property.

THEOREM 8.10 The feasible region M of system (8.13) is either empty or a convex set with at most a finite number of extreme points.

PROOF   We only prove that, if $M \neq \emptyset$, it is a convex set. Let $\mathbf{x}^1, \mathbf{x}^2 \in M$, i.e. we have

$$A\mathbf{x}^1 \ \mathbf{R} \ \mathbf{b}, \ \mathbf{x}^1 \geq \mathbf{0} \qquad \text{and} \qquad A\mathbf{x}^2 \ \mathbf{R} \ \mathbf{b}, \ \mathbf{x}^2 \geq \mathbf{0},$$

and we prove that $\lambda\mathbf{x}^1 + (1 - \lambda)\mathbf{x}^2 \in M$. Then

$$A\left[\lambda\mathbf{x}^1 + (1 - \lambda)\mathbf{x}^2\right] = \lambda A\mathbf{x}^1 + (1 - \lambda)A\mathbf{x}^2 \text{ R } \lambda\mathbf{b} + (1 - \lambda)\mathbf{b} = \mathbf{b} \qquad (8.14)$$

and due to $\lambda\mathbf{x}^1 \geq 0$ and $(1 - \lambda)\mathbf{x}^2 \geq 0$ for $0 \leq \lambda \leq 1$, we get

$$\lambda\mathbf{x}^1 + (1 - \lambda)\mathbf{x}^2 \geq 0. \qquad (8.15)$$

From (8.14) and (8.15), it follows that M is convex. ∎

A convex set with a finite number of extreme points is also called a *convex polyhedron*. One of the possible cases of Theorem 8.10 is illustrated in Figure 8.4. Here the values of both variables can be arbitrarily large provided that the given two inequalities are satisfied, i.e. the feasible region is unbounded. In this case, the feasible region has three extreme points P_1, P_2 and P_3. The following theorem characterizes the feasible region of a system of linear inequalities provided that this set is bounded.

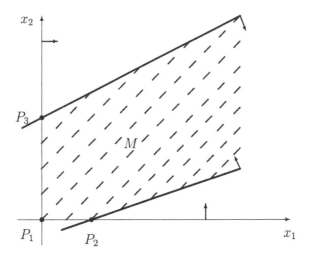

Figure 8.4 An unbounded set of solutions M.

THEOREM 8.11 Let the feasible region M of system (8.13) be bounded. Then it can be written as the set of all convex combinations of the extreme points $\mathbf{x}^1, \mathbf{x}^2, \ldots, \mathbf{x}^s$ of set M, i.e.

$$M = \left\{ \mathbf{x} \in \mathbb{R}^n \mid \mathbf{x} = \lambda_1\mathbf{x}^1 + \lambda_2\mathbf{x}^2 + \cdots + \lambda_s\mathbf{x}^s; \right.$$

$$\left. 0 \leq \lambda_i \leq 1, \quad i = 1, 2, \ldots, s, \quad \sum_{i=1}^{s} \lambda_i = 1 \right\}.$$

In the case of only two variables, we can graphically solve the problem. To illustrate the determination of set M, consider the following example.

Example 8.11 Two goods G_1 and G_2 are produced by means of two raw materials R_1 and R_2 with the capacities of 50 and 80 units, respectively. To produce 1 unit of G_1, 1 unit of R_1 and 1 unit of R_2 are required. To produce 1 unit of G_2, 1 unit of R_1 and 2 units of R_2 are required. The price of G_1 is 3 EUR per unit, the price of G_2 is 2 EUR per unit and at least 60 EUR worth of goods need to be sold.

Let x_i be the number of produced units of $G_i, i \in \{1, 2\}$. A feasible production programme has to satisfy the following constraints:

$$
\begin{array}{rlcll}
x_1 & + & x_2 & \le & 50 & \quad (I) & \text{(constraint for } R_1) \\
x_1 & + & 2x_2 & \le & 80 & \quad (II) & \text{(constraint for } R_2) \\
3x_1 & + & 2x_2 & \ge & 60 & \quad (III) & \text{(selling constraint)} \\
& & x_1, x_2 & \ge & 0 & & \text{(non-negativity constraints)}
\end{array}
$$

This is a system of linear inequalities with only two variables, which can be easily solved graphically. The feasible region is given in Figure 8.5. The convex set of feasible solutions has five extreme points described by the vectors $\mathbf{x}^i$ (or points P_i), $i = 1, 2, \ldots, 5$:

$$
\mathbf{x}^1 = \begin{pmatrix} 20 \\ 0 \end{pmatrix}, \quad \mathbf{x}^2 = \begin{pmatrix} 50 \\ 0 \end{pmatrix}, \quad \mathbf{x}^3 = \begin{pmatrix} 20 \\ 30 \end{pmatrix},
$$

$$
\mathbf{x}^4 = \begin{pmatrix} 0 \\ 40 \end{pmatrix} \quad \text{and} \quad \mathbf{x}^5 = \begin{pmatrix} 0 \\ 30 \end{pmatrix}.
$$

Therefore, the feasible region M is the set of all convex combinations of the above five extreme points:

$$
M = \left\{ \mathbf{x} \in \mathbb{R}^2_+ \;\middle|\; \mathbf{x} = \lambda_1 \begin{pmatrix} 20 \\ 0 \end{pmatrix} + \lambda_2 \begin{pmatrix} 50 \\ 0 \end{pmatrix} + \lambda_3 \begin{pmatrix} 20 \\ 30 \end{pmatrix} + \lambda_4 \begin{pmatrix} 0 \\ 40 \end{pmatrix} \right.
$$

$$
\left. + \lambda_5 \begin{pmatrix} 0 \\ 30 \end{pmatrix}, \quad \lambda_1, \lambda_2, \ldots, \lambda_5 \ge 0, \; \sum_{i=1}^5 \lambda_i = 1 \right\}.
$$

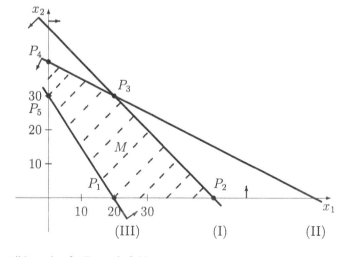

Figure 8.5 Feasible region for Example 8.11.

If the feasible region M is unbounded, there exist (unbounded) one-dimensional rays emanating from some extreme point of M on which points are feasible. Assume that there are u such rays and let $\mathbf{r}^1, \mathbf{r}^2, \ldots, \mathbf{r}^u$ denote those vectors pointing from a corresponding extreme point in the direction of such an unbounded one-dimensional ray. Then the feasible region M can be described as follows.

THEOREM 8.12   Let the feasible region M of system (8.13) be unbounded. Then it can be written as follows:

$$M = \left\{ \mathbf{x} \in \mathbb{R}^n \mid \mathbf{x} = \sum_{i=1}^{s} \lambda_i \mathbf{x}^i + \sum_{j=1}^{u} \mu_j \mathbf{r}^j; \right.$$

$$\left. 0 \le \lambda_i \le 1, \quad i = 1, 2, \ldots, s, \quad \sum_{i=1}^{s} \lambda_i = 1; \quad \mu_1, \mu_2, \ldots, \mu_u \ge 0 \right\},$$

where $\mathbf{x}^1, \mathbf{x}^2, \ldots, \mathbf{x}^s$ are the extreme points of set M and $\mathbf{r}^1, \mathbf{r}^2, \ldots, \mathbf{r}^u$ are the vectors of the unbounded one-dimensional rays of set M.

According to Theorem 8.12, any feasible solution of an unbounded feasible region M can be written as a convex combination of the extreme points and a linear combination of the vectors of the unbounded rays with non-negative scalars $\mu_j, j = 1, 2, \ldots, u$. Considering the example given in Figure 8.4, there are two unbounded one-dimensional rays emanating from points P_2 and P_3.

A relationship between extreme points and basic feasible solutions is given in the following theorem.

THEOREM 8.13   Any extreme point of the feasible region M of system (8.13) corresponds to at least one basic feasible solution, and conversely, any basic feasible solution corresponds exactly to one extreme point.

The latter theorem needs to be discussed in a bit more detail. We know from our previous considerations that in a basic solution, all non-basic variables are equal to zero. Thus, if the coefficient matrix A of the system of linear equations has rank m, at most m variables have positive values. We can distinguish the following two cases.

Definition 8.13   Let M be the feasible region of system (8.13) and let $r(A) = m$. If a basic feasible solution $\mathbf{x} \in M$ has m positive components, the solution is called *non-degenerate*. If the basic feasible solution $\mathbf{x}$ has less than m positive components, the solution is called *degenerate*.

As we discuss later in connection with linear programming problems in Chapter 9, degeneracy of solutions may cause computational problems. In the case where all basic feasible solutions are non-degenerate solutions, Theorem 8.13 can be strengthened.

THEOREM 8.14   Let all basic feasible solutions of system (8.13) be non-degenerate solutions. Then there is a one-to-one correspondence between basic feasible solutions and extreme points of the set M of feasible solutions of system (8.13).

8.2.3   A solution procedure

According to Theorem 8.11 we have to generate all extreme points in order to describe the feasible region of a system of linear inequalities. Using Theorems 8.13 and 8.14, respectively, this can be done by generating all basic feasible solutions of the given system.

In this section, we restrict ourselves to the case when system (8.13) is given in the special form

$$A\mathbf{x} \le \mathbf{b}, \quad \mathbf{x} \ge \mathbf{0}, \quad \text{with } \mathbf{b} \ge \mathbf{0}, \tag{8.16}$$

where A is an $m \times n$ matrix and we assume that $r(A) = m$. (The case of arbitrary constraints is discussed in detail in the next chapter when dealing with linear programming problems.) This particular situation often occurs in economic applications. For instance, it is necessary to determine feasible production programmes of n goods by means of m raw materials, where the coefficient matrix A describes the use of the particular raw materials per unit of each good and the non-negative vector $\mathbf{b}$ describes the capacity constraints on the use of the raw materials.

For a system (8.16), inequalities are transformed into equations by introducing a slack variable in each constraint, i.e. for the ith constraint

$$a_{i1}x_1 + a_{i2}x_2 + \cdots + a_{in}x_n \le b_i,$$

we write

$$a_{i1}x_1 + a_{i2}x_2 + \cdots + a_{in}x_n + u_i = b_i,$$

where $u_i \ge 0$ is a so-called *slack variable* $(i \in \{1, 2, \ldots, m\})$. In the following, we use the m-dimensional vector $\mathbf{u} = (u_1, u_2, \ldots, u_m)^\mathsf{T}$ for the slack variables introduced in the given m inequalities.

Letting

$$A^* = (A, I) \qquad \text{and} \qquad \overline{\mathbf{x}} = \begin{pmatrix} \mathbf{x} \\ \mathbf{u} \end{pmatrix},$$

the system turns into

$$A^*\mathbf{x}^* = \mathbf{b} \qquad \text{with} \qquad \mathbf{x}^* = \begin{pmatrix} \mathbf{x} \\ \mathbf{u} \end{pmatrix} \ge \mathbf{0}. \tag{8.17}$$

In this case, we can choose

$$\mathbf{x}^* = \begin{pmatrix} \mathbf{x} \\ \mathbf{u} \end{pmatrix} = \begin{pmatrix} \mathbf{0} \\ \mathbf{b} \end{pmatrix}$$

as an initial basic feasible solution, i.e. the variables $x_1, x_2, \ldots, x_n$ are the non-basic variables and the slack variables $u_1, u_2, \ldots, u_m$ are the basic variables. Moreover, we assume that $\overline{\mathbf{x}}$ is a non-degenerate basic feasible solution (which means that $\mathbf{b} > \mathbf{0}$).

Starting with this basic feasible solution, we systematically have to generate all other basic feasible solutions. To this end, we have to determine the pivot element in each step such that the new basic solution is again feasible. (Such pivoting steps are repeated until all basic feasible solutions have been visited.)

For an $m \times n$ matrix A with $r(A) = m$, there are at most $\binom{n+m}{m}$ different basic feasible solutions of system (8.17) and therefore at most $\binom{n+m}{m}$ extreme points $\mathbf{x}$ of the set $M = \{\mathbf{x} \in \mathbb{R}^n_+ \mid A\mathbf{x} \le \mathbf{b}\}$. Each pivoting step corresponds to a move from an extreme point (represented by some basic feasible solution) to another basic feasible solution (usually corresponding to another extreme point). We can use the tableau in Table 8.2 to perform a pivoting step. Rows 1 to m represent the initial basic feasible solution, and rows $m+1$ to $2m$ represent the basic solution obtained after the pivoting step. (As before, we could also add an additional column to describe the operations that have to be performed in order to get rows $m+1$ to $2m$.)

Table 8.2 Tableau for a pivoting step

Row	bv	x_1	$\cdots$	x_l	$\cdots$	x_n	u_1	$\cdots$	u_k	$\cdots$	u_m	b
1	u_1	a_{11}	$\cdots$	a_{1l}	$\cdots$	a_{1n}	1	$\cdots$	0	$\cdots$	0	b_1
$\vdots$	$\vdots$	$\vdots$		$\vdots$		$\vdots$	$\vdots$		$\vdots$		$\vdots$	$\vdots$
k	u_k	a_{k1}	$\cdots$	$\mathbf{a_{kl}}$	$\cdots$	a_{kn}	0	$\cdots$	1	$\cdots$	0	b_k
$\vdots$	$\vdots$	$\vdots$		$\vdots$		$\vdots$	$\vdots$		$\vdots$		$\vdots$	$\vdots$
m	u_m	a_{m1}	$\cdots$	a_{ml}	$\cdots$	a_{mn}	0	$\cdots$	0	$\cdots$	1	b_m
$m+1$	u_1	$a_{11}-\frac{a_{1l}}{a_{kl}}a_{k1}$	$\cdots$	0	$\cdots$	$a_{1n}-\frac{a_{1l}}{a_{kl}}a_{kn}$	1	$\cdots$	$-\frac{a_{1l}}{a_{kl}}$	$\cdots$	0	$b_1-\frac{a_{1l}}{a_{kl}}b_k$
$\vdots$	$\vdots$	$\vdots$		$\vdots$		$\vdots$	$\vdots$		$\vdots$		$\vdots$	$\vdots$
$m+k$	x_l	$\frac{a_{k1}}{a_{kl}}$	$\cdots$	1	$\cdots$	$\frac{a_{kn}}{a_{kl}}$	0	$\cdots$	$\frac{1}{a_{kl}}$	$\cdots$	0	$\frac{b_k}{a_{kl}}$
$\vdots$	$\vdots$	$\vdots$		$\vdots$		$\vdots$	$\vdots$		$\vdots$		$\vdots$	$\vdots$
$2m$	u_m	$a_{m1}-\frac{a_{ml}}{a_{kl}}a_{k1}$	$\cdots$	0	$\cdots$	$a_{mn}-\frac{a_{ml}}{a_{kl}}a_{kn}$	0	$\cdots$	$-\frac{a_{ml}}{a_{kl}}$	$\cdots$	1	$b_m-\frac{a_{ml}}{a_{kl}}b_k$

Assume that we choose $a_{kl} \ne 0$ as pivot element, i.e. we replace the basic variable u_k by the original non-basic variable x_l. Then we obtain the new basic solution

$$\bar{\mathbf{x}}^* = \begin{pmatrix} \bar{\mathbf{x}} \\ \bar{\mathbf{u}} \end{pmatrix}$$

with

$$\bar{\mathbf{x}}^T = (\bar{x}_1,\dots,\bar{x}_l,\dots,\bar{x}_n) = \left(0,\dots,\frac{b_k}{a_{kl}},\dots,0\right)$$

and

$$\bar{\mathbf{u}}^T = (\bar{u}_1,\dots,\bar{u}_k,\dots,\bar{u}_m) = \left(b_1 - \frac{a_{1l}}{a_{kl}}\cdot b_k,\dots,0,\dots,b_m - \frac{a_{ml}}{a_{kl}}\cdot b_k\right).$$

The new basic solution

$$\bar{\mathbf{x}}^* = \begin{pmatrix} \bar{\mathbf{x}} \\ \bar{\mathbf{u}} \end{pmatrix}$$

is feasible if all components of vector $\bar{\mathbf{x}}^*$ are non-negative, i.e.

$$\frac{b_k}{a_{kl}} \geq 0 \tag{8.18}$$

and

$$b_i - \frac{a_{il}}{a_{kl}} \cdot b_k \geq 0, \quad i = 1, 2, \ldots, m, \ i \neq k. \tag{8.19}$$

Next, we derive a condition such that both (8.18) and (8.19) are satisfied. First, since $\mathbf{b} > 0$ (remember that we are considering a non-degenerate solution) and $a_{kl} \neq 0$ by assumption, we have the equivalence

$$\frac{b_k}{a_{kl}} \geq 0 \quad \Longleftrightarrow \quad a_{kl} > 0.$$

This means that in the chosen pivot column, only positive elements are a candidate for the pivot element. However, we also have to ensure that (8.19) is satisfied. We get the following equivalence:

$$b_i - \frac{a_{il}}{a_{kl}} \cdot b_k \geq 0 \quad \Longleftrightarrow \quad \frac{b_i}{a_{il}} \geq \frac{b_k}{a_{kl}} \geq 0 \quad \text{for all } i \text{ with } a_{il} > 0.$$

This means that we have to take that row as pivot row, which yields the smallest quotient of the current right-hand side component and the corresponding element in the chosen pivot column among all rows with a positive element in the pivot column. Summarizing, we can replace the basic variable u_k by the non-basic variable x_l if

$$a_{kl} > 0 \quad \text{and} \quad q_k = \frac{b_k}{a_{kl}} = \min \left\{ \frac{b_i}{a_{il}} \ \middle| \ a_{il} > 0, \ i \in \{1, 2, \ldots, m\} \right\}.$$

If one of the above conditions is violated, i.e. if $a_{kl} < 0$ or quotient $q_k = b_k/a_{kl}$ is not minimal among all quotients in the rows with a positive element in the pivot column, we do *not* get a basic feasible solution, i.e. at least one component of the new right-hand side vector would be negative. Therefore, we add a column Q in the tableau in Table 8.2 to calculate the corresponding quotients q_i which have to be taken into account. Notice that in the new tableau, there is again an identity submatrix (with possibly interchanged columns) contained in the transformed matrix belonging to the basic variables.

In the case of a degenerate basic feasible solution, all the above formulas for a pivoting step remain valid. In such a case, we have a smallest quotient $q_k = 0$ (which, by the way, means that also in the case of choosing a negative pivot element a basic feasible solution results). We will discuss the difficulties that may arise in case of degeneracy later in Chapter 9 in a bit more detail.

The process of enumerating all basic feasible solutions of system (8.13) can be done by hand (without using a computer) only for very small problems. Consider the following example.

Example 8.12 Consider the system of linear inequalities presented in Example 8.10. Introducing slack variables u_1, u_2 and u_3, we obtain the initial tableau in rows 1 to 3 below. Now, the goal is to generate *all* basic feasible solutions of the given system of linear equations.

Since each basic feasible solution has three basic variables, there are at most $\binom{5}{3} = 10$ basic feasible solutions. We successively perform pivoting steps and might obtain for instance the following sequence of basic feasible solutions:

Row	bv	x_1	x_2	u_1	u_2	u_3	b	Q
1	u_1	-2	1	1	0	0	0	—
2	u_2	**1**	-1	0	1	0	4	4
3	u_3	1	1	0	0	1	30	30
4	u_1	0	-1	1	2	0	8	—
5	x_1	1	-1	0	1	0	4	—
6	u_3	0	**2**	0	-1	1	26	13
7	u_1	0	0	1	$\frac{3}{2}$	$\frac{1}{2}$	21	14
8	x_1	1	0	0	$\frac{1}{2}$	$\frac{1}{2}$	17	34
9	x_2	0	1	0	$-\frac{1}{2}$	$\frac{1}{2}$	13	—
10	u_2	0	0	$\frac{2}{3}$	1	$\frac{1}{3}$	14	42
11	x_1	1	0	$-\frac{1}{3}$	0	$\frac{1}{3}$	10	30
12	x_2	0	1	$\frac{1}{3}$	0	$\frac{2}{3}$	20	30

The corresponding pivot elements are printed in bold face. In the first tableau (rows 1 to 3), we can choose either the column belonging to x_1 or that belonging to x_2 as pivot column (since there is at least one positive element in each of these columns). Selecting the column belonging to x_1, the quotient is uniquely determined, and we have to replace the basic variable u_2 by the non-basic variable x_1. In the second pivoting step, we can choose only the column belonging to x_2 as pivot column (since, when choosing the column belonging to u_2, we come back to the basic feasible solution represented by rows 1 to 3).

From the last basic feasible solution, we cannot generate another extreme point. However, we can perform another pivoting step so that u_3 becomes a basic variable. Determining now the minimum quotient in the Q column, it is not uniquely determined: we can choose x_1 or x_2 as the variable that becomes a non-basic variable (since in both cases, we have the smallest quotient 30). If we choose x_1 as the variable that becomes a non-basic variable, we get the tableau:

Row	bv	x_1	x_2	u_1	u_2	u_3	b	Q
13	u_2	-1	0	1	1	0	4	
14	u_3	3	0	-1	0	1	30	
15	x_2	-2	1	1	0	0	0	

If we choose in the fourth tableau x_2 as the variable that becomes a non-basic variable, we get the following tableau:

Row	bv	x_1	x_2	u_1	u_2	u_3	b	Q
16	u_2	0	$-\frac{1}{2}$	$\frac{1}{2}$	1	0	4	
17	x_1	1	$-\frac{1}{2}$	$-\frac{1}{2}$	0	1	0	
18	u_3	0	$\frac{3}{2}$	$\frac{1}{2}$	0	1	30	

We still have to show that we have now indeed generated all basic feasible solutions. For the second, third and fourth tableaus we have discussed all the possibilities. We still have to check the remaining possibilities for the first, fifth and six tableaus. If in the first tableau we choose instead of the x_1 column the x_2 column as pivot column, we get the fifth tableau (rows 13 to 15). Checking all possibilities for generating a new basic feasible solution in the fifth and sixth tableaus, we see that we can generate only such basic feasible solutions as we have already found. This means that the remaining possible combinations for selecting basic variables which are x_2, u_1, u_3; x_2, u_1, u_2; x_1, u_1, u_2 and x_1, x_2, u_3 do not lead to a basic feasible solution. (One can check this by trying to find all basic solutions.)

Therefore, in the above example, there are six basic feasible solutions and four basic infeasible solutions. From rows 1 to 18, we get the following basic feasible solutions. (The basic variables are printed in bold face.)

(1) $x_1 = 0$, $\quad x_2 = 0$, $\quad \mathbf{u_1 = 0}$, $\quad \mathbf{u_2 = 4}$, $\quad \mathbf{u_3 = 30}$;
(2) $\mathbf{x_1 = 4}$, $\quad x_2 = 0$, $\quad \mathbf{u_1 = 8}$, $\quad u_2 = 0$, $\quad \mathbf{u_3 = 26}$;
(3) $\mathbf{x_1 = 17}$, $\quad \mathbf{x_2 = 13}$, $\quad \mathbf{u_1 = 21}$, $\quad u_2 = 0$, $\quad u_3 = 0$;
(4) $\mathbf{x_1 = 10}$, $\quad \mathbf{x_2 = 20}$, $\quad u_1 = 0$, $\quad \mathbf{u_2 = 14}$, $\quad u_3 = 0$;
(5) $x_1 = 0$, $\quad x_2 = 0$, $\quad u_1 = 0$, $\quad \mathbf{u_2 = 4}$, $\quad \mathbf{u_3 = 30}$;
(6) $x_1 = 0$, $\quad x_2 = 0$, $\quad \mathbf{u_1 = 0}$, $\quad \mathbf{u_2 = 4}$, $\quad \mathbf{u_3 = 30}$.

Deleting now the introduced slack variables u_1, u_2, u_3, we get the corresponding extreme points P_1 with the coordinates $(0, 0)$, P_2 with the coordinates $(4, 0)$, P_3 with the coordinates $(17, 13)$ and P_4 with the coordinates $(10, 20)$. The fifth and sixth basic feasible solutions correspond to extreme point P_1 again. (In each of them, exactly one basic variable has value zero.) Therefore, the first, fifth and sixth basic feasible solutions are degenerate solutions corresponding to the same extreme point.

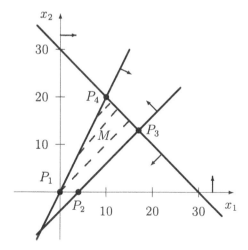

Figure 8.6 Feasible region for Example 8.12.

The feasible region together with the extreme points is given in Figure 8.6. It can be seen that our computations have started from point P_1, then we have moved to the adjacent extreme point P_2, then to the adjacent extreme point P_3 and finally to P_4. Then all extreme points have been visited and the procedure stops. According to Theorem 8.10, the feasible region is given by the set of all convex combinations of the four extreme points:

$$M = \left\{ \begin{pmatrix} x_1 \\ x_2 \end{pmatrix} \in \mathbb{R}^2 \;\middle|\; \begin{pmatrix} x_1 \\ x_2 \end{pmatrix} = \lambda_1 \begin{pmatrix} 0 \\ 0 \end{pmatrix} + \lambda_2 \begin{pmatrix} 4 \\ 0 \end{pmatrix} + \lambda_3 \begin{pmatrix} 17 \\ 13 \end{pmatrix} + \lambda_4 \begin{pmatrix} 10 \\ 20 \end{pmatrix}; \right.$$

$$\left. \lambda_i \geq 0, \; i \in \{1,2,3,4\}, \; \sum_{i=1}^{4} \lambda_i = 1 \right\}.$$

Since the columns belonging to the basic variables are always unit vectors, they can be omitted. All required information for performing a pivoting step is contained in the columns of the non-basic variables. Therefore, we can use a *short form* of the tableau, where the columns are assigned to the non-basic variables and the rows are assigned to the basic variables. Assume that altogether n variables denoted by $x_1, x_2, \ldots, x_n$ occur in the system of linear inequalities; among them there are m basic variables denoted by $x_{B1}, x_{B2}, \ldots, x_{Bm}$ and $n' = n - m$ non-basic variables denoted by $x_{N1}, x_{N2}, \ldots, x_{Nn'}$. In this case, the short form of the tableau is as in Table 8.3.

Table 8.3 Short form of the tableau for a pivoting step

Row	bv	x_{N1}	$\cdots$	x_{Nl}	$\cdots$	$x_{Nn'}$	
1	x_{B1}	a_{11}^*	$\cdots$	a_{1l}^*	$\cdots$	$a_{1n'}^*$	b_1^*
$\vdots$	$\vdots$	$\vdots$		$\vdots$		$\vdots$	$\vdots$
k	x_{Bk}	a_{k1}^*	$\cdots$	$\mathbf{a_{kl}^*}$	$\cdots$	$a_{kn'}^*$	b_k^*
$\vdots$	$\vdots$	$\vdots$		$\vdots$		$\vdots$	$\vdots$
m	x_{Bm}	a_{m1}^*	$\cdots$	a_{ml}^*	$\cdots$	$a_{mn'}^*$	b_m^*
		x_{N1}	$\cdots$	$\bar{x}_{Nl}$	$\cdots$	$x_{Nn'}$	
$m+1$	x_{B1}	$a_{11}^* - \frac{a_{1l}^*}{a_{kl}^*} a_{k1}^*$	$\cdots$	$-\frac{a_{1l}^*}{a_{kl}^*}$	$\cdots$	$a_{1n'}^* - \frac{a_{1l}^*}{a_{kl}^*} a_{kn'}^*$	$b_1^* - \frac{a_{1l}^*}{a_{kl}^*} b_k^*$
$\vdots$	$\vdots$	$\vdots$		$\vdots$		$\vdots$	$\vdots$
$m+k$	$\bar{x}_{Bk}$	$\frac{a_{k1}^*}{a_{kl}^*}$	$\cdots$	$\frac{1}{a_{kl}^*}$	$\cdots$	$\frac{a_{kn'}^*}{a_{kl}^*}$	$\frac{b_k^*}{a_{kl}^*}$
$\vdots$	$\vdots$	$\vdots$		$\vdots$		$\vdots$	$\vdots$
$2m$	x_{Bm}	$a_{m1}^* - \frac{a_{ml}^*}{a_{kl}^*} a_{k1}^*$	$\cdots$	$-\frac{a_{ml}^*}{a_{kl}^*}$	$\cdots$	$a_{mn'}^* - \frac{a_{ml}^*}{a_{kl}^*} a_{kn'}^*$	$b_m^* - \frac{a_{ml}^*}{a_{kl}^*} b_k^*$

In column bv, the basic variables $x_{B1}, x_{B2}, \ldots, x_{Bm}$ are given, and the next n' columns represent the non-basic variables $x_{N1}, x_{N2}, \ldots, x_{Nn'}$. After the first pivoting step, the kth basic variable is now the former lth non-basic variable: $\bar{x}_{Bk} = x_{Nl}$. Correspondingly, the lth non-basic variable in the new solution is the former kth basic variable: $\bar{x}_{Nl} = x_{Bk}$. Notice that in each transformation step we have to write the sequence of the non-basic variables since the variables in the columns do not appear according to the numbering.

It is worth emphasizing that, in contrast to the full form of the tableau, the new elements $\bar{a}_{il}$ of the pivot column are obtained as follows:

$$\bar{a}_{kl} = \frac{1}{a_{kl}^*},$$

$$\bar{a}_{il} = -\frac{a_{il}^*}{a_{kl}^*}, \quad i = 1, 2, \ldots, m, \ i \neq k,$$

i.e. the pivot element is replaced by its reciprocal value and the remaining elements of the pivot column are divided by the negative pivot element. Notice that this column also occurs in the full form of the tableau, though not in the pivot column (where after the pivoting step a unit vector occurs) but it appears in the column of the new non-basic variable $\bar{x}_{Nl}$ in the full form of the tableau after the pivoting step. The following example uses the short form of this tableau.

Example 8.13 We consider the following system of linear inequalities:

$$
\begin{array}{rcrcl}
 & & x_2 & \leq & 25 \\
2x_1 & + & x_2 & \leq & 30 \\
-x_1 & + & x_2 & \geq & -6 \\
 & & x_1, x_2 & \geq & 0.
\end{array}
$$

We multiply the third constraint by -1 and introduce in each inequality a slack variable denoted by u_1, u_2 and u_3. This yields a system with three equations and five variables.

$$
\begin{array}{rcrcrcrcl}
 & & x_2 & + & u_1 & & & & = 25 \\
2x_1 & + & x_2 & & & + & u_2 & & = 30 \\
x_1 & - & x_2 & & & & & + \ u_3 & = 6 \\
 & & & & \multicolumn{5}{l}{x_1, x_2, u_1, u_2, u_3 \geq 0.}
\end{array}
$$

There are at most $\binom{5}{3} = 10$ basic feasible solutions, each of them including exactly three basic variables. In order to describe the feasible region, we have to generate systematically all the possible basic feasible solutions. Starting from the first basic solution with $x_1 = 0, x_2 = 0, u_1 = 25, u_2 = 30, u_3 = 6$ and using the short form of the tableau, we obtain e.g. the following sequence of basic feasible solutions by subsequent pivoting steps. In the first step we have chosen the column belonging to x_1 as pivot column. (This is possible since at least one element in this column is positive; analogously one can also start with the column belonging to x_2 as the first pivot column.) Using the quotients in the Q column, we have found that row 3 must be the pivot row and, therefore, we have the pivot element 1. From the tableau given by rows 4 to 6, we have to select the column belonging to variable x_2 as pivot column (otherwise we come back to the basic feasible solution described by rows 1 to 3). The quotient rule determines row 5 as pivot row. Continuing in this way (note that the pivot column is now always uniquely determined since in the other case we would always go back to the previous basic feasible solution), we get the results in the following tableaus.

Row	bv	x_1	x_2	b	Q
1	u_1	0	1	25	—
2	u_2	2	1	30	15
3	u_3	1	−1	6	6
		u_3	x_2		
4	u_1	0	1	25	25
5	u_2	−2	3	18	6
6	x_1	1	−1	6	—
		u_3	u_2		
7	u_1	$\frac{2}{3}$	$-\frac{1}{3}$	19	$\frac{57}{2}$
8	x_2	$-\frac{2}{3}$	$\frac{1}{3}$	6	—
9	x_1	$\frac{1}{3}$	$\frac{1}{3}$	12	36
		u_1	u_2		
10	u_3	$\frac{3}{2}$	$-\frac{1}{2}$	$\frac{57}{2}$	—
11	x_2	1	0	25	—
12	x_1	$-\frac{1}{2}$	$\frac{1}{2}$	$\frac{5}{2}$	5
		u_1	x_1		
13	u_3	$-\frac{1}{2}$	1	31	
14	x_2	1	0	25	
15	u_2	−1	2	5	

Having generated the above five basic feasible solutions, we can stop our computations for the following reasons. There remain five other selections of three variables out of the variables x_1, x_2, u_1, u_2, u_3 which, however, do not yield basic feasible solutions. The variables x_2, u_1, u_3 as basic variables lead to a basic infeasible solution. (When selecting x_2 as the variable that becomes a basic variable in the next solution and row 2 as pivot row, this would violate the quotient rule since $q_2 = 30 > q_1 = 25$.) Similarly, the choice of x_1, u_1, u_3 as basic variables does not lead to a basic feasible solution since we would violate the quotient rule when choosing the x_1 column as pivot column and row 2 as pivot row. The choice of x_1, u_2, u_3 as basic variables is not possible because when choosing the x_1 column as pivot column and row 1 as pivot row in the first tableau, pivot zero would result and this is not allowed (i.e. the corresponding column vectors of matrix A belonging to these variables are linearly dependent). The choice of x_1, x_2, u_2 as basic variables leads to a basic infeasible solution because when choosing the u_2 column as pivot column and row 7 as pivot row in the third tableau, the resulting pivot element would be negative. For the same reason, the choice of variables x_2, u_1, u_2 as basic variables does not lead to a basic feasible solution. (Selecting the x_2 column as pivot column and row 3 as pivot row in the first tableau yields the negative pivot −1.) Therefore, in addition to the five basic feasible solutions determined from rows 1 to 15, there are four further basic infeasible solutions.

From the above five basic feasible solutions, we obtain the resulting five extreme points of the feasible region M (by dropping the slack variables u_1, u_2 and u_3). Therefore, set M is given by

$$M = \left\{ \begin{pmatrix} x_1 \\ x_2 \end{pmatrix} \in \mathbb{R}_+^2 \,\middle|\, \begin{pmatrix} x_1 \\ x_2 \end{pmatrix} = \lambda_1 \begin{pmatrix} 0 \\ 0 \end{pmatrix} + \lambda_2 \begin{pmatrix} 6 \\ 0 \end{pmatrix} + \lambda_3 \begin{pmatrix} 12 \\ 6 \end{pmatrix} \right.$$

$$\left. +\lambda_4 \begin{pmatrix} 5/2 \\ 25 \end{pmatrix} + \lambda_5 \begin{pmatrix} 0 \\ 25 \end{pmatrix}; \quad \lambda_i \geq 0, \ i = 1, 2, \ldots, 5, \ \sum_{i=1}^{5} \lambda_i = 1 \right\}.$$

The graphical solution of the system of linear inequalities is given in Figure 8.7. The ith basic feasible solution corresponds to extreme point P_i. In this example, there is exactly one basic feasible solution corresponding to each extreme point since each basic feasible solution is non-degenerate.

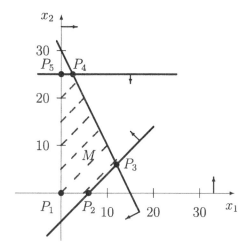

Figure 8.7 Feasible region for Example 8.13.

EXERCISES

8.1 Decide whether the following systems of linear equations are consistent and find the solutions. Apply both methods of Gaussian elimination and pivoting (use the rank criterion).

(a)
$$
\begin{array}{rcrcrcl}
x_1 &+& 2x_2 &+& 3x_3 &=& 5 \\
2x_1 &+& 3x_2 &+& x_3 &=& 8 \\
3x_1 &+& x_2 &+& 2x_3 &=& 5
\end{array}
$$

(b)
$$
\begin{array}{rcrcrcl}
3x_1 &+& x_2 &+& x_3 &=& 3 \\
4x_1 &-& x_2 &+& 2x_3 &=& 4 \\
x_1 &-& x_2 &+& x_3 &=& 1 \\
4x_1 &-& x_2 &+& 2x_3 &=& 5
\end{array}
$$

(c)
$$\begin{aligned}
3x_1 + 4x_2 + x_3 + 6x_4 &= 8 \\
3x_1 + 8x_2 + 6x_3 + 5x_4 &= 7 \\
8x_1 + 5x_2 + 6x_3 + 7x_4 &= 6 \\
6x_1 + 2x_2 + 5x_3 + 3x_4 &= 5
\end{aligned}$$

(d)
$$\begin{aligned}
x_1 + 2x_2 + 3x_3 &= 16 \\
8x_1 + 7x_2 + 6x_3 &= 74 \\
4x_1 + 5x_2 + 9x_3 &= 49 \\
5x_1 + 4x_2 + 2x_3 &= 43 \\
x_1 + 4x_2 + x_3 &= 22.
\end{aligned}$$

8.2 (a) Solve the following homogeneous systems of linear equations $Ax = 0$ with

(i) $A = \begin{pmatrix} 1 & 1 & 1 & 1 \\ 0 & 1 & 3 & 0 \\ 1 & 1 & 2 & 4 \\ 2 & 3 & 5 & 0 \end{pmatrix}$; (ii) $A = \begin{pmatrix} 1 & 1 & 1 & 1 \\ 2 & 3 & 6 & 3 \\ 1 & 1 & 2 & 4 \\ 2 & 3 & 5 & 0 \end{pmatrix}$

(b) Find the solutions of the non-homogeneous system $Ax = b$ with matrix A from (i) resp. (ii) and $b = (0, 0, -2, 2)^T$.

8.3 Find the general solutions of the following systems and specify two different basic solutions for each system:

(a)
$$\begin{aligned}
4x_1 - x_2 - 3x_3 + 5x_4 &= -2 \\
-x_1 + x_2 + x_3 - x_4 &= 4 \\
2x_1 - x_2 - x_3 - 2x_4 &= 1
\end{aligned}$$

(b)
$$\begin{aligned}
x_1 + 2x_2 - x_3 + 4x_5 &= 2 \\
x_1 + 4x_2 - 5x_3 + x_4 + 3x_5 &= 1 \\
2x_1 - 2x_2 + 10x_3 + x_4 - x_5 &= 11 \\
3x_1 + 2x_2 + 5x_3 + 2x_4 + 2x_5 &= 12
\end{aligned}$$

8.4 What restriction on the parameter a ensures the consistence of the following system? Find the solution depending on a.

$$\begin{aligned}
3x + 4y + 4z &= 2 \\
3x + 2y + 3z &= 3 \\
4x + 5y + az &= 4
\end{aligned}$$

8.5 Check the consistence of the following system as a function of the parameter λ:

$$\begin{aligned}
3x + 2y + z &= 0 \\
6x + 4y + \lambda z &= 0
\end{aligned}$$

Do the following cases exist?

(a) There is no solution.
(b) There is a unique solution.

Find the solution if possible.

8.6 Given is the system

$$\begin{pmatrix} 1 & 0 & 0 & -1 \\ 0 & 1 & -1 & 1 \\ 0 & 0 & a & 0 \\ 0 & 1 & 0 & b \end{pmatrix} \begin{pmatrix} x_1 \\ x_2 \\ x_3 \\ x_4 \end{pmatrix} = \begin{pmatrix} 2 \\ 3 \\ 1 \\ 0 \end{pmatrix}.$$

(a) Decide with respect to the parameters a and b in which cases the system is consistent or inconsistent. When does a unique solution exist?
(b) Find the general solution if possible.
(c) Calculate the solution for $a = 1$ and $b = 0$.

8.7 Decide whether the following system of linear equations is consistent and find the solution in dependence on parameters a and b:

$$\begin{aligned}
a x &+ (a+b) y &+ b z &= 3a + 5b \\
b x &+ ab y &+ a z &= a(2b+3) + b \\
a x &+ by &+ bz &= a + 5b
\end{aligned}$$

8.8 Find the kernel of the linear mapping described by matrix

$$A = \begin{pmatrix} 2 & -2 & 2 \\ 5 & -1 & 7 \\ 3 & -1 & 4 \end{pmatrix}.$$

8.9 Given are the two systems of linear equations

$$\begin{aligned}
x_1 &+ 2x_2 &+ 3x_3 &= 5 \\
3x_1 &+ 1x_2 &+ 2x_3 &= 5 \\
2x_1 &+ 3x_2 &+ 1x_3 &= 8
\end{aligned} \quad \text{and} \quad \begin{aligned}
x_1 &+ 2x_2 &+ 3x_3 &= 14 \\
2x_1 &+ 3x_2 &+ x_3 &= 13 \\
3x_1 &+ x_2 &+ 2x_3 &= 21.
\end{aligned}$$

Solve them in only one tableau.

8.10 Find the inverse matrices by pivoting:

(a) $A = \begin{pmatrix} 1 & 3 & 2 \\ 2 & 5 & 3 \\ -3 & -8 & -4 \end{pmatrix};$ (b) $B = \begin{pmatrix} 1 & 4 & 6 \\ 3 & 2 & 1 \\ 7 & 8 & 8 \end{pmatrix};$

(c) $C = \begin{pmatrix} 1 & 0 & -1 & 2 \\ 2 & -1 & -2 & 3 \\ -1 & 2 & 2 & -4 \\ 0 & 1 & 2 & -5 \end{pmatrix}.$

8.11 Find the solutions of the following matrix equations:

(a) $XA = B$ with

$$A = \begin{pmatrix} -1 & 3 & 2 \\ 2 & 5 & 3 \\ -3 & -8 & -4 \end{pmatrix} \quad \text{and} \quad B = \begin{pmatrix} 1 & -4 & 2 \\ 3 & 3 & -2 \end{pmatrix};$$

(b) $AXB = C$ with

$$A = \begin{pmatrix} 2 & 1 \\ -1 & 2 \end{pmatrix}, \qquad B = \begin{pmatrix} 3 & 2 \\ 4 & 3 \end{pmatrix} \qquad \text{and} \qquad C = \begin{pmatrix} 0 & 1 \\ 1 & 0 \end{pmatrix};$$

(c) $XA = -2(X + B)$ with

$$A = \begin{pmatrix} -3 & 1 & 1 \\ -1 & 0 & 1 \\ 2 & 1 & -2 \end{pmatrix} \qquad \text{and} \qquad B = \begin{pmatrix} 2 & -4 & 0 \\ 1 & 3 & -2 \end{pmatrix}.$$

8.12   Assume we have a four-industry economy. Each industry is to produce an output just
sufficient to meet its own input requirements and those of the other three industries
as well as the final demand of the open sector. That is, the output level x_i must satisfy
the following equation:

$$x_i = x_{i1} + x_{i2} + x_{i3} + x_{i4} + y_i, \quad i \in \{1, 2, 3, 4\},$$

where y_i is the final demand for industry i and x_{ij} is the amount of x_i needed as input
for the industry $j \in \{1, 2, 3, 4\}$. Let

$$X = (x_{ij}) = \begin{pmatrix} 5 & 15 & 15 & 10 \\ 25 & 25 & 10 & 30 \\ 10 & 20 & 20 & 20 \\ 10 & 30 & 15 & 25 \end{pmatrix} \qquad \text{and} \qquad y = \begin{pmatrix} 5 \\ 10 \\ 30 \\ 20 \end{pmatrix}.$$

(a) Find the input-coefficient matrix A which satisfies the equation

$$x = Ax + y$$

with $x = (x_1, x_2, x_3, x_4)^T$ and $x_{ij} = a_{ij} x_j$.
(b) How does the matrix X change if the final demand changes to $y = (10, 20, 20, 15)^T$ and the input coefficients are constant?

8.13   Given is the bounded convex set with the extreme points

$$x^1 = \begin{pmatrix} 2 \\ 1 \\ 0 \end{pmatrix}, \qquad x^2 = \begin{pmatrix} 1 \\ 1 \\ 1 \end{pmatrix}, \qquad x^3 = \begin{pmatrix} 2 \\ 0 \\ 1 \end{pmatrix} \qquad \text{and} \qquad x^4 = \begin{pmatrix} -1 \\ 3 \\ 1 \end{pmatrix}.$$

Do the points

$$a = \frac{1}{2} \begin{pmatrix} 2 \\ 3 \\ 1 \end{pmatrix} \qquad \text{and} \qquad b = \begin{pmatrix} 1 \\ 0 \\ 2 \end{pmatrix}$$

belong to the convex set given above?

8.14   In addition to its main production a firm produces two products A and B with two
machines I and II. The products are manufactured on machine I and packaged on
machine II. Machine I has a free capacity of 40 hours and machine II can be used for
20 hours. To produce 1 tonne of A takes 4 min on machine I and 6 min on machine II.

Producing 1 tonne of B takes 10 min on machine I and 3 min on machine II. What output combinations are possible?

(a) Model the problem by means of a system of linear inequalities.
(b) Solve the problem graphically.
(c) Find the general solution by calculation.

8.15 Solve the following problem graphically:

$$
\begin{array}{rrrcl}
-x_1 & + & x_2 & \leq & 4 \\
-2x_1 & + & x_2 & \leq & 3 \\
x_1 & - & 2x_2 & \leq & 1 \\
& & x_1, x_2 & \geq & 0
\end{array}
$$

8.16 Find the general solutions of the following systems by calculation:

$$
\text{(a)} \quad
\begin{array}{rrrrrcl}
2x_1 & + & 5x_2 & - & 2x_3 & \leq & 5 \\
4x_1 & + & x_2 & + & x_3 & \leq & 3 \\
& & & x_1, x_2, x_3 & & \geq & 0
\end{array}
$$

$$
\text{(b)} \quad
\begin{array}{rrrrcl}
x_1 & + & x_2 & + & x_3 & \leq & 3 \\
x_1 & & & + & 5x_3 & \leq & 3 \\
& & & x_1, x_2, x_3 & & \geq & 0
\end{array}
$$

9 Linear programming

In Chapter 8, we considered systems of linear inequalities and we discussed how to find the feasible region of such systems. In this chapter, any feasible solution of such a system is evaluated by the value of a linear objective function, and we are looking for a 'best' solution among the feasible ones. After introducing some basic notions, we discuss the simplex algorithm as a general method of solving such problems. Then we introduce a so-called dual problem, which is closely related to the problem originally given, and we present a modification of the simplex method based on the solution of this dual problem.

9.1 PRELIMINARIES

We start this section with an introductory example.

Example 9.1 A company produces a mixture consisting of three raw materials denoted as R_1, R_2 and R_3. Raw materials R_1 and R_2 must be contained in the mixture with a given minimum percentage, and raw material R_3 must not exceed a certain given maximum percentage. Moreover, the price of each raw material per kilogram is known. The data are summarized in Table 9.1.

We wish to determine a feasible mixture with the lowest cost. Let $x_i, i \in \{1, 2, 3\}$, be the percentage of raw material R_i. Then we get the following constraints. First,

$$x_1 + x_2 + x_3 = 100. \tag{9.1}$$

Equation (9.1) states that the sum of the percentages of all raw materials equals 100 per cent. Since the percentage of raw material R_3 should not exceed 30 per cent, we obtain the constraint

$$x_3 \leq 30. \tag{9.2}$$

Table 9.1 Data for Example 9.1

Raw material	Required (%)	Price in EUR per kilogram
R_1	at least 10	25
R_2	at least 50	17
R_3	at most 30	12

The percentage of raw material R_2 must be at least 50 per cent, or to put it another way, the sum of the percentages of R_1 and R_3 must be no more than 50 per cent:

$$x_1 + x_3 \leq 50. \tag{9.3}$$

Moreover, the percentage of R_1 must be at least 10 per cent, or equivalently, the sum of the percentages of R_2 and R_3 must not exceed 90 per cent, i.e.

$$x_2 + x_3 \leq 90. \tag{9.4}$$

Moreover, all variables should be non-negative:

$$x_1 \geq 0, \quad x_2 \geq 0, \quad x_3 \geq 0. \tag{9.5}$$

The cost of producing the resulting mixture should be minimized, i.e. the objective function is as follows:

$$z = 25x_1 + 17x_2 + 12x_3 \longrightarrow \text{min!} \tag{9.6}$$

The notation $z \longrightarrow$ min! indicates that the value of function z should become minimal for the desired solution. So we have formulated a problem consisting of an objective function (9.6), four constraints (three inequalities (9.2), (9.3) and (9.4) and one equation (9.1)) and the non-negativity constraints (9.5) for all three variables.

In general, a linear programming problem (abbreviated LPP) consists of constraints (a system of linear equations or linear inequalities), non-negativity constraints and a linear objective function. The general form of such an LPP can be given as follows.

General form of an LPP

$$z = c_1 x_1 + c_2 x_2 + \ldots + c_n x_n \longrightarrow \text{max!} \quad (\text{min!})$$

subject to (s.t.)

$$
\begin{array}{lll}
a_{11}x_1 + a_{12}x_2 + \ldots + a_{1n}x_n & R_1 & b_1 \\
a_{21}x_1 + a_{22}x_2 + \ldots + a_{2n}x_n & R_2 & b_2 \\
\quad \vdots & \vdots & \vdots \\
a_{m1}x_1 + a_{m2}x_2 + \ldots + a_{mn}x_n & R_m & b_m
\end{array}
$$

$$x_j \geq 0, \quad j \in J \subseteq \{1, 2, \ldots, n\},$$

where $R_i \in \{\leq, =, \geq\}$, $1 \leq i \leq m$.

An LPP considers either the maximization or the minimization of a linear function $z = \mathbf{c}^T \mathbf{x}$. In each constraint, we have exactly one of the signs $\leq, =$ or $\geq$, i.e. we may have both equations and inequalities as constraints, where we assume that at least one inequality occurs.

Alternatively, we can give the following *matrix representation* of an LPP:

$$z = c^T x \longrightarrow \max! \quad (\min!).$$

$$\text{s.t.} \quad Ax \; R \; b \tag{9.7}$$

$$x \geq 0,$$

where $R = (R_1, R_2, \ldots, R_m)^T$, $R_i \in \{\leq, =, \geq\}$, $i = 1, 2, \ldots, m$. Here matrix A is of order $m \times n$. The vector $c = (c_1, c_2, \ldots, c_n)^T$ is known as the vector of the coefficients in the objective function, and the vector $b = (b_1, b_2, \ldots, b_m)^T$ is the right-hand-side vector.

The feasibility of a solution of an LPP is defined in the same way as for a system of linear inequalities (see Definition 8.10).

Definition 9.1   A feasible solution $x = (x_1, x_2, \ldots, x_n)^T$, for which the objective function has an optimum (i.e. maximum or minimum) value is called an *optimal solution*, and $z_0 = c^T x$ is known as the *optimal objective function value*.

9.2   GRAPHICAL SOLUTION

Next, we give a geometric interpretation of an LPP with only two variables x_1 and x_2, which allows a graphical solution of the problem.

For fixed value z and $c_2 \neq 0$, the objective function $z = c_1 x_1 + c_2 x_2$ is a straight line of the form

$$x_2 = -\frac{c_1}{c_2} x_1 + \frac{z}{c_2},$$

i.e. for different values of z we get parallel lines all with slope $-c_1/c_2$. The vector

$$c = \begin{pmatrix} c_1 \\ c_2 \end{pmatrix}$$

points in the direction in which the objective function increases most. Thus, when maximizing the linear objective function z, we have to shift the line

$$x_2 = -\frac{c_1}{c_2} x_1 + \frac{z}{c_2}$$

in the direction given by vector c, while when minimizing z, we have to shift this line in the opposite direction, given by vector $-c$. This is illustrated in Figure 9.1.

Based on the above considerations, an LPP of the form (9.7) with two variables can be graphically solved as follows:

(1) Determine the feasible region as the intersection of all feasible half-planes with the first quadrant. (This has already been discussed when describing the feasible region of a system of linear inequalities in Chapter 8.2.2.)

(2) Draw the objective function $z = Z$, where Z is constant, and shift it either in the direction given by vector $\mathbf{c}$ (in the case of $z \to$ max!) or in the direction given by vector $-\mathbf{c}$ (in the case of $z \to$ min!). Apply this procedure as long as the line $z = constant$ has common points with the feasible region.

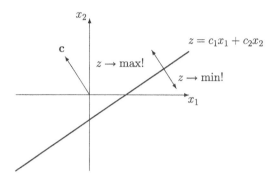

Figure 9.1 The cases $z \to$ max! and $z \to$ min!

Example 9.2 A firm can manufacture two goods G_1 and G_2 in addition to its current production programme, where Table 9.2 gives the available machine capacities, the processing times for one unit of good G_i and the profit per unit of each good G_i, $i \in \{1,2\}$.

Table 9.2 Data for Example 9.2

Process	Processing times per unit		Free machine capacity in minutes
	G_1	G_2	
Turning	0	8	640
Milling	6	6	720
Planing	6	3	600
Profit in EUR per unit	1	2	—

(1) First, we formulate the corresponding LPP. Let x_i denote the number of produced units of good G_i, $i \in \{1,2\}$. Then we obtain the following problem:

$$
\begin{aligned}
z = x_1 + 2x_2 &\to \text{max!}\\
\text{s.t.}\qquad 8x_2 &\le 640\\
6x_1 + 6x_2 &\le 720\\
6x_1 + 3x_2 &\le 600\\
x_1, x_2 &\ge 0.
\end{aligned}
$$

(2) We solve the above problem graphically (see Figure 9.2). We graph each of the three constraints as an equation and mark the corresponding half-spaces satisfying the given inequality constraint. The arrows on the constraints indicate the feasible half-space. The feasible region M is obtained as the intersection of these three feasible half-spaces with the first quadrant. (We get the dashed area in Figure 9.2.) We graph the objective function $z = Z$, where Z is constant. (In Figure 9.2 the parallel lines $z = 0$ and $z = 200$ are given.) The optimal extreme point is P_4 corresponding to $x_1^* = 40$ and $x_2^* = 80$.

The optimal objective function value is $z_0^{\max} = x_1^* + 2x_2^* = 40 + 2 \cdot 80 = 200$. Notice that for $z > 200$, the resulting straight line would not have common points with the feasible region M.

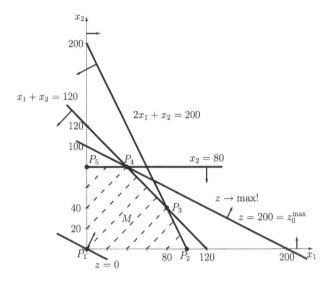

Figure 9.2 Graphical solution of Example 9.2.

Next, we determine the basic feasible solutions corresponding to the extreme points $P_1, P_2, \ldots, P_5$. To this end, we introduce a slack variable in each constraint which yields the following system of constraints:

$$
\begin{array}{rcrcrcrcrcl}
 & & 8x_2 & + & x_3 & & & & & = & 640 \\
6x_1 & + & 6x_2 & & & + & x_4 & & & = & 720 \\
6x_1 & + & 3x_2 & & & & & + & x_5 & = & 600 \\
\end{array}
$$
$$
x_1, x_2, x_3, x_4, x_5 \;\geq\; 0.
$$

Considering point P_1 with $x_1 = x_2 = 0$, the basic variables are x_3, x_4 and x_5 (and consequently the matrix of the basis vectors is formed by the corresponding column vectors, which is an identity matrix). The values of the basic variables are therefore

$$x_3 = 640, \quad x_4 = 720 \quad \text{and} \quad x_5 = 600,$$

and the objective function value is $z_1 = 0$. Considering point P_2, we insert $x_1 = 100$, $x_2 = 0$ into the system of constraints, which yields the basic feasible solution

$$x_1 = 100, \quad x_2 = 0, \quad x_3 = 640, \quad x_4 = 120, \quad x_5 = 0$$

with the objective function value $z_2 = 100$. Considering point P_3 with $x_1 = 80$, $x_2 = 40$, we obtain from the system of constraints the basic feasible solution

$$x_1 = 80, \quad x_2 = 40, \quad x_3 = 320, \quad x_4 = 0, \quad x_5 = 0$$

with the objective function value $z_3 = 160$. Considering point P_4 with $x_1 = 40$, $x_2 = 80$, we get from the system of constraints the basic feasible solution

$$x_1 = 40, \quad x_2 = 80, \quad x_3 = 0, \quad x_4 = 0, \quad x_5 = 120$$

with the objective function value $z_4 = 200$. Finally, considering point P_5 with $x_1 = 0$, $x_2 = 80$, we get the corresponding basic feasible solution

$$x_1 = 0, \quad x_2 = 80, \quad x_3 = 0, \quad x_4 = 240, \quad x_5 = 360$$

with the objective function value $z_5 = 160$.

The above considerations have confirmed our graphical solution that P_4 is the extreme point with the maximal objective function value $z_0^{max} = z_4 = 200$.

Example 9.3   We solve graphically the LPP

$$
\begin{aligned}
z = x_1 + x_2 &\rightarrow \max! \\
\text{s.t.} \quad x_1 + x_2 &\geq 6 \\
-ax_1 + x_2 &\geq 4 \\
x_1, x_2 &\geq 0,
\end{aligned}
$$

where a denotes some real parameter, i.e. $a \in \mathbb{R}$. We first draw the constraints as equations and check which of the resulting half-spaces satisfy the given inequality constraint. The arrows on the constraints indicate again the feasible half-spaces. This way, we get the feasible region M as the intersection of the half-spaces (see Figure 9.3). Since the second constraint contains parameter a, we graph in Figure 9.3 the line resulting from the second constraint as equation for some values of a, namely: $a = -2$, $a = 0$ and $a = 2$. (M is dashed for $a = 2$, while for $a = -2$ and $a = 0$ only the resulting constraint including parameter a is dashed.)

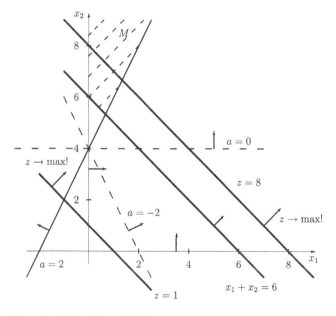

Figure 9.3 Feasible region M for Example 9.3.

It can be seen that all the resulting equations for different values of parameter a go through the point (0,4), and the slope of the line is given by $1/a$ for $a \neq 0$. To find the optimal solution, we now graph the objective function $z = Z$, where Z is constant. (In Figure 9.3 the lines $z = 1$ and $z = 8$ are shown.) The arrow on the lines $z = constant$ indicates in which direction the objective function value increases. (Remember that the dashed area gives the feasible region M of the problem for $a = 2$.) From Figure 9.3, we see that the function value can become arbitrarily large (independently of the value of parameter a), i.e. an optimal solution of the maximization problem does not exist.

We continue with some properties of an LPP.

9.3 PROPERTIES OF A LINEAR PROGRAMMING PROBLEM; STANDARD FORM

Let M be the feasible region and consider the maximization of the objective function $z = \mathbf{c}^{\mathrm{T}}\mathbf{x}$. We know already from Chapter 8.2.2 that the feasible region M of a system of linear inequalities is either empty or a convex polyhedron (see Theorem 8.10). Since the feasibility of a solution of an LPP is independent of the objective function, the latter property also holds for an LPP. We can further reinforce this and the following three cases for an LPP may occur:

(1) The feasible region is empty: $M = \varnothing$. In this case the constraints are inconsistent, i.e. there is no feasible solution of the LPP.
(2) M is a non-empty bounded subset of the n-space $\mathbb{R}^n$.
(3) M is an unbounded subset of the n-space $\mathbb{R}^n$, i.e. at least one variable may become arbitrarily large, or if some variables are not necessarily non-negative, at least one of them may become arbitrarily small.

In case (2), the feasible region M is also called *a convex polytope*, and there is always a solution of the maximization problem. In case (3), there are again two possibilities:

(3a) The objective function z is bounded from above. Then an optimal solution of the maximization problem under consideration exists.
(3b) The objective function z is not bounded from above. Then there does not exist an optimal solution for the maximization problem under consideration, i.e. there does not exist a *finite* optimal objective function value.

Cases (3a) and (3b) are illustrated in Figure 9.4. In case (3a), there are three extreme points, and P is an optimal extreme point. In case (3b), there exist four extreme points, however, the values of the objective function can become arbitrarily large.

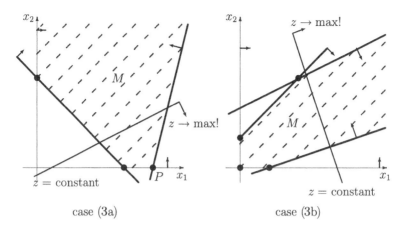

case (3a)   case (3b)

Figure 9.4 The cases (3a) and (3b).

THEOREM 9.1 If an LPP has an optimal solution, then there exists at least one extreme point, where the objective function has an optimum value.

According to Theorem 9.1, one can restrict the search for an optimal solution to the consideration of extreme points (represented by basic feasible solutions). The following theorem characterizes the set of optimal solutions.

THEOREM 9.2 Let $P_1, P_2, \ldots, P_r$ described by vectors $\mathbf{x}^1, \mathbf{x}^2, \ldots, \mathbf{x}^r$ be optimal extreme points. Then any convex combination

$$\mathbf{x}^0 = \lambda_1 \mathbf{x}^1 + \lambda_2 \mathbf{x}^2 + \ldots + \lambda_r \mathbf{x}^r, \quad \lambda_i \geq 0, \ i = 1, 2, \ldots, r, \quad \sum_{i=1}^{r} \lambda_i = 1 \qquad (9.8)$$

is also an optimal solution.

PROOF Let $\mathbf{x}^1, \mathbf{x}^2, \ldots, \mathbf{x}^r$ be optimal extreme points with

$$\mathbf{c}^T \mathbf{x}^1 = \mathbf{c}^T \mathbf{x}^2 = \ldots = \mathbf{c}^T \mathbf{x}^r = z_0^{\max}$$

and $\mathbf{x}^0$ be defined as in (9.8). Then $\mathbf{x}^0$ is feasible since the feasible region M is convex. Moreover,

$$\mathbf{c}^T \mathbf{x}^0 = \mathbf{c}^T (\lambda_1 \mathbf{x}^1 + \lambda_2 \mathbf{x}^2 + \ldots + \lambda_r \mathbf{x}^r) = \lambda_1 (\mathbf{c}^T \mathbf{x}^1) + \lambda_2 (\mathbf{c}^T \mathbf{x}^2) + \ldots + \lambda_r (\mathbf{c}^T \mathbf{x}^r)$$

$$= \lambda_1 z_0^{\max} + \lambda_2 z_0^{\max} + \ldots + \lambda_r z_0^{\max} = (\lambda_1 + \lambda_2 + \ldots + \lambda_r) z_0^{\max} = z_0^{\max},$$

i.e. point $\mathbf{x}^0$ is optimal. ∎

In Figure 9.5, Theorem 9.2 is illustrated for a problem with two variables having two optimal extreme points P_1 and P_2. Any point on the connecting line between the points P_1 and P_2 is optimal.

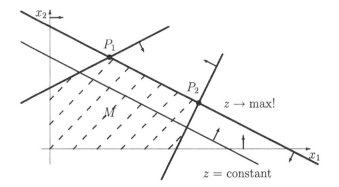

Figure 9.5 The case of several optimal solutions.

Standard form

In the next definition, we introduce a special form of an LPP.

Definition 9.2   An LPP of the form

$$z = \mathbf{c}^\mathsf{T}\mathbf{x} \longrightarrow \text{max!}$$

s.t. $A\mathbf{x} = \mathbf{b}, \quad \mathbf{x} \geq \mathbf{0},$

where $A = (A_N, I)$ and $\mathbf{b} \geq \mathbf{0}$ is called the *standard form* of an LPP.

According to Definition 9.2, matrix A can be partitioned into some matrix A_N and an identity submatrix I. Thus, the standard form of an LPP is characterized by the following properties:

(1) the LPP is a maximization problem;
(2) the constraints are given as a system of linear equations in canonical form with non-negative right-hand sides and
(3) all variables have to be non-negative.

It is worth noting that the standard form always includes a basic feasible solution for the corresponding system of linear inequalities (when the objective function is skipped). If no artificial variables are necessary when generating the standard form or when all artificial variables have value zero (this means that the right-hand sides of all constraints that contain an artificial variable are equal to zero), this solution is also feasible for the original problem and it corresponds to an extreme point of the feasible region M. However, it is an infeasible solution for the original problem if at least one artificial variable has a value greater than zero. In this case, the constraint from the original problem is violated and the basic solution does not correspond to an extreme point of set M. The generation of the standard form of an LPP plays an important role in finding a starting solution in the procedure that we present later for solving an LPP.

Any LPP can formally be transformed into the standard form by the following rules. We consider the possible violations of the standard form according to Definition 9.2.

(1) Some variable x_j is not necessarily non-negative, i.e. x_j may take arbitrary values. Then variable x_j is replaced by the difference of two non-negative variables, i.e. we set:

$$x_j = x_j^* - x_j^{**} \quad \text{with} \quad x_j^* \geq 0 \quad \text{and} \quad x_j^{**} \geq 0.$$

Then we get:

$$x_j^* > x_j^{**} \Longleftrightarrow x_j > 0$$

$$x_j^* = x_j^{**} \Longleftrightarrow x_j = 0$$

$$x_j^* < x_j^{**} \Longleftrightarrow x_j < 0.$$

(2) The given objective function has to be minimized:

$$z = c_1 x_1 + c_2 x_2 + \ldots + c_n x_n \rightarrow \text{min!}$$

The determination of a minimum of function z is equivalent to the determination of a maximum of function $\bar{z} = -z$:

$$z = c_1 x_1 + c_2 x_2 + \ldots + c_n x_n \rightarrow \text{min!}$$

$$\Longleftrightarrow \bar{z} = -z = -c_1 x_1 - c_2 x_2 - \ldots - c_n x_n \rightarrow \text{max!}$$

(3) For some right-hand sides, we have $b_i < 0$:

$$a_{i1} x_1 + a_{i2} x_2 + \ldots + a_{in} x_n = b_i < 0.$$

In this case, we multiply the above constraint by -1 and obtain:

$$-a_{i1} x_1 - a_{i2} x_2 - \ldots - a_{in} x_n = -b_i > 0.$$

(4) Let some constraints be inequalities:

$$a_{i1} x_1 + a_{i2} x_2 + \ldots + a_{in} x_n \leq b_i$$

or

$$a_{k1} x_1 + a_{k2} x_2 + \ldots + a_{kn} x_n \geq b_k.$$

Then by introducing a slack variable u_i and a surplus variable u_k, respectively, we obtain an equation:

$$a_{i1} x_1 + a_{i2} x_2 + \ldots + a_{in} x_n + u_i = b_i \quad \text{with} \quad u_i \geq 0$$

or

$$a_{k1} x_1 + a_{k2} x_2 + \ldots + a_{kn} x_n - u_k = b_k \quad \text{with} \quad u_k \geq 0.$$

(5) Let the given system of linear equations be not in canonical form, i.e. the constraints are given e.g. as follows:

$$a_{11}x_1 + a_{12}x_2 + \ldots + a_{1n}x_n = b_1$$
$$a_{21}x_1 + a_{22}x_2 + \ldots + a_{2n}x_n = b_2$$
$$\vdots$$
$$a_{m1}x_1 + a_{m2}x_2 + \ldots + a_{mn}x_n = b_m$$

with $b_i \geq 0$, $i = 1, 2, \ldots, m$; $x_j \geq 0$, $j = 1, 2, \ldots, n$.

In the above situation, there is no constraint that contains an eliminated variable with coefficient $+1$ (provided that all column vectors of matrix A belonging to variables $x_1, x_2, \ldots, x_n$ are different from the unit vector). Then we introduce in each equation an artificial variable x_{Ai} as basic variable and obtain:

$$a_{11}x_1 + a_{12}x_2 + \ldots + a_{1n}x_n + x_{A1} \qquad\qquad = b_1$$
$$a_{21}x_1 + a_{22}x_2 + \ldots + a_{2n}x_n \qquad + x_{A2} \qquad = b_2$$
$$\vdots \qquad\qquad\qquad \vdots$$
$$a_{m1}x_1 + a_{m2}x_2 + \ldots + a_{mn}x_n \qquad\qquad + x_{Am} = b_m$$

with $b_i \geq 0$, $i = 1, 2, \ldots, m$; $x_j \geq 0$, $j = 1, 2, \ldots, n$, and $x_{Ai} \geq 0$, $i = 1, 2, \ldots, m$.

At the end, we may renumber the variables so that they are successively numbered. (For the above representation by $x_1, x_2, \ldots, x_{n+m}$, but in the following, we always assume that the problem includes n variables $x_1, x_2, \ldots, x_n$ after renumbering.) In the way described above, we can transform any given LPP formally into the standard form. (It is worth noting again that a solution is only feasible for the original problem if all artificial variables have value zero.)

For illustrating the above transformation of an LPP into the standard form, we consider the following example.

Example 9.4   Given is the following LPP:

$$z = -x_1 + 3x_2 + x_4 \rightarrow \min!$$
$$\begin{array}{rcrcrcrcl}
\text{s.t.} \quad x_1 & - & x_2 & + & 3x_3 & - & x_4 & \geq & 8 \\
 & & x_2 & - & 5x_3 & + & 2x_4 & \leq & -4 \\
 & & & & x_3 & + & x_4 & \leq & 3 \\
 & & & & & & x_2, x_3, x_4 & \geq & 0.
\end{array}$$

First, we substitute for variable x_1 the difference of two non-negative variables x_1^* and x_1^{**}, i.e. $x_1 = x_1^* - x_1^{**}$ with $x_1^* \geq 0$ and $x_2^{**} \geq 0$. Further, we multiply the objective function z by -1 and obtain:

$$\bar{z} = -z = x_1^* - x_1^{**} - 3x_2 - x_4 \rightarrow \max!$$
$$\begin{array}{rcrcrcrcrcl}
\text{s.t.} \quad x_1^* & - & x_1^{**} & - & x_2 & + & 3x_3 & - & x_4 & \geq & 8 \\
 & & & & x_2 & - & 5x_3 & + & 2x_4 & \leq & -4 \\
 & & & & & & x_3 & + & x_4 & \leq & 3 \\
 & & & & & & x_1^*, x_1^{**}, x_2, x_3, x_4 & & & \geq & 0.
\end{array}$$

Multiplying the second constraint by -1 and introducing the slack variable x_7 in the third constraint as well as the surplus variables x_5 and x_6 in the first and second constraints, we obtain all constraints as equations with non-negative right-hand sides:

$$\bar{z} = -z = x_1^* - x_2^{**} - 3x_2 - x_4 \to \text{max!}$$

s.t.

$$
\begin{array}{rl}
x_1^* - x_1^{**} - x_2 + 3x_3 - x_4 - x_5 & = 8 \\
- x_2 + 5x_3 - 2x_4 - x_6 & = 4 \\
x_3 + x_4 + x_7 & = 3 \\
x_1^*, x_1^{**}, x_2, x_3, x_4, x_5, x_6, x_7 & \geq 0.
\end{array}
$$

Now we can choose variable x_1^* as eliminated variable in the first constraint and variable x_7 as the eliminated variable in the third constraint, but there is no variable that occurs only in the second constraint having coefficient $+1$. Therefore, we introduce the artificial variable x_{A1} in the second constraint and obtain:

$$\bar{z} = -z = x_1^* - x_2^{**} - 3x_2 - x_4 \to \text{max!}$$

s.t.

$$
\begin{array}{rl}
- x_1^{**} - x_2 + 3x_3 - x_4 - x_5 + \mathbf{x}_1^* & = 8 \\
- x_2 + 5x_3 - 2x_4 - x_6 + \mathbf{x}_{A1} & = 4 \\
x_3 + x_4 + \mathbf{x}_7 & = 3 \\
x_1^*, x_1^{**}, x_2, x_3, x_4, x_5, x_6, x_7, x_{A1} & \geq 0.
\end{array}
$$

Notice that we have written the variables in such a way that the identity submatrix (column vectors of variables x_1^*, x_{A1}, x_7) occurs at the end. So in the standard form, the problem has now $n = 9$ variables. A vector satisfying all constraints is only a feasible solution for the original problem if the artificial variable x_{A1} has value zero (otherwise the original second constraint would be violated).

9.4 SIMPLEX ALGORITHM

In this section, we always assume that the basic feasible solution resulting from the standard form of an LPP is feasible for the original problem. (In particular, we assume that no artificial variables are necessary to transform the given LPP into the standard form.) We now discuss a general method for solving linear programming problems, namely the *simplex method*. The basic idea of this approach is as follows. Starting with some initial extreme point (represented by a basic feasible solution resulting from the standard form of an LPP), we compute the value of the objective function and check whether the latter can be improved upon by moving to an adjacent extreme point (by applying the pivoting procedure). If so, we perform this move to the next extreme point and seek then whether further improvement is possible by a subsequent move. When finally an extreme point is attained that does not admit any further improvement, it will constitute an optimal solution. Thus, the simplex method is an iterative procedure which ends after a finite number of pivoting steps in an optimal extreme point (provided it is possible to move in each step to an adjacent extreme point). This idea is illustrated in Figure 9.6 for the case $z \to \text{max!}$ Starting from extreme point P_1, one can go via points P_2 and P_3 to the optimal extreme point P_4 or via $\overline{P}_2, \overline{P}_3, \overline{P}_4$ to P_4. In both cases, the objective function value increases from extreme point to extreme point.

In order to apply such an approach, a criterion to decide whether a move to an adjacent extreme point improves the objective function value is required, which we will derive in the following. We have already introduced the canonical form of a system of equations in Chapter 8.1.3 (see Definition 8.7). In the following, we assume that the rank of matrix A is equal to m, i.e. in the canonical form there are m basic variables among the n variables, and the number of non-basic variables is equal to $n' = n - m$. Consider a feasible canonical form with the basic variables x_{Bi} and the non-basic variables x_{Nj}:

$$x_{Bi} = b_i^* - \sum_{j=1}^{n'} a_{ij}^* x_{Nj} , \quad i = 1, 2, \ldots, m \qquad (n' = n - m). \tag{9.9}$$

Partitioning the set of variables into basic and non-basic variables, the objective function z can be written as follows:

$$\begin{aligned} z &= c_1 x_1 + c_2 x_2 + \ldots + c_n x_n \\ &= c_{B1} x_{B1} + c_{B2} x_{B2} + \ldots + c_{Bm} x_{Bm} + c_{N1} x_{N1} + c_{N2} x_{N2} + \ldots + c_{Nn'} x_{Nn'} \\ &= \sum_{i=1}^{m} c_{Bi} x_{Bi} + \sum_{j=1}^{n'} c_{Nj} x_{Nj}. \end{aligned}$$

Using equations (9.9), we can replace the basic variables and write the objective function only in dependence on the non-basic variables. We obtain

$$\begin{aligned} z &= \sum_{i=1}^{m} c_{Bi} \left(b_i^* - \sum_{j=1}^{n'} a_{ij}^* x_{Nj} \right) + \sum_{j=1}^{n'} c_{Nj} x_{Nj} \\ &= \sum_{i=1}^{m} c_{Bi} b_i^* - \sum_{j=1}^{n'} \left(\sum_{i=1}^{m} c_{Bi} a_{ij}^* - c_{Nj} \right) x_{Nj}. \end{aligned}$$

We refer to the latter row, where the objective function is written in terms of the current non-basic variables, as the *objective row*. Moreover, we define the following values:

$$g_j = \sum_{i=1}^{m} c_{Bi} a_{ij}^* - c_{Nj} \quad \text{(coefficient of variable } x_{Nj} \text{ in the objective row)}; \tag{9.10}$$

$$z_0 = \sum_{i=1}^{m} c_{Bi} b_i^* \qquad \text{(value of the objective function of the basic solution)}. \tag{9.11}$$

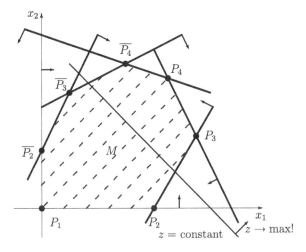

Figure 9.6 Illustration of the simplex method.

Table 9.3 Simplex tableau of the basic feasible solution (9.9)

No.	nbv	x_{N1}	x_{N2}	$\cdots$	$\mathbf{x_{Nl}}$	$\cdots$	$x_{Nn'}$		
bv	-1	c_{N1}	c_{N2}	$\cdots$	c_{Nl}	$\cdots$	$c_{Nn'}$	0	Q
x_{B1}	c_{B1}	a_{11}^*	a_{12}^*	$\cdots$	a_{1l}^*	$\cdots$	$a_{1n'}^*$	b_1^*	
x_{B2}	c_{B2}	a_{21}^*	a_{22}^*	$\cdots$	a_{2l}^*	$\cdots$	$a_{2n'}^*$	b_2^*	
$\vdots$	$\vdots$	$\vdots$	$\vdots$		$\vdots$		$\vdots$	$\vdots$	
$\mathbf{x_{Bk}}$	c_{Bk}	a_{k1}^*	a_{k2}^*	$\cdots$	$\mathbf{a_{kl}^*}$	$\cdots$	$a_{kn'}^*$	b_k^*	
$\vdots$	$\vdots$	$\vdots$	$\vdots$		$\vdots$		$\vdots$	$\vdots$	
x_{Bm}	c_{Bm}	a_{m1}^*	a_{m2}^*	$\cdots$	a_{ml}^*	$\cdots$	$a_{mn'}^*$	b_m^*	
z		g_1	g_2	$\cdots$	g_l	$\cdots$	$g_{n'}$	z_0	

Concerning the calculation of value z_0 according to formula (9.11), we remember that in a basic solution, all non-basic variables are equal to zero.

Then we get the following representation of the objective function in dependence on the non-basic variables x_{Nj}:

$$z = z_0 - g_1 x_{N1} - g_2 x_{N2} - \ldots - g_{n'} x_{Nn'}.$$

Here each coefficient g_j gives the change in the objective function value if the non-basic variable x_{Nj} is included in the set of basic variables (replacing some other basic variable) and if its value would increase by one unit. By means of the coefficients in the objective row, we can give the following optimality criterion.

THEOREM 9.3 (optimality criterion) If inequalities $g_j \geq 0$, $j = 1, 2, \ldots, n'$, hold for the coefficients of the non-basic variables in the objective row, the corresponding basic feasible solution is optimal.

From Theorem 9.3 we get the following obvious corollary.

COROLLARY 9.1 If there exists a column l with $g_l < 0$ in a basic feasible solution, the value of the objective function can be increased by including the column vector belonging to the non-basic variable x_{Nl} into the set of basis vectors, i.e. variable x_{Nl} becomes a basic variable in the subsequent basic feasible solution.

Assume that we have some current basic feasible solution (9.9). The corresponding simplex tableau is given in Table 9.3. It corresponds to the *short form* of the tableau of the pivoting procedure as introduced in Chapter 8.2.3 when solving systems of linear inequalities. An additional row contains the coefficients g_j together with the objective function value z_0 (i.e. the objective row) calculated as given above.

In the second row and column of the simplex tableau in Table 9.3, we write the coefficients of the corresponding variables in the objective function. Notice that the values of the objective row in the above tableau (i.e. the coefficients g_j and the objective function value z_0) are obtained as scalar products of the vector $(-1, c_{B1}, c_{B2}, \ldots, c_{Bm})^T$ of the second column and the vector of the corresponding column (see formulas (9.10) and (9.11)). Therefore, numbers -1 and 0 in the second row are fixed in each tableau. In the left upper box, we write the number of the tableau. Note that the basic feasible solution (9.9) represented by Table 9.3 is not necessarily the initial basic feasible solution resulting from the standard form (in the latter case, we would simply have $a_{ij}^* = a_{ij}$ and $b_i^* = b_i$ for $i = 1, 2, \ldots, m$ and $j = 1, 2, \ldots, n'$ provided that $A = (a_{ij})$ is the matrix formed by the column vectors belonging to the initial non-basic variables).

If in the simplex tableau given in Table 9.3 at least one of the coefficients g_j is negative, we have to perform a further pivoting step, i.e. we interchange one basic variable with a non-basic variable. This may be done as follows, where first the pivot column is determined, then the pivot row, and in this way the pivot element is obtained.

Determination of the pivot column *l*

Choose some column l, $1 \le l \le n'$, such that $g_l < 0$. Often, a column l is used with

$$g_l = \min\{g_j \mid g_j < 0, j = 1, 2, \ldots, n'\}.$$

It is worth noting that the selection of the smallest negative coefficient g_l does not guarantee that the algorithm terminates after the smallest possible number of iterations. It guarantees only that there is the biggest increase in the objective function value when going towards the resulting subsequent extreme point.

Determination of the pivot row *k*

We recall that after the pivoting step, the feasibility of the basic solution must be maintained. Therefore, we choose row k with $1 \le k \le m$ such that

$$\frac{b_k^*}{a_{kl}^*} = \min\left\{\frac{b_i^*}{a_{il}^*} \;\middle|\; a_{il}^* > 0, \; i = 1, 2, \ldots, m\right\}.$$

It is the same selection rule which we have already used for the solution of systems of linear inequalities (see Chapter 8.2.3). To determine the above quotients, we have added the last column Q in the tableau given in Table 9.3, where we enter the quotient in each row in which the corresponding element in the chosen pivot column is positive.

If column l is chosen as pivot column, the corresponding variable x_{Nl} becomes a basic variable in the next step. We also say that x_{Nl} is the *entering variable*, and the column of the initial matrix A belonging to variable x_{Nl} is entering the basis. Using row k as pivot row, the corresponding variable x_{Bk} becomes a non-basic variable in the next step. In this case, we say that x_{Bk} is the *leaving variable*, and the column vector of matrix A belonging to variable x_{Nl} is leaving the basis. Element a_{kl}^* is known as the *pivot* or *pivot element*. It has been printed in bold face in the tableau together with the leaving and the entering variables.

The following two theorems characterize situations when either an optimal solution does not exist or when an existing optimal solution is not uniquely determined.

THEOREM 9.4 If inequality $g_l < 0$ holds for a coefficient of a non-basic variable in the objective row and inequalities $a_{il}^* \le 0$, $i = 1, 2, \ldots, m$, hold for the coefficients in column l of the current tableau, then the LPP does not have an optimal solution.

In the latter case, the objective function value is unbounded from above, and we can stop our computations. Although there is a negative coefficient g_l, we cannot move to an adjacent extreme point with a better objective function value (there is no leaving variable which can be interchanged with the non-basic variable x_{Nl}).

THEOREM 9.5 If there exists a coefficient $g_l = 0$ in the objective row of the tableau of an optimal basic feasible solution and inequality $a_{il}^* > 0$ holds for at least one coefficient in column l, then there exists another optimal basic feasible solution, where x_{Nl} is a basic variable.

If the assumptions of Theorem 9.5 are satisfied, we can perform a further pivoting step with x_{Nl} as entering variable, and there is at least one basic variable which can be chosen as leaving variable. However, due to $g_l = 0$, the objective function value does not change. Based on the results above, we can summarize the simplex algorithm as follows.

Simplex algorithm

(1) Transform the LPP into the standard form, where the constraints are given in canonical form as follows (remember that it is assumed that no artificial variables are necessary to transform the given problem into standard form):

$$A_N \mathbf{x}_N + I \mathbf{x}_B = \mathbf{b}, \qquad \mathbf{x}_N \geq \mathbf{0}, \qquad \mathbf{x}_B \geq \mathbf{0}, \qquad \mathbf{b} \geq \mathbf{0},$$

where $A_N = (a_{ij})$ is of order $m \times n'$ and $\mathbf{b} = (b_1, b_2, \ldots, b_m)^{\mathrm{T}}$. The initial basic feasible solution is

$$\mathbf{x} = \begin{pmatrix} \mathbf{x}_N \\ \mathbf{x}_B \end{pmatrix} = \begin{pmatrix} \mathbf{0} \\ \mathbf{b} \end{pmatrix}$$

with the objective function value $z_0 = \mathbf{c}^{\mathrm{T}} \mathbf{x}$. Establish the corresponding initial tableau.

(2) Consider the coefficients g_j, $j = 1, 2, \ldots, n'$, of the non-basic variables x_{Nj} in the objective row.
If $g_j \geq 0$ for $j = 1, 2, \ldots, n'$, then the current basic feasible solution is optimal, STOP. Otherwise, there is a coefficient $g_j < 0$ in the objective row.

(3) Determine column l with

$$g_l = \min\{g_j \mid g_j < 0, j = 1, 2, \ldots, n'\}$$

as pivot column.

(4) If $a_{il} \leq 0$ for $i = 1, 2, \ldots, m$, then STOP. (In this case, there does not exist an optimal solution of the problem.) Otherwise, there is at least one element $a_{il} > 0$.

(5) Determine the pivot row k such that

$$\frac{b_k}{a_{kl}} = \min \left\{ \frac{b_i}{a_{il}} \, \middle| \, a_{il} > 0, \ i = 1, 2, \ldots, m \right\}.$$

(6) Interchange the basic variable x_{Bk} of row k with the non-basic variable x_{Nl} of column l and calculate the following values of the new tableau:

$$a_{kl}^* = \frac{1}{a_{kl}};$$

$$a_{kj}^* = \frac{a_{kj}}{a_{kl}}; \qquad b_k^* = \frac{b_k}{a_{kl}}; \qquad j = 1, 2, \ldots, n', j \neq l;$$

$$a_{il}^* = -\frac{a_{il}}{a_{kl}}; \qquad i = 1, 2, \ldots, m, \ i \neq k;$$

$$a_{ij}^* = a_{ij} - \frac{a_{il}}{a_{kl}} \cdot a_{kj}; \qquad b_i^* = b_i - \frac{a_{il}}{a_{kl}} \cdot b_k;$$

$$i = 1, 2, \ldots, m, \quad i \neq k; \quad j = 1, 2, \ldots, n', \quad j \neq l.$$

Moreover, calculate the values of the objective row in the new tableau:

$$g_l^* = -\frac{g_l}{a_{kl}};$$

$$g_j^* = g_j - \frac{g_l}{a_{kl}} \cdot a_{kj}; \qquad j = 1, 2, \ldots, n', j \neq l;$$

$$z_0^* = z_0 - \frac{g_l}{a_{kl}} \cdot b_k.$$

Consider the tableau obtained as a new starting solution and go to step 2.

It is worth noting that the coefficients g_j^* and the objective function value z_0^* in the objective row of new tableau can also be obtained by means of formulas (9.10) and (9.11), respectively, using the new values a_{ij}^* and b_i^*.

If in each pivoting step the objective function value improves, the simplex method certainly terminates after a finite number of pivoting steps. However, it is possible that the objective function value does not change after a pivoting step. Assume that, when determining the pivot row, the minimal quotient is equal to zero. This means that some component of the current right-hand side vector is equal to zero. (This always happens when in the previous tableau the minimal quotient was not uniquely defined.) In this case, the basic variable x_{Bk} has the value zero, and in the next pivoting step, one non-basic variable becomes a basic variable again with value zero. Geometrically, this means that we do not move to an adjacent extreme point in this pivoting step: there is only one non-basic variable interchanged with some basic variable having value zero, and so it may happen that after a finite number of such pivoting steps with unchanged objective function value, we come again to a basic feasible solution that has already been visited. So, a cycle occurs and the procedure would not stop after a finite number of steps. We only mention that there exist several rules for selecting the pivot row and column which prevent such a cycling. One of these rules can be given as follows.

Smallest subscript rule. If there are several candidates for the entering and/or leaving variables, always choose the corresponding variables having the *smallest subscript*.

The above rule, which is due to Bland, means that among the non-basic variables with negative coefficient g_j in the objective row, the variable with the smallest subscript is taken as the entering variable and if, in this case, the same smallest quotient is obtained for several rows, then among the corresponding basic variables again the variable with the smallest subscript is chosen. However, since cycling occurs rather seldom in practice, the remote possibility of cycling is disregarded in most computer implementations of the simplex algorithm.

Example 9.5   Let us consider again the data given in Example 9.2. We can immediately give the tableau for the initial basic feasible solution:

1	nbv	x_1	x_2		
bv	-1	1	2	0	Q
x_3	0	0	8	640	80
x_4	0	6	6	720	120
x_5	0	6	3	600	200
		-1	-2	0	

Choosing x_2 as entering variable, we get the quotients given in the last column of the above tableau and thus we choose x_3 as leaving variable. (Hereafter, both the entering and leaving variables are printed in bold face.) Number 8 (also printed in bold face) becomes the pivot element, and we get the following new tableau:

2	nbv	x_1	x_3		
bv	-1	1	0	0	Q
x_2	2	0	$\frac{1}{8}$	80	—
x_4	0	6	$-\frac{3}{4}$	240	40
x_5	0	6	$-\frac{3}{8}$	360	60
		-1	$\frac{1}{4}$	160	

Now, the entering variable x_1 is uniquely determined, and from the quotient column we find x_4 as the leaving variable. Then we obtain the following new tableau:

3	nbv	x_4	x_3		
bv	-1	0	0	0	Q
x_2	2	0	$\frac{1}{8}$	80	
x_1	1	$\frac{1}{6}$	$-\frac{1}{8}$	40	
x_5	0	-1	$\frac{3}{8}$	120	
		$\frac{1}{6}$	$\frac{1}{8}$	200	

From the last tableau, we get the optimal solution which has already been found in Example 9.2 geometrically, i.e. the basic variables x_1, x_2 and x_5 are equal to the corresponding values of the right-hand side: $x_1 = 40$, $x_2 = 80$, $x_5 = 120$, and the non-basic variables x_3 and x_4 are equal to zero. We also see that the simplex method moves in each step from an extreme point to an adjacent extreme point. Referring to Figure 9.2, the algorithm starts at point P_1, moves then to point P_5 and finally to the optimal extreme point P_4. If in the first tableau variable x_1 were chosen as the entering variable (which would also be allowed since the corresponding coefficient in the objective row is negative), the resulting path would be P_1, P_2, P_3, P_4, and in the latter case three pivoting steps would be necessary.

Example 9.6   A firm intends to manufacture three types of products P_1, P_2 and P_3 so that the total production cost does not exceed 32,000 EUR. There are 420 working hours possible and 30 units of raw materials may be used. Additionally, the data presented in Table 9.4 are given.

Table 9.4 Data for Example 9.6

Product	P_1	P_2	P_3
Selling price (EUR/piece)	1,600	3,000	5,200
Production cost (EUR/piece)	1,000	2,000	4,000
Required raw material (per piece)	3	2	2
Working time (hours per piece)	20	10	20

The objective is to determine the quantities of each product so that the profit is maximized. Let x_i be the number of produced pieces of P_i, $i \in \{1, 2, 3\}$. We can formulate the above problem as an LPP as follows:

$$
\begin{aligned}
z = 6x_1 + 10x_2 + 12x_3 &\rightarrow \max! \\
\text{s.t.}\quad x_1 + 2x_2 + 4x_3 &\leq 32 \\
3x_1 + 2x_2 + 2x_3 &\leq 30 \\
2x_1 + x_2 + 2x_3 &\leq 42 \\
x_1, x_2, x_3 &\geq 0.
\end{aligned}
$$

The objective function has been obtained by subtracting the production cost from the selling price and dividing the resulting profit by 100 for each product. Moreover, the constraint on the production cost has been divided by 1,000, and the constraint on the working time by 10.

Introducing now in the ith constraint the slack variable $x_{3+i} \geq 0$, we obtain the standard form together with the following initial tableau:

1	nbv	x_1	x_2	x_3		
bv	-1	6	10	12	0	Q
x_4	0	1	2	4	32	8
x_5	0	3	2	2	30	15
x_6	0	2	1	2	42	21
		-6	-10	-12	0	

Choosing x_3 now as the entering variable (since it has the smallest negative coefficient in the objective row), variable x_4 becomes the leaving variable due to the quotient rule. We obtain:

2	nbv	x_1	x_2	x_4		
bv	−1	6	10	0	0	Q
x_3	12	$\frac{1}{4}$	$\frac{1}{2}$	$\frac{1}{4}$	8	16
x_5	0	$\frac{5}{2}$	1	$-\frac{1}{2}$	14	14
x_6	0	$\frac{3}{2}$	0	$-\frac{1}{2}$	26	—
		−3	−4	3	96	

Choosing now x_2 as entering variable, x_5 becomes the leaving variable. We obtain the tableau:

3	nbv	x_1	x_5	x_4		
bv	−1	6	0	0	0	Q
x_3	12	−1	$-\frac{1}{2}$	$\frac{1}{2}$	1	
x_2	10	$\frac{5}{2}$	1	$-\frac{1}{2}$	14	
x_6	0	$\frac{3}{2}$	0	$-\frac{1}{2}$	26	
		7	4	1	152	

Since now all coefficients g_j are positive, we get the following optimal solution from the latter tableau:

$$x_1 = 0, \quad x_2 = 14, \quad x_3 = 1, \quad x_4 = 0, \quad x_5 = 0, \quad x_6 = 26.$$

This means that the optimal solution is to produce no piece of product P_1, 14 pieces of product P_2 and one piece of product P_3. Taking into account that the coefficients of the objective function were divided by 100, we get a total profit of 15,200 EUR.

Example 9.7 We consider the following LPP:

$$z = -2x_1 - 2x_2 \to \text{min!}$$
$$\begin{aligned}
\text{s.t.} \quad x_1 \ - \ x_2 &\geq -1 \\
-x_1 + 2x_2 &\leq 4 \\
x_1, x_2 &\geq 0.
\end{aligned}$$

First, we transform the given problem into the standard form, i.e. we multiply the objective function and the first constraint by -1 and introduce the slack variables x_3 and x_4. We obtain:

$$\bar{z} = 2x_1 + 2x_2 \to \text{max!}$$
$$\begin{aligned}
\text{s.t.} \quad -x_1 + \ x_2 + x_3 \qquad\ &= 1 \\
-x_1 + 2x_2 \qquad + x_4 &= 4 \\
x_1, x_2, x_3, x_4 &\geq 0.
\end{aligned}$$

Now we can establish the first tableau:

1	nbv	x_1	x_2			
bv	−1	2	2	0	Q	
x_3	0	−1	1	1	1	
x_4	0	−1	2	4	2	
		−2	−2	0		

Since there are only negative elements in the column of variable x_1, only variable x_2 can be the entering variable. In this case, we get the quotients given in the last column of the latter tableau and therefore, variable x_3 is the leaving variable. We obtain the following tableau:

2	nbv	x_1	x_3			
bv	−1	2	0	0	Q	
x_2	2	−1	1	1	−	
x_4	0	1	−2	2	2	
		−4	2	2		

In the latter tableau, there is only one negative coefficient of a non-basic variable in the objective row, therefore variable x_1 becomes the entering variable. Since there is only one positive element in the column belonging to x_1, variable x_4 becomes the leaving variable. We obtain the following tableau:

3	nbv	x_4	x_3			
bv	−1	0	0	0	Q	
x_2	2	1	−1	3		
x_1	2	1	−2	2		
		4	−6	10		

Since there is only one negative coefficient of a non-basic variable in the objective row, variable x_3 should be chosen as entering variable. However, there are only negative elements in the column belonging to x_3. This means that we cannot perform a further pivoting step, and so there does not exist an optimal solution of the maximization problem considered (i.e. the objective function value can become arbitrarily large, see Theorem 9.4).

Example 9.8   Given is the following LPP:

$$z = x_1 + x_2 + x_3 + x_4 + x_5 + x_6 \to \min!$$

$$\begin{aligned} \text{s.t.} \quad & 2x_1 + x_2 + x_3 && \geq 4,000 \\ & x_2 && + 2x_4 + x_5 && \geq 5,000 \\ & x_3 && + 2x_5 + 3x_6 \geq 3,000 \\ & x_1, x_2, x_3, x_4, x_5, x_6 \geq 0. \end{aligned}$$

To get the standard form, we notice that in each constraint there is one variable that occurs only in this constraint. (Variable x_1 occurs only in the first constraint, variable x_4 only in the second constraint and variable x_6 only in the third constraint.) Therefore, we divide the first constraint by the coefficient 2 of variable x_1, the second constraint by 2 and the third constraint by 3. Then, we introduce a surplus variable in each of the constraints, multiply the objective function by -1 and obtain the standard form. (Again the variables are written in such a way that the identity submatrix of the coefficient matrix occurs now at the end.)

$$\bar{z} = -z = -x_1 - x_2 - x_3 - x_4 - x_5 - x_6 \rightarrow \max!$$

$$\text{s.t.} \quad \tfrac{1}{2}x_2 + \tfrac{1}{2}x_3 \qquad - x_7 \qquad\quad + x_1 \qquad\qquad = 2{,}000$$

$$\tfrac{1}{2}x_2 \qquad + \tfrac{1}{2}x_5 \qquad - x_8 \qquad\quad + x_4 \qquad = 2{,}500$$

$$\tfrac{1}{3}x_3 + \tfrac{2}{3}x_5 \qquad\qquad - x_9 \qquad\quad + x_6 = 1{,}000$$

$$x_1, x_2, x_3, x_4, x_5, x_6, x_7, x_8, x_9 \geq 0.$$

This yields the following initial tableau:

1	nbv	x_2	x_3	x_5	x_7	x_8	x_9		
bv	-1	-1	-1	-1	0	0	0	0	Q
x_1	-1	$\tfrac{1}{2}$	$\tfrac{1}{2}$	0	-1	0	0	2,000	—
x_4	-1	$\tfrac{1}{2}$	0	$\tfrac{1}{2}$	0	-1	0	2,500	5,000
x_6	-1	0	$\tfrac{1}{3}$	$\tfrac{2}{3}$	0	0	-1	1,000	1,500
	0	$\tfrac{1}{6}$	$-\tfrac{1}{6}$	1	1	1		$-5{,}500$	

Choosing now x_5 as entering variable, we obtain the quotients given in the last column of the above tableau and therefore, x_6 is chosen as leaving variable. We obtain the following tableau:

2	nbv	x_2	x_3	x_6	x_7	x_8	x_9		
bv	-1	-1	-1	-1	0	0	0	0	Q
x_1	-1	$\tfrac{1}{2}$	$\tfrac{1}{2}$	0	-1	0	0	2,000	
x_4	-1	$\tfrac{1}{2}$	$-\tfrac{1}{4}$	$-\tfrac{3}{4}$	0	-1	$\tfrac{3}{4}$	1,750	
x_5	-1	0	$\tfrac{1}{2}$	$\tfrac{3}{2}$	0	0	$-\tfrac{3}{2}$	1,500	
	0	$\tfrac{1}{4}$	$\tfrac{1}{4}$	1	1	$\tfrac{3}{4}$	$-5{,}250$	

Now all coefficients of the non-basic variables in the objective row are non-negative and from the latter tableau we obtain the following optimal solution:

$$x_1 = 2{,}000, \quad x_2 = x_3 = 0, \quad x_4 = 1{,}750, \quad x_5 = 1{,}500, \quad x_6 = 0$$

with the optimal objective function value $\bar{z}_0^{\max} = -5{,}250$ which corresponds to $z_0^{\min} = 5{,}250$ (for the original minimization problem). Notice that the optimal solution is not uniquely determined. In the last tableau, there is one coefficient in the objective row equal to zero. Taking x_2 as the entering variable, the quotient rule determines x_4 as the leaving variable, and the following basic feasible solution with the same objective function value is obtained:

$$x_1 = 250, \quad x_2 = 3{,}500, \quad x_3 = x_4 = 0, \quad x_5 = 1{,}500, \quad x_6 = 0.$$

9.5   TWO-PHASE SIMPLEX ALGORITHM

In this section, we discuss the case when artificial variables are necessary to transform a given problem into standard form. In such a case, we have to determine a basic solution feasible for the original problem that can also be done by applying the simplex algorithm. This procedure is called *phase I* of the simplex algorithm. It either constructs an initial basic feasible solution or recognizes that the given LPP does not have a feasible solution at all. If a feasible starting solution has been found, *phase II* of the simplex algorithm starts, which corresponds to the simplex algorithm described in Chapter 9.4.

The introduction of artificial variables is necessary when at least one constraint is an equation with no eliminated variable that has coefficient $+1$, e.g. the constraints may have the following form:

$$a_{11} x_1 + a_{12}x_2 + \cdots + a_{1n}x_n = b_1$$
$$a_{21} x_1 + a_{22}x_2 + \cdots + a_{2n}x_n = b_2$$
$$\vdots \qquad \qquad \vdots$$
$$a_{m1}x_1 + a_{m2}x_2 + \cdots + a_{mn}x_n = b_m$$

$$x_j \geq 0, \ j = 1, 2, \ldots, n; \quad b_i \geq 0, \ i = 1, 2, \ldots, m.$$

As discussed in step (5) of generating the standard form, we introduce an artificial variable x_{Ai} in each equation. Additionally we replace the original objective function z by an objective function z_I minimizing the sum of all artificial variables (or equivalently, maximizing the negative sum of all artificial variables) since it is our goal that all artificial variables will get value zero to ensure feasibility for the original problem. This gives the following linear programming problem to be considered in phase I:

$$z_I = -x_{A1} - x_{A2} - \ldots - x_{Am} \longrightarrow \max!$$

$$
\begin{aligned}
\text{s.t.} \quad & a_{11} x_1 + a_{12}x_2 + \ldots + a_{1n}x_n + x_{A1} && = b_1 \\
& a_{21} x_1 + a_{22}x_2 + \ldots + a_{2n}x_n && + x_{A2} && = b_2 \\
& \qquad \qquad \vdots && \vdots \\
& a_{m1}x_1 + a_{m2}x_2 + \ldots + a_{mn}x_n && + x_{Am} = b_m
\end{aligned}
\tag{9.12}
$$

$$x_j \geq 0, \ j = 1, 2, \ldots, n \quad \text{and} \quad x_{Ai} \geq 0, \ i = 1, 2, \ldots, m.$$

The above problem is the standard form of an LPP with function z replaced by function z_I. The objective function z_I used in the first phase is also called the *auxiliary* objective function. When minimizing function z_I, the smallest possible objective function value is equal to zero. It is attained when all artificial variables have value zero, i.e. the artificial variables have become non-basic variables, or possibly some artificial variable is still a basic variable but it has value zero. When an artificial variable becomes the leaving variable (and therefore has value zero in the next tableau), it never will be the entering variable again, and therefore this variable together with the corresponding column can be dropped in the new tableau.

Assume that we have determined an optimal solution of the auxiliary problem (9.12) by the simplex method, i.e. the procedure stops with $g_j \geq 0$ for all coefficients of the non-basic variables in the objective row (for the auxiliary objection function z_I). Then, at the end of phase I, the following cases are possible:

(1) We have $z_I^{max} < 0$. Then the initial problem does not have a feasible solution, i.e. $M = \emptyset$.
(2) We have $z_I^{max} = 0$. Then one of the following cases occurs:

   (a) All artificial variables are non-basic variables. Then the basic solution obtained represents a feasible canonical form for the initial problem, and we can start with phase II of the simplex algorithm described in Chapter 9.4.

   (b) Among the basic variables, there is still an artificial variable in row k: $x_{Bk} = x_{AI} = 0$ (degeneration case). Then we have one of the two possibilities:

      (i) In the row belonging to the basic variable $x_{AI} = 0$, all coefficients are also equal to zero. In this case, the corresponding equation is superfluous and can be omitted.

      (ii) In the row belonging to the basic variable $x_{AI} = 0$, we have $a_{kj}^* \neq 0$ in the tableau according to Table 9.3 (with function z_I) for at least one coefficient. Then we can choose a_{kj}^* as pivot element and replace the artificial variable x_{AI} by the non-basic variable x_{Nj}.

We illustrate the two-phase simplex algorithm by the following three examples.

Example 9.9   Given is the following LPP:

$$z = x_1 - 2x_2 \rightarrow \max!$$
$$\text{s.t.} \quad \begin{array}{rcrcl} x_1 & + & x_2 & \leq & 4 \\ 2x_1 & - & x_2 & \geq & 1 \\ & & x_1, x_2 & \geq & 0. \end{array}$$

We transform the given problem into standard form by introducing a surplus variable (x_3) in the second constraint, a slack variable (x_4) in the first constraint and an artificial variable (x_{A1}) in the second constraint. Now we replace the objective function z by the auxiliary function z_I. Thus, in phase I of the simplex method, we consider the following LPP:

$$z_I = -x_{A1} \rightarrow \max!$$
$$\text{s.t.} \quad \begin{array}{rcrcrcrcrcl} x_1 & + & x_2 & & & + & x_4 & & & = & 4 \\ 2x_1 & - & x_2 & - & x_3 & & & + & x_{A1} & = & 1 \\ & & & & x_1, x_2, x_3, x_4, x_{A1} & & & & & \geq & 0. \end{array}$$

We start with the following tableau:

1	*nbv*	x_1	x_2	x_3		
bv	-1	0	0	0	0	Q
x_4	0	1	1	0	4	4
x_{A1}	-1	2	-1	-1	1	$\frac{1}{2}$
		-2	1	1	-1	

Choosing x_1 as entering variable gives the quotients presented in the last column of the above tableau, and variable x_{A1} becomes the leaving variable. This leads to the following tableau:

2	nbv	x_{A1}	x_2	x_3	Q
bv	-1	-1	0	0	0
x_4	0	$-\frac{1}{2}$	$\frac{3}{2}$	$\frac{1}{2}$	$\frac{7}{2}$
x_1	0	$\frac{1}{2}$	$-\frac{1}{2}$	$-\frac{1}{2}$	$\frac{1}{2}$
	1	0	0	0	

Now phase I is finished, we drop variable x_{A1} and the corresponding column, use the original objective function and determine the coefficients g_j of the objective row. This yields the following tableau:

2*	nbv	x_2	x_3		Q
bv	-1	-2	0	0	
x_4	0	$\frac{3}{2}$	$\frac{1}{2}$	$\frac{7}{2}$	7
x_1	1	$-\frac{1}{2}$	$-\frac{1}{2}$	$\frac{1}{2}$	—
	$\frac{3}{2}$	$-\frac{1}{2}$	$\frac{1}{2}$		

Due to the negative coefficient in the objective row, we choose x_3 as the entering variable in the next step and variable x_4 becomes the leaving variable. Then we obtain the following tableau

3	nbv	x_2	x_4		Q
bv	-1	-2	0	0	
x_3	0	3	2	7	
x_1	1	1	1	4	
	3	1	4		

Since all coefficients g_j are non-negative, the obtained solution is optimal: $x_1 = 4, x_2 = 0$. The introduced surplus variable x_3 is equal to seven while the introduced slack variable x_4 is equal to zero. The optimal objective function value is $z_0^{\max} = 4$. The graphical solution of this problem is illustrated in Figure 9.7. We see from Figure 9.7 that the origin of the coordinate system with variables x_1 and x_2 is not feasible since the second constraint is violated. This was the reason for introducing the artificial variable x_{A1} which has initially the value one. After the first pivoting step, we get a feasible solution for the original problem which corresponds to extreme point P_1 in Figure 9.7. Now phase II of the simplex algorithm starts, and after the next pivoting step we reach an adjacent extreme point P_2 which corresponds to an optimal solution of the considered LPP.

Let us consider another LPP assuming that the objective function changes now to

$$\tilde{z} = -x_1 + 3x_2 \longrightarrow \min!$$

Can we easily decide whether the optimal solution for the former objective function is also optimal for the new one? We replace only the coefficients c_1 and c_2 of the objective function in the last tableau (again for the maximization version of the problem), recompute the coefficients g_j of the objective row and obtain the following tableau:

$\bar{3}$	nbv	x_2	x_4		
bv	-1	-3	0	0	Q
x_3	0	3	2	7	
x_1	1	1	1	4	
	4	1	4		

Since also in this case all coefficients g_j in the objective row are non-negative, the solution $x_1 = 4$, $x_2 = 0$ is optimal for $\tilde{z} = -x_1 + 3x_2 \longrightarrow$ min as well with a function value $\tilde{z}_0^{min} = -4$. This can also be confirmed by drawing the objective function $\tilde{z}$ in Figure 9.7.

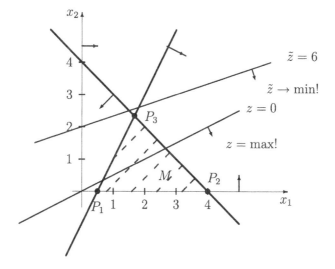

Figure 9.7 Graphical solution of Example 9.9.

Example 9.10 We consider the data given in Example 9.1 and apply the two-phase simplex method. Transforming the given problem into standard form we obtain:

$$\bar{z} = -25x_1 - 17x_2 - 12x_3 \rightarrow \text{max!}$$

$$
\begin{aligned}
\text{s.t.} \quad x_1 + x_2 + x_3 + x_{A1} &= 100 \\
x_3 + x_4 &= 30 \\
x_1 + x_3 + x_5 &= 50 \\
x_2 + x_3 + x_6 &= 90 \\
x_1, x_2, x_3, x_4, x_5, x_6, x_{A1} &\geq 0.
\end{aligned}
$$

Starting with phase I of the simplex method, we replace function z by the auxiliary objective function

$$z_I = -x_{A1} \longrightarrow \text{max!}$$

We obtain the following initial tableau:

1	nbv	x_1	x_2	x_3		
bv	−1	0	0	0	0	Q
x_{A1}	−1	1	1	1	100	100
x_4	0	0	0	1	30	−
x_5	0	1	0	1	50	50
x_6	0	0	1	1	90	−
		−1	−1	−1	−100	

Choosing x_1 as the entering variable, we get the quotients given above and select x_5 as the leaving variable. This leads to the following tableau:

2	nbv	x_5	x_2	x_3		
bv	−1	0	0	0	0	Q
x_{A1}	−1	−1	1	0	50	50
x_4	0	0	0	1	30	−
x_1	0	1	0	1	50	−
x_6	0	0	1	1	90	90
		1	−1	0	− 50	

Now x_2 becomes the entering variable and the artificial variable x_{A1} is the leaving variable. We get the following tableau, where the superfluous column belonging to x_{A1} is dropped.

3	nbv	x_5	x_3		
bv	−1	0	0	0	Q
x_2	0	−1	0	50	
x_4	0	0	1	30	
x_1	0	1	1	50	
x_6	0	1	1	40	
		0	0	0	

Now, phase I is finished, and we can consider the objective function

$$\bar{z} = -25x_1 - 17x_2 - 12x_3 \longrightarrow \text{max!}$$

We recompute the coefficients in the objective row and obtain the following tableau:

3*	nbv	x_5	x_3		
bv	-1	0	-12	0	Q
x_2	-17	-1	0	50	$-$
x_4	0	0	1	30	30
x_1	-25	1	1	50	50
x_6	0	1	1	40	40
	-8	-13	$-2,100$		

We choose x_3 as entering variable and based on the quotients given in the last column, x_4 is the leaving variable. After this pivoting step, we get the following tableau:

4	nbv	x_5	x_4		
bv	-1	0	0	0	Q
x_2	-17	-1	0	50	$-$
x_3	-12	0	1	30	$-$
x_1	-25	1	-1	20	20
x_6	0	1	-1	10	10
	-8	13	$-1,710$		

We choose x_5 as entering variable and x_6 as leaving variable which gives the following tableau:

5	nbv	x_6	x_4		
bv	-1	0	0	0	Q
x_2	-17	1	-1	60	
x_3	-12	0	1	30	
x_1	-25	-1	0	10	
x_5	0	1	-1	10	
	8	5	$-1,630$		

This last tableau gives the following optimal solution:

$$x_1 = 10, \quad x_2 = 60, \quad x_3 = 30, \quad x_4 = 0, \quad x_5 = 10, \quad x_6 = 0$$

with the objective function value $z_0^{\min} = 1,630$ for the minimization problem.

Example 9.11   Consider the following LPP:

$$z = x_1 + 2x_2 \rightarrow \max!$$
$$\text{s.t.} \quad x_1 \quad - \quad x_2 \quad \geq \quad 1$$
$$5x_1 \quad - \quad 2x_2 \quad \leq \quad 3$$
$$x_1, x_2 \quad \geq \quad 0.$$

Transforming the above problem into standard form, we obtain

$$z = x_1 + 2x_2 \rightarrow \text{max!}$$

$$\text{s.t.} \quad x_1 \quad - \quad x_2 \quad - \quad x_3 \quad + \quad x_{A1} \qquad\qquad = \quad 1$$

$$\qquad 5x_1 \quad - \quad 2x_2 \qquad\qquad\qquad + \quad x_4 \quad = \quad 3$$

$$\qquad\qquad\qquad\qquad\qquad x_1, x_2, x_3, x_4, x_{A1} \quad \geq \quad 0.$$

This leads to the following starting tableau for phase I with the auxiliary objective function $z_I = -x_{A1} \rightarrow \text{max!}$

1	nbv	x_1	x_2	x_3			
bv	-1	0	0	0	0	Q	
x_{A1}	-1	1	-1	-1	1	1	
x_4	0	5	-2	0	3	$\frac{3}{5}$	
		-1	1	1	-1		

We choose now variable x_1 as entering variable which gives the leaving variable x_4. This yields the following tableau:

2	nbv	x_4	x_2	x_3			
bv	-1	0	0	0	0	Q	
x_{A1}	-1	$-\frac{1}{5}$	$-\frac{3}{5}$	-1	$\frac{2}{5}$		
x_1	0	$\frac{1}{5}$	$-\frac{2}{5}$	0	$\frac{3}{5}$		
		$\frac{1}{5}$	$\frac{3}{5}$	1	$-\frac{2}{5}$		

So, we finish with case (1) described earlier, i.e. $z_I^{\max} < 0$. Consequently, the above LPP does not have a feasible solution. In fact, in the final tableau, variable x_{A1} is still positive (so the original constraint $x_1 - x_2 - x_3 = 1$ is violated). For the considered problem, the empty feasible region is given in Figure 9.8. (It can be seen that there are no feasible solutions for this problem, which confirms our computations.)

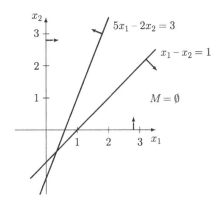

Figure 9.8 Empty feasible region M for Example 9.11.

Remark For many problems, it is possible to reduce the number of artificial variables to be considered in phase I. This is important since in any pivoting step, only one artificial variable is removed from the set of basic variables. Thus, in the case of introducing m artificial variables, at least m pivoting steps are required in phase I. Let us consider the following example.

Example 9.12 Consider the following constraints of an LPP:

$$
\begin{aligned}
2x_2 &- x_3 - x_4 + x_5 &\geq 1 \\
-2x_1 &+ 2x_3 - x_4 + x_5 &\geq 2 \\
x_1 - 2x_2 &- x_4 + x_5 &\geq 3 \\
x_1 + x_2 &+ x_3 &\geq 5 \\
& x_1, x_2, x_3, x_4, x_5 &\geq 0.
\end{aligned}
$$

First, we introduce in each constraint a surplus variable and obtain

$$
\begin{aligned}
2x_2 - x_3 - x_4 + x_5 - x_6 & = 1 \\
-2x_1 + 2x_3 - x_4 + x_5 - x_7 & = 2 \\
x_1 - 2x_2 - x_4 + x_5 - x_8 & = 3 \\
x_1 + x_2 + x_3 - x_9 & = 5 \\
x_1, x_2, x_3, x_4, x_5, x_6, x_7, x_8, x_9 & \geq 0.
\end{aligned}
$$

Now, we still need an eliminated variable with coefficient +1 in each equation constraint. Instead of introducing an artificial variable in each of the constraints, we subtract all other constraints from the constraint with the largest right-hand side. In this example, we get the first three equivalent constraints by subtracting the corresponding constraint from the fourth constraint:

$$
\begin{aligned}
x_1 - x_2 + 2x_3 + x_4 - x_5 - x_9 + x_6 & = 4 \\
3x_1 + x_2 - x_3 + x_4 - x_5 - x_9 + x_7 & = 3 \\
3x_2 + x_3 + x_4 - x_5 - x_9 + x_8 & = 2 \\
x_1 + x_2 + x_3 - x_9 & = 5 \\
x_1, x_2, x_3, x_4, x_5, x_6, x_7, x_8, x_9 & \geq 0.
\end{aligned}
$$

Now we have to introduce only *one* artificial variable x_{41} in the last constraint in order to start with phase I of the simplex algorithm.

9.6 DUALITY; COMPLEMENTARY SLACKNESS

We consider the following LPP denoted now as primal problem (P):

$$
\begin{aligned}
z = \mathbf{c}^T\mathbf{x} &\longrightarrow \max! \\
\text{s.t.} \quad A\mathbf{x} &\leq \mathbf{b} \qquad \text{(P)} \\
\mathbf{x} &\geq \mathbf{0},
\end{aligned}
$$

where $\mathbf{x} = (x_1, x_2, \ldots, x_n)^T \in \mathbb{R}^n$. By means of matrix A and vectors $\mathbf{c}$ and $\mathbf{b}$, we can define a dual problem (D) as follows:

$$w = \mathbf{b}^T \mathbf{u} \longrightarrow \min!$$
$$\text{s.t.} \quad A^T \mathbf{u} \geq \mathbf{c} \qquad\qquad \text{(D)}$$
$$\mathbf{u} \geq \mathbf{0},$$

where $\mathbf{u} = (u_{n+1}, u_{n+2}, \ldots, u_{n+m})^T \in \mathbb{R}^m$. Thus, the dual problem (D) is obtained by the following rules:

(1) The coefficient matrix A of problem (P) is transposed.
(2) The variables of the dual problem are denoted as $u_{n+1}, u_{n+2}, \ldots, u_{n+m}$, and they have to be non-negative.
(3) The vector $\mathbf{b}$ of the right-hand side of problem (P) is the vector of the objective function of problem (D).
(4) The vector $\mathbf{c}$ of the objective function of the primal problem (P) is the vector of the right-hand side of the dual problem (D).
(5) In all constraints of the dual problem (D), we have inequalities with the relation $\geq$.
(6) The objective function of the dual problem (D) has to be minimized.

By 'dualizing' problem (P), there is an assignment between the constraints of the primal problem (P) and the variables of the dual problem (D), and conversely, between the variables of problem (P) and the constraints of problem (D). Both problems can be described by the scheme given in Table 9.5 which has to be read row-wise for problem (P) and column-wise for problem (D).

Table 9.5 Relationships between problems (P) and (D)

	x_1	x_2	$\ldots$	x_n	$\leq$
u_{n+1}	a_{11}	a_{12}	$\ldots$	a_{1n}	b_1
u_{n+2}	a_{21}	a_{22}	$\ldots$	a_{2n}	b_2
$\vdots$	$\vdots$	$\vdots$		$\vdots$	$\vdots$
u_{n+m}	a_{m1}	a_{m2}	$\ldots$	a_{mn}	b_m
					min!
$\geq$	c_1	c_2	$\ldots$	c_n	max!

For instance, the first constraint of problem (P) reads as

$$a_{11}x_1 + a_{12}x_2 + \ldots + a_{1n}x_n \leq b_1$$

while e.g. the second constraint of problem (D) reads as

$$a_{12}u_{n+1} + a_{22}u_{n+2} + \ldots + a_{m2}u_{n+m} \geq c_2.$$

Notice that the variables are successively numbered, i.e. the n variables of the primal problem (P) are indexed by $1, 2, \ldots, n$ while the m variables of the dual problem (D) are indexed by $n+1, n+2, \ldots, n+m$. We get the following relationships between problems (P) and (D).

THEOREM 9.6   Let problem (D) be the dual problem of problem (P). Then the dual problem of problem (D) is problem (P).

THEOREM 9.7 Let $\mathbf{x} = (x_1, x_2, \ldots, x_n)^T$ be an arbitrary feasible solution of problem (P) and $\mathbf{u} = (u_{n+1}, u_{n+2}, \ldots, u_{n+m})^T$ be an arbitrary feasible solution of problem (D). Then

$$z_0 = \mathbf{c}^T \mathbf{x} \le \mathbf{b}^T \mathbf{u} = w_0.$$

From the latter theorem, it follows that the objective function value of some feasible solution for a maximization problem is always smaller than or equal to the function value of a feasible solution for the dual (minimization) problem.

THEOREM 9.8 If one of the problems (P) or (D) has an optimal solution, then the other one also has an optimal solution, and both optimal solutions $\mathbf{x}^*$ and $\mathbf{u}^*$ have the same objective function value, i.e.

$$z_0^{\max} = \mathbf{c}^T \mathbf{x}^* = \mathbf{b}^T \mathbf{u}^* = w_0^{\min}.$$

From Theorem 9.8 it follows that, if we know a feasible solution for the primal problem and a feasible solution for the dual problem, and both solutions have the same function value, then they must be optimal for the corresponding problems. The following theorem treats the case when one of the problems (P) and (D) has feasible solutions but the other one has not.

THEOREM 9.9 If one of the problems (P) or (D) has a feasible solution but no optimal solution, then the other one does not have a feasible solution at all.

Next, we consider the following question. Can we find the optimal solution for the dual problem immediately from the final tableau for the primal problem provided that both problems (P) and (D) have feasible solutions?

We introduce slack variables into the constraints of problem (P) and surplus variables into the constraints of problem (D), and we obtain for the constraints:

$$
\begin{aligned}
a_{11}\,x_1 + a_{12}\,x_2 + \cdots + a_{1n}\,x_n + x_{n+1} &= b_1 \\
a_{21}\,x_1 + a_{22}\,x_2 + \cdots + a_{2n}\,x_n \qquad\quad + x_{n+2} &= b_2 \\
\vdots \qquad\qquad\qquad \vdots \\
a_{m1}\,x_1 + a_{m2}\,x_2 + \cdots + a_{mn}\,x_n \qquad\qquad\quad + x_{n+m} &= b_m
\end{aligned}
\qquad (P^*)
$$

$$
\begin{aligned}
-u_1 \qquad\qquad +a_{11}\,u_{n+1} + a_{21}\,u_{n+2} + \cdots + a_{m1}\,u_{n+m} &= c_1 \\
-u_2 \qquad\quad +a_{12}\,u_{n+1} + a_{22}\,u_{n+2} + \cdots + a_{m2}\,u_{n+m} &= c_2 \\
\vdots \qquad\qquad\qquad \vdots \\
-u_n + a_{1n}\,u_{n+1} + a_{2n}\,u_{n+2} + \cdots + a_{mn}\,u_{n+m} &= c_n.
\end{aligned}
\qquad (D^*)
$$

Then we obtain the following relationship between the primal variables $x_j, j = 1, 2, \ldots, n+m$, and the dual variables $u_j, j = 1, 2, \ldots, n + m$, where the variables are numbered according to (P^*) and (D^*).

THEOREM 9.10 The coefficients of the non-basic variables x_j in the objective row of the optimal tableau of problem (P) are equal to the optimal values of the corresponding variables u_j of problem (D), and conversely: The optimal values of the basic variables of problem (P) are equal to the coefficients of the corresponding dual variables in the objective row of the optimal tableau of problem (P).

Thus, from the optimal tableau of one of the problems (P) or (D), we can determine the optimal solution of the other one.

Example 9.13   Consider the data given in Example 9.2. The optimal tableau for this problem has been determined in Example 9.5. Writing the dual problem with the constraints as in (D*), we obtain:

$$w = 640u_3 + 720u_4 + 600u_5 \longrightarrow \min!$$

$$
\begin{array}{rcccccl}
\text{s.t.} \quad -u_1 & & & + & 6u_4 & + & 6u_5 & = & 1 \\
& -u_2 & + 8u_3 & + & 6u_4 & + & 3u_5 & = & 2 \\
& & & & u_1, u_2, u_3, u_4, u_5 & \geq & 0.
\end{array}
$$

Applying Theorem 9.10, we get the following. Since in the optimal solution given in Example 9.5, variables x_4 and x_3 are non-basic variables, the corresponding variables u_4 and u_3 are basic variables in the optimal solution of the dual problem and their values are equal to the coefficients of x_4 and x_3 in the objective row of the optimal tableau of the primal problem, i.e. $u_4 = 1/6$ and $u_3 = 1/8$. Since x_2, x_1 and x_5 are basic variables in the optimal tableau of the primal problem, the variables u_2, u_1 and u_5 are non-basic variables in the optimal solution of the dual problem, i.e. their values are equal to zero. Accordingly, the values of the right-hand side in the optimal primal tableau correspond to the coefficients of the dual non-basic variables in the objective row, i.e. they are equal to 80, 40 and 120, respectively.

Finally, we briefly deal with a 'mixed' LPP (i.e. equations may occur as constraints, we may have inequalities both with $\leq$ and $\geq$ sign, and variables are not necessarily non-negative). Then we can establish a dual problem using the rules given in Table 9.6. To illustrate, let us consider the following example.

Example 9.14   Consider the LPP

$$z = 2x_1 + 3x_2 + 4x_3 \longrightarrow \max!$$

$$
\begin{array}{rcccccl}
\text{s.t.} \quad x_1 & - & x_2 & + & x_3 & \leq & 20 \\
-3x_1 & + & x_2 & + & x_3 & \geq & 3 \\
4x_1 & - & x_2 & + & 2x_3 & = & 10 \\
& & & & x_3 & \leq & 5 \\
\end{array}
$$

$$x_1 \geq 0 \qquad (x_2, x_3 \text{ arbitrary}).$$

Applying the rules given in Table 9.6, we obtain the following dual problem:

$$w = 20u_4 + 3u_5 + 10u_6 + 5u_7 \longrightarrow \min!$$

$$
\begin{array}{rcccccl}
\text{s.t.} \quad u_4 & - & 3u_5 & + & 4u_6 & & & \geq & 2 \\
-u_4 & + & u_5 & - & u_6 & & & = & 3 \\
u_4 & + & u_5 & + & 2u_6 & + & u_7 & = & 4 \\
\end{array}
$$

$$u_4 \geq 0, u_5 \leq 0, u_7 \geq 0 \qquad (u_6 \text{ arbitrary}).$$

Table 9.6 Rules for dualizing a problem

Primal problem	*Dual* problem
(P)	(D)
Maximization problem	*Minimization* problem
Constraints	Variables
ith constraint: inequality with sign $\leq$ $\implies$	$u_{n+i} \geq 0$
ith constraint: inequality with sign $\geq$ $\implies$	$u_{n+i} \leq 0$
ith constraint: equation $\implies$	u_{n+i} arbitrary
Variables	Constraints
$x_j \geq 0$ $\implies$	jth constraint: inequality with sign $\geq$
$x_j \leq 0$ $\implies$	jth constraint: inequality with sign $\leq$
x_j arbitrary $\implies$	jth constraint: equation

We finish this section with an economic interpretation of duality. Consider problem (P) as the problem of determining the optimal production quantities x_j^* of product $P_j, j = 1, 2, \ldots, n$, subject to given budget constraints, i.e. for the production there are m raw materials required and the right-hand side value b_i gives the available quantity of raw material R_i, $i \in \{1, 2, \ldots, m\}$. The objective is to maximize the profit described by the linear function $\mathbf{c}^T \mathbf{x}$. What is the economic meaning of the corresponding dual variables of problem (D) in this case? Assume that we increase the available amount of raw material R_i by one unit (i.e. we use the right-hand side component $b_i' = b_i + 1$) and all other values remain constant. Moreover, assume that there is an optimal non-degenerate basic solution (i.e. all basic variables have a value greater than zero) and that the same basic solution is optimal for the modified problem (P′) with b_i replaced by b_i' (i.e. no further pivoting step is required). In this case, the optimal function value $\bar{z}_0^{\max}$ of problem (P′) is given by

$$\bar{z}_0^{\max} = b_1 u_{n+1}^* + \ldots + (b_i + 1)u_{n+i}^* + \ldots + b_m u_{n+m}^* = w_0^{\min} + u_{n+i}^*,$$

where $\mathbf{u}^* = (u_{n+1}^*, u_{n+2}^*, \ldots, u_{n+m}^*)^T$ denotes the optimal solution of the dual problem (D). The latter equality holds due to Theorem 9.8 (i.e. $z_0^{\max} = w_0^{\min}$). This means that after 'buying' an additional unit of resource R_i, the optimal function value increases by the value u_{n+i}^*. Hence, the optimal values u_{n+i}^* of the variables of the dual problem can be interpreted as *shadow prices* for buying an additional unit of raw materials R_i, i.e. u_{n+i}^* characterizes the price that a company is willing to pay for an additional unit of raw material R_i, $i = 1, 2, \ldots, m$.

Next, we consider the primal problem (P) and the dual problem (D). We can formulate the following relationship between the optimal solutions of both problems.

THEOREM 9.11 Let $\mathbf{x}^* = (x_1^*, x_2^*, \ldots, x_n^*)^T$ be a feasible solution of problem (P) and $\mathbf{u}^* = (u_{n+1}^*, u_{n+2}^*, \ldots, u_{n+m}^*)^T$ be a feasible solution of problem (D). Then both solutions $\mathbf{x}^*$ and $\mathbf{u}^*$ are simultaneously optimal if and only if the following two conditions hold:

(1) $\displaystyle\sum_{i=1}^{m} a_{ij} u_{n+i}^* = c_j$ or $x_j^* = 0$ for $j = 1, 2, \ldots, n$;

(2) $\displaystyle\sum_{j=1}^{n} a_{ij} x_j^* = b_i$ or $u_{n+i}^* = 0$ for $i = 1, 2, \ldots, m$.

Theorem 9.11 means that vectors $\mathbf{x}^*$ and $\mathbf{u}^*$ are optimal for problems (P) and (D) if and only if

(1) either the jth constraint in problem (D) is satisfied with equality or the jth variable of problem (P) is equal to zero (or both) and

(2) either the ith constraint in problem (P) is satisfied with equality or the ith variable of problem (D) (i.e. $u_{n+i}^* = 0$) is equal to zero (or both).

Theorem 9.11 can be rewritten in the following form:

THEOREM 9.12   A feasible solution $\mathbf{x}^* = (x_1^*, x_2^*, \ldots, x_n^*)^T$ of problem (P) is optimal if and only if there exists a vector $\mathbf{u}^* = (u_{n+1}^*, u_{n+2}^*, \ldots, u_{n+m}^*)^T$ such that the following two properties are satisfied:

(1)   if $x_j^* > 0$, then $\displaystyle\sum_{i=1}^{m} a_{ij} u_{n+i}^* = c_j;$ $\hspace{3cm}$ (9.13)

(2)   if $\displaystyle\sum_{j=1}^{n} a_{ij} x_j^* < b_i$, then $u_{n+i}^* = 0,$

provided that vector $\mathbf{u}^*$ is feasible for problem (D), i.e.

$$\sum_{i=1}^{m} a_{ij} u_{n+i}^* \geq c_j \quad \text{for } j = 1, 2, \ldots, n;$$
$$u_{n+i}^* \geq 0 \qquad\qquad \text{for } i = 1, 2, \ldots, m.$$
$\hspace{11cm}$ (9.14)

Remark   Theorem 9.12 is often useful in checking the optimality of allegedly optimal solutions when no certificate of optimality is provided. Confronted with an allegedly optimal solution $\mathbf{x}^*$ of problem (P), we first set up the system of linear equations (9.13) and solve it for $\mathbf{u}^* = (u_{n+1}^*, u_{n+2}^*, \ldots, u_{n+m}^*)^T$. If the solution $\mathbf{u}^* = (u_{n+1}^*, u_{n+2}^*, \ldots, u_{n+m}^*)^T$ is *uniquely* determined, then vector $\mathbf{x}^*$ is optimal if and only if vector $\mathbf{u}^*$ is feasible, i.e. conditions (9.14) hold.

Example 9.15   Consider the LPP in Example 9.2 and the corresponding dual problem established in Example 9.13. Assume that we still do not know an optimal solution, but we want to use complementary slackness to verify that $x_1^* = 40$, $x_2^* = 80$ is indeed an optimal solution for problem (P). It follows from the second condition of (9.13) that $u_5^* = 0$ since, for the third constraint of the primal problem, we have

$$6x_1^* + 3x_2^* = 6 \cdot 40 + 3 \cdot 80 = 480 < 600$$

while the other two constraints of the primal problem are satisfied with equality, i.e.

$$8x_2^* = 8 \cdot 80 = 640 \quad\text{and}\quad 6x_1^* + 6x_2^* = 6 \cdot 40 + 6 \cdot 80 = 720.$$

Establishing the system of linear equations resulting from the first part of (9.13), we get

$$\begin{aligned} 6u_4^* &= 1 \\ 8u_3^* + 6u_4^* &= 2. \end{aligned}$$

(Notice that $x_1^* = 40 > 0$ and $x_2^* = 80 > 0$.) The latter system has the unique solution

$$u_3^* = \frac{1}{8} \quad\text{and}\quad u_4^* = \frac{1}{6}$$

which satisfies conditions (9.14). Therefore, without applying the simplex algorithm, we have confirmed that $x_1^* = 40$, $x_2^* = 80$ as well as $u_3^* = 1/8$, $u_4^* = 1/6$, $u_5^* = 0$ constitute optimal solutions for the primal and dual problems, respectively.

9.7 DUAL SIMPLEX ALGORITHM

Based on the relationships between problems (P) and (D), one can give an alternative variant of the simplex algorithm known as the *dual simplex algorithm*. While the simplex algorithm presented in Chapter 9.4 always works with (primal) feasible solutions and it stops when the current feasible solution is optimal (i.e. all coefficients g_j in the objective row are greater than or equal to zero), the dual algorithm operates with infeasible solutions (i.e. there are right-hand side components smaller than zero), but the coefficients g_j in the objective row are always all greater than or equal to zero (this means that the current solution satisfies the optimality criterion but it is not feasible). In the latter variant, all resulting basic solutions are feasible for the dual problem, but infeasible for the primal problem.

In the dual simplex method, first the pivot row and then the pivot column are determined. Using the notations introduced in Chapter 9.4 (see description of the simplex algorithm), the pivot row and column are determined as follows provided that the simplex tableau of the current basic solution is as given in Table 9.3.

Determination of the pivot row k

Choose row $k, 1 \leq k \leq m$, such that $b_k < 0$. Often, a row k is used with

$$b_k^* = \min\{b_i^* \mid b_i^* < 0, \quad i = 1, 2, \ldots, m\}.$$

Determination of the pivot column l

Choose column $l, 1 \leq l \leq n'$, such that

$$\frac{g_l}{|a_{kl}^*|} = \min\left\{\frac{g_i}{|a_{kj}^*|} \;\middle|\; a_{kj}^* < 0, \quad j = 1, 2, \ldots, n'\right\}.$$

Notice that, contrary to the (primal) simplex algorithm given in Chapter 9.4, the pivot element in the dual simplex algorithm is always negative. The determination of the pivot column as given above guarantees that after each pivoting step, all coefficients g_j in the objective row remain greater than or equal to zero. In order to find the smallest quotient that defines the pivot column, we can add a Q row under the objective row (instead of the Q column in the primal simplex algorithm). The remaining transformation formulas are the same as described in Chapter 9.4, see step (6) of the simplex algorithm. The procedure stops when all right-hand sides are non-negative. The application of the dual simplex algorithm is particularly favourable when the given problem has the form

$$z = \mathbf{c}^{\mathrm{T}}\mathbf{x} \longrightarrow \min!$$

$$\text{s.t.} \quad A\mathbf{x} \geq \mathbf{b}$$

$$\mathbf{x} \geq \mathbf{0}.$$

We illustrate the dual simplex algorithm by the following example.

Example 9.16   Let the following LPP be given:

$$z = 2x_1 + 3x_2 + 3x_3 \longrightarrow \text{min!}$$

$$
\begin{array}{rcrcrcrcl}
\text{s.t.} & x_1 & + & x_2 & - & 3x_3 & \geq & 6 \\
 & -x_1 & + & 2x_2 & + & x_3 & \geq & 2 \\
 & 4x_1 & + & 3x_2 & - & 2x_3 & \geq & 3 \\
 & & & & x_1, x_2, x_3 & \geq & 0.
\end{array}
$$

We rewrite the problem as a maximization problem, multiply all constraints by -1 and introduce a slack variable in each constraint. This gives the following LPP:

$$\bar{z} = -z = -2x_1 - 3x_2 - 3x_3 \longrightarrow \text{max!}$$

$$
\begin{array}{rcrcrcrcrcrcl}
\text{s.t.} & -x_1 & - & x_2 & + & 3x_3 & + & x_4 & & & & & = & -6 \\
 & x_1 & - & 2x_2 & - & x_3 & & & + & x_5 & & & = & -2 \\
 & -4x_1 & - & 3x_2 & + & 2x_3 & & & & & + & x_6 & = & -3 \\
 & & & & & & & x_1, x_2, x_3, x_4, x_5, x_6 & \geq & 0.
\end{array}
$$

Now we can start with the dual simplex algorithm (notice that the application of the dual simplex algorithm does not now require the introduction of artificial variables as in the case of applying the primal simplex algorithm), and the initial tableau is as follows:

1	nbv	x_1	x_2	x_3	
bv	-1	-2	-3	-3	0
x_4	0	-1	-1	3	-6
x_5	0	1	-2	-1	-2
x_6	0	-4	-3	2	-3
		2	3	3	0
Q		2	3	$-$	

In the above tableau, all coefficients g_j in the objective row are greater than or equal to zero (i.e. the optimality criterion is satisfied), but the current basic solution is infeasible (each basic variable is negative), which can be interpreted such that the current objective function value is 'better' than the optimal one. Applying the dual simplex algorithm, we first determine the pivot row. Choosing the smallest right-hand side component, x_4 becomes the leaving variable, and based on the values in the quotient row, x_1 becomes the entering variable, which gives the pivot element -1. This leads to the following tableau:

2	nbv	x_4	x_2	x_3	
bv	-1	0	-3	-3	0
x_1	-2	-1	1	-3	6
x_5	0	1	-3	2	-8
x_6	0	-4	1	-10	21
		2	1	9	-12
Q		$-$	$\frac{1}{3}$	$-$	

Choosing now x_5 as leaving variable, the only possible entering variable is x_2 which gives the pivot element -3. We obtain the following tableau:

3	nbv	x_4	x_5	x_3
bv	-1	0	0	-3	0
x_1	-2	$-\frac{2}{3}$	$\frac{1}{3}$	$-\frac{7}{3}$	$\frac{10}{3}$
x_2	-3	$-\frac{1}{3}$	$-\frac{1}{3}$	$-\frac{2}{3}$	$\frac{8}{3}$
x_6	0	$-\frac{11}{3}$	$\frac{1}{3}$	$-\frac{28}{3}$	$\frac{55}{3}$
		$\frac{7}{3}$	$\frac{1}{3}$	$\frac{29}{3}$	$-\frac{44}{3}$
	Q				

Now, all right-hand sides are non-negative, and so the procedure stops. The optimal solution is as follows:

$$x_1 = \frac{10}{3}, \quad x_2 = \frac{8}{3}, \quad x_3 = x_4 = x_5 = 0, \quad x_6 = \frac{55}{3},$$

and the optimal function value is

$$z_0^{\max} = -\bar{z}_0^{\min} = \frac{44}{3}.$$

EXERCISES

9.1 Solve the following problems graphically:

(a) $z = x_1 + x_2 \rightarrow$ min!
s.t.
$$\begin{array}{rcrcl}
x_1 & + & x_2 & \geq & 2 \\
x_1 & - & x_2 & \geq & 3 \\
x_1 & + & 2x_2 & \leq & 6 \\
x_1 & + & 4x_2 & \geq & 0 \\
& & x_1, x_2 & \geq & 0
\end{array}$$

(b) $z = -x_1 + 4x_2 \rightarrow$ min!
s.t.
$$\begin{array}{rcrcl}
x_1 & - & 2x_2 & \leq & 4 \\
-x_1 & + & 2x_2 & \leq & 4 \\
x_1 & + & 2x_2 & \leq & 8 \\
& & x_1, x_2 & \geq & 0
\end{array}$$

(c) $z = -x_1 - 2x_2 \rightarrow$ min!
s.t.
$$\begin{array}{rcrcl}
-x_1 & + & x_2 & \leq & 4 \\
x_1 & + & 2x_2 & \leq & 11 \\
2x_1 & + & x_2 & \leq & 10 \\
x_1 & & & \leq & 4 \\
& & x_1, x_2 & \geq & 0
\end{array}$$

(d) $z = x_1 + x_2 \rightarrow$ max!
s.t.
$$\begin{array}{rcrcl}
x_1 & - & x_2 & \geq & 0 \\
-x_1 & - & 2x_2 & \leq & 4 \\
& & x_1, x_2 & \geq & 0
\end{array}$$

9.2 A craftsman has a free capacity of 200 working hours which he wants to use for making two products A and B. Production of one piece uses up 1 hour for A and 4 hours for B.

The number of pieces of product A is at most 100. The number of pieces of product B must be at least 30, but more than three times the amount of A is impossible. Finally, products A and B can be sold at prices of 20 EUR and 27 EUR but the variable costs incurred amount to 10 EUR and 21 EUR per piece of products A and B, respectively. What output combination should the craftsman choose in order to maximize the total profit? Find the solution of the problem graphically.

9.3 Find the standard forms of the following linear programming problems:

(a) $z = x_1 - 2x_2 + x_3 \rightarrow$ min!

$$
\begin{array}{rcccccc}
\text{s.t.} & x_1 & + & x_2 & + & x_3 & \leq & 7 \\
& 3x_1 & - & x_2 & + & x_3 & \geq & -4 \\
& & & x_1, x_2, x_3 & & & \geq & 0
\end{array}
$$

(b) $z = -x_1 + 2x_2 - 3x_3 + x_4 \rightarrow$ min!

$$
\begin{array}{rcccccccc}
\text{s.t.} & 2x_1 & + & 2x_2 & - & x_3 & + & 3x_4 & = & 8 \\
& x_1 & & & + & 2x_3 & + & x_4 & \leq & 10 \\
& -2x_1 & & & + & 2x_3 & - & 3x_4 & \geq & 0 \\
\end{array}
$$
$$x_1 \leq 0, \quad x_2, x_3 \geq 0, \quad x_4 \in \mathbb{R}$$

9.4 (a) Find the optimal solution of the following problem graphically and by the simplex method:

$$z = x_1 + x_2 \rightarrow \text{max!}$$
$$
\begin{array}{rcccc}
\text{s.t.} & 3x_1 & + & 2x_2 & \leq & 6 \\
& x_1 & + & 4x_2 & \leq & 4 \\
& & & x_1, x_2 & \geq & 0
\end{array}
$$

(b) Solve problem 9.1 (c) by the simplex method.

9.5 Solve the following problems by the simplex method:

(a) $z = 7x_1 + 4x_2 + 5x_3 + 6x_4 \rightarrow$ max!

$$
\begin{array}{rcccccccc}
\text{s.t.} & 20x_1 & + & 10x_2 & + & 12x_3 & + & 16x_4 & \leq & 400 \\
& & & & & x_3 & & & \leq & 5 \\
& x_1 & + & x_2 & + & x_3 & + & x_4 & \leq & 30 \\
& & & & & x_1, x_2, x_3, x_4 & & & \geq & 0
\end{array}
$$

(b) $z = 2x_1 - 6x_2 \rightarrow$ min!

$$
\begin{array}{rcccccc}
\text{s.t.} & 2x_1 & - & x_2 & & & \leq & 10 \\
& x_1 & - & 3x_2 & & & \leq & 15 \\
& 3x_1 & & & + & x_3 & = & 12 \\
& & & x_1, x_2, x_3 & & & \geq & 0
\end{array}
$$

9.6 Solve problem 9.2 by the two-phase simplex algorithm.

9.7 Solve the following linear programming problems by the simplex method:

(a) $z = x_1 + x_2 - x_3 \rightarrow$ min!

$$
\begin{array}{rcccccc}
\text{s.t.} & 3x_1 & - & x_2 & - & 4x_3 & \leq & 0 \\
& x_1 & + & 2x_2 & & & \geq & 10 \\
& & & x_2 & + & 3x_3 & \geq & 4 \\
& & & x_1, x_2, x_3 & & & \geq & 0
\end{array}
$$

(b) $z = x_1 + 2x_2 + x_3 \rightarrow$ max!

$$
\begin{aligned}
\text{s.t.} \quad x_1 &+ x_2 && &&\leq 10 \\
&\ \ x_2 &+ x_3 && &&\leq 14 \\
x_1 &&+ x_3 && &&\geq 15 \\
&x_1, x_2, x_3 && &&\geq 0
\end{aligned}
$$

(c) $z = 2x_1 - x_2 + x_3 \rightarrow$ max!

$$
\begin{aligned}
\text{s.t.} \quad x_1 &+ x_2 &-\ x_3 &-\ x_4 &&= 4 \\
2x_1 &+ 3x_2 &-\ x_3 &-\ 2x_4 &&= 9 \\
&\ \ x_2 &+ 2x_3 && &&\geq 3 \\
&x_1, x_2, x_3, x_4 && &&\geq 0
\end{aligned}
$$

9.8 Formulate the dual problems of the three primal problems given in exercises 9.5 (a), 9.7 (a) and 9.7 (c).

9.9 Find the dual problems and both the primal and dual optimal solutions:

(a) $z = 2x_1 - 2x_2 - x_3 - x_4 \rightarrow$ min!

$$
\begin{aligned}
\text{s.t.} \quad -x_1 &+ 2x_2 &+ x_3 &-\ x_4 &&\leq 1 \\
x_1 &-\ x_2 && &+ x_4 &&\leq 4 \\
x_1 &+ 4x_2 && &-\ 2x_4 &&\leq 1 \\
&x_1, x_2, x_3, x_4 && &&\geq 0
\end{aligned}
$$

(b) $z = 3x_1 - x_2 + x_3 + x_4 \rightarrow$ max!

$$
\begin{aligned}
\text{s.t.} \quad 2x_1 &+ x_2 &-\ 3x_3 &-\ x_4 &&\leq 4 \\
-x_1 &+ x_2 &+ 3x_3 &-\ 2x_4 &&\leq 4 \\
-x_1 && &+ 2x_3 &+ x_4 &&\leq 4 \\
&x_1, x_2, x_3, x_4 && &&\geq 0
\end{aligned}
$$

9.10 A firm produces by means of three raw materials R_1, R_2 and R_3 two different products P_1 and P_2. Profits, raw material requirements and capacities are given in the following table:

	P_1	P_2	capacity
R_1	2	4	16
R_2	2	1	10
R_3	4	0	20
profit per unit P_i	2	3	

Use the method of complementary slackness to decide whether a production of four units of P_1 and two units of P_2 is optimal.

9.11 A publicity agency needs at least 100 paper strips that are 1 m long and 5 cm wide, 200 strips that are 1 m long and 3 cm wide and 400 strips that are 1 m long and 2 cm wide. The necessary strips can be cut from 1 m long and 10 cm wide strips.

(a) Find all cutting variants without waste strips.
(b) Formulate a linear programming problem, where the number of 10 cm wide strips needed is minimized. Denote by x_i the number of 10 cm wide strips that are cut by variant i.
(c) Solve the problem by the dual simplex algorithm.

10 Eigenvalue problems and quadratic forms

In this chapter, we deal with an application of homogeneous systems of linear equations, namely eigenvalue problems. Moreover, we introduce so-called quadratic forms and investigate their sign. Quadratic forms play an important role when determining extreme points and values of functions of several variables, which are discussed in Chapter 11.

10.1 EIGENVALUES AND EIGENVECTORS

An eigenvalue problem can be defined as follows.

> **Definition 10.1** Let A be an $n \times n$ matrix. Then the scalar λ is called an *eigenvalue* of matrix A if there exists a non-trivial solution $\mathbf{x} \in \mathbb{R}^n, \mathbf{x} \neq \mathbf{0}$, of the matrix equation
>
> $$A\mathbf{x} = \lambda\mathbf{x}. \tag{10.1}$$
>
> The solution $\mathbf{x}^{\mathrm{T}} = (x_1, x_2, \ldots, x_n) \neq (0, 0, \ldots, 0)$ is called an *eigenvector* of A (associated with scalar λ).

Equation (10.1) is equivalent to a homogeneous linear system of equations. From the matrix equation

$$A\mathbf{x} = \lambda\mathbf{x} = \lambda I \mathbf{x}$$

we obtain

$$(A - \lambda I)\mathbf{x} = \mathbf{0}, \tag{10.2}$$

where I is the identity matrix of order $n \times n$. Hence, system (10.2) includes n linear equations with n variables $x_1, x_2, \ldots, x_n$.

According to Definition 10.1, eigenvalue problems are defined only for *square* matrices A. In an eigenvalue problem, we look for all (real or complex) values λ such that the image of a non-zero vector $\mathbf{x}$ given by a linear mapping described by matrix A is a multiple $\lambda\mathbf{x}$ of this vector $\mathbf{x}$. It is worth noting that value zero is possible as an eigenvalue, while the zero vector is not possible as an eigenvector. Although eigenvalue problems arise mainly in engineering sciences, they also have some importance in economics. For example, as we show in the following two chapters, they are useful for deciding whether a function has an extreme point or for solving certain types of differential and difference equations.

The following theorem gives a necessary and sufficient condition for the existence of non-trivial (i.e. different from the zero vector) solutions of problem (10.1).

THEOREM 10.1 Problem (10.1) has a non-trivial solution $x \neq 0$ if and only if the determinant of matrix $A - \lambda I$ is equal to zero, i.e. $|A - \lambda I| = 0$.

The validity of the above theorem can easily be seen by taking into account that a homogeneous system (10.2) of linear equations has non-trivial solutions if and only if the rank of the coefficient matrix of the system (i.e. the rank of matrix $A - \lambda I$) is less than the number of variables. The latter condition is equivalent to the condition that the determinant of the coefficient matrix is equal to zero, which means that matrix $A - \lambda I$ has no inverse matrix.

We rewrite the determinant of matrix $A - \lambda I$ as a function of the variable λ. Letting $P(\lambda) = |A - \lambda I|$, where $A = (a_{ij})$ is a matrix of order $n \times n$, we get the following equation in λ:

$$
P(\lambda) = \begin{vmatrix}
a_{11} - \lambda & a_{12} & \cdots & a_{1n} \\
a_{21} & a_{22} - \lambda & \cdots & a_{2n} \\
\vdots & \vdots & & \vdots \\
a_{n1} & a_{n2} & \cdots & a_{nn} - \lambda
\end{vmatrix} = 0,
$$

which is known as the *characteristic equation* (or *eigenvalue equation*) of matrix A.

From the definition of a determinant, it follows that $P(\lambda)$ is a polynomial in λ which has degree n for a matrix A of order $n \times n$. The zeroes of this *characteristic polynomial*

$$
P(\lambda) = (-1)^n \lambda^n + b_{n-1} \lambda^{n-1} + \cdots + b_1 \lambda + b_0
$$

of degree n are the eigenvalues of matrix A. Thus, in order to find all eigenvalues of an $n \times n$ matrix A, we have to determine all zeroes of polynomial $P(\lambda)$ of degree n (i.e. all roots of the characteristic equation $P(\lambda) = 0$). Here we often have to apply numerical methods (described in Chapter 4) to find them (approximately). In general, the eigenvalues of a real matrix A can be complex numbers (and also the eigenvectors may contain complex components). The following theorem describes a case when all eigenvalues of matrix A are real.

THEOREM 10.2 If matrix A of order $n \times n$ is a symmetric matrix (i.e. $A = A^{\mathsf{T}}$), then all eigenvalues of A are real numbers.

For each eigenvalue $\lambda_i, i = 1, 2, \ldots, n$, we have to find the general solution of the homogeneous system of linear equations

$$
(A - \lambda_i I)x = 0 \tag{10.3}
$$

in order to get the corresponding eigenvectors. Since the rank of matrix A is smaller than n, the solution of the corresponding system of equations is not uniquely determined and for each eigenvalue λ_i, $i = 1, 2, \ldots, n$, of system (10.3), there is indeed a solution where not all variables are equal to zero.

We continue with some properties of the set of eigenvectors belonging to the *same* eigenvalue.

THEOREM 10.3 Let λ be an eigenvalue of multiplicity k (i.e. λ is k times a root of the characteristic equation $P(\lambda) = 0$) of a matrix A of order $n \times n$. Then:

(1) The number of linearly independent eigenvectors associated with eigenvalue λ is at least one and at most k.
(2) If A is a symmetric matrix, then there exist k linearly independent eigenvectors associated with λ. The set of linearly independent eigenvectors associated with eigenvalue λ forms a vector space.

As a consequence of Theorem 10.3, we mention that, if x^1 and x^2 are eigenvectors associated with eigenvalue λ, then also vector $sx^1 + tx^2$ with $s \in \mathbb{R}$ and $t \in \mathbb{R}$ is an eigenvector associated with λ. It also follows from Theorem 10.3 that for an arbitrary square matrix A there always exists exactly one linearly independent eigenvector associated with an eigenvalue of multiplicity one.

THEOREM 10.4 Let A be a matrix of order $n \times n$. Then:

(1) Eigenvectors associated with different eigenvalues of matrix A are linearly independent.
(2) If matrix A is symmetric, then eigenvectors associated with different eigenvalues are orthogonal.

Let us consider the following three examples to determine all eigenvalues and eigenvectors of a given matrix.

Example 10.1 We determine the eigenvalues and eigenvectors of matrix

$$A = \begin{pmatrix} 1 & 2 \\ 2 & 1 \end{pmatrix}.$$

The characteristic equation is given by

$$P(\lambda) = |A - \lambda I| = \begin{vmatrix} 1 - \lambda & 2 \\ 2 & 1 - \lambda \end{vmatrix} = (1 - \lambda)(1 - \lambda) - 4 = \lambda^2 - 2\lambda - 3 = 0$$

with the solutions

$$\lambda_1 = 1 + \sqrt{1 + 3} = 3 \qquad \text{and} \qquad \lambda_2 = 1 - \sqrt{1 + 3} = -1.$$

To determine the corresponding eigenvectors, we have to solve the matrix equation $(A - \lambda I)x = 0$ for $\lambda = \lambda_1$ and $\lambda = \lambda_2$. Thus, we get the following system for $\lambda = \lambda_1 = 3$:

$$-2x_1 + 2x_2 = 0$$
$$2x_1 - 2x_2 = 0.$$

The second equation may be obtained from the first equation by multiplying by -1 (i.e. both row vectors of the left-hand side are linearly dependent) and can therefore be dropped. The coefficient matrix of the above system has rank one and so we can choose one variable arbitrarily, say $x_2 = t, t \in \mathbb{R}$. This yields $x_1 = t$, and each eigenvector associated with the eigenvalue $\lambda_1 = 3$ can be described in the form

$$\mathbf{x}^1 = t \begin{pmatrix} 1 \\ 1 \end{pmatrix}, \qquad t \in \mathbb{R}.$$

Analogously, for $\lambda = \lambda_2 = -1$, we get the system:

$$2x_1 + 2x_2 = 0$$
$$2x_1 + 2x_2 = 0.$$

Again, we can drop one of the two identical equations and can choose one variable arbitrarily, say $x_2 = s, s \in \mathbb{R}$. Then we get $x_1 = -s$. Thus, all eigenvectors associated with $\lambda_2 = -1$ can be represented as

$$\mathbf{x}^2 = s \begin{pmatrix} -1 \\ 1 \end{pmatrix}, \qquad s \in \mathbb{R}.$$

This example illustrates Theorem 10.4. For arbitrary choice of $s, t \in \mathbb{R}$, the eigenvectors $\mathbf{x}^1$ and $\mathbf{x}^2$ are linearly independent and orthogonal (i.e. the scalar product of vectors $\mathbf{x}^1$ and $\mathbf{x}^2$ is equal to zero).

Example 10.2   We determine all eigenvalues and eigenvectors of matrix

$$A = \begin{pmatrix} 0 & -1 & 1 \\ -7 & 0 & 5 \\ -5 & -2 & 5 \end{pmatrix}.$$

To find the eigenvalues, we consider the characteristic equation $P(\lambda) = 0$:

$$P(\lambda) = |A - \lambda I| = \begin{vmatrix} -\lambda & -1 & 1 \\ -7 & -\lambda & 5 \\ -5 & -2 & 5 - \lambda \end{vmatrix} = \begin{vmatrix} -\lambda & -1 & 0 \\ -7 & -\lambda & 5 - \lambda \\ -5 & -2 & 3 - \lambda \end{vmatrix} = 0.$$

The above transformation is obtained by adding columns 2 and 3 to get the third element equal to zero in row 1. Expanding the latter determinant by row 1, we obtain

$$P(\lambda) = |A - \lambda I| = -\lambda \cdot \begin{vmatrix} -\lambda & 5 - \lambda \\ -2 & 3 - \lambda \end{vmatrix} + 1 \cdot \begin{vmatrix} -7 & 5 - \lambda \\ -5 & 3 - \lambda \end{vmatrix}$$

$$= \lambda^2 (3 - \lambda) - 2\lambda(5 - \lambda) + [(-7) \cdot (3 - \lambda) + 5(5 - \lambda)]$$

$$= (3\lambda^2 - \lambda^3 - 10\lambda + 2\lambda^2) + (-21 + 7\lambda + 25 - 5\lambda)$$

$$= -\lambda^3 + 5\lambda^2 - 8\lambda + 4.$$

Considering now the characteristic equation $P(\lambda) = 0$, we try to find a first root and use Horner's scheme (see Chapter 3.3.3) for the computation of the function value.

Checking $\lambda_1 = 1$, we get

$$
\lambda_1 = 1 \quad
\begin{array}{r|rrrr}
& -1 & 5 & -8 & 4 \\
& & -1 & 4 & -4 \\
\hline
& -1 & 4 & -4 & \underline{0}
\end{array}
$$

i.e. $\lambda_1 = 1$ is a root of the characteristic equation $P(\lambda) = 0$. From Horner's scheme (see last row), we obtain that dividing $P(\lambda)$ by the linear factor $\lambda - \lambda_1 = \lambda - 1$ gives the polynomial $P_2(\lambda)$ of degree two:

$$
P_2(\lambda) = -\lambda^2 + 4\lambda - 4.
$$

Setting $P_2(\lambda) = 0$, we obtain

$$
\lambda_2 = 2 + \sqrt{4-4} = 2 \qquad \text{and} \qquad \lambda_3 = 2 - \sqrt{4-4} = 2,
$$

i.e. $\lambda_2 = \lambda_3 = 2$ is an eigenvalue of multiplicity two.

In order to determine the eigenvectors associated with $\lambda_1 = 1$, we get the homogeneous system of linear equations

$$
\begin{array}{rcrcrcl}
-x_1 & - & x_2 & + & x_3 & = & 0 \\
-7x_1 & - & x_2 & + & 5x_3 & = & 0 \\
-5x_1 & - & 2x_2 & + & 4x_3 & = & 0.
\end{array}
$$

Applying Gaussian elimination, we get the following tableaus:

Row	x_1	x_2	x_3	b	Operation
1	-1	-1	1	0	
2	-7	-1	5	0	
3	-5	-2	4	0	
4	-1	-1	1	0	row 1
5	0	6	-2	0	row 2 $-$ 7 row 1
6	0	3	-1	0	row 3 $-$ 5 row 1
7	-1	-1	1	0	row 4
8	0	6	-2	0	row 5
9	0	0	0	0	row 6 $-\frac{1}{2}$ row 5

Since the rank of the coefficient matrix is equal to two, we can choose one variable arbitrarily. Setting $x_3 = 3$, we get $x_2 = 1$ and $x_1 = 2$ (here we set $x_3 = 3$ in order to get integer solutions for the other two variables), i.e. each eigenvector associated with $\lambda_1 = 1$ has the form

$$
\mathbf{x}^1 = s \begin{pmatrix} 2 \\ 1 \\ 3 \end{pmatrix}, \qquad s \in \mathbb{R}.
$$

Considering the eigenvalue $\lambda_2 = \lambda_3 = 2$, we get the following system of linear equations:

$$
\begin{array}{rrrrrl}
-2x_1 & - & x_2 & + & x_3 & = & 0 \\
-7x_1 & - & 2x_2 & + & 5x_3 & = & 0 \\
-5x_1 & - & 2x_2 & + & 3x_3 & = & 0.
\end{array}
$$

After applying Gaussian elimination or pivoting, we find that the coefficient matrix of this system of linear equations has rank two. Hence we can choose one variable arbitrarily. If we choose $x_3 = 1$, we get $x_2 = -1$ and $x_1 = 1$. Therefore, each eigenvector associated with eigenvalue $\lambda_2 = \lambda_3 = 2$ has the form

$$
\mathbf{x}^2 = t \begin{pmatrix} 1 \\ -1 \\ 1 \end{pmatrix}, \qquad t \in \mathbb{R}.
$$

In particular, for the latter eigenvalue of multiplicity two, there exists only one linearly independent eigenvector.

Example 10.3 We determine the eigenvalues and eigenvectors of matrix

$$
A = \begin{pmatrix} -4 & -3 & 3 \\ 2 & 3 & -6 \\ -1 & -3 & 0 \end{pmatrix}.
$$

The characteristic equation is given by

$$
\begin{aligned}
P(\lambda) = |A - \lambda I| &= \begin{vmatrix} -4 - \lambda & -3 & 3 \\ 2 & 3 - \lambda & -6 \\ -1 & -3 & -\lambda \end{vmatrix} \\
&= \begin{vmatrix} -4 - \lambda & 0 & 3 \\ 2 & -3 - \lambda & -6 \\ -1 & -3 - \lambda & -\lambda \end{vmatrix} \\
&= \begin{vmatrix} -4 - \lambda & 0 & 3 \\ 3 & 0 & -6 + \lambda \\ -1 & -3 - \lambda & -\lambda \end{vmatrix} \\
&= (-3 - \lambda) \cdot \begin{vmatrix} -4 - \lambda & 0 & 3 \\ 3 & 0 & -6 + \lambda \\ -1 & 1 & -\lambda \end{vmatrix} \\
&= -(-3 - \lambda) \cdot \begin{vmatrix} -4 - \lambda & 3 \\ 3 & -6 + \lambda \end{vmatrix} \\
&= -(-3 - \lambda) \cdot [(-4 - \lambda)(-6 + \lambda) - 9] \\
&= (3 + \lambda) \cdot (-\lambda^2 + 2\lambda + 15) = 0.
\end{aligned}
$$

In the transformations above, we have first added column 3 to column 2, and then we have added row 3 multiplied by -1 to row 2. In the next step, the term $(-3 - \lambda)$ has been factored out from the second column and finally, the resulting determinant has been expanded by column 2. From equation $3 + \lambda = 0$, we obtain the first eigenvalue

$$\lambda_1 = -3,$$

and from equation $-\lambda^2 + 2\lambda + 15 = 0$, we obtain the two eigenvalues

$$\lambda_2 = -3 \qquad \text{and} \qquad \lambda_3 = 5.$$

Next, we determine a maximum number of linearly independent eigenvectors for each eigenvalue. We first consider the eigenvalue $\lambda_1 = \lambda_2 = -3$ of multiplicity two and obtain the following system of linear equations:

$$-x_1 - 3x_2 + 3x_3 = 0$$
$$2x_1 + 6x_2 - 6x_3 = 0$$
$$-x_1 - 3x_2 + 3x_3 = 0.$$

Since equations one and three coincide and since equation two corresponds to equation one multiplied by -2, the rank of the coefficient matrix of the above system is equal to one, and therefore we can choose two variables arbitrarily. Consequently, there exist two linearly independent eigenvectors associated with this eigenvalue. Using our knowledge about the general solution of homogeneous systems of linear equations, we get linearly independent solutions by choosing for the first vector $\mathbf{x}^1$: $x_2^1 = 1, x_3^1 = 0$ and for the second vector $\mathbf{x}^2$: $x_2^2 = 0, x_3^2 = 1$ (i.e. we have taken x_2 and x_3 as the variables that can be chosen arbitrarily). Then the remaining variables are uniquely determined and we obtain e.g. from the first equation of the above system: $x_1^1 = -3, x_1^2 = 3$. Therefore, the set of all eigenvectors associated with $\lambda_2 = \lambda_3 = -3$ is given by

$$\left\{ \mathbf{x} \in \mathbb{R}^3 \ \middle| \ \mathbf{x} = s \begin{pmatrix} -3 \\ 1 \\ 0 \end{pmatrix} + t \begin{pmatrix} 3 \\ 0 \\ 1 \end{pmatrix}, \ s \in \mathbb{R}, \ t \in \mathbb{R} \right\}.$$

While in Example 10.2 only one linearly independent vector was associated with the eigenvalue of multiplicity two, in this example there are two linearly independent eigenvectors associated with an eigenvalue of multiplicity two. This is the maximal possible number since we know from Theorem 10.3 that at most k linearly independent eigenvectors are associated with an eigenvalue of multiplicity k. To finish this example, we still have to find the eigenvalue associated with $\lambda_3 = 5$ by the solution of the following system of linear equations:

$$-9x_1 - 3x_2 + 3x_3 = 0$$
$$2x_1 - 2x_2 - 6x_3 = 0$$
$$-x_1 - 3x_2 - 5x_3 = 0.$$

By applying Gaussian elimination or pivoting, we find that the coefficient matrix has rank two, and therefore one variable can be chosen arbitrarily. Choosing $x_3 = 1$, we finally get

$$x_2 = -2x_3 = -2 \qquad \text{and} \qquad x_1 = -3x_2 - 5x_3 = 1.$$

Therefore, an eigenvector associated with $\lambda_3 = 5$ can be written in the form

$$\mathbf{x}^3 = u \begin{pmatrix} 1 \\ -2 \\ 1 \end{pmatrix}, \quad u \in \mathbb{R}.$$

In the next chapters, we show how eigenvalues can be used for solving certain optimization problems as well as differential and difference equations. The problem of determining eigenvalues and the corresponding eigenvectors often arises in economic problems dealing with processes of proportionate growth or decline. We demonstrate this by the following example.

Example 10.4 Let x_t^M be the number of men and x_t^W the number of women in some population at time t. The relationship between the populations at some successive times t and $t + 1$ has been found to be as follows:

$$x_{t+1}^M = 0.8x_t^M + 0.4x_t^W$$
$$x_{t+1}^W = 0.3x_t^M + 0.9x_t^W.$$

Letting $\mathbf{x}^t = (x_t^M, x_t^W)^T$, we obtain the following relationship between the populations $\mathbf{x}^{t+1}$ and $\mathbf{x}^t$ at successive times:

$$\begin{pmatrix} x_{t+1}^M \\ x_{t+1}^W \end{pmatrix} = \begin{pmatrix} 0.8 & 0.4 \\ 0.3 & 0.9 \end{pmatrix} \begin{pmatrix} x_t^M \\ x_t^W \end{pmatrix}.$$

Moreover, we assume that the ratio of men and women is constant over time, i.e.

$$\begin{pmatrix} x_{t+1}^M \\ x_{t+1}^W \end{pmatrix} = \lambda \begin{pmatrix} x_t^M \\ x_t^W \end{pmatrix}.$$

Now, the question is: do there exist such values $\lambda \in \mathbb{R}_+$ and vectors $\mathbf{x}^t$ satisfying the above equations, i.e. can we find numbers λ and vectors $\mathbf{x}^t$ such that

$$A\mathbf{x}^t = \lambda\mathbf{x}^t \ ?$$

To answer this question, we have to find the eigenvalues of matrix

$$A = \begin{pmatrix} 0.8 & 0.4 \\ 0.3 & 0.9 \end{pmatrix}$$

and then for an appropriate eigenvalue the corresponding eigenvector. We obtain the characteristic equation

$$P(\lambda) = |A - \lambda I| = \begin{vmatrix} 0.8 - \lambda & 0.4 \\ 0.3 & 0.9 - \lambda \end{vmatrix} = (0.8 - \lambda)(0.9 - \lambda) - 0.3 \cdot 0.4 = 0.$$

This yields

$$P(\lambda) = \lambda^2 - 1.7\lambda + 0.6 = 0$$

and the eigenvalues

$$\lambda_1 = 0.85 + \sqrt{0.7225 - 0.6} = 0.85 + 0.35 = 1.2 \qquad \text{and} \qquad \lambda_2 = 0.85 - 0.35 = 0.5.$$

Since we are looking for a proportionate growth in the population, only the eigenvalue greater than one is of interest, i.e. we have to consider $\lambda_1 = 1.2$ and determine the corresponding eigenvector from the system

$$\begin{array}{rcl} -0.4x_t^M & + \quad 0.4x_t^W & = 0 \\ 0.3x_t^M & - \quad 0.3x_t^W & = 0. \end{array}$$

The coefficient matrix has rank one, we can choose one variable arbitrarily and get the eigenvector

$$x^t = u \begin{pmatrix} 1 \\ 1 \end{pmatrix}, \quad u \in \mathbb{R}.$$

In order to have a proportionate growth in the population, the initial population must consist of the same number of men and women, and the population grows by 20 per cent until the next time it is considered. This means that if initially at time $t = 1$, the population is given by vector $x^1 = (1,000; \ 1,000)^T$, then at time $t = 2$ the population is given by $x^2 = (1,200; \ 1,200)^T$, at time $t = 3$ the population is given by $x^3 = (1,440; \ 1,440)^T$, and so on.

10.2   QUADRATIC FORMS AND THEIR SIGN

We start with the following definition.

Definition 10.2   If $A = (a_{ij})$ is a matrix of order $n \times n$ and $x^T = (x_1, x_2, \ldots, x_n)$, then the term

$$Q(x) = x^T A x \tag{10.4}$$

is called a *quadratic form*.

Writing equation (10.4) explicitly, we have:

$$Q(\mathbf{x}) = Q(x_1, x_2, \ldots, x_n) = (x_1, x_2, \ldots, x_n) \begin{pmatrix} a_{11} & a_{12} & \cdots & a_{1n} \\ a_{21} & a_{22} & \cdots & a_{2n} \\ \vdots & \vdots & & \vdots \\ a_{n1} & a_{n2} & \cdots & a_{nn} \end{pmatrix} \begin{pmatrix} x_1 \\ x_2 \\ \vdots \\ x_n \end{pmatrix}$$

$$= a_{11}x_1x_1 + a_{12}x_1x_2 + \cdots + a_{1n}x_1x_n + a_{21}x_2x_1 + a_{22}x_2x_2 + \cdots + a_{2n}x_2x_n + \cdots$$

$$+ a_{n1}x_nx_1 + a_{n2}x_nx_2 + \cdots + a_{nn}x_nx_n$$

$$= \sum_{i=1}^{n}\sum_{j=1}^{n} a_{ij}x_ix_j.$$

THEOREM 10.5   Let A be a matrix of order $n \times n$. Then the quadratic form $\mathbf{x}^{\mathrm{T}}A\mathbf{x}$ can be written as a quadratic form $\mathbf{x}^{\mathrm{T}}A^{*}\mathbf{x}$ of a symmetric matrix A^{*} of order $n \times n$, i.e. we have $\mathbf{x}^{\mathrm{T}}A\mathbf{x} = \mathbf{x}^{\mathrm{T}}A^{*}\mathbf{x}$, where

$$A^{*} = \frac{1}{2} \cdot \left(A + A^{\mathrm{T}}\right).$$

As a result of Theorem 10.5, we can restrict ourselves to the consideration of quadratic forms of symmetric matrices, where all eigenvalues are real numbers (see Theorem 10.2). In the following definition, the sign of a quadratic form $\mathbf{x}^{\mathrm{T}}A\mathbf{x}$ is considered.

Definition 10.3   A square matrix A of order $n \times n$ and its associated quadratic form $Q(\mathbf{x})$ are said to be

(1) *positive definite* if $Q(\mathbf{x}) = \mathbf{x}^{\mathrm{T}}A\mathbf{x} > 0$ for all $\mathbf{x}^{\mathrm{T}} = (x_1, x_2, \ldots, x_n) \neq (0, 0, \ldots, 0)$;
(2) *positive semi-definite* if $Q(\mathbf{x}) = \mathbf{x}^{\mathrm{T}}A\mathbf{x} \geq 0$ for all $\mathbf{x} \in \mathbb{R}^{n}$;
(3) *negative definite* if $Q(\mathbf{x}) = \mathbf{x}^{\mathrm{T}}A\mathbf{x} < 0$ for all $\mathbf{x}^{\mathrm{T}} = (x_1, x_2, \ldots, x_n) \neq (0, 0, \ldots, 0)$;
(4) *negative semi-definite* if $Q(\mathbf{x}) = \mathbf{x}^{\mathrm{T}}A\mathbf{x} \leq 0$ for all $\mathbf{x} \in \mathbb{R}^{n}$;
(5) *indefinite* if they are neither positive semi-definite nor negative semi-definite.

The following example illustrates Definition 10.3.

Example 10.5   Let

$$A = \begin{pmatrix} 1 & -1 \\ -1 & 1 \end{pmatrix}.$$

We determine the sign of the quadratic form $Q(\mathbf{x}) = \mathbf{x}^{\mathrm{T}}A\mathbf{x}$ by applying Definition 10.3. Then

$$Q(\mathbf{x}) = \mathbf{x}^{\mathrm{T}}A\mathbf{x} = (x_1, x_2) \begin{pmatrix} 1 & -1 \\ -1 & 1 \end{pmatrix} \begin{pmatrix} x_1 \\ x_2 \end{pmatrix}$$

$$= x_1(x_1 - x_2) + x_2(-x_1 + x_2)$$

$$= x_1(x_1 - x_2) - x_2(x_1 - x_2)$$

$$= (x_1 - x_2)^2 \geq 0.$$

Therefore, matrix A is positive semi-definite. However, matrix A is not positive definite since there exist vectors $\mathbf{x}^T = (x_1, x_2) \neq (0, 0)$ such that $Q(\mathbf{x}) = \mathbf{x}^T A \mathbf{x} = 0$, namely if $x_1 = x_2$ but is different from zero.

The following theorem shows how we can decide by means of the eigenvalues of a symmetric matrix whether the matrix is positive or negative (semi-)definite.

THEOREM 10.6 Let A be a symmetric matrix of order $n \times n$ with the eigenvalues $\lambda_1, \lambda_2, \ldots, \lambda_n \in \mathbb{R}$. Then:

(1) A is positive definite if and only if all eigenvalues of A are positive (i.e. $\lambda_i > 0$ for $i = 1, 2, \ldots, n$).
(2) A is positive semi-definite if and only if all eigenvalues of A are non-negative (i.e. $\lambda_i \geq 0$ for $i = 1, 2, \ldots, n$).
(3) A is negative definite if and only if all eigenvalues of A are negative (i.e. $\lambda_i < 0$ for $i = 1, 2, \ldots, n$).
(4) A is negative semi-definite if and only if all eigenvalues of A are non-positive (i.e. $\lambda_i \leq 0$ for $i = 1, 2, \ldots, n$).
(5) A is indefinite if and only if A has at least two eigenvalues with opposite signs.

Example 10.6 Let us consider matrix A with

$$A = \begin{pmatrix} 0 & 2 \\ 1 & -1 \end{pmatrix}.$$

We determine the eigenvalues of A and obtain the characteristic equation

$$P(\lambda) = |A - \lambda I| = \begin{vmatrix} -\lambda & 2 \\ 1 & -1 - \lambda \end{vmatrix} = 0.$$

This yields

$$-\lambda(-1 - \lambda) - 2 = \lambda^2 + \lambda - 2 = 0.$$

The above quadratic equation has the solutions

$$\lambda_1 = -\frac{1}{2} + \sqrt{\frac{1}{4} + 2} = -\frac{1}{2} + \frac{3}{2} = 1 > 0 \qquad \text{and}$$

$$\lambda_2 = -\frac{1}{2} - \sqrt{\frac{1}{4} + 2} = -\frac{1}{2} - \frac{3}{2} = -2 < 0.$$

Since both eigenvalues have opposite signs, matrix A is indefinite according to part (5) of Theorem 10.6.

Next, we present another criterion to decide whether a given matrix A is positive or negative definite. To apply this criterion, we have to investigate the sign of certain minors introduced in the following definition.

Definition 10.4 The *leading principal minors* of matrix $A = (a_{ij})$ of order $n \times n$ are the determinants

$$D_k = \begin{vmatrix} a_{11} & a_{12} & \cdots & a_{1k} \\ a_{21} & a_{22} & \cdots & a_{2k} \\ \vdots & \vdots & & \vdots \\ a_{k1} & a_{k2} & \cdots & a_{kk} \end{vmatrix}, \quad k = 1, 2, \ldots, n,$$

i.e. D_k is obtained from $|A|$ by crossing out the last $n - k$ columns and rows.

By means of the leading principal minors, we can give a criterion to decide whether a matrix A is positive or negative definite.

THEOREM 10.7 Let matrix A be a symmetric matrix of order $n \times n$ with the leading principal minors $D_k, k = 1, 2, \ldots, n$. Then:

(1) A is positive definite if and only if $D_k > 0$ for $k = 1, 2, \ldots, n$.
(2) A is negative definite if and only if $(-1)^k \cdot D_k > 0$ for $k = 1, 2, \ldots, n$.

For a symmetric matrix A of order $n \times n$, we have to check the sign of n determinants to find out whether A is positive (negative) definite. If all leading principal minors are greater than zero, matrix A is positive definite according to part (1) of Theorem 10.7. If the signs of the n leading principal minors alternate, where the first minor is negative (i.e. element a_{11} is smaller than zero), then matrix A is negative definite according to part (2) of Theorem 10.7. The following two examples illustrate the use of Theorem 10.7.

Example 10.7 Let

$$A = \begin{pmatrix} -3 & 2 & 0 \\ 2 & -3 & 0 \\ 0 & 0 & -5 \end{pmatrix}.$$

For matrix A, we get the leading principal minors

$$D_1 = a_{11} = -3 < 0,$$

$$D_2 = \begin{vmatrix} a_{11} & a_{12} \\ a_{21} & a_{22} \end{vmatrix} = \begin{vmatrix} -3 & 2 \\ 2 & -3 \end{vmatrix} = 9 - 4 = 5 > 0,$$

$$D_3 = \begin{vmatrix} a_{11} & a_{12} & a_{13} \\ a_{21} & a_{22} & a_{23} \\ a_{31} & a_{32} & a_{33} \end{vmatrix} = \begin{vmatrix} -3 & 2 & 0 \\ 2 & -3 & 0 \\ 0 & 0 & -5 \end{vmatrix} = -5D_2 = -25 < 0.$$

Since the leading principal minors $D_k, k \in \{1, 2, 3\}$, alternate in sign, starting with a negative sign of D_1, matrix A is negative definite according to part (2) of Theorem 10.7.

Example 10.8   We check for which values $a \in \mathbb{R}$ the matrix

$$A = \begin{pmatrix} 2 & 1 & 0 & 0 \\ 1 & a & 1 & 0 \\ 0 & 1 & 3 & 1 \\ 0 & 0 & 1 & 2 \end{pmatrix}$$

is positive definite. We apply Theorem 10.7 and investigate for which values of a all leading principal minors are greater than zero. We obtain:

$$D_1 = 2 > 0,$$

$$D_2 = \begin{vmatrix} 2 & 1 \\ 1 & a \end{vmatrix} = 2a - 1 > 0.$$

The latter inequality holds for $a > 1/2$. Next, we obtain

$$D_3 = \begin{vmatrix} 2 & 1 & 0 \\ 1 & a & 1 \\ 0 & 1 & 3 \end{vmatrix} = 6a - 2 - 3 > 0,$$

which holds for $a > 5/6$. For calculating $D_4 = |A|$, we can expand $|A|$ by row 4:

$$D_4 = |A| = - \begin{vmatrix} 2 & 1 & 0 \\ 1 & a & 0 \\ 0 & 1 & 1 \end{vmatrix} + 2D_3 = -(2a - 1) + 12a - 10 = 10a - 9 > 0,$$

which holds for $a > 9/10$. Due to $1/2 < 5/6 < 9/10$, and since all leading principal minors must be positive, matrix A is positive definite for $a > 9/10$.

THEOREM 10.8   Let A be a symmetric matrix of order $n \times n$. Then:

(1) If matrix A is positive semi-definite, then each leading principal minor D_k, $k = 1, 2, \ldots, n$, is non-negative.

(2) If matrix A is negative semi-definite, then each leading principal minor D_k is either zero or has the same sign as $(-1)^k$, $k = 1, 2, \ldots, n$.

It is worth noting that Theorem 10.8 is only a necessary condition for a positive (negative) semi-definite matrix A. If all leading principal minors of a matrix A are non-negative, we cannot conclude that this matrix must be positive semi-definite. We only note that, in order to get a sufficient condition for a positive (negative) semi-definite matrix, we have to check *all* minors of matrix A and they have to satisfy the conditions of Theorem 10.8 concerning their signs.

EXERCISES

10.1 Given are the matrices

$$A = \begin{pmatrix} 2 & 1 \\ 4 & -1 \end{pmatrix}; \quad B = \begin{pmatrix} 2 & 1 \\ -2 & 4 \end{pmatrix};$$

$$C = \begin{pmatrix} -3 & -2 & 4 \\ -3 & 2 & 3 \\ -2 & -2 & 3 \end{pmatrix} \quad \text{and} \quad D = \begin{pmatrix} 2 & 0 & 0 \\ 2 & 1 & -2 \\ -1 & 0 & 2 \end{pmatrix}.$$

Find the eigenvalues and the eigenvectors of each of these matrices.

10.2 Let x_t be the consumption value of a national economy in period t and y_t the capital investment of the economy in this period. For the following period $t + 1$, we have

$$x_{t+1} = 0.7x_t + 0.6y_t$$

which describes the change in consumption from one period to the subsequent one depending on consumption and capital investment in the current period. Consumption increases by 70 per cent of consumption and by 60 per cent of the capital investment. The capital investment follows the same type of strategy:

$$y_{t+1} = 0.6x_t + 0.2y_t.$$

Thus, we have the system

$$\mathbf{u}_{t+1} = A\mathbf{u}_t \quad \text{with} \quad \mathbf{u}_t = (x_t, y_t)^T, \quad t = 1, 2, \dots.$$

(a) Find the greatest eigenvalue λ of the matrix A and the eigenvectors associated with this value.
(b) Interpret the result above with λ as a factor of proportionate growth.
(c) Let $10,000$ units be the sum of consumption value and capital investment in the first period. How does it have to be split for proportionate growth? Assume you have the same growth rate λ for the following two periods, what are the values of consumption and capital investment?

10.3 Given are the matrices

$$A = \begin{pmatrix} 1 & 2 & 0 & 1 \\ 0 & 3 & 2 & 0 \\ 0 & 0 & -2 & 0 \\ 0 & 0 & 0 & 5 \end{pmatrix} \quad \text{and} \quad B = \begin{pmatrix} 4 & 0 & 0 \\ 0 & 1 & 3 \\ 0 & 3 & 1 \end{pmatrix}.$$

(a) Find the eigenvalues and the eigenvectors of each of these matrices.
(b) Verify that the eigenvectors of matrix A form a basis of the space $\mathbb{R}^4$ and that the eigenvectors of matrix B are linearly independent and orthogonal.

10.4   Verify that the quadratic form $x^T Bx$ with matrix B from Exercise 10.1 is positive definite.

10.5   Given are the matrices:

$$A = \begin{pmatrix} 3 & -1 \\ -1 & 1 \end{pmatrix}; \qquad B = \begin{pmatrix} 2 & 2 \\ 2 & 1 \end{pmatrix};$$

$$C = \frac{1}{2} \begin{pmatrix} 5 & 1 & 0 \\ 1 & 5 & 0 \\ 0 & 0 & -8 \end{pmatrix} \qquad \text{and} \qquad D = \begin{pmatrix} 1 & 0 & 2 \\ 0 & 1 & 0 \\ 2 & 0 & 5 \end{pmatrix}.$$

(a) Find the eigenvalues of each of these matrices.
(b) Determine by the given criterion (see Theorem 10.7) which of the matrices A, B, C, D (and their associated quadratic forms, respectively) are positive definite and which are negative definite.
(c) Compare the results of (b) with the results of (a).

10.6   Let $x = (1, 1, 0)^T$ be an eigenvector associated with the eigenvalue $\lambda_1 = 3$ of the matrix

$$A = \begin{pmatrix} a_1 & 0 & 1 \\ 2 & a_2 & 0 \\ 1 & -1 & a_3 \end{pmatrix}.$$

(a) What can you conclude about the values of a_1, a_2 and a_3?
(b) Find another eigenvector associated with λ_1.
(c) Is it possible in addition to the answers concerning part (a) to find further conditions for a_1, a_2 and a_3 when A is positive definite?
(d) If your answer is affirmative for part (c), do you see a way to find a_1, a_2 and a_3 exactly when $\lambda_2 = -3$ is also an eigenvalue of the matrix A?

11 Functions of several variables

In Chapter 4, we considered such situations as when a firm produces some output $y = f(x)$ by means of a certain input x. However, in many economic applications, we have to deal with situations where several variables have to be included in the mathematical model, e.g. usually the output depends on a set of different input factors. Therefore, we deal in this chapter with functions depending on more than one independent variable.

11.1 PRELIMINARIES

If we have a pair $(x, y) \in \mathbb{R}^2$ of two input factors, the output may be measured by a function value $f(x, y) \in \mathbb{R}$ depending on two independent variables x and y. The notion of independent variables means that each variable can vary by itself without affecting the others. In a more general case, the input may be described by an n-tuple $(x_1, x_2, \ldots, x_n) \in \mathbb{R}^n$ of numbers, and the output is described by a value $f(x_1, x_2, \ldots, x_n) \in \mathbb{R}$. In the former case, $f : \mathbb{R}^2 \to \mathbb{R}$ is a function of two variables mapping the pair (x, y) into a real number $f(x, y)$. In the latter case, $f : \mathbb{R}^n \to \mathbb{R}$ is a function of n variables, or we also say that it is a function of an n-dimensional vector $\mathbf{x}$. Instead of $f(x_1, x_2, \ldots, x_n)$, we also use the notation $f(\mathbf{x})$ for the function value, where $\mathbf{x}$ denotes an n-vector. Similarly as in the case of a function of one variable, we denote the set of points for which the function value is defined as domain D_f of function f. Formally we can summarize: a function $f : D_f \to \mathbb{R}, D_f \subseteq \mathbb{R}^n$, depending on n variables $x_1, x_2, \ldots, x_n$ assigns a specified real number to each n-vector (or point) $(x_1, x_2, \ldots, x_n) \in D_f$. (Since vector $\mathbf{x}$ can also be considered as a point, we always skip the transpose sign 'T' of a row vector in this chapter.)

Example 11.1 (Cobb–Douglas production function) A production function assigns to each n-vector $\mathbf{x}$ with non-negative components, where the ith component represents the amount of input i $(i = 1, 2, \ldots, n;$ such a vector is also known as an input bundle), the maximal output $z = f(x_1, x_2, \ldots, x_n)$. Assume that an agricultural production function Q, where Q is the number of units produced, depends on three input variables K, L and R, where K stands for the capital invested, L denotes the labour input and R denotes the area of land that is used for the agricultural production. If the relationship between the independent variables K, L and R and the dependent variable (output) Q is described by the equation

$$Q(K, L, R) = A \cdot K^{a_1} \cdot L^{a_2} \cdot R^{a_3},$$

where A, a_1, a_2, a_3 are given parameters, we say that $Q = Q(K, L, R)$ is a *Cobb–Douglas* function. In this case, the output depends on the three variables K, L and R, but in general, it may depend on n variables, say $K_1, K_2, \ldots, K_n$.

The graph of a function f with $z = f(x, y)$, called a *surface*, consists of all points (x, y, z) for which $z = f(x, y)$ and (x, y) belongs to the domain D_f. The surface of a function f with $D_f \subseteq \mathbb{R}^2$ is illustrated in Figure 11.1. To get an overview on a specific function, we might be interested in knowing all points (x, y) of the domain D_f which have the same function value. A *level curve* for function f depending on two variables is a curve in the xy plane given by

$$z = f(x, y) = C,$$

where C is a constant. This curve is also called an *isoquant* (indicating 'equal quantity'). To illustrate, consider the following example.

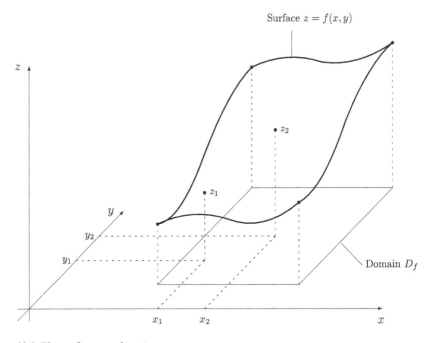

Figure 11.1 The surface $z = f(x, y)$.

Example 11.2 Let function $f : D_f \to \mathbb{R}$ with

$$f(x, y) = 4x^2 - y^2, \qquad D_f = \{(x, y) \in \mathbb{R}^2 \mid -1 \le x \le 1, \ -2 \le y \le 2\},$$

be given. We determine the isoquants for this function. Putting $4x^2 - y^2 = C$, where C denotes a constant, and eliminating variable y (and x, respectively), we get

$$y = \pm\sqrt{4x^2 - C} \qquad \left(\text{or, correspondingly,} \qquad x = \pm\frac{1}{2} \sqrt{y^2 + C} \right).$$

For $C = 0$, we get the two lines $y = 2x$ and $y = -2x$ intersecting at point $(0,0)$. For $C \neq 0$, each of the equations above gives two functions (namely one with sign '+' and one with sign '−') provided that C is chosen such that the term under the square root is non-negative. The level curves for some values of C are given in Figure 11.2.

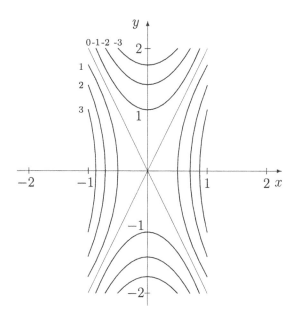

Figure 11.2 Level curves for function f with $f(x, y) = 4x^2 - y^2$.

Often, one wishes to determine to which value a function tends provided that the independent variables tend to a certain point $x^0 = (x_1^0, x_2^0, \dots, x_n^0)$. In the following, we introduce the limit of a function of n independent variables as a generalization of the limit of a function of one variable introduced in Chapter 4.1.1.

Definition 11.1 The real number L is called the *limit* of the function $f : D_f \to \mathbb{R}$, $D_f \subseteq \mathbb{R}^n$, as $x = (x_1, x_2, \dots, x_n)$ tends to point $x^0 = (x_1^0, x_2^0, \dots, x_n^0)$ if for any real number $\epsilon > 0$ there exists a real number $\delta = \delta(\epsilon) > 0$ such that

$$|f(x) - L| < \epsilon,$$

provided that

$$|x_1 - x_1^0| < \delta, \quad |x_2 - x_2^0| < \delta, \quad \dots, \quad |x_n - x_n^0| < \delta.$$

We can also write

$$\lim_{x \to x^0} f(x) = L.$$

The limit exists only if L is a finite number. The left-hand side in the above notation means that the variables x_i tend *simultaneously* and *independently* to x_i^0. In the case of a function of one variable, there were only two possibilities in which a variable x might tend to a value x_0, namely either from the right-hand side or from the left-hand side. Definition 11.1 requires that the limit has to exist for any possible path on which $\mathbf{x}$ approaches $\mathbf{x}^0$.

A specific path in the case of a function f depending on two variables is e.g. considered if we first determine the limit as x tends to x^0, and then the limit as y tends to y^0, i.e. we determine

$$L_1 = \lim_{y \to y^0} \lim_{x \to x^0} f(x, y).$$

Accordingly, we can interchange the order in which both limits above are determined. These specific limits are called *iterated* limits. However, from the existence and equality of both iterated limits one cannot conclude that the limit of function f exists, as the following example shows.

Example 11.3 Let function $f : D_f \to \mathbb{R}$ with

$$f(x, y) = \left(\frac{2x - 2y}{x + y} \right)^2, \qquad D_f = \{(x, y) \in \mathbb{R}^2 \mid y \neq -x\},$$

and $(x^0, y^0) = (0, 0)$ be given. For the iterated limits we get

$$L_1 = \lim_{y \to 0} \lim_{x \to 0} \left(\frac{2x - 2y}{x + y} \right)^2 = \lim_{y \to 0} (-2)^2 = \lim_{y \to 0} 4 = 4;$$

$$L_2 = \lim_{x \to 0} \lim_{y \to 0} \left(\frac{2x - 2y}{x + y} \right)^2 = \lim_{x \to 0} 2^2 = \lim_{x \to 0} 4 = 4.$$

Although the iterated limits exist and are equal, the limit of function f as (x, y) tends to $(0, 0)$ does not exist since e.g. along the path $y = x$, we get

$$f(x, y) = f(x, x) = \frac{0}{2x} = 0 \quad \text{for} \quad x \neq 0$$

and consequently

$$L_3 = \lim_{x \to 0} f(x, x) = 0,$$

which is different from L_1 and L_2.

Next, we deal with the generalization of the concept of a continuous function to the case of functions of more than one variable. Roughly speaking, a function f of n variables is continuous if small changes in the independent variables produce small changes in the function value. Formally we can introduce the following definition.

Definition 11.2   A function $f : D_f \to \mathbb{R}, D_f \subseteq \mathbb{R}^n$, is said to be *continuous* at point $x^0 = (x_1^0, x_2^0, \ldots, x_n^0) \in D_f$ if the limit of function f as x tends to x^0 exists and if this limit coincides with the function value $f(x^0)$, i.e.

$$\lim_{x \to x^0} f(x_1, x_2, \ldots, x_n) = f(x_1^0, x_2^0, \ldots, x_n^0).$$

If function f is continuous at all points $x^0 \in D_f$, then we also say that f is continuous. Analogously to functions of one variable (see Theorems 4.3 and 4.4 in Chapter 4.1.2), we get the following statement for functions of more than one variable: Any function of $n > 1$ independent variables that can be constructed from continuous functions by combining the operations of addition, subtraction, multiplication, division and functional composition is continuous wherever it is defined.

11.2   PARTIAL DERIVATIVES; GRADIENT

In Chapter 4, we considered the first-order and higher-order derivatives of a function of one independent variable. In the following, we discuss how to find derivatives of a function depending on more than one variable. First, we assume that function f with $z = f(x, y)$ depends on two independent variables x and y, and we introduce the notion of a partial derivative with respect to one of the two variables.

Before proceeding, we introduce the notion of an ϵ-neighbourhood $U_\epsilon(x^0)$ of a point $x^0 \in D_f \subseteq \mathbb{R}^n$. We denote by $U_\epsilon(x^0)$ with $\epsilon > 0$ the set of all points $x \in \mathbb{R}^n$ such that the Euclidean distance between the vectors x and x^0 is smaller than ϵ, i.e.

$$U_\epsilon(x^0) = \left\{ x \in \mathbb{R}^n \,\middle|\, \sqrt{\sum_{i=1}^{n} (x_i - x_i^0)^2} < \epsilon \right\}$$

In the case of $n = 2$ variables, the above definition of an ϵ-neighbourhood of point (x^0, y^0) includes all points that are within a circle (but not on the boundary of the circle) with the centre at point (x^0, y^0) and the radius ϵ.

Definition 11.3   Let function $f : D_f \to \mathbb{R}, D_f \subseteq \mathbb{R}^2$, be defined in some neighbourhood $U_\epsilon(x^0, y^0)$ of the point $(x^0, y^0) \in D_f$. Then function f is said to be *partially differentiable* with respect to x at point (x^0, y^0) if the limit

$$\lim_{\Delta x \to 0} \frac{f(x^0 + \Delta x, y^0) - f(x^0, y^0)}{\Delta x}$$

exists, and f is said to be partially differentiable with respect to y at the point (x^0, y^0) if the limit

$$\lim_{\Delta y \to 0} \frac{f(x^0, y^0 + \Delta y) - f(x^0, y^0)}{\Delta y}$$

exists.

The above limits are denoted as

$$f_x(x^0, y^0) \qquad \text{resp.} \qquad f_y(x^0, y^0)$$

and are called *partial derivatives* (of the first order) of function f with respect to x and with respect to y at point (x^0, y^0).

We also write

$$\frac{\partial f(x^0, y^0)}{\partial x} \qquad \text{and} \qquad \frac{\partial f(x^0, y^0)}{\partial y}$$

for the partial derivatives with respect to x and y at point (x^0, y^0). If for any point $(x, y) \in D_f$ the partial derivatives $f_x(x, y)$ and $f_y(x, y)$ exist, we get functions f_x and f_y by assigning to each point (x, y) the values $f_x(x, y)$ and $f_y(x, y)$, respectively. Finding the partial derivatives of function f with $z = f(x, y)$ means:

(1) to find $f_x(x, y)$, one has to differentiate function f with respect to variable x while treating variable y as a constant, and
(2) to find $f_y(x, y)$, one has to differentiate function f with respect to variable y while treating variable x as a constant.

The partial derivatives at some point (x^0, y^0) correspond again to slopes of certain straight lines. We can give the following interpretations.

(1) *Geometric interpretation of* $f_x(x^0, y^0)$. The slope of the tangent line $f_T(x)$ to the curve of intersection of the surface $z = f(x, y)$ and the plane $y = y^0$ at the point (x^0, y^0, z^0) on the surface is equal to $f_x(x^0, y^0)$.
(2) *Geometric interpretation of* $f_y(x^0, y^0)$. The slope of the tangent line $f_T(y)$ to the curve of intersection of the surface $z = f(x, y)$ and the plane $x = x^0$ at the point (x^0, y^0, z^0) on the surface is equal to $f_y(x^0, y^0)$.

The partial derivatives are illustrated in Figure 11.3. The slopes of the lines $f_T(x)$ and $f_T(y)$ are given by the partial derivatives of function f with $z = f(x, y)$ with respect to the variables x and y, respectively, at point (x^0, y^0). Both tangent lines span a plane $f_T(x, y)$, which is called the *tangent plane* to function $z = f(x, y)$ at point (x^0, y^0), and the equation of this tangent plane is given by

$$f_T(x, y) = f(x_0, y_0) + f_x(x_0, y_0) \cdot (x - x_0) + f_y(x_0, y_0) \cdot (y - y_0).$$

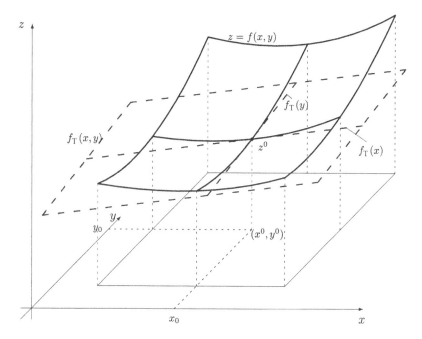

Figure 11.3 Partial derivatives of function f with $z = f(x,y)$ and the tangent plane $f_T(x,y)$.

Example 11.4   Assume that the total cost C of a firm manufacturing two products A and B depends on the produced quantities. Let x be the number of produced units of product A, y be the number of produced units of product B and let the total cost $C(x,y)$ be given as

$$C(x,y) = 200 + 22x + 16y^{3/2}, \qquad D_C = \mathbb{R}_+^2.$$

The first term of function C represents the fixed cost and the other two terms represent the variable cost in dependence on the values x and y. We determine the partial derivatives with respect to x and y and obtain

$$C_x(x,y) = 22 \qquad \text{and} \qquad C_y(x,y) = \frac{3}{2} \cdot 16 \cdot \sqrt{y} = 24\sqrt{y}.$$

$C_x(x,y)$ is known as the marginal cost of product A, and $C_y(x,y)$ as the marginal cost of product B. The partial derivative $C_x(x,y) = 22$ can be interpreted such that, when the quantity of product B does not change, an increase of the production of product A by one unit leads to an approximate increase of 22 units in total cost (independent of the concrete value of variable x). When keeping the production of A constant, the marginal cost of product B depends on the particular value of variable y. For instance, if $y = 16$, then an increase of the production of B by one unit causes approximately an increase of $C_y(x,16) = 24 \cdot \sqrt{16} = 96$ units in total cost, whereas for $y = 100$ the increase in total cost is approximately equal to $C_y(x,100) = 24 \cdot \sqrt{100} = 240$ units.

The previous considerations can be generalized to functions of more than two variables. In the following, we consider a function f with $z = f(x_1, x_2, \ldots, x_n)$ depending on the n independent variables $x_1, x_2, \ldots, x_n$.

Definition 11.4 Let function $f \colon D_f \to \mathbb{R}, D_f \subseteq \mathbb{R}^n$, be defined in some neighbourhood $U_\epsilon(\mathbf{x}^0)$ of point $\mathbf{x}^0 = (x_1^0, x_2^0, \ldots, x_n^0) \in D_f$. Then function f is said to be *partially differentiable* with respect to x_i at point $\mathbf{x}^0$ if the limit

$$\lim_{\Delta x_i \to 0} \frac{f(x_1^0, \ldots, x_{i-1}^0, x_i^0 + \Delta x_i, x_{i+1}^0, \ldots, x_n^0) - f(x_1^0, x_2^0, \ldots, x_n^0)}{\Delta x_i}$$

exists. The above limit is denoted by

$$f_{x_i}(\mathbf{x}^0)$$

and is called the *partial derivative* (of the first order) of function f with respect to x_i at point $\mathbf{x}^0 = (x_1^0, x_2^0, \ldots, x_n^0)$.

We also write

$$\frac{\partial f(x_1^0, x_2^0, \ldots, x_n^0)}{\partial x_i}$$

for the partial derivative of function f with respect to x_i at point $\mathbf{x}^0$. The partial derivative $f_{x_i}(\mathbf{x}^0)$ gives approximately the change in the function value that results from an increase of variable x_i by one unit (i.e. from x_i^0 to $x_i^0 + 1$) while holding all other variables constant. That is, partial differentiation of a function of n variables means that we must keep $n - 1$ independent variables constant while allowing only one variable to change. Since we know already from Chapter 4 how to handle constants in the case of a function of one independent variable, it is not a problem to perform the partial differentiation.

If function f is partially differentiable with respect to x_i at any point $\mathbf{x} \in D_f$, then function f is said to be partially differentiable with respect to x_i. If function f is partially differentiable with respect to all variables, f is said to be *partially differentiable*. If all partial derivatives are continuous, we say that function f is *continuously partially differentiable*.

Example 11.5 We consider the cost function $C \colon \mathbb{R}_+^3 \to \mathbb{R}$ of a firm producing three products given by

$$C(x_1, x_2, x_3) = 300 + 2x_1 + 0.5x_2^2 + \sqrt{x_1 x_2 + x_3},$$

where x_i denotes the produced quantity of product $i \in \{1, 2, 3\}$. For the partial derivatives, we obtain

$$C_{x_1}(x_1, x_2, x_3) = 2 + \frac{x_2}{2\sqrt{x_1 x_2 + x_3}};$$

$$C_{x_2}(x_1, x_2, x_3) = x_2 + \frac{x_1}{2\sqrt{x_1 x_2 + x_3}};$$

$$C_{x_3}(x_1, x_2, x_3) = \frac{1}{2\sqrt{x_1 x_2 + x_3}}.$$

Next, we introduce *higher-order partial derivatives*. Note that the first-order partial derivatives of a function $f : D_f \to \mathbb{R}, D_f \subseteq \mathbb{R}^n$, are in general again functions of n variables $x_1, x_2, \ldots, x_n$. Thus, we can again determine the partial derivatives of these functions describing the first-order partial derivatives and so on. To illustrate this, consider a function f with $z = f(x, y)$ depending on $n = 2$ variables. We obtain:

$$f_x(x, y) = \frac{\partial f(x, y)}{\partial x} = g(x, y); \qquad f_y(x, y) = \frac{\partial f(x, y)}{\partial y} = h(x, y);$$

$$f_{xx}(x, y) = \frac{\partial^2 f(x, y)}{\partial x^2} = \frac{\partial g(x, y)}{\partial x} = g_x(x, y); \quad f_{yy}(x, y) = \frac{\partial^2 f(x, y)}{\partial y^2} = \frac{\partial h(x, y)}{\partial y} = h_y(x, y);$$

$$f_{xy}(x, y) = \frac{\partial^2 f(x, y)}{\partial y \partial x} = \frac{\partial g(x, y)}{\partial y} = g_y(x, y); \quad f_{yx}(x, y) = \frac{\partial^2 f(x, y)}{\partial x \partial y} = \frac{\partial h(x, y)}{\partial x} = h_x(x, y).$$

Here the notation $f_{xy}(x, y)$ means that we *first* differentiate function f with respect to variable x and *then* with respect to variable y, i.e.

$$f_{xy}(x, y) = \frac{\partial}{\partial y} \left(\frac{\partial f(x, y)}{\partial x} \right) = \frac{\partial^2 f(x, y)}{\partial y \partial x}.$$

Analogously, we can consider mth-order partial derivatives that include m successive partial differentiations with respect to certain variables. If all partial derivatives of mth-order are continuous, we say that function f is m times continuously partially differentiable.

Example 11.6 Let function $f : \mathbb{R}^2 \to [-1, 1]$ with

$$z = f(x, y) = \sin(3x - y)$$

be a function of two independent variables. We determine all second-order partial derivatives of function f and obtain:

$$f_x(x, y) = 3\cos(3x - y); \quad f_{xx}(x, y) = -9\sin(3x - y); \quad f_{xy}(x, y) = 3\sin(3x - y);$$
$$f_y(x, y) = -\cos(3x - y); \quad f_{yx}(x, y) = 3\sin(3x - y); \quad f_{yy}(x, y) = -\sin(3x - y).$$

We observe that in Example 11.6 the mixed second-order partial derivatives $f_{xy}(x,y)$ and $f_{yx}(x,y)$ are identical. We now ask for a condition under which the latter statement is necessarily true. This question is answered by the following theorem.

THEOREM 11.1 (Young's theorem)   Let two mth-order partial derivatives of function $f: D_f \to \mathbb{R}$, $D_f \subseteq \mathbb{R}^n$, involve the same number of differentiations with respect to each of the variables and suppose that they are both continuous. Then the two partial derivatives are necessarily equal at all points $\mathbf{x} \in D_f$.

If in Theorem 11.1 only a particular point $\mathbf{x}^0 \in D_f$ is considered, it suffices to assume that both mth-order partial derivatives are continuous only in some neighbourhood $U_\epsilon(\mathbf{x}^0)$ with $\epsilon > 0$ to ensure equality of both mth-order partial derivatives at point $\mathbf{x}^0$. Hereafter, for the sake of simplicity in the presentation, we always suppose in what follows that the corresponding assumptions on partial differentiability or continuous partial differentiability of a function f hold for all points of the domain D_f.

A special case of Theorem 11.1 is obtained when $n = m = 2$. Then Young's theorem may be interpreted as follows: if functions f_{xy} and f_{yx} are continuous, then $f_{xy}(x,y) = f_{yx}(x,y)$ holds at all points $(x,y) \in D_f$.

Definition 11.5   Let function $f: D_f \to \mathbb{R}$, $D_f \subseteq \mathbb{R}^n$, be partially differentiable at point $\mathbf{x}^0 \in D_f$. Then the vector

$$\operatorname{\mathbf{grad}} f(\mathbf{x}^0) = \begin{pmatrix} f_{x_1}(\mathbf{x}^0) \\ f_{x_2}(\mathbf{x}^0) \\ \vdots \\ f_{x_n}(\mathbf{x}^0) \end{pmatrix}$$

is called the *gradient* of function f at point $\mathbf{x}^0 \in D_f$.

The vector $\operatorname{\mathbf{grad}} f(\mathbf{x}^0)$ has an important geometric interpretation. It gives the direction in which function f at point $\mathbf{x}^0 \in D_f$ increases most.

Example 11.7   We consider function $f : \mathbb{R}^2 \to \mathbb{R}$ with

$$f(x_1,x_2) = 2x_1^2 + \frac{1}{2}x_2^2$$

and determine the gradient at point $\mathbf{x}^0 = (x_1^0, x_2^0) = (1,2)$. First, we find the partial derivatives

$$f_{x_1}(x_1,x_2) = 4x_1 \qquad \text{and} \qquad f_{x_2}(x_1,x_2) = x_2$$

which yields

$$\operatorname{\mathbf{grad}} f(x_1,x_2) = \begin{pmatrix} 4x_1 \\ x_2 \end{pmatrix}$$

and thus

$$\mathbf{grad}\, f(1,2) = \begin{pmatrix} 4 \\ 2 \end{pmatrix}.$$

An illustration is given in Figure 11.4. Consider the level curve

$$f(x_1, x_2) = 2x_1^2 + \frac{1}{2}x_2^2 = c \qquad \text{with} \qquad c = f(x_1^0, x_2^0) = 4$$

which is an ellipse

$$\frac{x_1^2}{2} + \frac{x_2^2}{8} = 1$$

with the half axes $a = \sqrt{2}$ and $b = \sqrt{8}$. Consider the tangent line to this level curve at point $(1, 2)$. From Figure 11.4 we see that the gradient at this point, i.e. vector $(4, 2)^\mathrm{T}$, is orthogonal to the tangent line to the level curve at point $(x_1^0, x_2^0) = (1, 2)$. It can be shown that this property always holds.

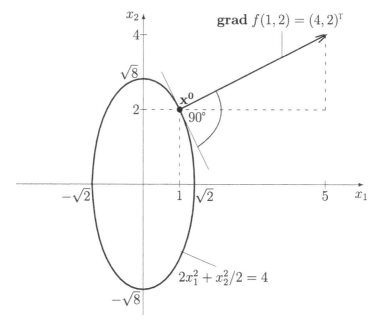

Figure 11.4 The gradient of function f with $f(x_1, x_2) = 2x_1^2 + x_2^2/2$ at point $(x_1^0, x_2^0) = (1, 2)$.

Example 11.8   Suppose that the production function $f : \mathbb{R}_+^2 \to \mathbb{R}_+$ of a firm is given by

$$f(x, y) = 3x^2 y + 0.5xe^y,$$

where x stands for the amount of labour used and y stands for the amount of capital used. We obtain the gradient

$$\operatorname{grad} f(x,y) = \begin{pmatrix} 6xy + 0.5e^y \\ 3x^2 + 0.5xe^y \end{pmatrix}.$$

If the current inputs are $x^0 = 10$ and $y^0 = \ln 12$, then, in order to get the biggest increase in output, we have to determine the gradient at point (x^0, y^0), and we obtain

$$\operatorname{grad} f(x^0, y^0) = \begin{pmatrix} 60 \ln 12 + 6 \\ 300 + 60 \end{pmatrix} \approx \begin{pmatrix} 155.09 \\ 360 \end{pmatrix}.$$

So, to increase the output as much as possible, the firm should increase labour and capital approximately in the ratio 155.09 : 360, i.e. in the ratio 1:2.32.

11.3 TOTAL DIFFERENTIAL

Partial derivatives consider only changes in one variable where the remaining ones are not changed. Now we investigate how a function value changes in the case of *simultaneous* changes of the independent variables. This leads to the concept of the total differential introduced in the following definition.

Definition 11.6 Let function $f : D_f \to \mathbb{R}, D_f \subseteq \mathbb{R}^n$, be continuously partially differentiable. Then the *total differential* of function f at point $\mathbf{x}^0 = (x_1^0, x_2^0, \ldots, x_n^0)$ is defined as

$$dz = f_{x_1}(\mathbf{x}^0)\, dx_1 + f_{x_2}(\mathbf{x}^0)\, dx_2 + \cdots + f_{x_n}(\mathbf{x}^0)\, dx_n.$$

The terms $f_{x_i}(\mathbf{x}^0)\, dx_i$, $i = 1, 2, \ldots, n$, are called *partial differentials*.

The partial differential $f_{x_i}(\mathbf{x}^0)\, dx_i$ gives approximately the change in the function value when changing variable x_i at point $\mathbf{x}^0$ by $\Delta x_i = dx_i$, $i \in \{1, 2, \ldots, n\}$. The total differential is illustrated in Figure 11.5 for the case $n = 2$. We consider the points $\mathbf{x}^0$ and $\mathbf{x}^0 + \Delta \mathbf{x}$, where $\Delta \mathbf{x} = (\Delta x_1, \Delta x_2)$. Then the total differential gives the difference in the function value of the tangent plane at point $\mathbf{x}^0 + \Delta \mathbf{x}$ and the function value of function f at point $\mathbf{x}^0$. If the changes Δx_1 and Δx_2 are small, then the total differential dz is approximately equal to Δz, which gives the difference in the function values of function f at the points $\mathbf{x}^0 + \Delta \mathbf{x}$ and $\mathbf{x}^0$.

In other words, in some neighbourhood of point $\mathbf{x}^0$, we can approximate function f by the tangent plane $f_T(x_1, x_2)$.

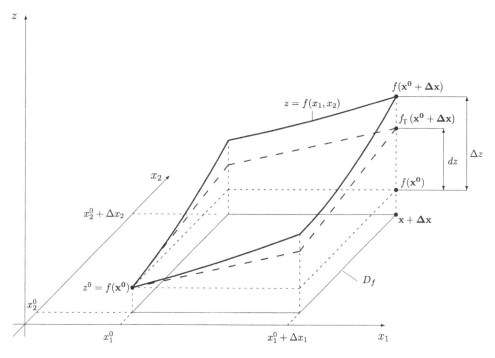

Figure 11.5 The total differential of function f with $z = f(x,y)$.

Example 11.9 We determine the total differential of function $f : D_f \to \mathbb{R}$ with

$$z = f(x,y) = \ln \tan \frac{x}{y}.$$

For the first-order partial derivatives we get

$$f_x(x,y) = \frac{\frac{1}{y}}{\tan \frac{x}{y} \cos^2 \frac{x}{y}} = \frac{1}{y \sin \frac{x}{y} \cos \frac{x}{y}} = \frac{2}{y \sin \frac{2x}{y}};$$

$$f_y(x,y) = \frac{-\frac{x}{y^2}}{\tan \frac{x}{y} \cos^2 \frac{x}{y}} = \frac{-x}{y^2 \sin \frac{x}{y} \cos \frac{x}{y}} = \frac{-2x}{y^2 \sin \frac{2x}{y}}.$$

In the above transformation, we have applied the addition theorem for the sine function (see property (1) of the trigonometric functions in Chapter 3.3.3). Then the total differential is given by

$$dz = f_x(x,y)\,dx + f_y(x,y)\,dy = \frac{2\,dx}{y \sin \frac{2x}{y}} - \frac{2x\,dy}{y^2 \sin \frac{2x}{y}} = \frac{2}{y^2 \sin \frac{2x}{y}} \cdot (y\,dx - x\,dy).$$

Next, we consider an application of the total differential, namely the estimation of the maximal error in the value of a function of n independent variables. Assume that we have to determine the function value $z = f(x_1, x_2, \ldots, x_n)$ at point $\mathbf{x}^0 = (x_1^0, x_2^0, \ldots, x_n^0)$, where some errors occur in the independent variables. The errors in these variables usually lead to an error Δz in the objective function value z. The aim is to estimate the error Δz provided that the errors $\Delta x_1, \Delta x_2, \ldots, \Delta x_n$ in the variables $x_1, x_2, \ldots, x_n$ are bounded by constants Δ_i, i.e. we have $|\Delta x_i| \le \Delta_i$. Since all errors $\Delta x_i, i = 1, 2, \ldots, m$, are assumed to be sufficiently small, we can apply Definition 11.6 and, because $\Delta z \approx dz$, we obtain

$$| \Delta z | \approx | dz | = \left| \sum_{i=1}^{n} f_{x_i}(\mathbf{x}^0) \cdot \Delta x_i \right| \le \sum_{i=1}^{n} |f_{x_i}(\mathbf{x}^0)| \cdot |\Delta x_i| \le \sum_{i=1}^{n} |f_{x_i}(\mathbf{x}^0)| \cdot \Delta_i.$$

By means of the last formula, we have an estimation for the maximal absolute error Δz in the function value at point $\mathbf{x}^0$ provided that the absolute error in the variable x_i at point x_i^0 is bounded by a sufficiently small constant Δ_i, $i \in \{1, 2, \ldots, n\}$.

Example 11.10 Assume that the radius R and the height H of a cylinder are measured with $R = 5.05 \pm 0.01$ cm and $H = 8.20 \pm 0.005$ cm, i.e. $R_0 = 5.05$ cm, $H_0 = 8.20$ cm, $|dR| \le 0.01$ cm and $|dH| \le 0.005$ cm. We wish to estimate the absolute and relative errors of the volume V of the cylinder. We have

$$V_0 = V(R_0, H_0) = \pi R_0^2 H_0 = 656.97 \text{ cm}^3.$$

For the total differential we get

$$dV = \frac{\partial V}{\partial R} dR + \frac{\partial V}{\partial H} dH = 2\pi R_0 H_0 \, dR + \pi R_0^2 \, dH.$$

Inserting $R_0 = 5.05$, $H_0 = 8.20$, $|dR| \le 0.01$ and $|dH| \le 0.005$, we obtain the estimate

$$|\Delta V| \approx |dV| \le |2\pi R_0 H_0| \cdot |dR| + |\pi R_0^2| \cdot |dH|$$
$$\le 82.82\pi \cdot 0.01 + 25.5025\pi \cdot 0.005 = 0.9557125\pi < 3.01 \text{ cm}^3$$

(note that we have used the triangle inequality to get the estimate given above), i.e.

$$V_0 \approx 656.97 \pm 3.01 \text{ cm}^3.$$

For the relative error we get

$$\left| \frac{\Delta V}{V_0} \right| \approx \left| \frac{dV}{V_0} \right| = \left| \frac{2\pi R_0 H_0 \, dR + \pi R_0^2 \, dH}{\pi R_0^2 H_0} \right| \le 2 \left| \frac{dR}{R} \right| + \left| \frac{dH}{H} \right| < 0.005.$$

Thus, for the given maximum errors in the variables, we get the estimate that the relative error for the volume is less than 0.5 per cent.

11.4 GENERALIZED CHAIN RULE; DIRECTIONAL DERIVATIVES

In Chapter 11.2, we investigated how a function value changes if one moves along one of the axes. The answer was obtained by means of the partial derivatives with respect to one of the variables. Now we investigate a more general question, namely how a function value changes if we move along an arbitrary curve C. First, we assume that this curve C is given in a parametric form

$$x = x(t) \qquad \text{and} \qquad y = y(t).$$

As an example, we mention the parametric representation of a function f with $z(t) = f(x(t), y(t))$, where the domain is a circle given by all points $(x, y) \in \mathbb{R}^2$ with

$$x = r \sin t, \quad y = r \cos t; \qquad t \in [0, 2\pi],$$

where r denotes the (constant) radius and t is the (variable) angle measured in radians. We only note that parameter t often represents the time in mathematical models of economical problems, i.e. we describe the variables x and y in representation $f(x, y)$ as functions of time t.

A point (x^0, y^0) corresponds to some value t^0 in the above parametric representation, i.e.

$$x^0 = x(t^0) \qquad \text{and} \qquad y^0 = y(t^0).$$

Accordingly, for some point on the curve C with parameter value $t = t^0 + \Delta t$, we get the representation

$$x = x(t^0 + \Delta t) \qquad \text{and} \qquad y = y(t^0 + \Delta t).$$

Now we can conclude that, if $\Delta t \to 0$, then also $\Delta x \to 0$ and $\Delta y \to 0$. Therefore, the derivative

$$\frac{df(x, y)}{dt} = \lim_{\Delta t \to 0} \frac{f(x(t^0 + \Delta t), y(t^0 + \Delta t)) - f(x(t^0), y(t^0))}{\Delta t}$$

$$= \lim_{\Delta t \to 0} \frac{f(x^0 + \Delta x, y^0 + \Delta y) - f(x^0, y^0)}{\Delta t}$$

characterizes the approximate change of function f if parameter t increases from t_0 by one unit, i.e. one moves along curve C from point $(x(t_0), y(t_0))$ to point $(x(t_0 + 1), y(t_0 + 1))$ provided that the above limit exists.

To obtain a formula for the derivative of such a function with variables depending on a parameter t, we consider the general case of n independent variables and assume that all variables x_i, $i = 1, 2, \ldots, n$, are also functions of a variable t, i.e.

$$x_1 = x_1(t), \qquad x_2 = x_2(t), \qquad \ldots, \qquad x_n = x_n(t).$$

Thus, we have:

$$z = f(x_1, x_2, \ldots, x_n) = f(x_1(t), x_2(t), \ldots, x_n(t)) = z(t).$$

To determine the derivative dz/dt, we have to compute

$$\frac{dz(x_1, x_2, \ldots, x_n)}{dt} = \lim_{\Delta t \to 0} \frac{f(x_1 + \Delta x_1, x_2 + \Delta x_2, \ldots, x_n + \Delta x_n) - f(x_1, x_2, \ldots, x_n)}{\Delta t},$$

provided that this limit exists, where $\Delta x_i = x_i(t + \Delta t) - x_i(t)$. This can be done by applying the following theorem.

THEOREM 11.2 (chain rule)   Let function $f : D_f \to \mathbb{R}, D_f \subseteq \mathbb{R}^n$, be continuously partially differentiable, where

$$x_1 = x_1(t), \quad x_2 = x_2(t), \quad \ldots, \quad x_n = x_n(t),$$

and functions $x_i(t) : D_{x_i} \to \mathbb{R}, \ i = 1, 2, \ldots, n$, are differentiable with respect to t. Then function

$$z(t) = f(\mathbf{x}(t)) = f(x_1(t), x_2(t), \ldots, x_n(t))$$

is also differentiable with respect to t, and we get

$$\frac{dz}{dt} = f_{x_1}(\mathbf{x}) \frac{dx_1}{dt} + f_{x_2}(\mathbf{x}) \frac{dx_2}{dt} + \ldots + f_{x_n}(\mathbf{x}) \frac{dx_n}{dt}.$$

The above result can be generalized to the case when the functions x_i depend on two variables u and v. We get the following theorem.

THEOREM 11.3 (generalized chain rule)   Let function $f : D_f \to \mathbb{R}, D_f \subseteq \mathbb{R}^n$, be continuously partially differentiable, where

$$x_1 = x_1(u, v), \quad x_2 = x_2(u, v), \quad \ldots, \quad x_n = x_n(u, v),$$

and functions $x_i(u, v) : D_{x_i} \to \mathbb{R}, \ D_{x_i} \subseteq \mathbb{R}^2, \ i = 1, 2, \ldots, n$, are continuously partially differentiable with respect to both variables u and v. Then function z with

$$z(u, v) = f(x_1(u, v), x_2(u, v), \ldots, x_n(u, v))$$

is also partially differentiable with respect to both variables u and v, and

$$\frac{\partial z}{\partial u} = f_{x_1}(\mathbf{x}) \frac{\partial x_1}{\partial u} + f_{x_2}(\mathbf{x}) \frac{\partial x_2}{\partial u} + \ldots + f_{x_n}(\mathbf{x}) \frac{\partial x_n}{\partial u},$$

$$\frac{\partial z}{\partial v} = f_{x_1}(\mathbf{x}) \frac{\partial x_1}{\partial v} + f_{x_2}(\mathbf{x}) \frac{\partial x_2}{\partial v} + \ldots + f_{x_n}(\mathbf{x}) \frac{\partial x_n}{\partial v}.$$

Example 11.11   Consider function $f : \mathbb{R}^3 \to \mathbb{R}_+$ with

$$z = f(x_1, x_2, x_3) = e^{x_1^2 + 2x_2 - x_3},$$

where $x_1 = uv$, $x_2 = u^2 + v^2$ and $x_3 = u + 2v$. We determine the partial derivatives of function f with respect to variables u and v. We obtain

$$f_{x_1} = 2x_1 e^{x_1^2 + 2x_2 - x_3}, \qquad f_{x_2} = 2e^{x_1^2 + 2x_2 - x_3}, \qquad f_{x_3} = -e^{x_1^2 + 2x_2 - x_3},$$

$$\frac{\partial x_1}{\partial u} = v, \qquad \frac{\partial x_1}{\partial v} = u, \qquad \frac{\partial x_2}{\partial u} = 2u, \qquad \frac{\partial x_2}{\partial v} = 2v, \qquad \frac{\partial x_3}{\partial u} = 1, \qquad \frac{\partial x_3}{\partial v} = 2,$$

and thus for function $z = z(u, v)$

$$\frac{\partial z}{\partial u} = f_{x_1} \frac{\partial x_1}{\partial u} + f_{x_2} \frac{\partial x_2}{\partial u} + f_{x_3} \frac{\partial x_3}{\partial u} = (2uv^2 + 4u - 1)e^{u^2 v^2 + 2(u^2 + v^2) - (u + 2v)},$$

$$\frac{\partial z}{\partial v} = f_{x_1} \frac{\partial x_1}{\partial v} + f_{x_2} \frac{\partial x_2}{\partial v} + f_{x_3} \frac{\partial x_3}{\partial v} = (2u^2 v + 4v - 2)e^{u^2 v^2 + 2(u^2 + v^2) - (u + 2v)}.$$

Note that the same result is obtained if we first substitute the functions $x_1 = x_1(u, v)$, $x_2 = x_2(u, v)$ and $x_3 = x_3(u, v)$ into function f with $z = f(x_1, x_2, x_3)$ and determine the partial derivatives with respect to variables u and v.

So far we have considered partial derivatives which characterize changes of the function value $f(\mathbf{x})$ in a direction of the axis of one variable. In the following, we deal with the change of $f(\mathbf{x})$ in a direction given by a certain vector $\mathbf{r} = (r_1, r_2, \ldots, r_n)^{\mathrm{T}}$. This leads to the introduction of the directional derivative given in the following definition.

Definition 11.7 Let function $f : D_f \to \mathbb{R}$, $D_f \subseteq \mathbb{R}^n$, be continuously partially differentiable and $\mathbf{r} = (r_1, r_2, \ldots, r_n)^{\mathrm{T}}$ be a directional vector with $|\mathbf{r}| = 1$. The term

$$[\mathbf{grad}\, f(\mathbf{x}^0)]^{\mathrm{T}} \cdot \mathbf{r} = [\nabla f(\mathbf{x}^0)]^{\mathrm{T}} \cdot \mathbf{r}$$

$$= f_{x_1}(\mathbf{x}^0) \cdot r_1 + f_{x_2}(\mathbf{x}^0) \cdot r_2 + \ldots + f_{x_n}(\mathbf{x}^0) \cdot r_n$$

is called the *directional derivative* of function f at point $\mathbf{x}^0 = (x_1^0, x_2^0, \ldots, x_n^0) \in D_f$ in the direction $\mathbf{r}$.

Notice that the directional derivative is a *scalar* obtained as the scalar product of the vectors $\mathbf{grad}\, f(\mathbf{x}^0)$ and $\mathbf{r}$. The assumption $|\mathbf{r}| = 1$ allows us to interpret the value of the directional derivative as the approximate change of the function value if we move one unit from point $\mathbf{x}^0$ in direction $\mathbf{r}$. The directional derivative of function f at point $\mathbf{x}^0$ in the direction $\mathbf{r}$ is illustrated in Figure 11.6. It gives the slope of the tangent line $f_{\mathrm{T}}(\mathbf{r})$ to function f at point $\mathbf{x}^0$ along the direction given by vector $\mathbf{r}$. If vector $\mathbf{r}^i = (0, \ldots, 0, 1, 0, \ldots, 0)^{\mathrm{T}}$ is the ith unit vector, the directional derivative in direction $\mathbf{r}^i$ corresponds to the ith partial derivative, i.e.

$$[\mathbf{grad}\, f(\mathbf{x}^0)]^{\mathrm{T}} \cdot \mathbf{r}^i = f_{x_i}(\mathbf{x}^0).$$

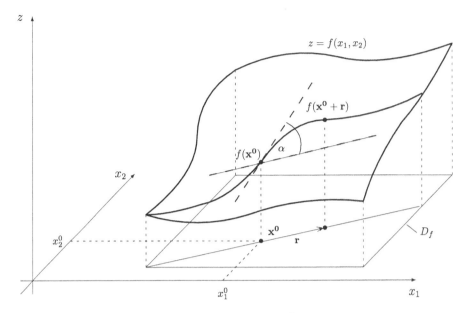

Figure 11.6 The directional derivative of function f at point $\mathbf{x}^0$ in the direction $\mathbf{r}$.

Thus, partial derivatives are special directional derivatives. For the directional derivative of function f at point $\mathbf{x}^0$ in the direction $\mathbf{r}$, we also write

$$\frac{\partial f}{\partial \mathbf{r}}(\mathbf{x}^0).$$

Example 11.12   The total cost function $f: \mathbb{R}^3_+ \to \mathbb{R}$ of a firm producing the quantities x_1, x_2 and x_3 of the three goods 1, 2 and 3 is given by

$$f(x_1, x_2, x_3) = 20 + x_1^2 + 2x_2 + 3x_3 + 2x_1x_3.$$

Let the current production quantities of the three goods be $x_1^0 = 10, x_2^0 = 15$ and $x_3^0 = 20$. Assume that the firm has the chance to increase its production by exactly five units, namely two products by two units and the remaining product by one unit, and we want to know which change results in the smallest increase in cost. This means that we have to investigate the directional derivatives in the direction of the vectors

$$\bar{\mathbf{r}}^1 = \begin{pmatrix} 1 \\ 2 \\ 2 \end{pmatrix}, \qquad \bar{\mathbf{r}}^2 = \begin{pmatrix} 2 \\ 1 \\ 2 \end{pmatrix} \qquad \text{and} \qquad \bar{\mathbf{r}}^3 = \begin{pmatrix} 2 \\ 2 \\ 1 \end{pmatrix}.$$

First, we consider direction $\bar{\mathbf{r}}^1 = (1, 2, 2)^{\mathrm{T}}$, and determine the directional derivative. The gradient of function f is given by

$$\mathbf{grad}\, f(x_1, x_2, x_3) = \begin{pmatrix} 2x_1 + 2x_3 \\ 2 \\ 3 + 2x_1 \end{pmatrix}$$

and therefore at point $\mathbf{x}^0 = (x_1^0, x_2^0, x_3^0) = (10, 15, 20)$ by

$$\mathbf{grad}\, f(\mathbf{x}^0) = \begin{pmatrix} 60 \\ 2 \\ 23 \end{pmatrix}.$$

Since $\bar{\mathbf{r}}^1$ is not a unit vector, we have to divide vector $\bar{\mathbf{r}}^1$ by its length

$$|\bar{\mathbf{r}}^1| = \sqrt{1^2 + 2^2 + 2^2} = \sqrt{9} = 3,$$

which yields vector

$$\mathbf{r}^1 = \frac{\bar{\mathbf{r}}^1}{|\bar{\mathbf{r}}^1|} = \frac{1}{3} \begin{pmatrix} 1 \\ 2 \\ 2 \end{pmatrix}.$$

We obtain for the directional derivative $[\mathbf{grad}\, f(\mathbf{x}^0)]^T (\mathbf{r}^1) = (\partial f)/(\partial \mathbf{r}^1)$ at point $\mathbf{x}^0$ in the direction $\mathbf{r}^1$:

$$\frac{\partial f}{\partial \mathbf{r}^1}(\mathbf{x}^0) = [\mathbf{grad}\, f(\mathbf{x}^0)]^T \cdot \frac{\bar{\mathbf{r}}^1}{|\bar{\mathbf{r}}^1|} = \frac{1}{3} \cdot (60, 2, 23) \cdot \begin{pmatrix} 1 \\ 2 \\ 2 \end{pmatrix}$$

$$= \frac{1}{3} \cdot (60 \cdot 1 + 2 \cdot 2 + 23 \cdot 2) = \frac{110}{3}.$$

How can we interpret the directional derivative $(\partial f)/(\partial \mathbf{r}^1)$? When moving from $\mathbf{x}^0$ one unit in the direction of $\bar{\mathbf{r}}^1$ (i.e. from $\mathbf{x}^0$ to $\mathbf{x}^0 + \mathbf{r}^1$), the increase in cost is approximately equal to $110/3$. Equivalently, if we increase the production of good 1 by one unit, and of goods 2 and 3 by two units each, the approximate increase in the cost is $3 \cdot 110/3 = 110$ units.

For the directional vector $\bar{\mathbf{r}}^2 = (2, 1, 2)^T$, we obtain first the unit vector

$$\mathbf{r}^2 = \frac{\bar{\mathbf{r}}^2}{|\bar{\mathbf{r}}^2|} = \frac{1}{3} \begin{pmatrix} 2 \\ 1 \\ 2 \end{pmatrix}$$

and then the directional derivative at point $\mathbf{x}^0$ in the direction $\mathbf{r}^2$:

$$\frac{\partial f}{\partial \mathbf{r}^2}(\mathbf{x}^0) = [\mathbf{grad}\, f(\mathbf{x}^0)]^T \cdot \frac{\bar{\mathbf{r}}^2}{|\bar{\mathbf{r}}^2|} = \frac{1}{3} \cdot (60, 2, 23) \cdot \begin{pmatrix} 2 \\ 1 \\ 2 \end{pmatrix}$$

$$= \frac{1}{3} \cdot (60 \cdot 2 + 2 \cdot 1 + 23 \cdot 2) = \frac{1}{3} \cdot 168 = 56.$$

Accordingly, for the directional vector $\mathbf{r}^3 = (2, 2, 1)^T$, we obtain first the unit vector

$$\mathbf{r}^3 = \frac{\bar{\mathbf{r}}^3}{|\bar{\mathbf{r}}^3|} = \frac{1}{3} \begin{pmatrix} 2 \\ 2 \\ 1 \end{pmatrix}$$

and then the directional derivative at point $x^0 = (10, 15, 20)$ in the direction r^3:

$$\frac{\partial f}{\partial r^3}(x^0) = [\mathbf{grad}\, f(x^0)]^T \cdot \frac{\bar{r}^3}{|\bar{r}^3|} = \frac{1}{3} \cdot (60, 2, 23) \cdot \begin{pmatrix} 2 \\ 2 \\ 1 \end{pmatrix}$$

$$= \frac{1}{3} \cdot (60 \cdot 2 + 2 \cdot 2 + 23 \cdot 1) = \frac{1}{3} \cdot 147 = 49.$$

Thus the cost has the smallest increase in the case of extending production of the three goods according to vector $\bar{r}^1 = (1, 2, 2)^T$, and the firm would choose this variant. It is worth noting that we can here compare immediately the values of the directional derivatives since all vectors $\bar{r}^1, \bar{r}^2, \bar{r}^3$ considered have the same length 3.

11.5   PARTIAL RATE OF CHANGE AND ELASTICITY; HOMOGENEOUS FUNCTIONS

As in the case of a function of one variable, we can define a rate of change and the elasticity of a function with respect to exactly one of the n variables keeping the remaining ones constant. For instance, if price and income may be considered as the two independent variables of a demand function, we can distinguish between price and income elasticity of demand. In this section, we generalize the rate of change and elasticity of a function of one variable introduced in Chapter 4.4 to the case of functions of several variables.

Definition 11.8   Let function $f : D_f \to \mathbb{R}, D_f \subseteq \mathbb{R}^n$, be partially differentiable and $x^0 = (x_1^0, x_2^0, \dots, x_n^0) \in D_f$ with $f(x^0) \neq 0$. The term

$$\rho_{f, x_i}(x^0) = \frac{f_{x_i}(x^0)}{f(x^0)}$$

is called the *partial rate of change* of function f with respect to x_i at point x^0.

Example 11.13   Let function $f : \mathbb{R}_+^2 \setminus \{(0,0)\} \to \mathbb{R}$ with

$$f(x, y) = x \cdot \sqrt{y} \cdot e^{x/2 + y/4}$$

be given. We determine the partial rates of change of function f at point (20,4) and obtain

$$f_x(x, y) = \sqrt{y} \cdot e^{x/2 + y/4} + \frac{1}{2} \cdot x \cdot \sqrt{y} \cdot e^{x/2 + y/4} = \sqrt{y} \cdot e^{x/2 + y/4} \cdot \left(1 + \frac{x}{2}\right)$$

$$f_y(x, y) = \frac{x}{2\sqrt{y}} \cdot e^{x/2 + y/4} + \frac{x\sqrt{y}}{4} \cdot e^{x/2 + y/4} = \frac{2 + y}{4\sqrt{y}} \cdot x \cdot e^{x/2 + y/4}.$$

This yields

$$\rho_{f,x}(x,y) = \frac{f_x(x,y)}{f(x,y)} = \frac{1 + \frac{x}{2}}{x} = \frac{2 + x}{2x}$$

and

$$\rho_{f,y}(x,y) = \frac{f_y(x,y)}{f(x,y)} = \frac{2 + y}{4y}.$$

For point (20, 4), we obtain

$$\rho_{f,x}(20, 4) = 0.55 \qquad \text{and} \qquad \rho_{f,y}(20, 4) = 0.375.$$

Definition 11.9   Let function $f : D_f \to \mathbb{R}, D_f \subseteq \mathbb{R}^n$, be partially differentiable and $\mathbf{x}^0 = (x_1^0, x_2^0, \ldots, x_n^0) \in D_f$ with $f(\mathbf{x}^0) \neq 0$. The term

$$\varepsilon_{f,x_i}(\mathbf{x}^0) = x_i^0 \cdot \rho_{f,x_i}(\mathbf{x}^0) = x_i^0 \cdot \frac{f_{x_i}(\mathbf{x}^0)}{f(\mathbf{x}^0)}$$

is called the *partial elasticity of function f* with respect to x_i at point $\mathbf{x}^0$.

For point $\mathbf{x}^0 \in D_f$, the value $\varepsilon_{f,x_i}(\mathbf{x}^0)$ is approximately equal to the percentage change in the function value caused by an increase of one per cent in the variable x_i (i.e. from x_i^0 to $1.01x_i^0$) while holding the other variables constant.

One important class of functions in economics is the class of so-called homogeneous functions, which we introduce next.

Definition 11.10   A function $f : D_f \to \mathbb{R}, D_f \subseteq \mathbb{R}^n$, is said to be *homogeneous of degree* k on D_f if $t > 0$ and $(x_1, x_2, \ldots, x_n) \in D_f$ imply $(tx_1, tx_2, \ldots, tx_n) \in D_f$ and

$$f(tx_1, tx_2, \ldots, tx_n) = t^k \cdot f(x_1, x_2, \ldots, x_n) \qquad \text{for all } t > 0,$$

where k can be positive, zero or negative.

According to the latter definition, a function is said to be homogeneous of degree k if the multiplication of each of its independent variables by a constant t changes the function value by the proportion t^k. The value t can take any positive value provided that $(tx_1, tx_2, \ldots, tx_n)$ belongs to the domain D_f of function f. We mention that many functions are not homogeneous of a certain degree. The notion of a homogeneous function is related to the notion 'returns to scale' in economics. It is said that for any production function, there are constant returns to scale, if a proportional increase in all independent variables $x_1, x_2, \ldots, x_n$ leads to the same proportional increase in the function value (output) $f(x_1, x_2, \ldots, x_n)$. Thus, constant returns to scale would mean that we have a homogeneous function of degree $k = 1$. If there is a larger increase in the function value than in the independent variables (i.e. taking t times the original values of the input variables, the production output increases by a bigger factor than t), it is said that there are increasing returns to scale. This corresponds to a homogeneous function of degree $k > 1$. Finally, if there is a production function with decreasing returns to scale, then it corresponds to a homogeneous function of degree $k < 1$. We continue with two properties of homogeneous functions.

THEOREM 11.4 (Euler's theorem)   Let function $f : D_f \to \mathbb{R}$, $D_f \subseteq \mathbb{R}^n$, be continuously partially differentiable, where inequality $t > 0$ and inclusion $\mathbf{x} = (x_1, x_2, \ldots, x_n) \in D_f$ imply inclusion $(tx_1, tx_2, \ldots, tx_n) \in D_f$. Then function f is homogeneous of degree k on D_f if and only if the equation

$$x_1 \cdot f_{x_1}(\mathbf{x}) + x_2 \cdot f_{x_2}(\mathbf{x}) + \ldots + x_n \cdot f_{x_n}(\mathbf{x}) = k \cdot f(\mathbf{x}) \tag{11.1}$$

holds for all points $\mathbf{x} = (x_1, x_2, \ldots, x_n) \in D_f$.

In gradient notation, we can express equation (11.1) in Euler's theorem as

$$\mathbf{x}^T \cdot \mathbf{grad}\, f(x) = k \cdot f(\mathbf{x}).$$

Applying the definition of partial elasticity, we obtain the following corollary.

COROLLARY 11.1   Let function $f : D_f \to \mathbb{R}$, $D_f \subseteq \mathbb{R}^n$, with $f(\mathbf{x}) \neq 0$ for $\mathbf{x} \in D_f$, be homogeneous of degree k. Then the sum of the partial elasticities of function f is equal to k, i.e.

$$\varepsilon_{f,x_1}(\mathbf{x}) + \varepsilon_{f,x_2}(\mathbf{x}) + \ldots + \varepsilon_{f,x_n}(\mathbf{x}) = k.$$

Example 11.14   Consider a Cobb–Douglas production function $Q : \mathbb{R}^2_+ \to \mathbb{R}$ with two variables K and L given by

$$Q(K, L) = A \cdot K^{a_1} \cdot L^{a_2},$$

where K denotes the quantity of capital and L denotes the quantity of labour used for the production of a firm. Determining the partial derivatives, we get

$$Q_K(K, L) = A \cdot a_1 \cdot K^{a_1 - 1} \cdot L^{a_2} \qquad \text{and} \qquad Q_L(K, L) = A \cdot K^{a_1} \cdot a_2 \cdot L^{a_2 - 1}.$$

The partial derivative Q_K is also known as the marginal product of capital and Q_L as the marginal product of labour.
For the partial elasticities at point (K, L), we obtain

$$\epsilon_{f,K}(K, L) = K \cdot \frac{A \cdot a_1 \cdot K^{a_1 - 1} L^{a_2}}{A \cdot K^{a_1} \cdot L^{a_2}} = a_1 \qquad \text{and}$$

$$\epsilon_{f,L}(K,L) = L \cdot \frac{A \cdot K^{a_1} \cdot a_2 \cdot L^{a_2-1}}{A \cdot K^{a_1} \cdot L^{a_2}} = a_2.$$

Thus, the exponents in the Cobb–Douglas function correspond to the partial elasticities. Finally, from the equalities

$$Q(tK, tL) = A \cdot (tK)^{a_1} \cdot (tL)^{a_2} = t^{a_1+a_2} \cdot Q(K, L),$$

we find that function Q is homogeneous of degree $a_1 + a_2$.

11.6 IMPLICIT FUNCTIONS

In our previous considerations, the function was defined explicitly, i.e. we considered a function of one or several variables in the form: $y = f(x)$ and $z = f(x_1, x_2, \ldots, x_n)$, respectively. However, in many applications a relationship between two variables x and y may be given in the form: $F(x, y) = 0$, or correspondingly, $F(x_1, x_2, \ldots, x_n; z) = 0$. In the former case, we say that y is *implicitly* defined as a function of x if there exists a function f with $y = f(x)$ such that $F(x, y) = F(x, f(x)) = 0$.

Example 11.15 Let

$$\frac{x^2}{a^2} + \frac{y^2}{b^2} - 1 = 0.$$

The above equation defines two elementary functions of the form

$$y_1 = +b \sqrt{1 - \frac{x^2}{a^2}}$$

and

$$y_2 = -b \sqrt{1 - \frac{x^2}{a^2}}.$$

Each of the above functions defines half an ellipse (see Figure 11.7) with the half-axes a and b and the centre at point $(0, 0)$. Inserting the above explicit representations into $F(x, y) = 0$, we obtain in both cases the identity

$$\frac{x^2}{a^2} + \left(1 - \frac{x^2}{a^2}\right) - 1 = 0.$$

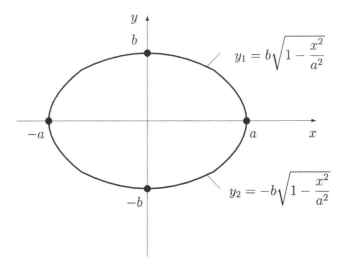

$y_1 = b\sqrt{1 - \dfrac{x^2}{a^2}}$

$y_2 = -b\sqrt{1 - \dfrac{x^2}{a^2}}$

Figure 11.7 The ellipse $x^2/a^2 + y^2/b^2 - 1 = 0$.

However, not every implicit representation $F(x, y) = 0$ can be solved for y, e.g. the equation $2y^6 + y - \ln x = 0$ cannot be solved for y.

Similarly, one can consider an explicitly given system in the form:

$$\begin{aligned}
y_1 &= f_1(x_1, x_2, \ldots, x_n) \\
y_2 &= f_2(x_1, x_2, \ldots, x_n) \\
&\;\;\vdots \\
y_m &= f_m(x_1, x_2, \ldots, x_n).
\end{aligned} \tag{11.2}$$

Such a system of equations might be given in implicit form:

$$\begin{aligned}
F_1(x_1, x_2, \ldots, x_n; y_1, y_2, \ldots, y_m) &= 0 \\
F_2(x_1, x_2, \ldots, x_n; y_1, y_2, \ldots, y_m) &= 0 \\
&\;\;\vdots \\
F_m(x_1, x_2, \ldots, x_n; y_1, y_2, \ldots, y_m) &= 0.
\end{aligned} \tag{11.3}$$

Representation (11.2) is referred to as the *reduced form* of system (11.3). The next problem we deal with is to find an answer to the following question: under what conditions is it possible to put a system given in the form (11.3) into its reduced form (11.2)? This question is answered by the following theorem.

THEOREM 11.5 Let a system (11.3) of implicitly defined functions with continuous partial derivatives with respect to all variables be given and $(\mathbf{x}^0; \mathbf{y}^0) = (x_1^0, x_2^0, \ldots, x_n^0; y_1^0, y_2^0, \ldots, y_m^0)$ be a point that satisfies system (11.3). If the matrix of the partial derivatives of the functions

F_j with respect to the variables y_k is regular at point $(\mathbf{x}^0; \mathbf{y}^0)$, i.e.

$$|J(\mathbf{x}^0; \mathbf{y}^0)| = \det \left(\frac{\partial F_j(\mathbf{x}^0; \mathbf{y}^0)}{\partial y_k} \right) \neq 0 \quad (|J(\mathbf{x}^0; \mathbf{y}^0)| \text{ is the Jacobian determinant}),$$

then system (11.3) can be uniquely put into its reduced form (11.2) in a neighbourhood of point $(\mathbf{x}^0; \mathbf{y}^0)$.

We illustrate the use of Theorem 11.5 by the following example.

Example 11.16 Consider the implicitly given system

$$F_1(x_1, x_2; y_1, y_2) = x_1 y_1 - x_2 - y_2 = 0 \tag{11.4}$$

$$F_2(x_1, x_2; y_1, y_2) = x_2^2 y_2 - x_1 + y_1 = 0, \tag{11.5}$$

and we investigate whether this system can be put into its reduced form. Checking the Jacobian determinant, we obtain

$$|J(x_1, x_2; y_1, y_2)| = \begin{vmatrix} \dfrac{\partial F_1}{\partial y_1} & \dfrac{\partial F_1}{\partial y_2} \\ \dfrac{\partial F_2}{\partial y_1} & \dfrac{\partial F_2}{\partial y_2} \end{vmatrix} = \begin{vmatrix} x_1 & -1 \\ 1 & x_2^2 \end{vmatrix} = x_1 x_2^2 + 1.$$

According to Theorem 11.5, we can conclude that the given system can be put into its reduced form for all points $(\mathbf{x}, \mathbf{y}) = (x_1, x_2; y_1, y_2)$ with $x_1 x_2^2 + 1 \neq 0$ satisfying equations (11.4) and (11.5). Indeed, from equation (11.4) we obtain

$$y_2 = x_1 y_1 - x_2.$$

Substituting the latter term into equation (11.5), we get

$$x_2^2 (x_1 y_1 - x_2) - x_1 + y_1 = 0$$

which can be rewritten as

$$(x_1 x_2^2 + 1) y_1 = x_1 + x_2^3.$$

For $x_1 x_2^2 + 1 \neq 0$, we get

$$y_1 = \frac{x_1 + x_2^3}{x_1 x_2^2 + 1}$$

and

$$y_2 = x_1 \cdot \left(\frac{x_1 + x_2^3}{x_1 x_2^2 + 1} \right) - x_2 = \frac{x_1^2 + x_1 x_2^3 - x_1 x_2^3 - x_2}{x_1 x_2^2 + 1} = \frac{x_1^2 - x_2}{x_1 x_2^2 + 1}.$$

If in the representation $F(x_1, x_2, \ldots, x_n; z) = 0$ variable z can be eliminated, we can perform the partial differentiation with respect to some variable as discussed before. However, if it is not possible to solve this equation for z, we can nevertheless perform the partial differentiations by applying the following theorem about the *differentiation of an implicitly given function*.

THEOREM 11.6 (implicit-function theorem) Let function $F : D_F \to \mathbb{R}$, $D_F \subseteq \mathbb{R}^{n+1}$, with $F(\mathbf{x}; z) = F(x_1, x_2, \ldots, x_n; z)$ be continuous and $F(\mathbf{x}^0, z^0) = 0$. Moreover, let function F be continuously partially differentiable with $F_z(\mathbf{x}^0, z^0) \neq 0$. Then in some neighbourhood $U_\epsilon(\mathbf{x}^0)$ of point $\mathbf{x}^0$ with $\epsilon > 0$, there exists a function f with $z = f(x_1, x_2, \ldots, x_n)$, and function f is continuously partially differentiable for $\mathbf{x} \in U_\epsilon(\mathbf{x}^0)$ with

$$f_{x_i}(\mathbf{x}) = f_{x_i}(x_1, x_2, \ldots, x_n) = -\frac{F_{x_i}(\mathbf{x}; z)}{F_z(\mathbf{x}; z)}.$$

For the special case of $n = 2$, the above implicit-function theorem can be formulated as follows. Let $F(x, y) = 0$ be an implicitly given function and (x^0, y^0) be a point that satisfies the latter equation. Then, under the assumptions of Theorem 11.6, for points x of some interval (a, b) containing point x_0 the first derivative of function f with $y = f(x)$ is given by

$$y'(x) = -\frac{F_x(x, y)}{F_y(x, y)}. \tag{11.6}$$

Example 11.17 We determine the derivative of function f with $y = f(x)$ implicitly defined by

$$F(x, y) = 8x^3 - 24xy^2 + 16y^3 = 0$$

by means of the implicit-function theorem. This yields

$$F_x(x, y) = 24x^2 - 24y^2, \qquad F_y(x, y) = -48xy + 48y^2,$$

and we obtain

$$y' = -\frac{24(x^2 - y^2)}{-48y(x - y)} = \frac{x + y}{2y}.$$

As an alternative, we can also directly differentiate function $F(x, y(x)) = 0$ with respect to x. In this case, using the product rule for the second term of $F(x, y)$ (since y depends on x), we get

$$24x^2 - 24y^2 - 48xyy' + 48y^2 y' = 0,$$

from which we obtain

$$48y'y(y - x) = 24(y^2 - x^2)$$

and consequently

$$y' = \frac{x + y}{2y}.$$

The latter formula allows us to determine the value of the first derivative at some point (x, y) with $F(x, y) = 0$ *without* knowing the corresponding explicit representation $y = f(x)$. This is particularly favourable when it is impossible or difficult to transform the implicit representation $F(x, y) = 0$ into an explicit representation $y = f(x)$ of function f.

Using formula (11.6), we can determine the second derivative y'' of function f. By applying the chain rule, we get

$$y'' = -\frac{\left[F_{xx}(x, y) + F_{xy}(x, y) \cdot y_x\right] \cdot F_y(x, y) - \left[F_{yx}(x, y) + F_{yy}(x, y) \cdot y_x\right] \cdot F_x(x, y)}{F_y^2(x, y)}$$

with $y_x = y' = f'$. Substituting $y_x = y'$ according to formula (11.6), we obtain

$$y'' = -\frac{F_{xx}(x, y) \cdot F_y^2(x, y) - 2 \cdot F_{xy}(x, y) \cdot F_x(x, y) \cdot F_y(x, y) + F_{yy}(x, y) \cdot F_x^2(x, y)}{F_y^3(x, y)}.$$

$$(11.7)$$

Example 11.18 Let function $f : D_f \to \mathbb{R}$ be implicitly given by

$$F(x, y) = x - y + 2 \sin y = 0.$$

(Note that we cannot solve the above equation for y.) We determine y' and y''. Using

$$F_x(x, y) = 1 \qquad \text{and} \qquad F_y(x, y) = -1 + 2 \cos y,$$

we get

$$y' = -\frac{1}{-1 + 2 \cos y} = \frac{1}{1 - 2 \cos y}.$$

Moreover, using

$$F_{xx}(x, y) = F_{xy}(x, y) = 0 \qquad \text{and} \qquad F_{yy}(x, y) = -2 \sin y,$$

we obtain by means of formula (11.7)

$$y'' = -\frac{0 \cdot (-1 + 2 \cos y)^2 - 2 \cdot 0 \cdot 0 \cdot (-1 + 2 \cos y) + (-2 \sin y \cdot 1)}{(-1 + 2 \cos y)^3}$$

$$= \frac{2 \sin y}{(2 \cos y - 1)^3}.$$

11.7 UNCONSTRAINED OPTIMIZATION

11.7.1 Optimality conditions

In economics, one often has to look for a minimum or a maximum of a function depending on several variables. If no additional constraints have to be considered, it is an *unconstrained optimization problem*. As an introductory example we consider the maximization of the profit of a firm.

Example 11.19   For simplicity assume that a firm produces only two goods X and Y. Suppose that this firm can sell each piece of product X for 45 EUR and each piece of product Y for 55 EUR. Denoting by x the quantity of product X and by y the quantity of product Y, we get revenue $R = R(x, y)$ with

$$R(x, y) = 45x + 55y, \qquad D_R = \mathbb{R}_+^2.$$

The production cost of the firm also depends on the quantities x and y. Suppose that this cost function $C = C(x, y)$ is known as

$$C(x, y) = 300 + x^2 + 1.5y^2 - 25x - 35y, \qquad D_C = \mathbb{R}_+^2.$$

One can easily show that inequality $C(x, y) \geq 0$ holds for all non-negative values of variables x and y, and for the fixed cost of production one has $C(0, 0) = 300$. In order to determine the profit $P = f(x, y)$ in dependence on both values x and y, we have to subtract the cost from revenue, i.e.

$$\begin{aligned} f(x, y) &= R(x, y) - C(x, y) \\ &= 45x + 55y - (300 + x^2 + 1.5y^2 - 25x - 35y) \\ &= -x^2 + 70x - 1.5y^2 + 90y - 300. \end{aligned}$$

To maximize the profit, we have to determine the point $(x, y) \in \mathbb{R}_+^2$ that maximizes function f. We say that we look for an extreme point (in this case for a maximum point) of function f. This is an unconstrained optimization problem since no other constraints of the production have to be taken into consideration (except the usual non-negativity constraints of variables x and y).

In the following, we look for necessary and sufficient conditions for the existence of extreme points of a function depending on n variables. First, we formally define a local extreme (i.e. minimum or maximum) point as well as a global extreme point.

Definition 11.11   A function $f : D_f \to \mathbb{R}, D_f \subseteq \mathbb{R}^n$, has a *local maximum (minimum)* at point x^0 if there exists an ϵ-neighbourhood $U_\epsilon(x^0) \in D_f$ of point x^0 with $\epsilon > 0$ such that

$$f(x) \leq f(x^0) \qquad (f(x) \geq f(x^0), \text{ respectively}) \tag{11.8}$$

for all points $x \in U_\epsilon(x^0)$. Point x^0 is called a *local maximum (minimum) point* of function f.

If inequality (11.8) holds for all points $x \in D_f$, function f has at point x^0 a *global maximum (minimum)*, and x^0 is called a *global maximum (minimum) point* of function f.

THEOREM 11.7 (necessary first-order conditions) Let function $f : D_f \to \mathbb{R}, D_f \subseteq \mathbb{R}^n$, be partially differentiable. If $x^0 = (x_1^0, x_2^0, \ldots, x_n^0) \in D_f$ is a local extreme point (i.e. a local maximum or minimum point) of function f, then

$$\operatorname{grad} f(x^0) = \operatorname{grad} f(x_1^0, x_2^0, \ldots, x_n^0) = 0,$$

i.e.

$$f_{x_1}(x^0) = 0, \quad f_{x_2}(x^0) = 0, \quad \ldots, \quad f_{x_n}(x^0) = 0.$$

A point x^0 with $\operatorname{grad} f(x^0) = 0$ is called a *stationary point*. According to Theorem 11.7 we can conclude that, if $x^0 = (x_1^0, x_2^0, \ldots, x_n^0)$ is a local extreme point (assuming that function f is partially differentiable at this point), then it has to be a stationary point.

In order to determine a global minimum or maximum point of a function f depending on n variables, we have to search among the stationary points (which can be found by means of differential calculus), the points on the boundary of the domain and possibly among the points where the partial derivatives do not exist. This is similar to the case of a function of one variable (see Chapter 4). In the following, we focus on the determination of local extreme points by differential calculus.

Next, we are looking for a sufficient condition for the existence of a local extreme point. In the case of a function of one variable, we have presented one criterion which uses higher-order derivatives. In the following, we present a criterion that uses the second-order partial derivatives to answer the question of whether a given stationary point is indeed a local extreme point. We begin with the definition of the Hessian.

Definition 11.12   The matrix

$$H_f(x^0) = \left(f_{x_i x_j}(x^0)\right) = \begin{pmatrix} f_{x_1 x_1}(x^0) & f_{x_1 x_2}(x^0) & \cdots & f_{x_1 x_n}(x^0) \\ f_{x_2 x_1}(x^0) & f_{x_2 x_2}(x^0) & \cdots & f_{x_2 x_n}(x^0) \\ \vdots & \vdots & & \vdots \\ f_{x_n x_1}(x^0) & f_{x_n x_2}(x^0) & \cdots & f_{x_n x_n}(x^0) \end{pmatrix}$$

is called the *Hessian*, or the *Hessian matrix*, of function f at point $x^0 = (x_1^0, x_2^0, \ldots, x_n^0) \in D_f \subseteq \mathbb{R}^n$.

Under the assumptions of Theorem 11.1 (i.e. continuous second-order partial derivatives), the Hessian is a symmetric matrix. The following theorems present sufficient conditions for so-called isolated (local or global) extreme points, for which in Definition 11.11, the strict inequality holds in inequality (11.8) for $x \neq x^0$.

THEOREM 11.8 (local and global sufficient second-order conditions)   Let function $f : D_f \to \mathbb{R}, D_f \subseteq \mathbb{R}^n$, be twice continuously partially differentiable and let $x^0 = (x_1^0, x_2^0, \ldots, x_n^0) \in D_f$ be a stationary point of function f. Then:

(1) If the Hessian $H_f(x^0)$ is negative (positive) definite, then x^0 is a local maximum (minimum) point of function f.

(2) If $H_f(\mathbf{x}^0)$ is indefinite, then function f does not have a local extremum at point $\mathbf{x}^0$.
(3) If $H_f(\mathbf{x})$ is negative (positive) definite for all points $\mathbf{x} \in D_f$, then $\mathbf{x}^0$ is a global maximum (minimum) point of function f over D_f.

A stationary point $\mathbf{x}^0$ of $f(\mathbf{x})$ that is neither a local maximum nor a local minimum point is called a *saddle point* of function f. Next, we give a sufficient condition for a stationary point to be either a local minimum, a local maximum or a saddle point in the case of a function of two variables.

THEOREM 11.9 (special case of Theorem 11.8 for the case $n = 2$)   Let function f: $D_f \to \mathbb{R}, D_f \subseteq \mathbb{R}^2$, be twice continuously partially differentiable and let $(x^0, y^0) \in D_f$ be a stationary point of function f. Then:

(1) If

$$|H_f(x^0, y^0)| = f_{xx}(x^0, y^0) \cdot f_{yy}(x^0, y^0) - \left[f_{xy}(x^0, y^0)\right]^2 > 0,$$

then (x^0, y^0) is a local extreme point of function f, in particular:

(a) If $f_{xx}(x^0, y^0) < 0$ or $f_{yy}(x^0, y^0) < 0$, it is a local maximum point.
(b) If $f_{xx}(x^0, y^0) > 0$ or $f_{yy}(x^0, y^0) > 0$, it is a local minimum point.

(2) If $|H_f(x^0, y^0)| < 0$, then (x^0, y^0) is a saddle point of function f.
(3) If $|H_f(x^0, y^0)| = 0$, then (x^0, y^0) could be a local maximum point, a local minimum point, or a saddle point of function f.

In case (1) of Theorem 11.9, we can take *any* of the second-order partial derivatives f_{xx} or f_{yy} in order to decide whether a stationary point is a local maximum or minimum point. Notice also that in the case of a function f depending on two variables, at a saddle point function f has a local maximum with respect to one of the independent variables and a local minimum with respect to the other independent variable.

Now we are able to find the extreme point of the profit function f with $z = f(x, y)$ given in Example 11.1.

Example 11.1 (continued)   We first establish the necessary first-order conditions according to Theorem 11.7 and obtain

$$f_x(x, y) = -2x + 70 = 0$$
$$f_y(x, y) = -3y + 90 = 0.$$

Thus, we get the only stationary point $P_1 : (x_1, y_1) = (35, 30)$. In order to check the sufficient second-order conditions according to Theorem 11.9, we determine the second-order partial derivatives

$$f_{xx}(x, y) = -2, \qquad f_{yy}(x, y) = -3, \qquad f_{xy}(x, y) = f_{yx}(x, y) = 0,$$

set up the Hessian

$$H_f(x,y) = \begin{pmatrix} f_{xx}(x,y) & f_{xy}(x,y) \\ f_{yx}(x,y) & f_{yy}(x,y) \end{pmatrix} = \begin{pmatrix} -2 & 0 \\ 0 & -3 \end{pmatrix},$$

and obtain

$$|H_f(35,30)| = f_{xx}(35,30) \cdot f_{yy}(35,30) - \left[f_{xy}(35,30)\right]^2 = (-2) \cdot (-3) - 0^2 = 6 > 0.$$

Therefore, point P_1 is a local extreme point. It is worth noting that in this case the value of the determinant of the Hessian does not even depend on the concrete values of variables x and y. Moreover, since

$$f_{xx}(35,30) = -2 < 0 \qquad \text{(or analagously since} \quad f_{yy}(35,30) = -3 < 0)$$

point P_1 with $(x_1,y_1) = (35,30)$ is a local maximum point of the profit function. The maximal profit is equal to $P = f(35,30) = 2,275$ units.

Example 11.20   Let function $f : \mathbb{R}^2 \to \mathbb{R}$ with

$$f(x,y) = \frac{1}{3}x^3 + \frac{1}{2}x^2y + x^2 - 4x - \frac{1}{6}y^3$$

be given. We determine all local extreme points of function f. The necessary first-order conditions according to Theorem 11.7 for an extreme point are

$$f_x(x,y) = x^2 + xy + 2x - 4 = 0 \tag{11.9}$$

$$f_y(x,y) = \frac{1}{2}x^2 - \frac{1}{2}y^2 = 0. \tag{11.10}$$

From equation (11.10) we obtain $x^2 = y^2$ which yields $|x| = |y|$ or equivalently $y = \pm x$.

Case 1 $y = x$. Substituting $y = x$ into equation (11.9), we get $2x^2 + 2x - 4 = 0$, or accordingly,

$$x^2 + x - 2 = 0.$$

This quadratic equation has the two solutions $x_1 = -2$ and $x_2 = 1$. Thus, we get $y_1 = -2$ and $y_2 = 1$, which gives the two stationary points

$$P_1 : (x_1,y_1) = (-2,-2) \qquad \text{and} \qquad P_2 : (x_2,y_2) = (1,1).$$

Case 2 $y = -x$. In this case, we obtain from condition (11.9) the equation $2x - 4 = 0$ which yields $x_3 = 2$ and $y_3 = -2$ and the stationary point

$$P_3 : (x_3,y_3) = (2,-2).$$

Thus, we have three stationary points, $P_1 : (-2,-2), P_2 : (1,1)$ and $P_3 : (2,-2)$. Next, we check the sufficient second-order conditions according to Theorem 11.9 and obtain

$$f_{xx}(x,y) = 2x + y + 2, \qquad f_{yy}(x,y) = -y, \qquad f_{xy}(x,y) = f_{yx}(x,y) = x.$$

This yields the Hessian

$$H_f(x,y) = \begin{pmatrix} f_{xx}(x,y) & f_{xy}(x,y) \\ f_{yx}(x,y) & f_{yy}(x,y) \end{pmatrix} = \begin{pmatrix} 2x+y+2 & x \\ x & -y \end{pmatrix},$$

and for the stationary points, we obtain

$$|H_f(-2,-2)| = f_{xx}(-2,-2) \cdot f_{yy}(-2,-2) - \left[f_{xy}(-2,-2) \right]^2$$
$$= (-4) \cdot 2 - (-2)^2 = -12 < 0,$$

$$|H_f(1,1)| = f_{xx}(1,1) \cdot f_{yy}(1,1) - \left[f_{xy}(1,1) \right]^2 = 5 \cdot (-1) - 1^2 = -6 < 0,$$

$$|H_f(2,-2)| = f_{xx}(2,-2) \cdot f_{yy}(2,-2) - \left[f_{xy}(2,-2) \right]^2 = 4 \cdot 2 - 2^2 = 4 > 0.$$

Thus, according to Theorem 11.9, P_1 and P_2 are saddle points, and only point P_3 is a local extreme point of function f. Due to $f_{xx}(2,-2) = 4 > 0$ (or accordingly $f_{yy}(2,-2) = 2 > 0$), point P_3 is a local minimum point with the function value $f(x_3,y_3) = -4$.

We continue with another example of finding the maximum profit of a firm.

Example 11.21 We consider the profit maximization problem of a firm that uses two input factors F_1 and F_2 to produce one output. Let P be the constant price per unit of the output, x be the quantity of factor F_1, y be the quantity of factor F_2 as well as p_1 and p_2 be the prices per unit of the two inputs. Moreover, function $f : \mathbb{R}_+^2 \to \mathbb{R}$ is the production function of the firm, i.e. $f(x,y)$ gives the quantity produced in dependence on quantities x and y. Thus, $Pf(x,y)$ denotes revenue from which cost of production, i.e. $p_1 x + p_2 y$, has to be subtracted to get the profit function $g : \mathbb{R}_+^2 \to \mathbb{R}$. This problem can be written as

$$g(x,y) = P \cdot f(x,y) - p_1 x - p_2 y \to \max!$$

The necessary first-order conditions (according to Theorem 11.7) for the optimal choices of the input factors F_1 and F_2 are as follows:

$$g_x(x,y) = Pf_x(x,y) - p_1 = 0$$
$$g_y(x,y) = Pf_y(x,y) - p_2 = 0.$$

Let us suppose that the production function f is of the Cobb–Douglas type given by

$$f(x,y) = x^{1/2} \cdot y^{1/3}.$$

Then the two necessary first-order optimality conditions according to Theorem 11.7 become

$$g_x(x,y) = \frac{1}{2} \cdot P \cdot x^{-1/2} \cdot y^{1/3} - p_1 = 0$$

$$g_y(x,y) = \frac{1}{3} \cdot P \cdot x^{1/2} \cdot y^{-2/3} - p_2 = 0.$$

Multiplying the first equation by x and the second one by y, we get

$$\frac{1}{2} \cdot P \cdot x^{1/2} \cdot y^{1/3} - p_1 x = 0$$

$$\frac{1}{3} \cdot P \cdot x^{1/2} \cdot y^{1/3} - p_2 y = 0.$$

Moreover, multiplying the first equation by $-2/3$ and summing up both equations gives

$$\frac{2}{3} p_1 x - p_2 y = 0$$

and therefore

$$y = \frac{2p_1}{3p_2} \cdot x.$$

Substituting this into the above equation $g_x(x, y) = 0$, we get

$$\frac{1}{2} \cdot P \cdot x^{-1/2} \cdot \left(\frac{2p_1}{3p_2} \cdot x\right)^{1/3} - p_1 = 0.$$

The latter can be rewritten as the equation

$$\frac{1}{2} \cdot P \cdot x^{(-1/2+1/3)} \cdot \left(\frac{2p_1}{3p_2}\right)^{1/3} = p_1$$

from which we obtain

$$x = \left[\frac{P}{2p_1} \cdot \left(\frac{2p_1}{3p_2}\right)^{1/3}\right]^6$$

and thus

$$x_0 = \left(\frac{P}{2p_1}\right)^6 \cdot \left(\frac{2p_1}{3p_2}\right)^2 = \frac{P^6}{(2p_1)^4 \cdot (3p_2)^2} = \left(\frac{P}{2p_1}\right)^4 \cdot \left(\frac{P}{3p_2}\right)^2.$$

Furthermore, we get

$$y_0 = \frac{2p_1}{3p_2} \cdot x_0 = \frac{2p_1}{3p_2} \cdot \left(\frac{P}{2p_1}\right)^4 \cdot \left(\frac{P}{3p_2}\right)^2 = \left(\frac{P}{2p_1}\right)^3 \cdot \left(\frac{P}{3p_2}\right)^3.$$

Thus, we have obtained the only stationary point

$$(x_0, y_0) = \left(\left(\frac{P}{2p_1}\right)^4 \cdot \left(\frac{P}{3p_2}\right)^2, \left(\frac{P}{2p_1}\right)^3 \cdot \left(\frac{P}{3p_2}\right)^3\right)$$

for the quantities of both input factors F_1 and F_2.

Checking the sufficient second-order conditions according to Theorem 11.9, we get

$$g_{xx}(x,y) = -\frac{1}{4} \cdot P \cdot x^{-3/2} \cdot y^{1/3}, \qquad g_{yy}(x,y) = -\frac{2}{9} \cdot P \cdot x^{1/2} \cdot y^{-5/3},$$

$$g_{xy}(x,y) = g_{yx}(x,y) = \frac{1}{6} \cdot P \cdot x^{-1/2} \cdot y^{-2/3}$$

and consequently,

$$|H_f(x,y)| = \begin{vmatrix} g_{xx}(x,y) & g_{xy}(x,y) \\ g_{yx}(x,y) & g_{yy}(x,y) \end{vmatrix} = \begin{vmatrix} -\dfrac{1}{4} \cdot P \cdot x^{-3/2} \cdot y^{1/3} & \dfrac{1}{6} \cdot P \cdot x^{-1/2} \cdot y^{-2/3} \\ \dfrac{1}{6} \cdot P \cdot x^{-1/2} \cdot y^{-2/3} & -\dfrac{2}{9} \cdot P \cdot x^{1/2} \cdot y^{-5/3} \end{vmatrix}$$

$$= \frac{2}{36} \cdot P^2 \cdot x^{-1} \cdot y^{-4/3} - \frac{1}{36} \cdot P^2 \cdot x^{-1} \cdot y^{-4/3} = \frac{1}{36} \cdot P^2 \cdot x^{-1} \cdot y^{-4/3}.$$

Since x_0 and y_0 are positive, we have obtained inequality $|H_f(x_0,y_0)| > 0$ which means that (x_0,y_0) is a local extreme point. Since $g_{xx}(x_0,y_0) < 0$, (x_0,y_0) is a local maximum point according to Theorem 11.9, and the corresponding supply is given by

$$f(x_0,y_0) = x_0^{1/2} \cdot y_0^{1/3} = \left[\left(\frac{P}{2p_1}\right)^4 \cdot \left(\frac{P}{3p_2}\right)^2 \right]^{1/2} \cdot \left[\left(\frac{P}{2p_1}\right)^3 \cdot \left(\frac{P}{3p_2}\right)^3 \right]^{1/3}$$

$$= \left(\frac{P}{2p_1}\right)^3 \cdot \left(\frac{P}{3p_2}\right)^2 .$$

Example 11.22 Each of two enterprises E_1 and E_2 offers a single product. The relationships between the outputs x_1, x_2 and the prices p_1, p_2 are as follows:

$$x_1 = 110 - 2p_1 - p_2,$$
$$x_2 = 140 - p_1 - 3p_2.$$

The total costs for the enterprises are given by

$$C_1(x_1) = 120 + 2x_1,$$
$$C_2(x_2) = 140 + 2x_2.$$

(a) First, we determine the profit functions f_1, f_2 of both enterprises and the total profit function $f = f_1 + f_2$. We recall that the profit function f_i, $i \in \{1,2\}$, is obtained as the difference of the revenue $x_i p_i$ and the cost $C_i = C_i(x_i)$, i.e.

$$f_1(p_1,p_2) = x_1 p_1 - C_1(x_1)$$
$$= (110 - 2p_1 - p_2) \cdot p_1 - [120 + 2(110 - 2p_1 - p_2)]$$
$$= 110p_1 - 2p_1^2 - p_1 p_2 - 120 - 220 + 4p_1 + 2p_2$$
$$= -340 - 2p_1^2 - p_1 p_2 + 114p_1 + 2p_2,$$

$$f_2(p_1, p_2) = x_2 p_2 - C_2(x_2)$$
$$= (140 - p_1 - 3p_2) \cdot p_2 - [140 + 2(140 - p_1 - 3p_2)]$$
$$= 140p_2 - p_1 p_2 - 3p_2^2 - 140 - 280 + 2p_1 + 6p_2$$
$$= -420 - p_1 p_2 - 3p_2^2 + 2p_1 + 146p_2.$$

This yields the total profit function $f = f_1 + f_2$ as follows:

$$f(p_1, p_2) = -760 - 2p_1^2 - 2p_1 p_2 - 3p_2^2 + 116p_1 + 148p_2.$$

(b) Next, we determine prices p_1 and p_2 such that the total profit is maximized. Applying Theorem 11.7, we get

$$f_{p_1}(p_1, p_2) = -4p_1 - 2p_2 + 116 = 0$$
$$f_{p_2}(p_1, p_2) = -2p_1 - 6p_2 + 148 = 0.$$

The latter system can be rewritten as

$$2p_1 + p_2 = 58$$
$$p_1 + 3p_2 = 74,$$

which has the unique solution

$$p_1^0 = 20 \quad \text{and} \quad p_2^0 = 18.$$

Checking the sufficient conditions according to Theorem 11.9, we obtain

$$|H_f(20, 18)| = f_{p_1 p_1}(20, 18) \cdot f_{p_2 p_2}(20, 18) - [f_{p_1 p_2}(20, 18)]^2$$
$$= (-4) \cdot (-6) - (-2)^2 = 20 > 0.$$

Therefore, the stationary point obtained is a local extreme point. From

$$f_{p_1 p_1}(20, 18) = -4 < 0,$$

we have found that $(p_1^0, p_2^0) = (20, 18)$ is a local maximum point with the total profit

$$f(20, 18) = -760 - 2 \cdot (20)^2 - 2 \cdot 20 \cdot 18 - 3 \cdot (18)^2 + 116 \cdot 20 + 148 \cdot 18 = 1,732.$$

(c) Following some problems between the enterprises, E_2 fixes the price $p_2^* = 22$. Next, we want to investigate which price p_1^* would ensure a maximum profit f_1 for enterprise E_1 under this additional condition. In this case, we get a function f_1 with

$$f_1(p_1, 22) = -340 - 2p_1^2 - 22p_1 + 114p_1 + 44 = -296 - 2p_1^2 + 92p_1$$

which depends now only on variable p_1. We denote $F(p_1) = f_1(p_1, 22)$. Applying the criterion given in Chapter 4.5 (see Theorem 4.12), we get as a necessary condition

$$F'(p_1) = -4p_1 + 92 = 0$$

which yields the only solution

$$p_1^* = 23.$$

Since

$$F''(p_1^*) = -4 < 0,$$

the solution obtained is indeed the maximum profit of enterprise E_1 under the above constraint $p_2^* = 22$. From equality $f(23, 22) = 1,642$, the total profit of both enterprises reduces by $1,732 - 1,642 = 90$ units in this case.

Example 11.23   The profit function $P : \mathbb{R}_+^3 \to \mathbb{R}$ for some agricultural product is a function of the quantities used of three sorts of chemical fertilizers. Let function P with

$$P(C_1, C_2, C_3) = 340 - \frac{1}{3}C_1^3 + \frac{1}{2}C_1^2 + 2C_1C_2 - C_2^2 + C_2C_3 - \frac{1}{2}C_3^2 + 2C_1 + C_2 + C_3$$

be given, where C_i denotes the quantity of the ith chemical fertilizer used in tons, $i \in \{1, 2, 3\}$, and $P(C_1, C_2, C_3)$ is the resulting profit in thousands of EUR. We determine the quantities of the three sorts of chemical fertilizers such that the profit is maximized. Checking the necessary conditions of Theorem 11.7, we get

$$P_{C_1} = -C_1^2 + C_1 + 2C_2 + 2 = 0 \tag{11.11}$$

$$P_{C_2} = 2C_1 - 2C_2 + C_3 + 1 = 0 \tag{11.12}$$

$$P_{C_3} = C_2 - C_3 + 1 = 0. \tag{11.13}$$

Adding equations (11.12) and (11.13), we obtain

$$C_2 = 2C_1 + 2 = 2(C_1 + 1). \tag{11.14}$$

Inserting equation (11.14) into equation (11.11), we get

$$-C_1^2 + C_1 + 4C_1 + 4 + 2 = 0$$

which can be rewritten as

$$C_1^2 - 5C_1 - 6 = 0.$$

The latter quadratic equation has two solutions denoted by $C_1^{(1)}$ and $C_1^{(2)}$:

$$C_1^{(1)} = 6 \qquad \text{and} \qquad C_1^{(2)} = -1.$$

The second root $C_1^{(2)} = -1$ does not lead to a point belonging to the domain of function P. Using $C_1^{(1)} = 6$, we get from equations (11.14) and (11.13)

$$C_2^{(1)} = 2(C_1^{(1)} + 1) = 14 \qquad \text{and} \qquad C_3^{(1)} = C_2^{(1)} + 1 = 15.$$

So, we have only one stationary point $(6, 14, 15)$ within the domain of function P. Next, we check whether this point is indeed a local maximum point. We get the Hessian

$H_P(C_1, C_2, C_3)$ of function P as follows:

$$H_P(C_1, C_2, C_3) = \begin{pmatrix} -2C_1 + 1 & 2 & 0 \\ 2 & -2 & 1 \\ 0 & 1 & -1 \end{pmatrix}.$$

Thus, for point $(C_1^{(1)}, C_2^{(1)}, C_3^{(1)}) = (6, 14, 15)$, we obtain

$$H_P(6, 14, 15) = \begin{pmatrix} -11 & 2 & 0 \\ 2 & -2 & 1 \\ 0 & 1 & -1 \end{pmatrix}.$$

Checking the sufficient conditions for a local maximum point, we determine the leading principal minors D_1, D_2 and D_3 of matrix $H_P(6, 14, 15)$ and test the conditions of Theorem 11.8 using Theorem 10.7:

$$D_1 = -11 < 0, \quad D_2 = \begin{vmatrix} -11 & 2 \\ 2 & -2 \end{vmatrix} = 18 > 0,$$

$$D_3 = \begin{vmatrix} -11 & 2 & 0 \\ 2 & -2 & 1 \\ 0 & 1 & -1 \end{vmatrix} = -7 < 0.$$

Thus, from Theorem 10.7 the Hessian $H_P(C_1, C_2, C_3)$ is negative definite for point $(6, 14, 15)$, and this point is indeed the only local maximum point found by differential calculus. Since function P has no larger objective function values on the boundary of the domain and since function P has continuous partial derivatives, the obtained solution is also the global maximum point with function value $P(6, 14, 15) = 396.5$, which means that the resulting profit is equal to $396, 500$ EUR.

11.7.2 Method of least squares

We consider the following application of the determination of an extreme point of a function of several variables. Assume that n points (x_i, y_i), $i = 1, 2, \ldots, n$, are given. The aim is to determine a function f of a specified class (for instance a linear function or a quadratic function) such that $f(x)$ describes the relationship between x and y 'as well as possible' for this class of functions.

We apply the criterion of minimizing the sum of the squared differences between y_i and $f(x_i)$ (denoted as the *Gaussian method of least squares*), i.e.

$$Q = \sum_{i=1}^{n} (f(x_i) - y_i)^2 \to \text{ min!}$$

It is assumed that we look for an approximation by a linear function f with $f(x) = ax + b$ (see Figure 11.8).

The objective formulated above leads to the following optimization problem with function $Q = Q(a, b)$ depending on two arguments a and b:

$$Q(a, b) = \sum_{i=1}^{n} (ax_i + b - y_i)^2 \to \text{ min!}$$

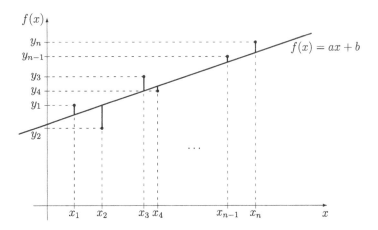

Figure 11.8 Approximation by a linear funtion f with $f(x) = ax + b$.

As we know from Theorem 11.7, the necessary conditions for an extreme point are that the first-order partial derivatives with respect to the variables a and b are equal to zero:

$$Q_a(a, b) = 0 \qquad \text{and} \qquad Q_b(a, b) = 0.$$

Determining the first-order partial derivatives of function Q, we get:

$$Q_a(a, b) = 2 \sum_{i=1}^{n} x_i \, (ax_i + b - y_i) = 0$$

$$Q_b(a, b) = 2 \sum_{i=1}^{n} (ax_i + b - y_i) = 0.$$

This is a system of two linear equations with two variables a and b which can be solved by applying Gaussian elimination or the pivoting procedure (we leave this calculation to the reader). As a result, we obtain the values of parameters a and b as follows:

$$a = \frac{n \cdot \sum_{i=1}^{n} x_i y_i - \left(\sum_{i=1}^{n} x_i \right) \cdot \left(\sum_{i=1}^{n} y_i \right)}{n \cdot \sum_{i=1}^{n} x_i^2 - \left(\sum_{i=1}^{n} x_i \right)^2}, \qquad (11.15)$$

$$b = \frac{\left(\sum_{i=1}^{n} x_i^2 \right) \cdot \left(\sum_{i=1}^{n} y_i \right) + \left(\sum_{i=1}^{n} x_i \right) \cdot \left(\sum_{i=1}^{n} x_i y_i \right)}{n \cdot \sum_{i=1}^{n} x_i^2 - \left(\sum_{i=1}^{n} x_i \right)^2}.$$

The latter equality can be rewritten in terms of parameter a calculated according to equation (11.15) as follows:

$$b = \frac{\sum_{i=1}^{n} y_i - a \cdot \sum_{i=1}^{n} x_i}{n}. \qquad (11.16)$$

One can derive that the denominator of value a is equal to zero if and only if all x_i values are equal. Thus, we get a unique candidate for a local extreme point of function Q provided that not all x_i are equal. Applying Theorem 11.9, we get that a and b obtained above satisfy the sufficient conditions for a local minimum point and we have determined the 'best' straight line 'through' this set of n points according to the criterion of minimizing the sum of the squared differences between y_i and the function values $f(x_i)$ of the determined straight line.

Example 11.24 The steel production of Nowhereland during the period 1996–2001 (in millions of tons) was as shown in Table 11.1:

Table 11.1 Data for Example 11.24

Year	1995	1996	1997	1998	1999	2000	2001
Production	450.2	458.6	471.4	480.9	492.6	501.0	512.2

Drawing the relationship between the year and the steel production in a coordinate system, we see that it is justified to describe it by a linear function, and therefore we set

$$f(x) = ax + b.$$

First, we assign a value x_i to each year as follows and compute the required values.

Table 11.2 Application of Gaussian method of least squares

Year	i	x_i	y_i	x_i^2	$x_i y_i$
1995	1	−3	450.2	9	−1,350.6
1996	2	−2	458.6	4	− 917.2
1997	3	−1	471.4	1	− 471.4
1998	4	0	480.9	0	0
1999	5	1	492.6	1	492.6
2000	6	2	501.0	4	1,002.0
2001	7	3	512.2	9	1,536.6
Sum		0	3,366.9	28	292.0

In Table 11.2, we have taken the year 1998 as origin $x_4 = 0$ (of course, this can be arbitrarily chosen). With our choice, the x values are integer values ranging from -3 to 3, i.e. they are symmetric around zero. Such a choice is recommended for cases when n is odd and distances between successive x values are equal since it simplifies that calculations.

Using $n = 7$, we can immediately apply formulas (11.15) and (11.16), and we obtain

$$a = \frac{n \cdot \sum_{i=1}^{n} x_i y_i - \left(\sum_{i=1}^{n} x_i\right) \cdot \left(\sum_{i=1}^{n} y_i\right)}{n \cdot \sum_{i=1}^{n} x_i^2 - \left(\sum_{i=1}^{n} x_i\right)^2} = \frac{7 \cdot 292.0 - 0 \cdot 3366.9}{7 \cdot 28 - 0^2} \approx 10.429,$$

$$b = \frac{\sum_{i=1}^{n} y_i - a \cdot \sum_{i=1}^{n} x_i}{n} = \frac{3366.9 - 10.429 \cdot 0}{7} \approx 480.986.$$

The above values for parameters a and b have been rounded to three decimal places. Thus, we have found function f with

$$f(x) \approx 10.429x + 480.986$$

describing steel production between 1995 and 2001 (i.e. for the years $x \in \{-3, -2, -1, 0, 1, 2, 3\}$). Using this trend function, we obtain for the expected steel production for the year 2006 (i.e. for $x = 8$):

$$f(8) \approx 10.429 \cdot 8 + 480.986 = 564.418$$

millions of tons.

Example 11.25 Assume that the following ordered pairs have been found experimentally:

$$(x_1, y_1) = (1, 3), \quad (x_2, y_2) = (2, 8), \quad (x_3, y_3) = (-1, 0) \quad \text{and} \quad (x_4, y_4) = (-2, 1).$$

Further, we assume that the relationship between variables x and y should be expressed by a function f of the class

$$f(x) = ax + bx^2.$$

In this case, we cannot use formulas (11.15) and (11.16) for calculating the parameters a and b, but we have to derive formulas for determining the parameters of the required function. This may be done in a quite similar way as for the approximation by a linear function. According to the Gaussian method of least squares we get the problem

$$Q(a, b) = \sum_{i=1}^{n} \left(ax_i + bx_i^2 - y_i\right)^2 \to \min!$$

Using $n = 4$, we obtain the first-order necessary conditions for a local extreme point:

$$Q_a(a, b) = 2 \sum_{i=1}^{4} x_i \left(ax_i + bx_i^2 - y_i\right) = 0$$

$$Q_b(a, b) = 2 \sum_{i=1}^{4} x_i^2 \left(ax_i + bx_i^2 - y_i\right) = 0.$$

We can rewrite the above conditions as follows:

$$\sum_{i=1}^{4} x_i y_i = a \sum_{i=1}^{4} x_i^2 + b \sum_{i=1}^{4} x_i^3 \tag{11.17}$$

$$\sum_{i=1}^{4} x_i^2 y_i = a \sum_{i=1}^{4} x_i^3 + b \sum_{i=1}^{4} x_i^4. \tag{11.18}$$

From equation (11.17) we obtain $10a = 17$ and thus $a = 1.7$, and from equation (11.18) we obtain $34b = 39$ and thus $b = 39/34 \approx 1.147$. Hence the best function f of the chosen class according to Gauss's method of least squares is approximately given by

$$f(x) \approx 1.7x + 1.147x^2.$$

11.7.3   Extreme points of implicit functions

Next, we give a criterion for the determination of local extreme points in the case of an implicitly defined function $F(x,y) = 0$.

THEOREM 11.10   Let function $f : D_f \rightarrow \mathbb{R}, D_f \subseteq \mathbb{R}$, be implicitly given by $F(x,y)=0$ and let function F be twice continuously partially differentiable. If the four conditions (1) $F(x^0,y^0) = 0$, (2) $F_x(x^0,y^0) = 0$, (3) $F_y(x^0,y^0) \neq 0$ and (4) $F_{xx}(x^0,y^0) \neq 0$ are satisfied, then (x^0,y^0) is a local extreme point of function f, in particular:

(1) If

$$\frac{F_{xx}(x^0,y^0)}{F_y(x^0,y^0)} > 0,$$

then (x^0,y^0) is a local maximum point.

(2) If

$$\frac{F_{xx}(x^0,y^0)}{F_y(x^0,y^0)} < 0,$$

then (x^0,y^0) is a local minimum point.

Example 11.26   Given

$$F(x,y) = -x^2 + 2y^2 - 8 = 0,$$

we determine local extreme points by applying Theorem 11.10. To find the stationary points, we check

$$F_x(x,y) = -2x = 0,$$

from which we obtain the only solution $x_1 = 0$. Substituting this into $F(x,y) = 0$, we get the two solutions $y_1 = 2$ and $y_2 = -2$, i.e. there are two stationary points

$$P_1 : (x_1, y_1) = (0, 2) \qquad \text{and} \qquad P_2 : (x_1, y_2) = (0, -2).$$

Furthermore we get

$$F_y(x, y) = 4y \qquad \text{and} \qquad F_{xx}(x, y) = -2.$$

For point P_1, this yields

$$\frac{F_{xx}(0, 2)}{F_y(0, 2)} = \frac{-2}{8} < 0,$$

i.e. point $(0, 2)$ is a local minimum point. For point P_2, we get

$$\frac{F_{xx}(0, -2)}{F_y(0, -2)} = \frac{-2}{-8} > 0,$$

i.e. point $(0, -2)$ is a local maximum point. It should be noted that $F(x, y) = 0$ is not a function but only a relation here, nevertheless we can apply Theorem 11.10 for finding local extreme points.

11.8   CONSTRAINED OPTIMIZATION

So far, we have dealt with optimization problems without additional constraints. However, in economics, one often looks for an optimum of a function subject to certain constraints. Such problems are known as *constrained optimization problems*. We have already considered problems where the output of a firm depends on the quantity of the used capital K and the used quantity of labour L. A natural question in economics is the maximization of the output subject to certain budget constraints. This means that the quantity of capital and labour which can be used in production is bounded by some given constants. In the case of a linear objective function and linear inequality constraints, we have already considered such problems in Chapter 9. Here, we consider the minimization (or maximization) of a function subject to some *equality constraints*. Such a problem may arise if the output $Q(K, L)$ has to be equal to a given isoquant Q^*. The production cost depends on the quantity of used capital and labour. In that case, the objective is the minimization of a cost function f depending on variables K and L subject to the constraint $Q(K, L) = Q^*$. In this section, we consider the general case of constrained optimization problems with non-linear objective and/or constraint functions.

11.8.1   Local optimality conditions

We consider the problem

$$z = f(x_1, x_2, \ldots, x_n) \rightarrow \min! \ (\max!)$$

$$\text{s.t.} \qquad g_1(x_1, x_2, \ldots, x_n) = 0$$
$$g_2(x_1, x_2, \ldots, x_n) = 0$$
$$\vdots$$
$$g_m(x_1, x_2, \ldots, x_n) = 0,$$

and we assume that $m < n$. Next, we discuss two basic methods of solving the above problem.

Optimization by substitution

If the constraints

$$g_i(x_1, x_2, \ldots, x_n) = 0, \quad i = 1, 2, \ldots, m,$$

can be put into its reduced form (according to Theorem 11.6, this means that the Jacobian determinant at point $\mathbf{x} = (x_1, x_2, \ldots, x_n)$ is different from zero)

$$x_1 = w_1(x_{m+1}, x_{m+2}, \ldots, x_n)$$
$$x_2 = w_2(x_{m+1}, x_{m+2}, \ldots, x_n)$$

$$\vdots$$

$$x_m = w_m(x_{m+1}, x_{m+2}, \ldots, x_n),$$

we can substitute the terms for $x_1, x_2, \ldots, x_m$ into the given function f. As a result, we get an unconstrained problem with $n - m$ variables

$$W(x_{m+1}, x_{m+2}, \ldots, x_n) \to \text{min! (max!)},$$

which we can already treat. Necessary and sufficient optimality conditions for the case $n = m + 1$ (i.e. function W depends on one variable) have been given in Chapter 4.5 and for the case $n > m + 1$ (function W depends on several variables) in Chapter 11.7. This method is known as *optimization by substitution*. The procedure can also be applied to non-linear functions provided that it is possible to eliminate the variables $x_1, x_2, \ldots, x_m$. This approach is mainly applied in the case when we look for the optimum of a function of two variables subject to one constraint (i.e. $n = 2$, $m = 1$).

Lagrange multiplier method

In what follows, we discuss an alternative general procedure known as the *Lagrange multiplier method* which works as follows. We introduce a new variable λ_i for each constraint i ($1 \le i \le m$), and add all the constraints together with the multipliers to the function to be minimized (or maximized). Thus, we obtain:

$$L(\mathbf{x}, \lambda) = L(x_1, x_2, \ldots, x_n; \lambda_1, \lambda_2, \ldots, \lambda_m)$$

$$= f(x_1, x_2, \ldots, x_n) + \sum_{i=1}^{m} \lambda_i \cdot g_i(x_1, x_2, \ldots, x_n).$$

The above function L is called the *Lagrangian function* and the parameters λ_i are known as *Lagrangian multipliers*.

Note that function L depends on $n + m$ variables. The following theorem gives a necessary condition for the existence of a local extreme point of the Lagrangian function.

THEOREM 11.11 (Lagrange's theorem) Let functions $f : D_f \to \mathbb{R}$, $D_f \subseteq \mathbb{R}^n$, and $g_i : D_{g_i} \to \mathbb{R}$, $D_{g_i} \subseteq \mathbb{R}^n$, $i = 1, 2, \ldots, m$, $m < n$, be continuously partially differentiable and let $\mathbf{x}^0 = (x_1^0, x_2^0, \ldots, x_n^0) \in D_f$ be a local extreme point of function f subject to the constraints $g_i(x_1, x_2, \ldots, x_n) = 0$, $i = 1, 2, \ldots, m$. Moreover, let $|J(x_1^0, x_2^0, \ldots x_n^0)| \neq 0$. Then

$$\mathbf{grad}\, L(\mathbf{x}^0, \lambda^0) = \mathbf{0}. \tag{11.19}$$

Condition (11.19) in Theorem 11.11 means that the Lagrangian function $L(\mathbf{x}; \lambda)$ has a stationary point at $(x_1^0, x_2^0, \ldots, x_n^0; \lambda_1^0, \lambda_2^0, \ldots, \lambda_m^0)$ and can be written in detail as follows:

$$L_{x_j}(\mathbf{x}^0; \lambda^0) = f_{x_j}(x_1^0, x_2^0, \ldots, x_n^0) + \sum_{i=1}^{m} \lambda_i^0 \cdot \frac{\partial g_i(x_1^0, x_2^0, \ldots, x_n^0)}{\partial x_j} = 0 \; ; \quad j = 1, 2, \ldots, n,$$

$$L_{\lambda_i}(\mathbf{x}^0; \lambda^0) = g_i(x_1^0, x_2^0, \ldots, x_n^0) = 0 \; ; \quad i = 1, 2, \ldots, m.$$

The second part of the formula above, which requires that the partial derivatives of function L with respect to all variables λ_i must be equal to zero, guarantees that only vectors $\mathbf{x}$ are obtained which satisfy all constraints. Assume that we have determined the stationary points of the Lagrangian function by means of Theorem 11.11. We now look for sufficient conditions to verify whether a stationary point is indeed an (isolated) local extreme point and if so, whether it is a local minimum or a local maximum point. In the case of an unconstrained optimization problem, a criterion has already been given, where the Hessian has to be checked. As the following theorem shows, in the case of constrained optimization problems, we have to check the so-called *bordered Hessian* which includes the usual Hessian in its lower right part as a submatrix.

THEOREM 11.12 (local sufficient conditions)   Let functions $f : D_f \to \mathbb{R}$, $D_f \subseteq \mathbb{R}^n$, and $g_i : D_{g_i} \to \mathbb{R}$, $D_{g_i} \subseteq \mathbb{R}^n$, $i = 1, 2, \ldots, m$, $m < n$, be twice continuously partially differentiable and let $(\mathbf{x}^0; \lambda^0)$ with $\mathbf{x}^0 \in D_f$ be a solution of the system $\mathbf{grad}\, L(\mathbf{x}; \lambda) = \mathbf{0}$. Moreover, let

$$H_L(\mathbf{x}; \lambda) = \left(\begin{array}{ccc|ccc} 0 & \cdots & 0 & L_{\lambda_1 x_1}(\mathbf{x}; \lambda) & \cdots & L_{\lambda_1 x_n}(\mathbf{x}; \lambda) \\ \vdots & & \vdots & \vdots & & \vdots \\ 0 & \cdots & 0 & L_{\lambda_m x_1}(\mathbf{x}; \lambda) & \cdots & L_{\lambda_m x_n}(\mathbf{x}; \lambda) \\ \hline L_{x_1 \lambda_1}(\mathbf{x}; \lambda) & \cdots & L_{x_1 \lambda_m}(\mathbf{x}; \lambda) & L_{x_1 x_1}(\mathbf{x}; \lambda) & \cdots & L_{x_1 x_n}(\mathbf{x}; \lambda) \\ \vdots & & \vdots & \vdots & & \vdots \\ L_{x_n \lambda_1}(\mathbf{x}; \lambda) & \cdots & L_{x_n \lambda_m}(\mathbf{x}; \lambda) & L_{x_n x_1}(\mathbf{x}; \lambda) & \cdots & L_{x_n x_n}(\mathbf{x}; \lambda) \end{array} \right)$$

be the bordered Hessian and consider its leading principal minors $|\bar{H}_j(\mathbf{x}^0; \lambda^0)|$ of the order $j = 2m + 1, 2m + 2, \ldots, n + m$ at point $(\mathbf{x}^0; \lambda^0)$. Then:

(1) If all leading principal minors $|\bar{H}_j(\mathbf{x}^0; \lambda^0)|$, $2m + 1 \leq j \leq n + m$, have the sign $(-1)^m$, then $\mathbf{x}^0 = (x_1^0, x_2^0, \ldots, x_n^0)$ is a local minimum point of function f subject to the constraints $g_i(\mathbf{x}) = 0, i = 1, 2, \ldots, m$.

(2) If all leading principal minors $|\bar{H}_j(\mathbf{x}^0; \lambda^0)|$, $2m + 1 \leq j \leq n + m$, alternate in sign, the sign of $|\bar{H}_{n+m}(\mathbf{x}^0; \lambda^0)| = |H_L(\mathbf{x}^0; \lambda^0)|$ being that of $(-1)^n$, then $\mathbf{x}^0 = (x_1^0, x_2^0, \ldots, x_n^0)$ is a local maximum point of function f subject to the constraints $g_i(\mathbf{x}) = 0, i = 1, 2, \ldots, m$.

(3) If neither the conditions of (1) nor those of (2) are satisfied, then $\mathbf{x}^0$ is not a local extreme point of function f subject to the constraints $g_i(\mathbf{x}) = 0, i = 1, 2, \ldots, m$. Here the case

when one or several leading principal minors have value zero is not considered a violation of condition (1) or (2).

It is worth noting that the order of the bordered Hessian H_L is $n + m$ if the problem includes n variables x_i and m constraints. We illustrate the application of Theorem 11.12 for two special cases. Assume first that $n = 2$ and $m = 1$, i.e. function f depends on two variables x, y and there is one constraint $g(x, y) = 0$. In that case, we have $2m + 1 = 3$ and only the determinant $|\bar{H}_3(x^0, y^0; \lambda^0)| = |H_L(x^0, y^0; \lambda^0)|$ has to be evaluated at the stationary point $(x^0, y^0; \lambda^0)$. If $|\bar{H}_3(x^0, y^0; \lambda^0)| < 0$, i.e. the sign of the determinant is equal to $(-1)^m = (-1)^1 = -1$, then (x^0, y^0) is a local minimum point according to part (1) of Theorem 11.12. If $|\bar{H}_3(x^0, y^0; \lambda^0)| > 0$, i.e. the sign of the determinant is equal to $(-1)^n = (-1)^2 = 1$, then (x^0, y^0) is a local maximum point according to part (2) of Theorem 11.12. However, if $|\bar{H}_3(x^0, \lambda^0)| = 0$, then a decision cannot be made on the basis of Theorem 11.12 since, due to the comment at the end of part (3), none of the conditions (1), (2) or (3) is satisfied (i.e. neither condition (1) nor condition (2) holds, nor does the violation mentioned in (3) occur).

Assume now that $n = 3$ and $m = 1$, i.e. function f depends on three variables and there is one constraint $g(\mathbf{x}) = 0$. In this case, the bordered Hessian H_L has order $n + m = 4$. Moreover, we have $2m + 1 = 3$, i.e. the leading principal minors $|\bar{H}_3(\mathbf{x}^0; \lambda^0)|$ and $|\bar{H}_4(\mathbf{x}^0; \lambda^0)| = |H_L(\mathbf{x}^0; \lambda^0)|$ have to be evaluated. From Theorem 11.12 we can draw the following conclusions. If $|\bar{H}_4(\mathbf{x}^0; \lambda^0)| > 0$, point $\mathbf{x}^0$ is not a local extreme point due to part (3) of Theorem 11.12, since the sign of $|\bar{H}_4(\mathbf{x}^0; \lambda^0)|$ is different both from $(-1)^m = (-1)^1 = -1$ and $(-1)^n = (-1)^3 = -1$.

If $|\bar{H}_4(\mathbf{x}^0, \lambda^0)| < 0$ and $|\bar{H}_3(\mathbf{x}^0, \lambda^0)| < 0$, i.e. both leading principal minors have the sign $(-1)^m = (-1)^1 = -1$, then according to part (1) of Theorem 11.12, $\mathbf{x}^0$ is a local minimum point.

If $|\bar{H}_4(\mathbf{x}^0; \lambda^0)| < 0$, but $|\bar{H}_3(\mathbf{x}^0; \lambda^0)| > 0$, i.e. both leading principal minors have alternative signs, and the sign of $|\bar{H}_4(\mathbf{x}^0; \lambda^0)|$ is equal to $(-1)^n = (-1)^3 = -1$, then according to part (2) of Theorem 11.12, $\mathbf{x}^0$ is a local maximum point. If $|\bar{H}_4(\mathbf{x}^0; \lambda^0)| < 0$, but $|\bar{H}_3(\mathbf{x}^0; \lambda^0)| = 0$, then a decision about an extreme point cannot be made. (Neither condition (1) nor condition (2) is satisfied, but according to the remark at the end of part (3), none of the conditions are violated.)

If $|\bar{H}_4(\mathbf{x}^0; \lambda^0)| = 0$, then a decision about an extreme point cannot be made independently of the value of the leading principal minor $|\bar{H}_3(\mathbf{x}^0; \lambda^0)|$ (see remark at the end of part (3) of Theorem 11.12).

Example 11.27   We want to determine all local minima and maxima of function $f : \mathbb{R}^2 \to \mathbb{R}$ with

$$z = f(x, y) = x^2 + 4y^2 - 3$$

subject to the constraint

$$g(x, y) = 4x^2 - 16x + y^2 + 12 = 0.$$

We formulate the Lagrangian function depending on the three variables x, y and λ, which gives

$$L(x, y, \lambda) = x^2 + 4y^2 - 3 + \lambda(4x^2 - 16x + y^2 + 12).$$

The necessary conditions for a local optimum according to Theorem 11.11 are

$$L_x(x, y, \lambda) = 2x + 8\lambda x - 16\lambda = 0 \tag{11.20}$$

$$L_y(x, y, \lambda) = 8y + 2\lambda y = 0 \tag{11.21}$$

$$L_\lambda(x, y, \lambda) = 4x^2 - 16x + y^2 + 12 = 0. \tag{11.22}$$

Factoring out $2y$ in equation (11.21), we obtain

$$2y \cdot (4 + \lambda) = 0.$$

Thus, we have to consider the following two cases.

Case 1 $y = 0$. Substituting $y = 0$ into equation (11.22), we get $4x^2 - 16x + 12 = 0$ and

$$x^2 - 4x + 3 = 0,$$

respectively. This yields

$$x_1 = 2 + \sqrt{4 - 3} = 3 \qquad \text{and} \qquad x_2 = 2 - \sqrt{4 - 3} = 1,$$

and from equation (11.20) we obtain the corresponding λ-values $\lambda_1 = -3/4$ and $\lambda_2 = 1/4$. Hence point

$$P_1 : (x_1, y_1) = (3, 0); \qquad \text{with } \lambda_1 = -\frac{3}{4}$$

and point

$$P_2 : (x_2, y_2) = (1, 0); \qquad \text{with } \lambda_2 = \frac{1}{4}$$

are candidates for a local extreme point in this case.

Case 2 $4 + \lambda = 0$, i.e. $\lambda = -4$. Then we obtain from (11.20) the equation $2x - 32x + 64 = 0$, which has the solution

$$x_3 = \frac{64}{30} = \frac{32}{15}.$$

Substituting the latter into equation (11.22), we obtain the equation

$$4\left(\frac{32}{15}\right)^2 - 16 \cdot \frac{32}{15} + y^2 + 12 = 0$$

which yields

$$y^2 = \frac{884}{225}.$$

Hence we obtain also the following points

$$P_3 : (x_3, y_3) = \left(\frac{32}{15}, \frac{\sqrt{884}}{15} \approx 1.98\right); \qquad \text{with } \lambda_3 = -4$$

and

$$P_4 : (x_3, y_4) = \left(\frac{32}{15}, -\frac{\sqrt{884}}{15} \approx -1.98 \right) ; \qquad \text{with } \lambda_4 = -4$$

as candidates for a local extreme point.

Next, we check the local sufficient conditions for a local extreme point given in Theorem 11.12. We first determine

$$L_{xx}(x, y; \lambda) = 2 + 8\lambda$$

$$L_{xy}(x, y; \lambda) = 0 \qquad = L_{yx}(x, y; \lambda)$$

$$L_{yy}(x, y; \lambda) = 8 + 2\lambda$$

$$L_{x\lambda}(x, y; \lambda) = 8x - 16 \qquad = L_{\lambda x}(x, y; \lambda)$$

$$L_{y\lambda}(x, y; \lambda) = 2y \qquad = L_{\lambda y}(x, y; \lambda).$$

Thus, we obtain the bordered Hessian

$$H_L(x, y; \lambda) = \left(\begin{array}{c|cc} 0 & 8x - 16 & 2y \\ \hline 8x - 16 & 2 + 8\lambda & 0 \\ 2y & 0 & 8 + 2\lambda \end{array} \right).$$

Since we have two variables in function f and one constraint, we apply Theorem 11.12 with $n = 2$ and $m = 1$. This means that we have to check only the leading principal minor for $j = 2m + 1 = n + m = 3$, i.e. we have to check only the determinant of the bordered Hessian to apply Theorem 11.12. We obtain for the four stationary points of the Lagrangian function:

$$\left| H_L \left(3, 0; -\frac{3}{4} \right) \right| = \begin{vmatrix} 0 & 8 & 0 \\ 8 & -4 & 0 \\ 0 & 0 & 6.5 \end{vmatrix} = 6.5 \cdot \begin{vmatrix} 0 & 8 \\ 8 & -4 \end{vmatrix} = 6.5 \cdot (-64) < 0,$$

$$\left| H_L \left(1, 0; \frac{1}{4} \right) \right| = \begin{vmatrix} 0 & -8 & 0 \\ -8 & 4 & 0 \\ 0 & 0 & 8.5 \end{vmatrix} = 8.5 \cdot \begin{vmatrix} 0 & -8 \\ -8 & 4 \end{vmatrix}$$

$$= 8.5 \cdot (-64) < 0,$$

$$\left| H_L \left(\frac{32}{15}, \frac{\sqrt{884}}{15}; -4 \right) \right| = \begin{vmatrix} 0 & \dfrac{16}{15} & \dfrac{2\sqrt{884}}{15} \\ \dfrac{16}{15} & -30 & 0 \\ \dfrac{2\sqrt{884}}{15} & 0 & 0 \end{vmatrix} > 0,$$

$$\left| H_L \left(\frac{32}{15}, -\frac{\sqrt{884}}{15}; -4 \right) \right| = \begin{vmatrix} 0 & \dfrac{16}{15} & -\dfrac{2\sqrt{884}}{15} \\ \dfrac{16}{15} & -30 & 0 \\ -\dfrac{2\sqrt{884}}{15} & 0 & 0 \end{vmatrix} > 0.$$

Hence, due to part (1) of Theorem 11.12, the points $(3,0)$ and $(1,0)$ are local minimum points, since in both cases the sign of the determinant of the bordered Hessian is equal to $(-1)^m = (-1)^1 = -1$, and due to part (2) of Theorem 11.12, the points $(32/15, \sqrt{884}/15)$ and $(32/15, -\sqrt{884}/15)$ are local maximum points, since the sign of the determinant of the bordered Hessian is equal to $(-1)^n = (-1)^2 = +1$. Notice that in this problem the check of the conditions of Theorem 11.12 reduces to the determination of the sign of the determinant of the bordered Hessian (however, in the general case of an unconstrained optimization problem we might have to check the signs of several leading principal minors in order to verify a local extreme point).

Some level curves of function f together with the constraint $g(x,y) = 0$ are given in Figure 11.9. To illustrate, we consider equation $f(x,y) = 0$ corresponding to $x^2 + 4y^2 = 3$ which can be rewritten as

$$\left(\frac{x}{\sqrt{3}}\right)^2 + \left(\frac{y}{\frac{\sqrt{3}}{2}}\right)^2 = 1.$$

The latter is the equation of an ellipse with the half-axes $a = \sqrt{3}$ and $b = \sqrt{3}/2$. If we consider now $f(x,y) = C$ with $C \neq 0$, we also get ellipses with different half-axes, where the half-axes are larger the bigger the value of C is. The constraint $g(x,y) = 0$ can be rewritten as

$$4(x^2 - 4x) + y^2 + 12 = 4(x^2 - 4x + 4) + y^2 - 4 = 0$$

which also corresponds to the equation of an ellipse:

$$(x-2)^2 + \left(\frac{y}{2}\right)^2 = 1.$$

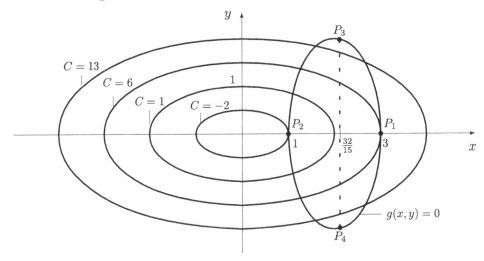

Figure 11.9 Level curves for function f with $f(x,y) = x^2 + 4y^2 - 3$.

From Figure 11.9 one can confirm that $z_1 = 6$ and $z_2 = -2$ are indeed the local minima and $z_3 = z_4 = 3885/225 \approx 17.3$ are the local maxima of function f subject to the given constraint.

Example 11.28 A utility function assigns to each point (vector) $\mathbf{x} = (x_1, x_2, \ldots, x_n)$, where the ith component x_i describes the amount of the ith commodity ($i = 1, 2, \ldots, n$; such a vector is also called a commodity bundle), a real number $u(x_1, x_2, \ldots, u_n)$ which measures the degree of satisfaction or utility of the consumer with the given commodity bundle. Assume that a Cobb–Douglas utility function $u : \mathbb{R}^2_+ \to \mathbb{R}_+$ depending on the amounts of two commodities with

$$u(x, y) = x^2 y^3$$

is given. We wish to maximize this utility function subject to the constraint

$$7x + 5y = 105.$$

The equation $7x + 5y = 105$ describes the budget constraint. It is convenient to take logarithms of function u and work with equation

$$\ln u(x, y) = 2 \ln x + 3 \ln y.$$

This is possible since, due to the strict monotonicity of the logarithmic function, it has no influence on the optimal solution whether we minimize function u or function $\ln u$. Thus, we have to solve the following constrained optimization problem:

$$f(x, y) = 2 \ln x + 3 \ln y \to \max!$$

subject to

$$7x + 5y = 105.$$

We consider two approaches to solving this problem.
The first way is to apply optimization by substitution, which means that we have to substitute the budget constraint into the function of the maximization problem from the beginning. This yields

$$y = w(x) = 21 - \frac{7}{5}x.$$

If we do this, the problem is now as follows:

$$W(x) = 2 \ln x + 3 \ln \left(21 - \frac{7}{5}x \right) \to \max!$$

The necessary optimality condition of first order for this problem is as follows (see Chapter 4.5, Theorem 4.12):

$$W'(x) = \frac{2}{x} - 3 \cdot \frac{5}{105 - 7x} \cdot \frac{7}{5} = 0$$

which can be written as

$$W'(x) = \frac{2}{x} - \frac{21}{105 - 7x} = 0.$$

The latter equation yields the solution $x^0 = 6$. Substituting this back into the budget constraint, we get

$$y^0 = 21 - \frac{7}{5} \cdot 6 = 12.6.$$

To check the sufficient condition given in Chapter 4.5 (see Theorem 4.14), we determine

$$W''(x) = -\frac{2}{x^2} + 21 \cdot \frac{-7}{(105 - 7x)^2}.$$

Thus, we obtain for $x^0 = 6$

$$W''(6) = -\frac{2}{36} - \frac{147}{63^2} \approx -0.0926 < 0,$$

and we can conclude that $x^0 = 6$ is a local maximum point of function W.
The second way to solve this example is to apply Lagrange's multiplier method. We set up the Lagrangian function L with

$$L(x, y; \lambda) = 2 \ln x + 3 \ln y + \lambda(7x + 5y - 105)$$

and differentiate it to get the three necessary optimality conditions of the first order according to Theorem 11.11:

$$L_x(x, y; \lambda) = \frac{2}{x} + 7\lambda = 0 \tag{11.23}$$

$$L_y(x, y; \lambda) = \frac{3}{y} + 5\lambda = 0 \tag{11.24}$$

$$L_\lambda(x, y, \lambda) = 7x + 5y - 105 = 0. \tag{11.25}$$

Let us solve this system first for λ and then for x and y. After cross-multiplying, we get from equations (11.23) and (11.24)

$$7\lambda x = -2 \quad \text{and} \quad 5\lambda y = -3.$$

We now add these two equations and use equation (11.25):

$$-5 = \lambda(7x + 5y) = 105\lambda$$

which gives

$$\lambda^0 = -\frac{1}{21}.$$

Just as before, substituting this back into the equations (11.23) and (11.24) and solving them for x and y, we get

$$x^0 = 6 \quad \text{and} \quad y^0 = 12.6.$$

To check the sufficient conditions given in Theorem 11.12, we determine

$$L_{xx}(x, y; \lambda) = -\frac{2}{x^2}$$

$$L_{xy}(x, y; \lambda) = 0 \qquad = L_{xy}(x, y; \lambda)$$

$$L_{yy}(x, y; \lambda) = -\frac{3}{y^2}$$

$$L_{x\lambda}(x, y; \lambda) = 7 \qquad = L_{\lambda x}(x, y; \lambda)$$

$$L_{y\lambda}(x, y; \lambda) = 5 \qquad = L_{\lambda y}(x, y; \lambda),$$

and obtain the bordered Hessian:

$$H_L(x, y; \lambda) = \begin{pmatrix} 0 & 7 & 5 \\ \hline 7 & -\dfrac{2}{x^2} & 0 \\ 5 & 0 & -\dfrac{3}{y^2} \end{pmatrix}.$$

Since $n = 2$ and $m = 1$, according to Theorem 11.12 we have to check only one leading principal minor for $j = 2m + 1 = 3$ which corresponds to the determinant of the bordered Hessian. For point $(x^0, y^0; \lambda^0) = (6, 12.6; -1/21)$, we get

$$H_L\left(6, 12.6; -\frac{1}{21}\right) = \begin{pmatrix} 0 & 7 & 5 \\ \hline 7 & -\dfrac{1}{18} & 0 \\ 5 & 0 & -\dfrac{3}{12.6^2} \end{pmatrix}.$$

Thus, we obtain

$$\left| H_L\left(6, 12.6; -\frac{1}{21}\right) \right| = (-7) \cdot \begin{vmatrix} 7 & 5 \\ 0 & -\dfrac{3}{12.6^2} \end{vmatrix} + 5 \cdot \begin{vmatrix} 7 & 5 \\ -\dfrac{1}{18} & 0 \end{vmatrix} > 0.$$

Notice that we do not have to calculate the values of both determinants above, we only need the sign of the left-hand-side determinant. Since the first summand obviously has a value greater than zero (the first determinant has a negative value multiplied by a negative factor) and the second determinant also has a value greater than zero, the determinant of the bordered Hessian is greater than zero. According to part (2) of Theorem 11.12, we find that $P^0 : (x^0, y^0) = (6, 12.6)$ is a local maximum point since the sign of the determinant of the bordered Hessian is equal to $(-1)^n = (-1)^2 = +1$.

11.8.2 Global optimality conditions

Next, we present a sufficient condition for a global maximum and a global minimum point.

THEOREM 11.13 (global sufficient conditions) Let functions $f : D_f \to \mathbb{R}$, $D_f \subseteq \mathbb{R}^n$, and $g_i : D_{g_i} \to \mathbb{R}$, $D_{g_i} \subseteq \mathbb{R}^n$, $i = 1, 2, \ldots, m$, $m < n$, be twice continuously partially differentiable and let $(\mathbf{x}^0; \lambda^0)$ with $\mathbf{x}^0 \in D_f$ be a solution of the system $\mathbf{grad}\, L(\mathbf{x}; \lambda) = \mathbf{0}$. If the Hessian

$$H(\mathbf{x}; \lambda^0) = \begin{pmatrix} L_{x_1 x_1}(\mathbf{x}; \lambda^0) & \cdots & L_{x_1 x_n}(\mathbf{x}; \lambda^0) \\ \vdots & & \vdots \\ L_{x_n x_1}(\mathbf{x}; \lambda^0) & \cdots & L_{x_n x_n}(\mathbf{x}; \lambda^0) \end{pmatrix}$$

(1) is negative definite for all points $\mathbf{x} \in D_f$, then $\mathbf{x}^0$ is the global maximum point of function f subject to the constraints $g_i(\mathbf{x}) = 0, i = 1, 2, \ldots, m$;
(2) is positive definite for all points $\mathbf{x} \in D_f$, then $\mathbf{x}^0$ is the global minimum point of function f subject to the constraints $g_i(\mathbf{x}) = 0, i = 1, 2, \ldots, m$.

Theorem 11.13 requires us to check whether the Hessian is positive or negative definite for *all* points $\mathbf{x} \in D_f$, which can be rather complicated or even impossible. If function f and the constraints g_j contain only quadratic and linear terms, the second-order partial derivatives are constant, and evaluation of the determinant of $H_L(\mathbf{x}, \lambda^0)$ is possible. We look at the following example.

Example 11.29 Given is the function $f : D_f \to \mathbb{R}, D_f \subseteq \mathbb{R}^3$, with

$$f(x, y, z) = 2x + 2y - 2z$$

s.t.

$$x^2 + 2y^2 + 3z^2 = 66.$$

We determine all global optimum points by applying Theorems 11.11 and 11.13. Setting up the Lagrangian function, we get

$$L(x, y, z; \lambda) = 2x + 2y - 2z + \lambda(x^2 + 2y^2 + 3z^2 - 66).$$

Applying Theorem 11.11, we get the necessary optimality conditions

$$L_x = 2 + 2\lambda x = 0 \tag{11.26}$$

$$L_y = 2 + 4\lambda y = 0 \tag{11.27}$$

$$L_z = -2 + 6\lambda z = 0 \tag{11.28}$$

$$L_\lambda = x^2 + 2y^2 + 3z^2 - 66 = 0. \tag{11.29}$$

From equations (11.26) to (11.28), we obtain

$$x = -\frac{1}{\lambda^2}, \qquad y = -\frac{1}{2\lambda} \qquad \text{and} \qquad z = \frac{1}{3\lambda}.$$

Substituting the above terms for x, y and z into equation (11.29), we obtain

$$x^2 + 2y^2 + 3z^2 = \frac{1}{\lambda^2} + 2 \cdot \frac{1}{4\lambda^2} + 3 \cdot \frac{1}{9\lambda^2} = 66$$

which can be rewritten as

$$\frac{1}{\lambda^2} \cdot \left(1 + \frac{1}{2} + \frac{1}{3}\right) = \frac{1}{\lambda^2} \cdot \frac{11}{6} = 66.$$

This yields

$$\lambda^2 = \frac{1}{36}$$

which gives the two solutions

$$\lambda_1 = \frac{1}{6} \quad \text{and} \quad \lambda_2 = -\frac{1}{6}.$$

Using equations (11.26) to (11.28), we obtain the following two stationary points of the Lagrangian function:

$$P_1: \quad x_1 = -6, \quad y_1 = -3, \quad z_1 = 2; \quad \text{with} \quad \lambda_1 = \frac{1}{6}$$

and

$$P_2: \quad x_2 = 6, \quad y_2 = 3, \quad z_2 = -2; \quad \text{with} \quad \lambda_2 = -\frac{1}{6}.$$

We now check the sufficient conditions for a global optimum point given in Theorem 11.13. For point P_1, we obtain

$$H(x,y,z;\lambda_1) = \begin{pmatrix} 2\lambda_1 & 0 & 0 \\ 0 & 4\lambda_1 & 0 \\ 0 & 0 & 6\lambda_1 \end{pmatrix} = \begin{pmatrix} \frac{1}{3} & 0 & 0 \\ 0 & \frac{2}{3} & 0 \\ 0 & 0 & 1 \end{pmatrix}.$$

(Notice that the above matrix does not depend on variables x, y and z since the mixed second-order partial derivatives are constant.) All leading principal minors have a positive sign and by Theorem 10.7, matrix $H(x,y,z;\lambda_1)$ is positive definite. Therefore, P_1 is the global minimum point.

$$H(x,y,z;\lambda_2) = \begin{pmatrix} 2\lambda_1 & 0 & 0 \\ 0 & 4\lambda_1 & 0 \\ 0 & 0 & 6\lambda_1 \end{pmatrix} = \begin{pmatrix} -\frac{1}{3} & 0 & 0 \\ 0 & -\frac{2}{3} & 0 \\ 0 & 0 & -1 \end{pmatrix}.$$

The leading principal minors alternate in sign beginning with a negative sign of the first leading principal minor. Therefore, matrix $H(x,y,z;\lambda_2)$ is negative definite, and P_2 is the global maximum point.

11.9   DOUBLE INTEGRALS

In this section, we extend the notion of the definite integral to functions of more than one variable. Integrals of functions of more than one variable have to be evaluated in probability theory and statistics in connection with multi-dimensional probability distributions. In the following, we restrict ourselves to continuous functions f depending on two independent variables.

Definition 11.13   The integrals

$$I_1 = \int_a^b \left(\int_{y_1(x)}^{y_2(x)} f(x,y)\,dy \right) dx = \int_a^b F_2(x)\,dx$$

and

$$I_2 = \int_c^d \left(\int_{x_1(y)}^{x_2(y)} f(x,y)\,dx \right) dy = \int_c^d F_1(y)\,dy$$

are called (*iterated*) *double integrals*.

For simplicity, the parentheses for the inner integral are often dropped. To evaluate integral I_1, we first determine

$$F_2(x) = \int_{y_1(x)}^{y_2(x)} f(x,y)\,dy,$$

where x is (in the same way as for partial differentiation) treated as a constant, and then we evaluate the integral

$$I_1 = \int_a^b F_2(x)\,dx$$

as in the case of integrating a function of one variable. To evaluate both integrals, we can apply the rules presented in Chapter 5.

Example 11.30   We evaluate the integral

$$I = \int \int_R 2xy\,dy\,dx$$

where R contains all points (x, y) with $x^2 \le y \le 3x$ and $0 \le x \le 1$ (see Figure 11.10). We obtain

$$I = \int\limits_0^1 \left(\int\limits_{x^2}^{3x} 2xy \, dy \right) dx = \int\limits_0^1 \left[xy^2 \Big|_{x^2}^{3x} \right] dx$$

$$= \int\limits_0^1 \left[x \cdot (3x)^2 - x \cdot (x^2)^2 \right] dx = \int\limits_0^1 (9x^3 - x^5) \, dx$$

$$= \left(\frac{9}{4} x^4 - \frac{1}{6} x^6 \right) \Big|_0^1 = \frac{9}{4} - \frac{1}{6} = \frac{25}{12}.$$

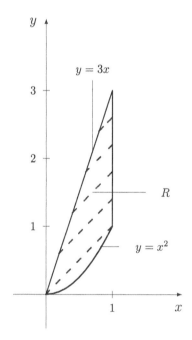

Figure 11.10 Region R considered in Example 11.30.

Finally, we consider the case where both the lower and the upper limits of integration of both variables are constant. In this case, we get the following property.

THEOREM 11.14 Let function $f : D_f \to \mathbb{R}$, $D_f \subseteq \mathbb{R}^2$, be continuous in the rectangle given by $a \le x \le b$ and $c \le y \le d$. Then

$$\int\limits_a^b \left(\int\limits_c^d f(x, y) \, dy \right) dx = \int\limits_c^d \left(\int\limits_a^b f(x, y) \, dx \right) dy.$$

One application of double integrals is the determination of the volume of a solid bounded above by the surface $z = f(x, y)$, below by the xy plane, and on the sides by the vertical walls defined by the region R. If we have $f(x, y) \geq 0$ over a region R, then the volume of the solid under the graph of function f and over the region R is given by

$$V = \iint\limits_{R} f(x, y) \, dx \, dy.$$

Example 11.31   We want to find the volume V under the graph of function f with

$$f(x, y) = 2x^2 + y^2$$

and over the rectangular region given by $0 \leq x \leq 2$ and $0 \leq y \leq 1$. We obtain:

$$V = \iint\limits_{R} (2x^2 + y^2) \, dy \, dx = \int\limits_{0}^{2} \left(\int\limits_{0}^{1} (2x^2 + y^2) \, dy \right) dx$$

$$= \int\limits_{0}^{2} \left(2x^2 y + \frac{y^3}{3} \right) \Bigg|_{0}^{1} dx$$

$$= \int\limits_{0}^{2} \left(2x^2 + \frac{1}{3} \right) dx = \left(\frac{2x^3}{3} + \frac{x}{3} \right) \Bigg|_{0}^{2} = \frac{16}{3} + \frac{2}{3} = \frac{18}{3} = 6.$$

Thus, the volume of the solid described above is equal to six volume units.

The considerations in this section can be extended to the case of integrals depending on more than two variables.

EXERCISES

11.1   (a)   Given is a Cobb–Douglas function $f : \mathbb{R}_+^2 \to \mathbb{R}$ with

$$z = f(x, y) = A \cdot x_1^{\alpha_1} \cdot x_2^{\alpha_2},$$

where $A = 1, \alpha_1 = 1/2$ and $\alpha_2 = 1/2$. Graph isoquants for $z = 1$ and $z = 2$ and illustrate the surface in $\mathbb{R}^3$.

(b)   Given are the following functions $f : D_f \to \mathbb{R}$, $D_f \subseteq \mathbb{R}^2$, with $z = f(x, y)$:

(i) $z = \sqrt{9 - x^2 - y^2}$;   (ii) $z = \frac{xy}{x - y}$;   (iii) $z = x^2 + 4x + 2y$.

Graph the domain of the function and isoquants for $z = 1$ and $z = 2$.

11.2 Find the first-order partial derivatives for each of the following functions:

(a) $z = f(x,y) = x^2 \sin^2 y$; (b) $z = f(x,y) = x^{(y^2)}$;

(c) $z = f(x,y) = x^y + y^x$; (d) $z = f(x,y) = \ln(\sqrt{x}\sqrt{y})$;

(e) $z = f(x_1,x_2,x_3) = 2xe^{x_1^2+x_2^2+x_3^2}$; (f) $z = f(x_1,x_2,x_3) = \sqrt{x_1^2 + x_2^2 + x_3^2}$.

11.3 The variable production cost C of two products P_1 and P_2 depends on the outputs x and y as follows:

$$C(x,y) = 120x + \frac{1,200,000}{x} + 800y + \frac{32,000,000}{y},$$

where $(x,y) \in \mathbb{R}^2$ with $x \in [20,200]$ and $y \in [50,400]$.

(a) Determine the marginal production cost of products P_1 and P_2.
(b) Compare the marginal cost of P_1 for $x_1 = 80$ and $x_2 = 120$ and of P_2 for $y_1 = 160$ and $y_2 = 240$. Give an interpretation of the results.

11.4 Find all second-order partial derivatives for each of the following functions:

(a) $z = f(x_1,x_2,x_3) = x_1^3 + 3x_1x_2^2x_3^3 + 2x_2 + \ln(x_1x_3)$;

(b) $z = f(x,y) = \frac{1+xy}{1-xy}$; (c) $z = f(x,y) = \ln\frac{x+y}{x-y}$.

11.5 Determine the gradient of function $f : D_f \rightarrow \mathbb{R}, D_f \subseteq \mathbb{R}^2$, with $z = f(x,y)$ and specify it at the points $(x_0,y_0) = (1,0)$ and $(x_1,y_1) = (1,2)$:

(a) $z = ax + by$; (b) $z = x^2 + xy^2 + \sin y$; (c) $z = \sqrt{9 - x^2 - y^2}$.

11.6 Given is the surface

$$z = f(x,y) = x^2 \sin^2 y$$

and the domain $D_f = \mathbb{R}^2$, where the xy plane is horizontal. Assume that a ball is located on the surface at point $(x,y,z) = (1,1,z)$. If the ball begins to roll, what is the direction of its movement?

11.7 Determine the total differential for the following functions:

(a) $z = f(x,y) = \sin\frac{x}{y}$; (b) $z = f(x,y) = x^2 + xy^2 + \sin y$;

(c) $z = f(x,y) = e^{(x^2+y^2)}$; (d) $z = f(x,y) = \ln(xy)$.

11.8 Find the surface of a circular cylinder with radius $r = 2$ m and height $h = 5$ m. Assume that measurements of radius and height may change as follows: $r = 2\pm0.05$ and $h = 5\pm0.10$. Use the total differential for an approximation of the change of the surface in this case. Find the absolute and relative (percentage) error of the surface.

11.9 Let $f : D_f \rightarrow \mathbb{R}, D_f \subseteq \mathbb{R}^2$, be a function with

$$z = f(x_1,x_2) = x_1^2 e^{x_2},$$

where $x_1 = x_1(t)$ and $x_2 = x_2(t)$.

(a) Find the derivative dz/dt.
(b) Use the chain rule to find $z'(t)$ if (i) $x_1 = t^2$; $x_2 = \ln t^2$; (ii) $x_1 = \ln t^2$; $x_2 = t^2$.

(c) Find $z'(t)$ by substituting the functions of (i) and (ii) for x_1 and x_2, and then differentiate them.

11.10   Given is the function

$$z = f(x, y) = \sqrt{9 - x^2 - y^2}.$$

Find the directional derivatives in direction $\mathbf{r}^1 = (1, 0)^T, \mathbf{r}^2 = (1, 1)^T, \mathbf{r}^3 = (-1, -2)^T$ at point $(1, 2)$.

11.11   Assume that

$$C(x_1, x_2, x_3) = 20 + 2x_1x_2 + 8x_3 + x_2 \ln x_3 + 4x_1$$

is the total cost function of three products, where x_1, x_2, x_3 are the outputs of these three products.

(a) Find the gradient and the directional derivative with the directional vector $\mathbf{r} = (1, 2, 3)^T$ of function C at point $(3, 2, 1)$. Compare the growth of the cost (marginal cost) in the direction of fastest growth with the directional marginal cost in the direction $\mathbf{r}$. Find the percentage rate of cost reduction at point $(3, 2, 1)$.

(b) The owner of the firm wants to increase the output by 6 units altogether. The owner can do it in the ratio of 1:2:3 or of 3:2:1 for the products $x_1 : x_2 : x_3$. Further conditions are $x_1 \geq 1$, $x_3 \geq 1$, and the output x_2 must be at least 4 units. Which ratio leads to a lower cost for the firm?

11.12   Success of sales z for a product depends on a promotion campaign in two media. Let x_1 and x_2 be the funds invested in the two media. Then the following function is to be used to reflect the relationship:

$$z = f(x_1, x_2) = 10\sqrt{x_1} + 20 \ln(x_2 + 1) + 50; \qquad x_1 \geq 0, x_2 \geq 0.$$

Find the partial rates of change and the partial elasticities of function f at point $(x_1, x_2) = (100, 150)$.

11.13   Determine whether the following function is homogeneous. If it is, what is the degree of homogeneity? Use Euler's theorem and interpret the result.

(a) $z = f(x_1, x_2) = \sqrt{x_1^3 + 2x_1x_2^2 + x_2^3}$;   (b) $z = f(x_1, x_2) = x_1x_2^2 + x_1^2$.

11.14   Let $F(x, y) = 0$ be an implicitly defined function. Find dy/dx by the implicit-function rule.

(a) $F(x, y) = \dfrac{x^2}{a^2} - \dfrac{y^2}{b^2} - 1 = 0$ $(y \geq 0)$;   (b) $F(x, y) = xy - \sin 3x = 0$;

(c) $F(x, y) = x^y - \ln xy + x^2y = 0$.

11.15   We consider the representation of variables x and y by so-called polar coordinates:

$$x = r \cos \varphi, \qquad y = r \sin \varphi,$$

or equivalently, the implicitly given system

$$F_1(x,y;r,\varphi) = r\cos\varphi - x = 0$$
$$F_2(x,y;r,\varphi) = r\sin\varphi - y = 0.$$

Check by means of the Jacobian determinant whether this system can be put into its reduced form, i.e. whether variables r and φ can be expressed in terms of x and y.

11.16   Check whether the following function $f : D_f \to \mathbb{R}, D_f \subseteq \mathbb{R}^2$, with

$$z = f(x,y) = x^3 y^2 (1 - x - y)$$

has a local maximum at point $(x_1, y_1) = (1/2, 1/3)$ and a local minimum at point $(x_2, y_2) = (1/7, 1/7)$.

11.17   Find the local extrema of the following functions $f : D_f \to \mathbb{R}, D_f \subseteq \mathbb{R}^2$:
(a) $z = f(x,y) = x^2 y + y^2 - y$;   (b) $z = f(x,y) = x^2 y - 2xy + \frac{3}{4}e^y$.

11.18   The variable cost of two products P_1 and P_2 depends on the production outputs x and y as follows:

$$C(x,y) = 120x + \frac{1,200,000}{x} + 800y + \frac{32,000,000}{y},$$

where $D_C = \mathbb{R}_+^2$. Determine the outputs x^0 and y^0 which minimize the cost function and determine the minimum cost.

11.19   Given is the function $f : \mathbb{R}^3 \to \mathbb{R}$ with

$$f(x,y,z) = x^2 - 2x + y^2 - 2z^3 y + 3z^2.$$

Find all stationary points and check whether they are local extreme points.

11.20   Find the local extrema of function $C : D_C \to \mathbb{R}, D_C \subseteq \mathbb{R}^3$, with

$$C(\mathbf{x}) = C(x_1, x_2, x_3) = 20 + 2x_1 x_2 + 8x_3 + x_2 \ln x_3 + 4x_1$$

and $\mathbf{x} \in \mathbb{R}^3$, $x_3 > 0$.

11.21   The profit P of a firm depends on three positive input factors x_1, x_2, x_3 as follows:

$$P(x_1, x_2, x_3) = 90x_1 x_2 - x_1^2 x_2 - x_1 x_2^2 + 60 \ln x_3 - 4x_3.$$

Determine input factors which maximize the profit function and find the maximum profit.

11.22   Sales y of a firm depend on the expenses x of advertising. The following values x_i of expenses and corresponding sales y_i of the last 10 months are known:

x_i	20	20	24	25	26	28	30	30	33	34
y_i	180	160	200	250	220	250	250	310	330	280

(a)   Find a linear function f by applying the criterion of minimizing the sum of the squared differences between y_i and $f(x_i)$.
(b)   What sales can be expected if $x = 18$, and if $x = 36$?

11.23   Given the implicitly defined functions

$$F(x,y) = \frac{(x-1)^2}{4} + \frac{(y-2)^2}{9} - 1 = 0,$$

verify that F has local extrema at points $P_1 : (x = 1, y = 5)$ and $P_2 : (x = 1, y = -1)$. Decide whether they are a local maximum or a local minimum point and graph the function.

11.24   Find the constrained optima of the following functions:

(a) $z = f(x,y) = x^2 + xy - 2y^2$,
    s.t.

$$2x + y = 8;$$

(b) $z = f(x_1, x_2, x_3) = 3x_1^2 + 2x_2 - x_1x_2x_3$,
    s.t.

$$x_1 + x_2 = 3 \quad \text{and} \quad 2x_1x_3 = 5.$$

11.25   Check whether the function $f : D_f \to \mathbb{R}, D_f \subseteq \mathbb{R}^3$, with

$$z = f(x,y,z) = x^2 - xz + y^3 + y^2z - 2z,$$

s.t.

$$x - y^2 - z - 2 = 0 \quad \text{and} \quad x + z - 4 = 0$$

has a local extremum at point $(11, -4, -7)$.

11.26   Find the dimensions of a box for washing powder with a double bottom so that the surface is minimal and the volume amounts to $3,000 \text{ cm}^3$. How much cardboard is required if glued areas are not considered?

11.27   Find all the points (x,y) of the ellipse $4x^2 + y^2 = 4$ which have a minimal distance from point $P_0 = (2,0)$. (Use the Lagrange multiplier method first, then substitute the constraint into the objective function.)

11.28   The cost function $C : \mathbb{R}_+^3 \to \mathbb{R}$ of a firm producing the quantities x_1, x_2 and x_3 is given by

$$C(x_1, x_2, x_3) = x_1^2 + 2x_2^2 + 2x_3^2 - 2x_1x_2 - 3x_1x_3 + 500.$$

The firm has to fulfil the constraint

$$2x_1 + 4x_2 + 3x_3 = 125.$$

Find the minimum cost. (Use the Langrange multiplier method as well as optimization by substitution.)

11.29   Evaluate

$$\iint\limits_R (2x + y + 1)\, dR,$$

where the region R is the triangle with the corners $P_1 : (1,1), P_2 : (5,3)$ and $P_3 :$ $(5,5)$.

11.30   Find the volume of a solid of constant height $h = 2$ which is bounded below by the xy plane, and the sides by the vertical walls defined by the region R: R is the set of all points (x,y) of the xy plane which are enclosed by the curves $y^2 = 2x + 1$ and $x - y - 1 = 0$.

11.31   Evaluate the double integral

$$\iint\limits_{R} \frac{x}{y} \, dx \, dy,$$

where R is the region between the two parabolas $x = y^2$ and $y = x^2$.

12 Differential equations and difference equations

We start our considerations with the following introductory example.

Example 12.1 We wish to determine all functions $y = y(x)$ having the rate of change $3\sqrt{2x+1}$. Using the definition of the rate of change (see Chapter 4.4), this problem can be formulated as follows. Determine all functions $y = y(x)$ which satisfy the equation

$$\varrho_y(x) = \frac{y'(x)}{y(x)} = 3\sqrt{2x+1}.$$

The latter equation can be rewritten as

$$y'(x) = 3\sqrt{2x+1} \cdot y(x).$$

This is a relationship between the independent variable x, the function $y(x)$ and its derivative $y'(x)$. Such an equation is called a differential equation.

In this chapter, we discuss some methods for solving special types of differential equations (and later related difference equations). In what follows, as well as $y(x), y'(x), \ldots$ we also use the short forms $y, y', \ldots$ First, we formally define a so-called *ordinary differential equation*.

Definition 12.1 A relationship

$$F(x, y, y', y'', \ldots, y^{(n)}) = 0 \tag{12.1}$$

between the independent variable x, a function $y(x)$ and its derivatives is called an *ordinary differential equation*. The *order* of the differential equation is determined by the highest order of the derivatives appearing in the differential equation.

If representation (12.1) can be solved for $y^{(n)}$, we get the explicit representation

$$y^{(n)} = f(x, y, y', y'', \ldots, y^{(n-1)})$$

of a differential equation of order n. The notion of an ordinary differential equation indicates that only functions depending on one variable can be involved, in contrast to partial differential equations which include also functions of several variables and their partial derivatives. Hereafter, we consider only ordinary differential equations and for brevity, we skip 'ordinary'.

Definition 12.2 A function $y(x)$ for which relationship $F(x, y, y', y'', \ldots, y^{(n)}) = 0$ holds for all values $x \in D_y$ is called a *solution* of the differential equation. The set

$$S = \{y(x) \mid F(x, y, y', y'', \ldots, y^{(n)}) = 0 \quad \text{for all } x \in D_y\}$$

is called the *set of solutions* or *general solution* of the differential equation.

12.1 DIFFERENTIAL EQUATIONS OF THE FIRST ORDER

A *first-order differential equation* can be written in implicit form:

$$F(x, y, y') = 0$$

or in explicit form:

$$y' = f(x, y).$$

In the rest of this section, we first discuss the graphical solution of a differential equation of the first order and then a special case of such an equation which can be solved using integration.

12.1.1 Graphical solution

Consider the differential equation $y' = f(x, y)$. At any point (x_0, y_0), the value $y' = f(x_0, y_0)$ of the derivative is given, which corresponds to the slope of the tangent at point (x_0, y_0). This means that we can roughly approximate the function in a rather small neighbourhood by the tangent line. To get an overview on the solution, we can determine the curves, along which the derivative y' has the same value (i.e. we can consider the isoquants with y' being constant). In this way, we can 'approximately' graph the (infinitely many) solutions of the given differential equation by drawing the so-called *direction field*. This procedure is illustrated in Figure 12.1 for the differential equation $y' = x/2$.

For any $x \in \mathbb{R}$ (and arbitrary y), the slope of the tangent line is equal to $x/2$, i.e. for $x = 1$, we have $y' = 1/2$ (that is, along the line $x = 1$ we have the slope $1/2$ of the tangent line), for $x = 2$, we have $y' = 1$ (i.e. along the line $x = 2$ the slope of the tangent line is equal to 1) and so on. Similarly, for $x = -1$, we get $y' = -1/2$, for $x = -2$ we get $y' = -1$ and so on. Looking at many x values, we get a rough impression of the solution. In the given case, we get parabolas (in Figure 12.1 the parabolas going through the points $(0, -2), (0, 0)$ and $(0, 2)$ are drawn), and the *general solution* is given by

$$y = \frac{x^2}{4} + C$$

with C being an arbitrary real constant.

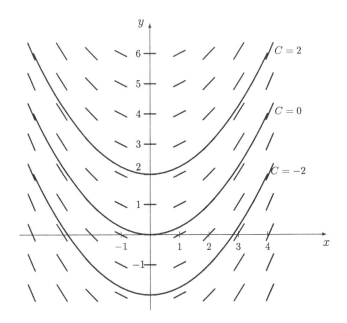

Figure 12.1 The direction field for $y' = x/2$.

Example 12.2   We apply this graphical approach to the differential equation

$$y' = -\frac{x}{y}.$$

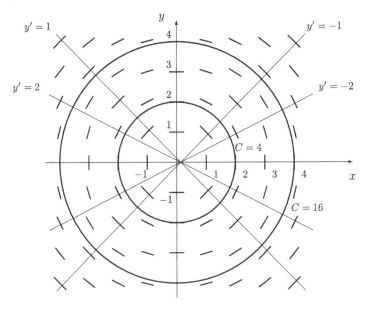

Figure 12.2 The direction field for $y' = -x/y$.

For each point on the line $y = x$, we have $y' = -1$, and for each point on the line $y = -x$, we have $y' = 1$. Analogously, for each point on the line $y = x/2$, we get $y' = -2$ and for each point on the line $y = -x/2$, we get $y' = 2$. Continuing in this way, we get the direction field given in Figure 12.2. Thus, the solutions are circles with the origin of the coordinate system as the centre, i.e.

$$x^2 + y^2 = C \qquad (C > 0).$$

(In Figure 12.2 the circles with radius 2, i.e. $C = 4$, and radius 4, i.e. $C = 16$, are given.) Implicit differentiation of $y^2 = C - x^2$ confirms the result:

$$2y \cdot y' = -2x$$

$$y' = -\frac{x}{y}.$$

12.1.2 Separable differential equations

We continue with such a type of differential equation, where a solution can be found by means of integration. Let us consider the following special case of a first-order differential equation:

$$y' = f(x, y) = g(x) \cdot h(y),$$

i.e. function $f(x, y)$ can be written as the product of a function g depending only on variable x and a function h depending only on variable y. In this case, function $f = f(x, y)$ has a special structure, and we say that this is a *differential equation with separable variables*. Solving this type of differential equation requires only integration techniques. First, we rewrite

$$y' = \frac{dy}{dx} = g(x) \cdot h(y)$$

in the form

$$\frac{dy}{h(y)} = g(x)\, dx.$$

That is, all terms containing function y are now on the left-hand side and all terms containing variable x are on the right-hand side (i.e. we have separated the variables). Then we integrate both sides

$$\int \frac{dy}{h(y)} = \int g(x)\, dx$$

and obtain

$$H(y) = G(x) + C.$$

If we eliminate y, we get the *general solution* $y(x)$ including constant C. A *particular (or definite) solution* $y^P(x)$ is obtained when C is assigned a particular value by considering an initial condition $y(x_0) = y_0$. The latter problem of finding a particular solution is also referred to as an *initial value problem*. We illustrate this approach by the following two examples.

Example 12.1 (continued)   If we rewrite equation

$$\frac{dy}{dx} = y' = 3\sqrt{2x+1} \cdot y$$

in the form

$$\frac{dy}{y} = 3\sqrt{2x+1}\ dx,$$

we get a differential equation with separable variables. Let us integrate both sides:

$$\int \frac{dy}{y} = 3 \int \sqrt{2x+1}\ dx. \tag{12.2}$$

Now we get on the left-hand side the natural logarithm function as an antiderivative. Considering the integral on the right-hand side, we apply the substitution

$$u = 2x + 1$$

which yields

$$\frac{du}{dx} = 2.$$

Now we can find the integral:

$$\int \sqrt{2x+1}\ dx = \frac{1}{2} \int \sqrt{u}\ du = \frac{1}{3} \cdot u^{3/2} + C^*.$$

Therefore, equation (12.2) leads to

$$\ln |y| = (2x+1)^{3/2} + C^*.$$

Thus, we get the general solution

$$|y| = e^{(2x+1)^{3/2}+C^*} = e^{(2x+1)^{3/2}} \cdot e^{C^*}$$

which can be rewritten as

$$|y| = C \cdot e^{(2x+1)^{3/2}}, \qquad C > 0.$$

Here we have written the term e^{C^*} as a new constant $C > 0$. By dropping the absolute values (they are now not necessary since the logarithmic terms have been removed), we can rewrite the solution obtained as

$$y(x) = C \cdot e^{(2x+1)^{3/2}}$$

with C being an arbitrary constant. If now the initial condition $y(0) = e$ is also given, we obtain from the general solution

$$e = C \cdot e^1,$$

i.e. $C = 1$. This yields the particular solution

$$y^P(x) = e^{(2x+1)^{3/2}}$$

of the initial value problem.

Example 12.3   We wish to determine all functions $y = y(x)$ whose elasticity is given by the function

$$\epsilon_y(x) = 2x + 1.$$

Applying the definition of elasticity (see Chapter 4.4), we obtain

$$\epsilon_y(x) = x \cdot \frac{y'(x)}{y(x)} = 2x + 1.$$

We can rewrite this equation as

$$\frac{\frac{dy}{dx}}{y} = \frac{2x+1}{x}$$

or, equivalently,

$$\frac{dy}{y} = \left(2 + \frac{1}{x}\right) dx.$$

This is a differential equation with separable variables and we can integrate the functions on both sides. This yields:

$$\int \frac{dy}{y} = \int \left(2 + \frac{1}{x}\right) dx$$
$$\ln|y| = 2x + \ln|x| + C^*$$
$$|y| = e^{2x+\ln|x|+C^*} = e^{2x} \cdot e^{\ln|x|} \cdot e^{C^*}$$
$$|y| = e^{2x} \cdot |x| \cdot C, \quad C > 0.$$

By dropping the absolute values, we can rewrite the solution obtained as

$$y(x) = C \cdot x \cdot e^{2x}$$

with C being an arbitrary real constant.

Finally, we mention that we can reduce certain types of differential equations by an appropriate substitution to a differential equation with separable variables. This can be done e.g. for a differential equation of the type:

$$y' = g(ax + by + c).$$

We illustrate this procedure by an example.

Example 12.4 Let the differential equation

$$y' = (9x + y + 2)^2$$

be given. (In this form it is not a differential equation with separable variables.) Applying the substitution

$$u = 9x + y + 2,$$

we get by differentiation

$$\frac{du}{dx} = 9 + \frac{dy}{dx}$$

which can be rewritten as

$$y' = \frac{dy}{dx} = \frac{du}{dx} - 9.$$

After substituting the corresponding terms, we get

$$\frac{du}{dx} - 9 = u^2.$$

This is a differential equation with separable variables, and we get

$$\frac{du}{u^2 + 9} = dx.$$

Integrating both sides, we obtain

$$\frac{1}{3} \cdot \arctan \frac{u}{3} = x + C$$

or, correspondingly,

$$\arctan \frac{u}{3} = 3(x + C)$$

$$\frac{u}{3} = \tan 3(x + C)$$

$$u = 3 \tan 3(x + C).$$

After substituting back, we get

$$9x + y + 2 = 3 \tan 3(x + C)$$

$$y = 3 \tan 3(x + C) - 9x - 2.$$

12.2 LINEAR DIFFERENTIAL EQUATIONS OF ORDER n

Definition 12.3 A differential equation

$$y^{(n)} + a_{n-1}(x) y^{(n-1)} + \cdots + a_1(x) y' + a_0(x) y = q(x) \tag{12.3}$$

is called a *linear differential equation of order n.*

Equation (12.3) is called linear since function y and all derivatives occur only in the first power and there are no products of the $n+1$ functions $y, y', \ldots, y^{(n)}$. The functions $a_i(x), i = 1, 2, \ldots, n - 1$, depend on variable x and can be arbitrary.

Definition 12.4 If $q(x) \equiv 0$ in equation (12.3), i.e. function $q = q(x)$ is identically equal to zero for all values x, the differential equation (12.3) is called *homogeneous.* Otherwise, i.e. if $q(x) \neq 0$, the differential equation is called *non-homogeneous.*

Function $q = q(x)$ is also known as the *forcing term.* Next, we present some properties of solutions of a linear differential equation of order n and then we discuss an important special case.

12.2.1 Properties of solutions

Homogeneous differential equations

First, we present some properties of solutions of linear homogeneous differential equations of order n.

THEOREM 12.1 Let $y_1(x), y_2(x), \ldots, y_m(x)$ be solutions of the linear homogeneous differential equation

$$y^{(n)} + a_{n-1}(x)y^{(n-1)} + \cdots + a_0(x)y = 0. \tag{12.4}$$

Then the linear combination

$$y(x) = C_1 y_1(x) + C_2 y_2(x) + \cdots + C_m y_m(x)$$

is a solution as well with $C_1, C_2, \ldots, C_m \in \mathbb{R}$.

Definition 12.5   The solutions $y_1(x), y_2(x), \ldots, y_m(x)$, $m \le n$, of the linear homogeneous differential equation (12.4) are said to be *linearly independent* if

$$C_1 y_1(x) + C_2 y_2(x) + \cdots + C_m y_m(x) = 0 \quad \text{for all } x \in D_y$$

is possible only for $C_1 = C_2 = \cdots = C_m = 0$. Otherwise, the solutions are said to be *linearly dependent*.

The next theorem gives a criterion to decide whether a set of solutions for the homogeneous differential equation is linearly independent.

THEOREM 12.2   The solutions $y_1(x), y_2(x), \ldots, y_m(x)$, $m \le n$, of the linear homogeneous differential equation (12.4) are linearly independent if and only if

$$W(x) = \begin{vmatrix} y_1(x) & y_2(x) & \cdots & y_m(x) \\ y_1'(x) & y_2'(x) & \cdots & y_m'(x) \\ \vdots & \vdots & & \vdots \\ y_1^{(m-1)}(x) & y_2^{(m-1)}(x) & \cdots & y_m^{(m-1)}(x) \end{vmatrix} \ne 0 \quad \text{for } x \in D_y.$$

$W(x)$ is called *Wronski's* determinant. It can be proved that, if $W(x_0) \ne 0$ for some $x_0 \in D_y = (a, b)$, then we also have $W(x) \ne 0$ for all $x \in (a, b)$. Under the assumptions of Theorem 12.2, it is therefore sufficient to investigate $W(x_0)$ for some particular point $x_0 \in D_y$. If we have $W(x_0) \ne 0$, the m solutions $y_1(x), y_2(x), \ldots, y_m(x)$ are linearly independent.

Example 12.5   We check whether the functions

$$y_1(x) = x, \qquad y_2(x) = \frac{1}{x} \qquad \text{and} \qquad y_3(x) = \frac{\ln x}{x}$$

are linearly independent solutions of the differential equation

$$x^3 y''' + 4x^2 y'' + xy' - y = 0.$$

We first determine the required derivatives of the function y_i, $i \in \{1, 2, 3\}$, and obtain

$$y_1'(x) = 1, \qquad y_1''(x) = 0, \qquad y_1'''(x) = 0,$$

$$y_2'(x) = -\frac{1}{x^2}, \qquad y_2''(x) = \frac{2}{x^3}, \qquad y_2'''(x) = -\frac{6}{x^4},$$

$$y_3'(x) = \frac{1 - \ln x}{x^2}, \qquad y_3''(x) = \frac{-3 + 2\ln x}{x^3}, \qquad y_3'''(x) = \frac{11 - 6\ln x}{x^4}.$$

Next, we check whether each $y_i, i \in \{1, 2, 3\}$, is a solution of the differential equation. We get

$$y_1(x) : x^3 \cdot 0 + 4x^2 \cdot 0 + x \cdot 1 - x \qquad\qquad = 0,$$

$$y_2(x) : x^3 \cdot \left(-\frac{6}{x^4}\right) + 4x^2 \cdot \left(\frac{2}{x^3}\right) + x \cdot \left(-\frac{1}{x^2}\right) - \frac{1}{x} = -\frac{6}{x} + \frac{8}{x} - \frac{1}{x} - \frac{1}{x} = 0,$$

$$y_3(x) : x^3 \cdot \frac{11 - 6\ln x}{x^4} + 4x^2 \cdot \frac{-3 + 2\ln x}{x^3} + x \cdot \frac{1 - \ln x}{x^2} - \frac{\ln x}{x}$$

$$= \frac{11 - 6\ln x}{x} + \frac{-12 + 8\ln x}{x} + \frac{1}{x} - \frac{\ln x}{x} - \frac{\ln x}{x} = 0.$$

We now investigate Wronski's determinant $W(x)$:

$$W(x) = \begin{vmatrix} x & \dfrac{1}{x} & \dfrac{\ln x}{x} \\[2mm] 1 & -\dfrac{1}{x^2} & \dfrac{1 - \ln x}{x^2} \\[2mm] 0 & \dfrac{2}{x^3} & \dfrac{-3 + 2\ln x}{x^3} \end{vmatrix}$$

$$= x \cdot \begin{vmatrix} -\dfrac{1}{x^2} & \dfrac{1 - \ln x}{x^2} \\[2mm] \dfrac{2}{x^3} & \dfrac{-3 + 2\ln x}{x^3} \end{vmatrix} - 1 \cdot \begin{vmatrix} \dfrac{1}{x} & \dfrac{\ln x}{x} \\[2mm] \dfrac{2}{x^3} & \dfrac{-3 + 2\ln x}{x^3} \end{vmatrix}$$

$$= x \cdot \frac{3 - 2\ln x - 2 + 2\ln x}{x^5} - \frac{-3 + 2\ln x - 2\ln x}{x^4} = \frac{4}{x^4}.$$

Thus, Wronski's determinant is different from zero for all $x \in D_y = (0, \infty)$. Therefore, the solutions $y_1(x), y_2(x)$ and $y_3(x)$ are linearly independent.

Now we can describe the general solution of a linear homogeneous differential equation of order n as follows.

THEOREM 12.3 Let $y_1(x), y_2(x), \dots, y_n(x)$ be n linearly independent solutions of a linear homogeneous differential equation (12.4) of order n. Then the *general solution* can be written as

$$S_H = \{y^H(x) \mid y^H(x) = C_1 y_1(x) + C_2 y_2(x) + \cdots + C_n y_n(x), \ C_1, C_2, \dots, C_n \in \mathbb{R}\}. \tag{12.5}$$

The solutions $y_1(x), y_2(x), \dots, y_n(x)$ in equation (12.5) constitute a *fundamental system* of the differential equation, and we also say that $y^H(x)$ is the *complementary function*. Referring to Example 12.5, the general solution of the given homogeneous differential equation is

$$y^H(x) = C_1 x + C_2 \cdot \frac{1}{x} + C_3 \cdot \frac{\ln x}{x}.$$

Non-homogeneous differential equations

We present the following theorem that describes the structure of the general solution of a linear non-homogeneous differential equation.

THEOREM 12.4   Let

$$S_H = \{y^H(x) \mid y^H(x) = C_1\, y_1(x) + C_2\, y_2(x) + \cdots + C_n\, y_n(x),\ C_1, C_2, \ldots, C_n \in \mathbb{R}\}$$

be the general solution of the homogeneous equation (12.4) of order n and $y^N(x)$ a particular solution of the non-homogeneous equation (12.3). Then the general solution of a linear non-homogeneous differential equation (12.3) can be written as

$$S = \{y(x) \mid y(x) = y^H(x) + y^N(x),\quad y^H(x) \in S_H\}$$
$$= \{y(x) \mid y(x) = C_1\, y_1(x) + C_2\, y_2(x) + \ldots + C_n\, y_n(x) + y^N(x),\ C_1, C_2, \ldots, C_n \in \mathbb{R}\}.$$

Initial value problems

If an initial value problem of the linear differential equation of order n is considered, n initial conditions

$$y(x_0) = y_0, \qquad y'(x_0) = y_1, \qquad \ldots, \qquad y^{(n-1)}(x_0) = y_{n-1} \tag{12.6}$$

are given from which the n constants $C_1, C_2, \ldots, C_n$ of the general solution can be determined. This yields the *particular solution* $y^P(x)$. The following theorem gives a sufficient condition for the existence and uniqueness of a particular solution.

THEOREM 12.5 (existence and uniqueness of a solution)   Let n initial conditions according to (12.6) for a linear differential equation of order n

$$y^{(n)} + a_{n-1}\, y^{(n-1)} + \cdots + a_1\, y' + a_0\, y = q(x) \tag{12.7}$$

with constant coefficients a_i, $i \in \{0, 1, \ldots, n-1\}$, be given. Then this linear differential equation has exactly one particular solution $y^P(x)$.

We only note that for arbitrary functions $a_i(x)$, $i \in \{1, 2, \ldots, n-1\}$, an initial value problem does not necessarily have a solution.

12.2.2   Differential equations with constant coefficients

Now, we consider the special case of a linear differential equation of order n with *constant coefficients*, i.e. $a_i(x) = a_i$ for $i = 0, 1, 2, \ldots, n-1$.

Homogeneous equations

First, we consider the homogeneous differential equation. Thus, we have

$$y^{(n)} + a_{n-1}\, y^{(n-1)} + \cdots + a_1\, y' + a_0\, y = 0. \tag{12.8}$$

This type of homogeneous differential equation can be easily solved. We set

$$y(x) = e^{\lambda x} \qquad \text{with } \lambda \in \mathbb{C}$$

and obtain

$$y'(x) = \lambda e^{\lambda x}, \qquad y''(x) = \lambda^2 e^{\lambda x}, \qquad \dots, \qquad y^{(n)}(x) = \lambda^n e^{\lambda x}.$$

Substituting the latter terms into equation (12.8) and taking into account that the exponential function $e^{\lambda x}$ has no zeroes, we obtain the so-called *characteristic equation*:

$$P_n(\lambda) = \lambda^n + a_{n-1} \lambda^{n-1} + \cdots + a_1 \lambda + a_0 = 0. \tag{12.9}$$

We know from Chapter 3 that polynomial $P_n(\lambda)$ has n (real or complex) zeroes $\lambda_1, \lambda_2, \dots, \lambda_n$. For each root λ_j of the characteristic equation, we determine a specific solution of the homogeneous differential equation (Table 12.1).

It can be shown by Theorem 12.2 that the resulting set of solutions forms a fundamental system, i.e. the n solutions obtained in this way are linearly independent. It is worth noting that also in the case of complex roots of the characteristic equation, the solutions used are real. The solutions in cases (c) and (d) are based on Euler's formula:

$$e^{(\alpha_j + i\beta_j)x} = e^{\alpha_j x} \cdot e^{i\beta_j x} = e^{\alpha_j x}(\cos \beta_j + i \sin \beta_j)$$
$$= e^{\alpha_j x} \cos \beta_j x + i e^{\alpha_j x} \sin \beta_j x = y_j(x) + i y_{j+1}(x)$$

with $y_j(x)$ and $y_{j+1}(x)$ being real solutions.

Table 12.1 Specific solutions in dependence on the roots of the characteristic equation

Roots of the characteristic equation	Solution
(a) λ_j is a real root of multiplicity one	$y_j(x) = e^{\lambda_j x}$
(b) $\lambda_j = \lambda_{j+1} = \dots = \lambda_{j+k-1}$ is a real root of multiplicity $k > 1$	$y_j(x) = e^{\lambda_j x}$ $y_{j+1}(x) = x \cdot e^{\lambda_j x}$ $\vdots$ $y_{j+k-1}(x) = x^{k-1} \cdot e^{\lambda_j x}$
(c) $\lambda_j = \alpha_j + \beta_j i$ $\lambda_{j+1} = \alpha_j - \beta_j i$ is a pair of conjugate complex roots of multiplicity one	$y_j(x) = e^{\alpha_j x} \cdot \cos \beta_j x$ $y_{j+1}(x) = e^{\alpha_j x} \cdot \sin \beta_j x$
(d) $\lambda_j = \lambda_{j+2} = \cdots = \lambda_{j+2k-2} = \alpha_j + \beta_j i$ $\lambda_{j+1} = \lambda_{j+3} = \cdots = \lambda_{j+2k-1} = \alpha_j - \beta_j i$ is a pair of conjugate complex roots of multiplicity $k > 1$	$y_j(x) = e^{\alpha_j x} \cdot \cos \beta_j x$ $y_{j+1}(x) = e^{\alpha_j x} \cdot \sin \beta_j x$ $y_{j+2}(x) = x \cdot e^{\alpha_j x} \cdot \cos \beta_j x$ $y_{j+3}(x) = x \cdot e^{\alpha_j x} \cdot \sin \beta_j x$ $\vdots$ $y_{j+2k-2}(x) = x^{k-1} \cdot e^{\alpha_j x} \cdot \cos \beta_j x$ $y_{j+2k-1}(x) = x^{k-1} \cdot e^{\alpha_j x} \cdot \sin \beta_j x$

Table 12.2 Settings for special forcing terms

Forcing term $q(x)$	Setting $y^N(x)$
$a \cdot e^{bx}$	(a) $A \cdot e^{bx}$ if b is not a root of the characteristic equation (12.9) (b) $A \cdot x^k \cdot e^{bx}$ if b is a root of multiplicity k of the characteristic equation (12.9)
$b_n x^n + b_{n-1} x^{n-1} + \cdots + b_1 x + b_0$	(a) $A_n x^n + A_{n-1} x^{n-1} + \cdots + A_1 x + A_0$ if $a_0 \neq 0$ in equation (12.9) (b) $x^k \cdot (A_n x^n + A_{n-1} x^{n-1} + \cdots + A_1 x + A_0)$ if $a_0 = a_1 = \cdots = a_{k-1} = 0$ in (12.9), i.e. $y, y', \ldots, y^{(k-1)}$ do not occur in (12.8)
$a \cdot \cos \alpha x + b \cdot \sin \alpha x$ (a or b can be equal to zero)	(a) $A \cos \alpha x + B \sin \alpha x$ if αi is not a root of the characteristic equation (12.9) (b) $x^k \cdot (A \cdot \cos \alpha x + B \cdot \sin \alpha x)$ if αi is a root of multiplicity k of the characteristic equation (12.9)

Non-homogeneous differential equations

Now we discuss how a particular solution $y^N(x)$ of a non-homogeneous linear differential equation with constant coefficients can be found. We describe the *method of undetermined coefficients* and consider here only three specific types of forcing terms q. In these cases, a solution can be found by a special setting of the solution function $y^N(x)$ in dependence on function q in equation (12.7).

It can be seen that the setting in Table 12.2 follows the structure of the forcing term given on the left-hand side. We emphasize that the small letters in Table 12.2 are fixed in dependence on the given function $q = q(x)$ whereas capital letters are parameters which have to be determined such that $y^N(x)$ is a solution of the given differential equation. Using the corresponding setting above, we now insert $y^N(x)$ into the non-homogeneous differential equation and determine the coefficients A, B and A_i, respectively, by a comparison of the coefficients of the left-hand and right-hand sides. It should be noted that the modifications described in cases (b) of Table 12.2 are necessary to guarantee that the suggested approach works (otherwise we would not be able to determine all coefficients by this approach). Moreover, we mention that this approach can also be used for sums or products of the above forcing terms, where the setting has to follow the structure of the given forcing term. We illustrate the determination of the general solution by the following examples.

Example 12.6   We solve the differential equation

$$y'' + y' - 2y = 2x^2 - 4x + 1. \tag{12.10}$$

First, we consider the homogeneous differential equation

$$y'' + y' - 2y = 0.$$

By setting $y(x) = e^{\lambda x}$ and substituting $y(x), y'(x)$ and $y''(x)$ into the homogeneous equation, we get the characteristic equation

$$\lambda^2 + \lambda - 2 = 0$$

which has the two solutions

$$\lambda_1 = -\frac{1}{2} + \sqrt{\frac{1}{4} + 2} = 1 \qquad \text{and} \qquad \lambda_2 = \frac{1}{2} - \sqrt{\frac{1}{4} + 2} = -2.$$

Since we have two real roots of multiplicity one, case (a) in Table 12.1 applies for each root and we get the complementary function

$$y^H(x) = C_1 e^x + C_2 e^{-2x}.$$

Since function q is a polynomial of degree two, we use the setting

$$y^N(x) = Ax^2 + Bx + C$$

in order to find a particular solution of the non-homogeneous differential equation. We obtain

$$(y^N)'(x) = 2Ax + B$$
$$(y^N)''(x) = 2A.$$

Substituting the above terms into the differential equation (12.10), we get

$$2A + (2Ax + B) - 2(Ax^2 + Bx + C) = 2x^2 - 4x + 1.$$

Comparing now the coefficients of all powers of x, i.e. the coefficients of x^2, x^1 and x^0, we obtain:

$$
\begin{array}{rcccccc}
x^2: & -2A & & & & = & 2 \\
x^1: & 2A & - & B & & = & -4 \\
x^0: & 2A & + & 2B & - 2C & = & 1.
\end{array}
$$

From the first equation we obtain $A = -1$, and then from the second equation $B = 1$ and, finally, from the third equation $C = -1/2$. The general solution of the non-homogeneous differential equation (12.10) is given by

$$y(x) = y^H(x) + y^N(x) = C_1 e^x + C_2 e^{-2x} - x^2 + 2x - \frac{1}{2}.$$

Example 12.7 We solve the non-homogeneous differential equation

$$y''' + 2y'' + 5y' = \cos 2x.$$

Considering the homogeneous equation and again setting $y = e^{\lambda x}$, we obtain the characteristic equation

$$\lambda^3 + 2\lambda^2 + 5\lambda = \lambda \cdot (\lambda^2 + 2\lambda + 5) = 0.$$

From the latter equation we get

$$\lambda_1 = 0, \quad \lambda_2 = -1 + \sqrt{1-5} = -1 + 2i \quad \text{and} \quad \lambda_3 = -1 - \sqrt{1-5} = -1 - 2i,$$

i.e. there is a real root of multiplicity one and two (conjugate) complex roots of multiplicity one. Therefore, we get the complementary function

$$y^H(x) = C_1 e^{0x} + C_2 e^{-x} \cos 2x + C_3 e^{-x} \sin 2x$$
$$= C_1 + e^{-x}(C_2 \cos 2x + C_3 \sin 2x).$$

To find a particular solution y^N of the non-homogeneous differential equation, we use the setting

$$y^N(x) = A \cos 2x + B \sin 2x$$

and obtain

$$(y^N)'(x) = -2A \sin 2x + 2B \cos 2x$$
$$(y^N)''(x) = -4A \cos 2x - 4B \sin 2x$$
$$(y^N)'''(x) = 8A \sin 2x - 8B \cos 2x.$$

Substituting $y^N(x)$ and its derivatives into the non-homogeneous differential equation, we obtain

$$(8A \sin 2x - 8B \cos 2x) + 2(-4A \cos 2x - 4B \sin 2x)$$
$$+ 5(-2A \sin 2x + 2B \cos 2x) = \cos 2x.$$

Comparing now the coefficients of $\sin 2x$ and $\cos 2x$ on both sides of the above equation, we obtain:

$$\sin 2x : 8A - 8B - 10A = -2A - 8B = 0$$
$$\cos 2x : -8B - 8A + 10B = -8A + 2B = 1.$$

This is a system of two linear equations with the two variables A and B which can be easily solved:

$$A = -\frac{2}{17} \quad \text{and} \quad B = \frac{1}{34}.$$

Therefore, we obtain the general solution

$$y(x) = y^H(x) + y^N(x) = C_1 + e^{-x}(C_2 \cos 2x + C_3 \sin 2x) - \frac{2}{17} \cos 2x + \frac{1}{34} \sin 2x.$$

Example 12.8   Let the non-homogeneous differential equation

$$y''' - 4y' = e^{2x}$$

be given. In order to solve the homogeneous equation, we consider the characteristic equation

$$\lambda^3 - 4\lambda = \lambda(\lambda^2 - 4) = 0$$

which has the three roots

$$\lambda_1 = 0, \qquad \lambda_2 = 2 \qquad \text{and} \qquad \lambda_3 = -2.$$

This yields the complementary function

$$y^H(x) = C_1 e^{0x} + C_2 e^{2x} + C_3 e^{-2x}$$
$$= C_1 + C_2 e^{2x} + C_3 e^{-2x}.$$

To get a particular solution of the non-homogeneous equation, we have to use case (b) with $k = 1$ for an exponential forcing term e^{2x} in Table 12.2 since $\lambda_2 = 2$ is a root of multiplicity one of the characteristic equation, and we set

$$y^N(x) = Axe^{2x}.$$

(Indeed, if we used case (a) and set $y^N(x) = Ae^{2x}$, we would be unable to determine coefficient A, as the reader can check.) We obtain

$$(y^N)'(x) = Ae^{2x} + 2Axe^{2x} = Ae^{2x}(1 + 2x)$$
$$(y^N)''(x) = 2Ae^{2x}(1 + 2x) + 2Ae^{2x} = 2Ae^{2x}(2 + 2x) = 4Ae^{2x}(1 + x)$$
$$(y^N)'''(x) = 8Ae^{2x}(1 + x) + 4Ae^{2x} = 4Ae^{2x}(3 + 2x).$$

Substituting the terms for $(y^N)'$ and $(y^N)'''$ into the non-homogeneous equation, we get

$$4Ae^{2x}(3 + 2x) - 4Ae^{2x}(1 + 2x) = e^{2x}.$$

Comparing the coefficients of e^{2x} on both sides, we get

$$12A - 4A = 1$$

which gives $A = 1/8$ and thus the particular solution

$$y^N(x) = \frac{1}{8}xe^{2x}.$$

Therefore, the general solution of the given differential equation is given by

$$y(x) = y^H(x) + y^N(x) = C_1 + C_2 e^{2x} + C_3 e^{-2x} + \frac{1}{8}xe^{2x}.$$

Example 12.9   Consider the homogeneous differential equation

$$y''' - 4y'' + 5y' - 2y = 0$$

with the initial conditions

$$y(0) = 1, \qquad y'(0) = 1 \qquad \text{and} \qquad y''(0) = 2.$$

The characteristic equation is given by

$$\lambda^3 - 4\lambda^2 + 5\lambda - 2 = 0.$$

Since the sum of the above four coefficients is equal to zero we immediately get the root $\lambda_1 = 1$. Applying now Horner's scheme for dividing the polynomial of degree three by $\lambda - 1$, we get the equation

$$\lambda^3 - 4\lambda^2 + 5\lambda - 2 = (\lambda - 1) \cdot (\lambda^2 - 3\lambda + 2).$$

The quadratic equation $\lambda^2 - 3\lambda + 2 = 0$ has the two roots

$$\lambda_2 = 1 \qquad \text{and} \qquad \lambda_3 = 2,$$

i.e. there is one real root $\lambda_1 = \lambda_2 = 1$ of multiplicity two and one real root $\lambda_3 = 2$ of multiplicity one. Therefore, the following complementary function is obtained:

$$y^H(x) = C_1 e^x + C_2 x e^x + C_3 e^{2x}.$$

Now we determine a particular solution satisfying the given initial conditions. We obtain:

$$y^H(0) = 1: \ C_1 \qquad\quad + C_3 = 1$$
$$(y^H)'(0) = 1: \ C_1 + C_2 + 2C_3 = 1$$
$$(y^H)''(0) = 2: \ C_1 + 2C_2 + 4C_3 = 2.$$

This is a system of three linear equations with three variables which has the unique solution

$$C_1 = 0, \qquad C_2 = -1 \qquad \text{and} \qquad C_3 = 1.$$

Hence we get the particular solution

$$y^P(x) = -xe^x + e^{2x}$$

of the initial value problem.

12.3 SYSTEMS OF LINEAR DIFFERENTIAL EQUATIONS OF THE FIRST ORDER

In this section, we briefly discuss the solution of systems of linear differential equations of the first order. We consider the system

$$y_1'(x) = a_{11} y_1(x) + a_{12} y_2(x) + \cdots + a_{1n} y_n(x) + q_1(x)$$
$$y_2'(x) = a_{21} y_1(x) + a_{22} y_2(x) + \cdots + a_{2n} y_n(x) + q_2(x)$$

$$\vdots \qquad\qquad \vdots$$

$$y_n'(x) = a_{n1} y_1(x) + a_{n2} y_2(x) + \cdots + a_{nn} y_n(x) + q_n(x)$$

or, equivalently, in matrix representation

$$
\begin{pmatrix} y_1'(x) \\ y_2'(x) \\ \vdots \\ y_n'(x) \end{pmatrix}
=
\begin{pmatrix} a_{11} & a_{12} & \cdots & a_{1n} \\ a_{21} & a_{22} & \cdots & a_{2n} \\ \vdots & \vdots & & \vdots \\ a_{n1} & a_{n2} & \cdots & a_{nn} \end{pmatrix}
\cdot
\begin{pmatrix} y_1(x) \\ y_2(x) \\ \vdots \\ y_n(x) \end{pmatrix}
+
\begin{pmatrix} q_1(x) \\ q_2(x) \\ \vdots \\ q_n(x) \end{pmatrix}.
\tag{12.11}
$$

A system is called *homogeneous* if *all* functions $q_i(x)$, $i = 1, 2, \ldots, n$, are identically equal to zero. We define a *solution* and the *general solution* of a system of linear differential equations of the first order in an analogous way as for a differential equation. Now a solution is characterized by a vector $\mathbf{y}(x) = (y_1(x), y_2(x), \ldots, y_n(x))^T$ of n functions satisfying the given system of n linear differential equations. For an *initial value problem*, also given are the n initial conditions

$$y_1(x_0) = y_{10}, \qquad y_2(x_0) = y_{20}, \qquad \ldots, \qquad y_n(x_0) = y_{n0},$$

and a *particular solution* $\mathbf{y}^P(x)$ is a vector of n functions in which the n constants from the general solution have been determined by means of the initial values.

The following theorem gives a relationship between one linear differential equation of order n and a system of n linear differential equations of first order.

THEOREM 12.6 Let a differential equation of order n be linear with constant coefficients. Then it can be written as a system of n linear differential equations of first order with constant coefficients.

PROOF If a linear differential equation of order n is given in the form (12.7), we set

$$y_1(x) = y(x), \qquad y_2(x) = y'(x), \qquad \ldots, \qquad y_n(x) = y^{(n-1)}(x)$$

and obtain a system of differential equations:

$$y_1'(x) = y_2(x)$$
$$y_2'(x) = y_3(x)$$

$$\vdots \qquad \vdots$$

$$y_{n-1}'(x) = y_n(x)$$
$$y_n'(x) = -a_0 y_1(x) - a_1 y_2(x) - \cdots - a_{n-1} y_{n-1}(x) + q(x).$$

∎

Under additional assumptions (which we do not discuss here), there is even an equivalence between a linear differential equation of order n with constant coefficients and a system of n linear differential equations of the first order with constant coefficients, i.e. a linear differential equation of order n can be transformed into a system of n linear differential equations of the first order with constant coefficients, and vice versa. By the following example, we illustrate this process of transforming a given system into a differential equation of order n by eliminating systematically all but one of the functions $y_i(x)$.

Example 12.10   Let the system of first-order differential equations

$$y_1' = y_2 + y_3 + e^x$$
$$y_2' = y_1 - y_3$$
$$y_3' = y_1 + y_2 + e^x$$

be given with the initial conditions

$$y_1(0) = 0, \qquad y_2(0) = 0 \qquad \text{and} \qquad y_3(0) = 1.$$

We transform this system containing three functions $y_1(x), y_2(x), y_3(x)$ and its derivatives into one linear differential equation of order three. To this end, we differentiate the first equation which yields

$$y_1'' = y_2' + y_3' + e^x$$
$$= (y_1 - y_3) + (y_1 + y_2 + e^x) + e^x$$
$$= 2y_1 + y_2 - y_3 + 2e^x.$$

Differentiating the latter equation again, we get

$$y_1''' = 2y_1' + y_2' - y_3' + 2e^x$$
$$= 2(y_2 + y_3 + e^x) + (y_1 - y_3) - (y_1 + y_2 + e^x) + 2e^x$$
$$= y_2 + y_3 + 3e^x.$$

Now we can replace the sum $y_2 + y_3$ by means of the first equation of the given system, and we obtain the third-order differential equation

$$y_1''' - y_1' = 2e^x.$$

Using the setting

$$y_1(x) = e^{\lambda x},$$

we obtain the characteristic equation

$$\lambda^3 - \lambda = 0$$

with the roots

$$\lambda_1 = 0, \qquad \lambda_2 = 1 \qquad \text{and} \qquad \lambda_3 = -1.$$

This yields the following general solution of the homogeneous equation:

$$y_1^H = C_1 + C_2 \cdot e^x + C_3 \cdot e^{-x}.$$

In order to find a particular solution of the non-homogeneous equation, we set

$$y_1^N = Axe^x$$

(notice that number 1 is a root of the characteristic equation, therefore we use case (b) in the setting for an exponential forcing term according to Table 12.2), which yields

$$(y_1^N)' = Ae^x(x+1), \qquad (y_1^N)'' = Ae^x(x+2) \qquad \text{and} \qquad (y_1^N)''' = Ae^x(x+3)$$

and therefore

$$Ae^x(x+3) - Ae^x(x+1) = 2e^x,$$

i.e. we obtain

$$e^x(Ax + 3A - Ax - A) = 2e^x$$

from which we obtain

$$A = 1.$$

Therefore, the general solution is given by

$$y_1(x) = C_1 + C_2 e^x + C_3 e^{-x} + xe^x.$$

We still have to determine the solutions $y_2(x)$ and $y_3(x)$. This can be done by using the first equation of the given system and the equation obtained for y_1''. We get

$$y_2 + y_3 = y_1' - e^x$$
$$y_2 - y_3 = y_1'' - 2y_1 - 2e^x.$$

Adding both equations and dividing the resulting equation by 2 gives

$$y_2(x) = \frac{1}{2}y_1' + \frac{1}{2}y_1'' - y_1 - \frac{3}{2}e^x.$$

Inserting the general solution y_1 and its derivatives

$$y_1' = C_2 e^x - C_3 e^{-x} + e^x + x e^x$$
$$y_1'' = C_2 e^x + C_3 e^{-x} + 2 e^x + x e^x,$$

we get

$$y_2(x) = -C_1 - C_3 e^{-x}.$$

Subtracting the above equations for $y_2 + y_3$ and $y_2 - y_3$ and dividing the resulting equation by two yields

$$y_3(x) = \frac{1}{2} y_1' - \frac{1}{2} y_1'' + y_1 + \frac{1}{2} e^x.$$

Inserting the above terms for y_1' and y_1'' we obtain

$$y_3(x) = C_1 + C_2 e^x + x e^x.$$

Next, we consider the initial conditions and obtain

$$
\begin{array}{llrrrrrl}
y_1(0) = 0: & \quad & C_1 & + & C_2 & + & C_3 & = & 0 \\
y_2(0) = 0: & \quad & -C_1 & & & - & C_3 & = & 0 \\
y_3(0) = 1: & \quad & C_1 & + & C_2 & & & = & 1.
\end{array}
$$

This system of three linear equations with three variables has the unique solution

$$C_1 = 1, \qquad C_2 = 0 \qquad \text{and} \qquad C_3 = -1.$$

The particular solution of the initial value problem is

$$y_1^P(x) = 1 - e^{-x} + x e^x$$
$$y_2^P(x) = -1 + e^{-x}$$
$$y_3^P(x) = 1 + x e^x.$$

It is worth noting that not every initial value problem necessarily has a solution. Next, we discuss an approach for solving a system of linear differential equations of the first order directly, without using the comment after Theorem 12.6.

Homogeneous systems

We first discuss the solution of the homogeneous system, i.e. we assume $q_i(x) \equiv 0$ for $i = 1, 2, \ldots, n$. We set

$$
\mathbf{y}(x) = \begin{pmatrix} y_1(x) \\ y_2(x) \\ \vdots \\ y_n(x) \end{pmatrix} = \begin{pmatrix} z_1 \\ z_2 \\ \vdots \\ z_n \end{pmatrix} e^{\lambda x}. \tag{12.12}
$$

Using (12.12) and

$$\begin{pmatrix} y_1'(x) \\ y_2'(x) \\ \vdots \\ y_n'(x) \end{pmatrix} = \lambda \begin{pmatrix} z_1 \\ z_2 \\ \vdots \\ z_n \end{pmatrix} e^{\lambda x},$$

we obtain from system (12.11) the eigenvalue problem:

$$\begin{pmatrix} a_{11} - \lambda & a_{12} & \cdots & a_{1n} \\ a_{21} & a_{22} - \lambda & \cdots & a_{2n} \\ \vdots & \vdots & & \vdots \\ a_{n1} & a_{n2} & \cdots & a_{nn} - \lambda \end{pmatrix} \begin{pmatrix} z_1 \\ z_2 \\ \vdots \\ z_n \end{pmatrix} = \mathbf{0}.$$

We determine the eigenvalues $\lambda_1, \lambda_2, \ldots, \lambda_n$ and the corresponding eigenvectors

$$\mathbf{z}^1 = \begin{pmatrix} z_1^1 \\ z_2^1 \\ \vdots \\ z_n^1 \end{pmatrix}, \quad \mathbf{z}^2 = \begin{pmatrix} z_1^2 \\ z_2^2 \\ \vdots \\ z_n^2 \end{pmatrix}, \quad \ldots, \quad \mathbf{z}^n = \begin{pmatrix} z_1^n \\ z_2^n \\ \vdots \\ z_n^n \end{pmatrix}.$$

As in the case of one differential equation, we have to distinguish several cases.
In the case of real eigenvalues λ_i of multiplicity one, the general solution $\mathbf{y}^H(x)$ is as follows:

$$\mathbf{y}^H(x) = \begin{pmatrix} y_1^H(x) \\ y_2^H(x) \\ \vdots \\ y_n^H(x) \end{pmatrix} = C_1 \begin{pmatrix} z_1^1 \\ z_2^1 \\ \vdots \\ z_n^1 \end{pmatrix} e^{\lambda_1 x} + C_2 \begin{pmatrix} z_1^2 \\ z_2^2 \\ \vdots \\ z_n^2 \end{pmatrix} e^{\lambda_2 x} + \cdots + C_n \begin{pmatrix} z_1^n \\ z_2^n \\ \vdots \\ z_n^n \end{pmatrix} e^{\lambda_n x}.$$

The other cases can be handled analogously to the case of a differential equation of order n.
We give only a few comments for the cases of a real eigenvalue of multiplicity $k > 1$ and of complex eigenvalues $\alpha_j \pm i\beta_j$ of multiplicity one.

If λ_i is a real eigenvalue of multiplicity k, the eigenvectors $\mathbf{z}^i, \mathbf{z}^{i+1}, \ldots, \mathbf{z}^{i+k-1}$ associated with this eigenvalue are not necessarily linearly independent. Therefore, we have to modify the term for finding the corresponding part of the general solution, and we use the following setting for an eigenvalue λ_i of multiplicity k:

$$C_i \begin{pmatrix} z_1^1 \\ z_2^1 \\ \vdots \\ z_n^1 \end{pmatrix} e^{\lambda_i x} + C_{i+1} \begin{pmatrix} z_1^2 \\ z_2^2 \\ \vdots \\ z_n^2 \end{pmatrix} xe^{\lambda_i x} + \ldots + C_{i+k-1} \begin{pmatrix} z_1^k \\ z_2^k \\ \vdots \\ z_n^k \end{pmatrix} x^{k-1} e^{\lambda_i x}.$$

In the case of complex eigenvalues $\alpha_j \pm i\beta_j$ of multiplicity one, we take one of them, e.g. $\alpha_j + i\beta_j$, and determine the corresponding eigenvector $\mathbf{z}^j$ (which contains in general complex numbers as components in this case). Then we calculate the product

$$\mathbf{z}^j \cdot e^{\alpha_j + i\beta_j} = \mathbf{z}^j \cdot e^{\alpha_j} \cdot (\cos \beta_j + i \sin \beta_j) = \mathbf{z}^*.$$

The real part $\bar{z}^j$ and the imaginary part $\bar{z}^{j+1}$ of the resulting vector z^*, respectively, yield two linearly independent solutions, and we use the setting

$$C_j \bar{z}^j + C_{j+1} \bar{z}^{j+1}$$

for the corresponding part of the general solution of the homogeneous differential equation.

Non-homogeneous systems

A particular solution of the non-homogeneous system can be determined in a similar way as for a differential equation of order n. However, *all* occurring specific forcing terms $q_1(x), q_2(x), \ldots, q_n(x)$ have to be considered in *any* function $y_i^N(x)$.

We finish this section with two systems of linear differential equations of the first order.

Example 12.11 We consider the system

$$
\begin{array}{rcrcrcr}
y_1' & = & 2y_1 & & & & \\
y_2' & = & y_1 & + & y_2 & + & 2y_3 \\
y_3' & = & -2y_1 & - & 4y_2 & - & 3y_3
\end{array}
$$

with the initial conditions

$$y_1(0) = 13, \qquad y_2(0) = 0 \qquad \text{and} \qquad y_3(0) = -4.$$

We use the setting

$$
\begin{pmatrix} y_1(x) \\ y_2(x) \\ y_3(x) \end{pmatrix} = \begin{pmatrix} z_1 \\ z_2 \\ z_3 \end{pmatrix} e^{\lambda x}.
$$

Inserting this and the derivatives y_i', $i \in \{1,2,3\}$, into the system, we get the eigenvalue problem

$$
\begin{pmatrix} 2-\lambda & 0 & 0 \\ 1 & 1-\lambda & 2 \\ -2 & -4 & -3-\lambda \end{pmatrix} \cdot \begin{pmatrix} z_1 \\ z_2 \\ z_3 \end{pmatrix} = 0,
$$

i.e. we have to look for the eigenvalues of matrix

$$
A = \begin{pmatrix} 2 & 0 & 0 \\ 1 & 1 & 2 \\ -2 & -4 & -3 \end{pmatrix}.
$$

Determining the eigenvalues of matrix A, we investigate

$$
|A - \lambda I| = \begin{vmatrix} 2-\lambda & 0 & 0 \\ 1 & 1-\lambda & 2 \\ -2 & -4 & -3-\lambda \end{vmatrix} = (2-\lambda) \cdot \begin{vmatrix} 1-\lambda & 2 \\ -4 & -3-\lambda \end{vmatrix}
$$

$$
= (2-\lambda) \cdot [(1-\lambda)(-3-\lambda) + 8] = (2-\lambda) \cdot (\lambda^2 + 2\lambda + 5) = 0.
$$

This yields the eigenvalues

$$\lambda_1 = 2, \quad \lambda_2 = -1 + \sqrt{1-5} = -1 + 2i \quad \text{and} \quad \lambda_3 = -1 - \sqrt{1-5} = -1 - 2i.$$

Next, we determine for each eigenvalue a corresponding eigenvector. For $\lambda_1 = 2$, we obtain the system

$$
\begin{array}{rrrcl}
0z_1^1 & + \; 0z_2^1 & + \; 0z_3^1 & = & 0 \\
z_1^1 & - \; z_2^1 & + \; 2z_3^1 & = & 0 \\
-2z_1^1 & - \; 4z_2^1 & - \; 5z_3^1 & = & 0.
\end{array}
$$

The coefficient matrix of this system has rank two (row 1 can be dropped and the third equation is not a multiple of the second equation), so we can choose one variable arbitrarily. Choosing $z_3^1 = 6$, all components in the resulting solution are integer and the corresponding eigenvector is

$$
\mathbf{z}^1 = \begin{pmatrix} -13 \\ -1 \\ 6 \end{pmatrix}.
$$

For $\lambda_2 = -1 + 2i$, we get the following system:

$$
\begin{array}{rrrcl}
(3 - 2i)z_1^2 & + \quad 0z_2^2 & + \quad 0z_3^2 & = & 0 \\
z_1^2 & + \; (2 - 2i)z_2^2 & + \quad 2z_3^2 & = & 0 \\
-2z_1^2 & - \quad 4z_2^2 & + \; (-2 - 2i)z_3^2 & = & 0.
\end{array}
$$

From the first equation, we get

$$z_1^2 = 0.$$

This reduces the original system to the following system with two equations and two variables:

$$
\begin{array}{rrcl}
(2 - 2i)z_2^2 & + \quad 2z_3^2 & = & 0 \\
-4z_2^2 & + \; (-2 - 2i)z_3^2 & = & 0.
\end{array}
$$

Obviously, both equations (i.e. both row vectors) are linearly dependent (since multiplying the first equation by $(-1 - i)$ yields the second equation). Choosing $z_2^2 = 2$, we obtain $z_3^2 = -2 + 2i$. Thus, we have obtained the eigenvector

$$
\mathbf{z}^2 = \begin{pmatrix} 0 \\ 2 \\ -2 + 2i \end{pmatrix}.
$$

Finally, for $\lambda_3 = -1 - 2i$, we get the system

$$
\begin{array}{rrrcl}
(3 + 2i)z_1^3 & + \quad 0z_2^3 & + \quad 0z_3^3 & = & 0 \\
z_1^3 & + \; (2 + 2i)z_2^3 & + \quad 2z_3^3 & = & 0 \\
-2z_1^3 & - \quad 4z_2^3 & + \; (-2 + 2i)z_3^3 & = & 0.
\end{array}
$$

As in the above case, we get from the first equation $z_1^3 = 0$. Solving the reduced system with two equations and two variables, we finally get the eigenvector

$$z^3 = \begin{pmatrix} 0 \\ 2 \\ -2 - 2i \end{pmatrix}.$$

Thus, we have obtained the solutions

$$\begin{pmatrix} -13 \\ -1 \\ 6 \end{pmatrix} e^{2x}, \qquad \begin{pmatrix} 0 \\ 2 \\ -2 + 2i \end{pmatrix} e^{(-1+2i)x} \qquad \text{and} \qquad \begin{pmatrix} 0 \\ 2 \\ -2 - 2i \end{pmatrix} e^{(-1+2i)x}.$$

In order to get two linearly independent *real* solutions, we rewrite the second specific solution as follows. (Alternatively, we could also use the third solution resulting from the complex conjugate.)

$$\begin{pmatrix} 0 \\ 2 \\ -2 + 2i \end{pmatrix} e^{(-1+2i)x} = \begin{pmatrix} 0 \\ 2(\cos 2x + i \sin 2x) \\ (-2 + 2i)(\cos 2x + i \sin 2x) \end{pmatrix} e^{-x}$$

$$= \begin{pmatrix} 0 \\ 2(\cos 2x + i \sin 2x) \\ -2\cos 2x - 2\sin 2x + i(2\cos 2x - 2\sin 2x) \end{pmatrix} e^{-x}.$$

In the first transformation step above, we have applied Euler's formula for complex numbers. As we know from the consideration of differential equations of order n, the real part and the imaginary part form specific independent solutions (see also Table 12.1), i.e. we obtain

$$\begin{pmatrix} 0 \\ 2\cos 2x \\ -2(\cos 2x + \sin 2x) \end{pmatrix} e^{-x} \qquad \text{and} \qquad \begin{pmatrix} 0 \\ 2\sin 2x \\ -2(-\cos 2x + \sin 2x) \end{pmatrix} e^{-x}.$$

From the considerations above, we obtain the general solution as follows:

$$\begin{pmatrix} y_1(x) \\ y_2(x) \\ y_3(x) \end{pmatrix} = C_1 \begin{pmatrix} -13 \\ -1 \\ 6 \end{pmatrix} e^{2x} + C_2^* \begin{pmatrix} 0 \\ 2\cos 2x \\ -2(\cos 2x + \sin 2x) \end{pmatrix} e^{-x}$$

$$+ C_3^* \begin{pmatrix} 0 \\ 2\sin 2x \\ -2(-\cos 2x + \sin 2x) \end{pmatrix} e^{-x}.$$

Using $-2C_i^* = C_i$ for $i \in \{2,3\}$, we can simplify the above solution and obtain

$$
\begin{pmatrix} y_1(x) \\ y_2(x) \\ y_3(x) \end{pmatrix} = C_1 \begin{pmatrix} -13 \\ -1 \\ 6 \end{pmatrix} e^{2x} + C_2 \begin{pmatrix} 0 \\ -\cos 2x \\ \cos 2x + \sin 2x \end{pmatrix} e^{-x}
$$

$$
+ C_3 \begin{pmatrix} 0 \\ -\sin 2x \\ -\cos 2x + \sin 2x \end{pmatrix} e^{-x}.
$$

It remains to determine a solution satisfying the initial conditions. We obtain

$$
\begin{array}{rrccccccc}
y_1(0) = & 13: & -13C_1 & + & 0 & + & 0 & = & 13 \\
y_2(0) = & 0: & -C_1 & - & C_2 & + & 0 & = & 0 \\
y_3(0) = & -4: & 6C_1 & + & C_2 & - & C_3 & = & -4
\end{array}
$$

which has the solution:

$$
C_1 = -1; \qquad C_2 = 1 \qquad \text{and} \qquad C_3 = -1.
$$

Thus, the particular solution $\mathbf{y}^P(x)$ of the given initial value problem for a system of differential equations is as follows:

$$
\mathbf{y}^P(x) = \begin{pmatrix} y_1^P(x) \\ y_2^P(x) \\ y_3^P(x) \end{pmatrix}
$$

$$
= \begin{pmatrix} 13 \\ 1 \\ -6 \end{pmatrix} e^{2x} + \begin{pmatrix} 0 \\ -\cos 2x \\ \cos 2x + \sin 2x \end{pmatrix} e^{-x} + \begin{pmatrix} 0 \\ \sin 2x \\ \cos 2x - \sin 2x \end{pmatrix} e^{-x}.
$$

Example 12.12   We consider the following system of linear differential equations of the first order:

$$
\begin{aligned}
y_1' &= y_1 & + e^{2x} \\
y_2' &= 4y_1 - 3y_2 + 3x.
\end{aligned}
$$

In order to determine a solution of the homogeneous system

$$
\begin{aligned}
y_1' &= y_1 \\
y_2' &= 4y_1 - 3y_2,
\end{aligned}
$$

we use the setting

$$
\begin{pmatrix} y_1(x) \\ y_2(x) \end{pmatrix} = \begin{pmatrix} z_1 \\ z_2 \end{pmatrix} e^{\lambda x}.
$$

and determine first the eigenvalues of matrix

$$A = \begin{pmatrix} 1 & 0 \\ 4 & -3 \end{pmatrix}$$

formed by the coefficients of the functions $y_i(x)$ in the given system of differential equations. We obtain

$$|A - \lambda I| = \begin{vmatrix} 1 - \lambda & 0 \\ 4 & -3 - \lambda \end{vmatrix}$$

$$= (1 - \lambda)(-3 - \lambda) = 0$$

which has the solutions

$$\lambda_1 = 1 \quad \text{and} \quad \lambda_2 = -3.$$

We determine the eigenvectors associated with both eigenvalues. For $\lambda_1 = -1$, we obtain the system

$$0z_1^1 + 0z_2^1 = 0$$
$$4z_1^1 - 4z_2^1 = 0,$$

which yields the eigenvector

$$\mathbf{z}^1 = \begin{pmatrix} z_1^1 \\ z_2^1 \end{pmatrix} = \begin{pmatrix} 1 \\ 1 \end{pmatrix}.$$

For $\lambda_2 = -3$, we obtain the system

$$4z_1^2 + 0z_2^2 = 0$$
$$4z_1^2 + 0z_2^2 = 0,$$

which yields the eigenvector

$$\mathbf{z}^2 = \begin{pmatrix} z_1^2 \\ z_2^2 \end{pmatrix} = \begin{pmatrix} 0 \\ 1 \end{pmatrix}.$$

Thus, we obtain the following general solution $\mathbf{y}^H(x)$ of the homogeneous system:

$$\mathbf{y}^H(x) = \begin{pmatrix} y_1^H(x) \\ y_2^H(x) \end{pmatrix} = C_1 \begin{pmatrix} 1 \\ 1 \end{pmatrix} e^x + C_2 \begin{pmatrix} 0 \\ 1 \end{pmatrix} e^{-3x}.$$

Next, we determine a particular solution $\mathbf{y}^N(x)$ of the non-homogeneous system. Since we have the two forcing terms $3x$ and e^{2x}, we use the setting

$$\mathbf{y}^N(x) = \begin{pmatrix} y_1^N(x) \\ y_2^N(x) \end{pmatrix} = \begin{pmatrix} A_1 x + A_0 + A e^{2x} \\ B_1 x + B_0 + B e^{2x} \end{pmatrix}.$$

Notice that, according to the earlier comment, we have to use both the linear term and the exponential term in each of the functions $y_1^N(x)$ and $y_2^N(x)$. Differentiating the above functions, we get

$$\begin{pmatrix} (y_1^N)'(x) \\ (y_2^N)'(x) \end{pmatrix} = \begin{pmatrix} A_1 + 2Ae^{2x} \\ B_1 + 2Be^{2x} \end{pmatrix}.$$

Inserting now y_i^N and $(y_i^N)'$, $i \in \{1, 2\}$, into the initial non-homogeneous system, we obtain

$$A_1 + 2Ae^{2x} = A_1 x + A_0 + Ae^{2x} \qquad\qquad\qquad + e^{2x}$$

$$B_1 + 2Be^{2x} = 4A_1 x + 4A_0 + 4Ae^{2x} - 3B_1 x - 3B_0 - 3Be^{2x} + 3x,$$

which can be rewritten as

$$A_1 + 2Ae^{2x} = A_0 \qquad\qquad + A_1 x \qquad\qquad + (A + 1)e^{2x}$$

$$B_1 + 2Be^{2x} = (4A_0 - 3B_0) + (4A_1 - 3B_1 + 3)x + (4A - 3B)e^{2x}.$$

Now we compare in each of the two equations the coefficients of the terms x^1, x^0 and e^{2x} which must coincide on the left-hand and right-hand sides:

$$x^1 : 0 = A_1$$

$$0 = 4A_1 - 3B_1 + 3,$$

which yields the solution

$$A_1 = 0 \qquad \text{and} \qquad B_1 = 1.$$

Furthermore,

$$x^0 : A_1 = A_0$$

$$B_1 = 4A_0 - 3B_0.$$

Using the solution obtained for B_1 and A_1, we find

$$A_0 = 0 \qquad \text{and} \qquad B_0 = -\frac{1}{3}.$$

Finally,

$$e^{2x} : 2A = A + 1$$

$$2B = 4A - 3B,$$

which yields the solution

$$A = 1 \qquad \text{and} \qquad B = \frac{4}{5}.$$

Combining the results above, we get the particular solution $\mathbf{y}^N(x)$ of the non-homogeneous system:

$$
\mathbf{y}^N(x) = \begin{pmatrix} y_1^N(x) \\ y_2^N(x) \end{pmatrix} = \begin{pmatrix} e^{2x} \\ x - \dfrac{1}{3} + \dfrac{4}{5} e^{2x} \end{pmatrix}.
$$

Finally, the general solution $\mathbf{y}(x)$ of the non-homogeneous system of differential equations is given by the sum of $\mathbf{y}^H(x)$ and $\mathbf{y}^N(x)$:

$$
\mathbf{y}(x) = \begin{pmatrix} y_1(x) \\ y_2(x) \end{pmatrix} = \mathbf{y}^H(x) + \mathbf{y}^N(x)
$$

$$
= C_1 \begin{pmatrix} 1 \\ 1 \end{pmatrix} e^x + C_2 \begin{pmatrix} 0 \\ 1 \end{pmatrix} e^{-3x} + \begin{pmatrix} e^{2x} \\ x - \dfrac{1}{3} + \dfrac{4}{5} e^{2x} \end{pmatrix}.
$$

12.4   LINEAR DIFFERENCE EQUATIONS

12.4.1   Definitions and properties of solutions

While differential equations consider a relationship between a function of a *real* variable and its derivatives, in several economic applications differences in function values have to be considered. A typical situation arises when the independent variable is a *discrete* variable which may take e.g. only integer values. (This applies when a function value is observed every day, every month or every year or any other discrete periods.) For instance, if time is considered as the independent variable, and we are interested in the development of a certain function value over time, then it is often desired to give a relationship between several successive values $y_t, y_{t+1}, \ldots, y_{t+k}$ which are described by a so-called *difference equation*. For the solution of special types of such equations we use a similar concept as for the solution of differential equations. We begin with the formal definition of a difference equation of order n.

Definition 12.6   If for arbitrary successive $n+1$ terms of a sequence $\{y_t\}$ the equation

$$
y_{t+n} = a_{n-1}(t) y_{t+n-1} + \cdots + a_1(t) y_{t+1} + a_0(t) y_t + q(t) \tag{12.13}
$$

holds, it is called a *linear difference equation*. If $a_0(t) \neq 0$, the number n denotes the *order* of the difference equation.

Definition 12.7   If we have $q(t) \equiv 0$ in equation (12.13), the difference equation is called *homogeneous*, otherwise it is called *non-homogeneous*.

Definition 12.8   Any sequence $\{y_t\}$ satisfying the difference equation (12.13) is called a *solution*. The set S of all sequences $\{y_t\}$ satisfying the difference equation (12.13) is called the *set of solutions* or the *general solution*.

First, we present some properties of solutions of linear homogeneous differential equations of order n.

Homogeneous difference equations

For a homogeneous difference equation of order n, we can state analogous results as for a homogeneous differential equation of order n given in Chapter 12.2.1.

THEOREM 12.7   Let $\{y_t^1\}, \{y_t^2\}, \dots, \{y_t^m\}$ with $t = 0, 1, \dots$ be solutions of the linear homogeneous difference equation

$$y_{t+n} = a_{n-1}(t)\, y_{t+n-1} + \cdots + a_1(t)\, y_{t+1} + a_0(t)\, y_t \tag{12.14}$$

Then the linear combination

$$y_t = C_1 y_t^1 + C_2 y_t^2 + \cdots + C_m y_t^m$$

is a solution as well with $C_1, C_2, \dots, C_m \in \mathbb{R}$.

Definition 12.9   The solutions $\{y_t^1\}, \{y_t^2\}, \dots, \{y_t^m\}$, $m \le n$, $t = 0, 1, \dots$, of the linear homogeneous difference equation (12.14) are said to be *linearly independent* if

$$C_1 y_t^1 + C_2 y_t^2 + \cdots + C_m y_t^m = 0 \quad \text{for } t = 0, 1, \dots.$$

is only possible for $C_1 = C_2 = \cdots = C_m = 0$. Otherwise, the solutions are said to be *linearly dependent*.

The next theorem gives a criterion to decide whether a set of solutions for the homogeneous difference equation is linearly independent.

THEOREM 12.8   The solutions $\{y_t^1\}, \{y_t^2\}, \dots, \{y_t^m\}$, $m \le n$, $t = 0, 1, \dots$, of the linear homogeneous difference equation (12.14) are linearly independent if and only if

$$C(t) = \begin{vmatrix} y_t^1 & y_t^2 & \cdots & y_t^m \\ y_{t+1}^1 & y_{t+1}^2 & \cdots & y_{t+1}^m \\ \vdots & \vdots & & \vdots \\ y_{t+m-1}^1 & y_{t+m-1}^2 & \cdots & y_{t+m-1}^m \end{vmatrix} \ne 0 \quad \text{for } t = 0, 1, \dots.$$

$C(t)$ is known as *Casorati's* determinant. Now we can describe the general solution of a linear homogeneous difference equation of order n as follows.

THEOREM 12.9   Let $\{y_t^1\}, \{y_t^2\}, \ldots, \{y_t^n\}$ with $t = 0, 1, \ldots$, be n linearly independent solutions of a linear homogeneous difference equation (12.14) of order n. Then the general solution can be written as

$$S_H = \left\{ \{y_t^H\} \mid y_t^H = C_1 y_t^1 + C_2 y_t^2 + \cdots + C_n y_t^n, \quad C_1, C_2, \ldots, C_n \in \mathbb{R} \right\}.$$

Non-homogeneous difference equations

The following theorem describes the structure of the general solution of a linear non-homogeneous difference equation of order n.

THEOREM 12.10   Let

$$S_H = \left\{ \{y_t^H\} \mid y_t^H = C_1 y_t^1 + C_2 y_t^2 + \cdots + C_n y_t^n, \quad C_1, C_2, \ldots, C_n \in \mathbb{R} \right\}$$

be the general solution of a linear homogeneous difference equation (12.14) of order n and $\{y_t^N\}$ be a particular solution of the non-homogeneous equation (12.13). Then the general solution of a linear non-homogeneous difference equation (12.13) can be written as

$$S = \left\{ \{y_t\} \mid y_t = y_t^H + y_t^N, \quad \{y_t^H\} \in S_H \right\}$$
$$= \left\{ \{y_t\} \mid y_t = C_1 y_t^1 + C_2 y_t^2 + \cdots + C_n y_t^n + y_t^N, \quad C_1, C_2, \ldots, C_n \in \mathbb{R} \right\}.$$

Initial value problems

If in addition n initial values of a linear difference equation of order n are given, we can present the following sufficient condition for the existence and uniqueness of a solution of the inital value problem.

THEOREM 12.11 (existence and uniqueness of a solution)   Let n successive initial values $y_0, y_1, \ldots, y_{n-1}$ for a linear difference equation

$$y_{t+n} = a_{n-1} y_{t+n-1} + \cdots + a_1 y_{t+1} + a_0 y_t + q(t)$$

with constant coefficients a_i, $i \in \{0, 1, \ldots, n-1\}$, be given. Then this linear difference equation has exactly one particular solution $\{y_t^P\}$ with $t = 0, 1, \ldots$.

We note that, if the n initial values are given for not successive values of t, it is possible that no solution or no unique solution exists. In the following, we describe the solution process for difference equations of the first order with arbitrary coefficients and for difference equations of the second order with constant coefficients.

12.4.2   Linear difference equations of the first order

Next, we consider a *linear difference equation of the first order* such as:

$$y_{t+1} = a_0(t) y_t + q(t).$$

This difference equation can be alternatively written by means of the differences of two successive terms y_{t+1} and y_t as

$$\Delta y = y_{t+1} - y_t = a(t)y_t + q(t),$$

where $a(t) = a_0(t) - 1$. (The latter notation illustrates why such an equation is called a difference equation.)

Constant coefficient and forcing term

First, we consider the special cases of a constant coefficient $a_0(t) = a_0$ and a constant forcing term $q(t) = q$.

If we consider the homogeneous equation $y_{t+1} = a_0 y_t$, we obtain the following:

$$y_1 = a_0 y_0$$

$$y_2 = a_0 y_1 = a_0^2 y_0$$

$$\vdots \quad \vdots$$

$$y_t = a_0 y_{t-1} = a_0^t y_0.$$

If there is a constant forcing term, i.e. the non-homogeneous difference equation has the form $y_{t+1} = a_0 y_t + q$, we obtain:

$$y_1 = a_0 y_0 + q$$

$$y_2 = a_0 y_1 + q = a_0(a_0 y_0 + q) + q = a_0^2 y_0 + a_0 q + q$$

$$y_3 = a_0 y_2 + q = a_0(a_0^2 y_0 + a_0 q + q) + q = a_0^3 y_0 + q(a_0^2 + a_0 + 1)$$

$$\vdots \quad \vdots$$

$$y_t = a_0^t y_0 + q(a_0^{t-1} + \cdots + a_0 + 1).$$

For $a_0 \neq 1$, the latter equation can be written as follows (using the formula for the $(t+1)$th partial sum of a geometric sequence):

$$y_t = a_0^t y_0 + q \cdot \frac{a_0^t - 1}{a_0 - 1}, \quad t = 1, 2, \ldots.$$

For $a_0 = 1$, we obtain

$$y_t = y_0 + q \cdot t.$$

Variable coefficient and forcing term

Now we consider the general case, where the coefficient and the forcing term depend on variable t, i.e. the difference equation has the form

$$y_{t+1} = a_0(t)y_t + q(t).$$

Then we obtain:

$$y_1 = a_0(0)y_0 + q(0)$$

$$y_2 = a_0(1)y_1 + q(1) = a_0(1)[a_0(0)y_0 + q(0)] + q(1)$$

$$= a_0(1)a_0(0)y_0 + a_0(1)q(0) + q(1)$$

$$y_3 = a_0(2)y_2 + q(2) = a_0(2)[a_0(1)a_0(0)y_0 + a_0(1)q(0) + q(1)] + q(2)$$

$$= a_0(2)a_0(1)a_0(0)y_0 + a_0(2)a_0(1)q(0) + a_0(2)q(1) + q(2)$$

$$\vdots \qquad \vdots$$

$$y_t = \prod_{i=0}^{t-1} a_0(i)y_0 + \sum_{j=0}^{t-2}\left(\prod_{i=j+1}^{t-1} a_0(i)\right)q(j) + q(t-1), \quad t = 1, 2, \ldots$$

Example 12.13   Consider the following instance of the so-called cobweb model:

$$y_{t+1} = 1 + 2p_t$$

$$x_t = 15 - p_t$$

$$y_t = x_t,$$

where for a certain product, y_t denotes the supply, x_t denotes the demand and p_t the price at time t. Moreover, let the initial price $p_0 = 4$ at time zero be given.

The first equation expresses that a producer fixes the supply of a product in period $t + 1$ in dependence on the price p_t of this product in the previous period (i.e. the supply increases with the price). The second equation means that the demand for a product in period t depends on its price in that period (i.e. the demand decreases with increasing price). The third equation gives the equilibrium condition: all products are sold, i.e. supply is equal to demand.

Using the equilibrium $y_{t+1} = x_{t+1}$, we obtain

$$1 + 2p_t = 15 - p_{t+1}$$

$$p_{t+1} = 14 - 2p_t$$

which is a linear difference equation of the first order (with constant coefficient and forcing term). Applying the above formula and taking into account that $a_0 = -2 \neq 1$, we obtain

$$p_t = a_0^t p_0 + q \cdot \frac{a_0^t - 1}{a_0 - 1} = (-2)^t p_0 + 14 \cdot \frac{(-2)^t - 1}{-2 - 1}$$

$$= (-2)^t p_0 + \frac{14}{3} \cdot [1 - (-2)^t] = \frac{14}{3} + (-2)^t \left(p_0 - \frac{14}{3}\right).$$

Using the initial condition $p_0 = 4$, we get the solution

$$p_t = \frac{14}{3} + (-2)^t \cdot \left(4 - \frac{14}{3}\right) = \frac{14}{3} + (-2)^t \cdot \left(-\frac{2}{3}\right).$$

For the initial value problem, we get the terms

$$p_0 = 4, \qquad p_1 = \frac{14}{3} + (-2)^1 \cdot \left(-\frac{2}{3}\right) = \frac{18}{3} = 6,$$

$$p_2 = \frac{14}{3} + (-2)^2 \cdot \left(-\frac{2}{3}\right) = \frac{6}{3} = 2, \qquad p_3 = \frac{14}{3} + (-2)^3 \cdot \left(-\frac{2}{3}\right) = \frac{30}{3} = 10,$$

$$p_4 = \frac{14}{3} + (-2)^4 \cdot \left(-\frac{2}{3}\right) = -\frac{18}{3} = -6, \ldots,$$

which is illustrated in Figure 12.3.

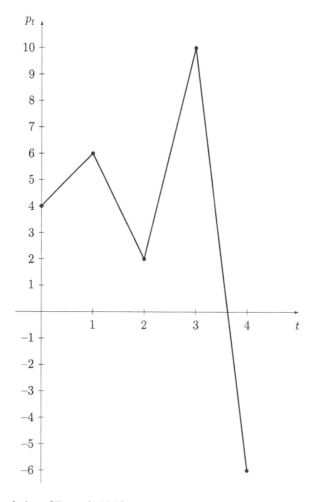

Figure 12.3 The solution of Example 12.13.

12.4.3   Linear difference equations of the second order

Next, we consider a *linear difference equation of the second order with constant coefficients*:

$$y_{t+2} = a_1 y_{t+1} + a_0 y_t + q(t), \qquad\qquad (12.15)$$

or equivalently,

$$y_{t+2} - a_1 y_{t+1} - a_0 y_t = q(t).$$

First, we consider the homogeneous equation, i.e. $q(t) \equiv 0$:

$$y_{t+2} - a_1 y_{t+1} - a_0 y_t = 0.$$

In order to determine two linearly independent solutions and then the general solution y_t^H according to Theorem 12.10, we use the setting

$$y_t^H = m^t,$$

which is substituted in the homogeneous difference equation. This yields the quadratic characteristic equation

$$m^2 - a_1 m - a_0 = 0 \qquad\qquad (12.16)$$

which has the two solutions

$$m_{1,2} = \frac{a_1}{2} \pm \sqrt{\frac{a_1^2}{4} + a_0}.$$

We can distinguish the following three cases. Treatment of these cases is done in the same way as in Chapter 12.2.2 for linear differential equations (where the term y_{t+i} in a difference equation 'corresponds' to the derivative $y^{(i)}(x)$ in the differential equation).

(1) Inequality

$$\frac{a_1^2}{4} + a_0 > 0$$

holds, i.e. there are two different real roots of the quadratic equation. Then we set

$$y_t = C_1 m_1^t + C_2 m_2^t.$$

(2) Equation

$$\frac{a_1^2}{4} + a_0 = 0$$

holds, i.e. there is a real root $m_1 = m_2$ of multiplicity two. Analogously to the treatment of a differential equation, we set

$$y_t = (C_1 + C_2 t) m_1^t.$$

(3) Inequality

$$\frac{a_1^2}{4} + a_0 < 0$$

holds, i.e. there are two complex roots

$$m_{1,2} = a \pm bi$$

of the quadratic characteristic equation with

$$a = \frac{a_1}{2} \quad \text{and} \quad b = \frac{\sqrt{-a_1^2 - 4a_0}}{2}.$$

Then we get the real solutions

$$y_t = r^t \cdot (C_1 \cos \alpha t + C_2 \sin \alpha t),$$

where

$$r = \sqrt{-a_0} = \sqrt{a^2 + b^2}$$

and α is such that

$$\cos \alpha = -\frac{a_1}{2a_0} \cdot \sqrt{-a_0} = \frac{a}{r} \quad \text{and} \quad \sin \alpha = -\frac{\sqrt{4a_0^2 + a_0 a_1^2}}{2a_0} = \frac{b}{r}.$$

Next, we consider the non-homogeneous equation, i.e. we have $q(t) \neq 0$ in equation (12.15). A particular solution y_t^N of a non-homogeneous differential equation of the second order can be found for a polynomial or an exponential function as forcing term $q(t)$ by a specific setting as shown in Table 12.3.

Analogously to differential equations, we can also use this *method of undetermined coefficients* for a sum or product of the above forcing terms. In such a case, the setting has to

Table 12.3 Settings for specific forcing terms in the difference equation

Forcing term $q(t)$	Setting y_t^N
$b_n t^n + b_{n-1} t^{n-1} + \cdots + b_1 t + b_0$	(a) $y_t^N = A_n t^n + A_{n-1} t^{n-1} + \cdots + A_1 t + A_0$ if $a_0 + a_1 \neq 1$ in the characteristic equation (12.16)
	(b) $y_t^N = t \cdot (A_n t^n + A_{n-1} t^{n-1} + \cdots + A_1 t + A_0)$ if $a_0 + a_1 = 1$ and $a_1 \neq 2$ in (12.16)
	(c) $y_t^N = t^2 \cdot (A_n t^n + A_{n-1} t^{n-1} + \cdots + A_1 t + A_0)$ if $a_0 + a_1 = 1$ and $a_1 = 2$ in (12.16)
ab^t	(a) $y_t^N = Ab^t$ if b is not a root of the characteristic equation (12.16)
	(b) $y_t^N = A \cdot t^k \cdot b^t$ if b is a root of multiplicity k of the characteristic equation (12.16)

represent the corresponding sum or product structure of the forcing term. We now apply this method as before in the case of differential equations.

Example 12.14   As an example, we consider an instance of the *multiplier-accelerator model* by Samuelson, which describes a relationship between the national income y_t, the total consumption c_t and the total investment i_t of a country at some time t. Let the following equations

$$y_t = c_t + i_t + 50$$

$$c_t = \frac{1}{2} \cdot y_{t-1}$$

$$i_t = c_t - c_{t-1}$$

together with the initial conditions

$$y_0 = 55 \qquad \text{and} \qquad y_1 = 48$$

be given. The first equation shows that the income at time t is equal to the consumption plus the investment plus the constant government expenditure (in our example they are equal to 50 units). The second equation shows that the consumption at time t is half the income at time $t - 1$. Finally, the third equation shows that the total investment at time t is equal to the difference in the total consumption at times t and $t - 1$.

Substituting the equations for c_t and i_t into the first equation, we obtain

$$y_t = \frac{1}{2} \cdot y_{t-1} + (c_t - c_{t-1}) + 50$$

$$= \frac{1}{2} \cdot y_{t-1} + \left(\frac{1}{2} \cdot y_{t-1} - \frac{1}{2} \cdot y_{t-2} \right) + 50$$

$$= y_{t-1} - \frac{1}{2} \cdot y_{t-2} + 50.$$

Using the terms y_{t+2}, y_{t+1}, y_t instead of y_t, y_{t-1}, y_{t-2}, we obtain the second-order difference equation

$$y_{t+2} = y_{t+1} - \frac{1}{2} \cdot y_t + 50.$$

To find the general solution of the homogeneous equation, we use the setting

$$y_t = m^t$$

which, after substituting into the difference equation, yields the quadratic equation

$$m^2 - m + \frac{1}{2} = 0.$$

We get the two solutions

$$m_1 = \frac{1}{2} + \sqrt{\frac{1}{4} - \frac{1}{2}} = \frac{1}{2} + \frac{i}{2} \qquad \text{and} \qquad m_2 = \frac{1}{2} - \sqrt{\frac{1}{4} - \frac{1}{2}} = \frac{1}{2} - \frac{i}{2}.$$

Using $a_1 = 1$ and $a_0 = -1/2$, we get the following equations for r and α:

$$r = \sqrt{\frac{1}{2}} = \frac{1}{2}\sqrt{2}$$

and α is such that

$$\cos \alpha = -\frac{a_1}{2a_0} \cdot \sqrt{-a_0} = -\frac{1}{2(-\frac{1}{2})} \cdot \sqrt{\frac{1}{2}} = \frac{1}{2}\sqrt{2}$$

and

$$\sin \alpha = -\frac{\sqrt{4a_0^2 + a_0a_1^2}}{2a_0} = -\frac{\sqrt{4 \cdot (-\frac{1}{2})^2 + (-\frac{1}{2}) \cdot 1}}{2(-\frac{1}{2})} = \sqrt{1 - \frac{1}{2}} = \frac{1}{2}\sqrt{2}.$$

The latter two equations give $\alpha = \pi/4$. Thus, the general solution of the homogeneous difference equation is given by

$$y_t^H = \left(\frac{1}{2}\sqrt{2}\right)^t \cdot \left(C_1 \cos \frac{\pi}{4}t + C_2 \sin \frac{\pi}{4}t\right).$$

To find a particular solution of the non-homogeneous equation, we use the setting $y_t = A$. Using $y_{t+2} = y_{t+1} = A$ and substituting the latter into the difference equation, we obtain

$$A = A - \frac{1}{2}A + 50.$$

This yields

$$\frac{1}{2} \cdot A = 50$$

$$A = 100,$$

and thus the particular solution of the non-homogeneous equation is

$$y_t^N = 100.$$

Therefore, the general solution of the non-homogeneous difference equation is given by

$$y_t = y_t^H + y_t^N = \left(\frac{1}{2}\sqrt{2}\right)^t \cdot \left(C_1 \cos \frac{\pi}{4}t + C_2 \sin \frac{\pi}{4}t\right) + 100.$$

Finally, we determine the solution satisfying the initial conditions. We get

$$y_0 = C_1 \cdot 1 + C_2 \cdot 0 + 100 = 55 \quad \Longrightarrow \quad C_1 = -45$$

$$y_1 = C_1 \cdot \frac{1}{2}\sqrt{2} \cdot \frac{1}{2}\sqrt{2} + C_2 \frac{1}{2}\sqrt{2} \cdot \frac{1}{2}\sqrt{2} + 100$$

$$= C_1 \cdot \frac{1}{2} + C_2 \cdot \frac{1}{2} + 100$$

$$= -\frac{45}{2} + C_2 \cdot \frac{1}{2} + 100 = 48 \quad \Longrightarrow \quad C_2 = -59.$$

Thus, we get the particular solution y_t^P of the initial value problem

$$y_t^P = y_t^H + y_t^N = -45 \left(\frac{1}{2}\sqrt{2}\right)^t \cos\frac{\pi}{4}t - 59 \left(\frac{1}{2}\sqrt{2}\right)^t \sin\frac{\pi}{4}t + 100.$$

Using this solution, we get

$$y_2 = -45 \cdot \left(\frac{1}{2}\sqrt{2}\right)^2 \cos\frac{\pi}{2} - 59 \cdot \left(\frac{1}{2}\sqrt{2}\right)^2 \sin\frac{\pi}{2} + 100 = 70.5,$$

$$y_3 = 96.5, \quad y_4 = 111.25, \ldots$$

The solution is illustrated in Figure 12.4.

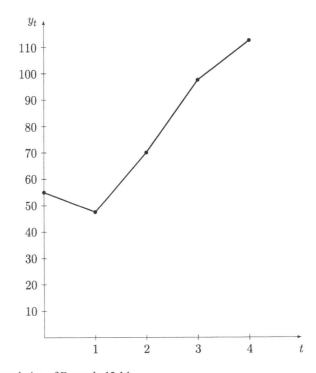

Figure 12.4 The solution of Example 12.14.

Example 12.15   Let the non-homogeneous differential equation

$$y_{t+2} = 2y_{t+1} - y_t + 1$$

be given. To solve the homogeneous equation, we get the characteristic equation

$$m^2 - 2m + 1 = 0$$

(notice that we have $a_1 = 2$ and $a_0 = -1$ in the characteristic equation) which has a real solution of multiplicity two:

$$m_1 = m_2 = 1.$$

Therefore, the general solution of the homogeneous equation is given by

$$y_t^H = C_1 \cdot 1^t + C_2 \cdot t \cdot 1^t = C_1 + C_2 t.$$

To get a particular solution of the non-homogeneous equation, we have to use case (c) of a polynomial forcing term (since $a_0 + a_1 = -1 + 2 = 1$ and $a_1 = 2$) which means that we set

$$y_t^N = At^2.$$

(In fact, the reader may check that both settings $y_t^N = A$ and $y_t^N = At$ are not appropriate to determine coefficient A). Substituting y_t^N into the non-homogeneous equation, we get

$$A(t+2)^2 = 2A(t+1)^2 - At^2 + 1$$

which can be rewritten as

$$A(t^2 + 4t + 4) - 2A(t^2 + 2t + 1) + At^2 = 1.$$

Comparing now the terms on both sides, we get

$$4A - 2A = 1$$

which gives the solution

$$A = \frac{1}{2}.$$

Therefore, we have obtained

$$y_t^N = \frac{1}{2} t^2,$$

and the general solution of the difference equation is given by

$$y_t = y_t^H + y_t^N = C_1 + C_2 t + \frac{1}{2} t^2.$$

EXERCISES

12.1 Consider the differential equation $y' = y$.

(a) Draw the direction field.
(b) Solve the differential equation by calculation. Find the particular solution satisfying $y(0) = 1$ resp. $y(0) = -1$.
(c) Draw the particular solution from part (b) in the direction field of part (a).

12.2   Find the solutions of the following differential equations:

(a) $y' = e^{x-y}$;   (b) $(1 + e^x)yy' = e^x$ with $y(0) = 1$.

12.3   Let a curve go through the point $P : (1, 1)$. The slope (of the tangent line) of the function at any point of the curve should be proportional to the squared function value at this point. Find all curves which satisfy this condition.

12.4   The elasticity $\varepsilon_f(x)$ of a function $f : D_f \to \mathbb{R}$ is given by

$$\varepsilon_f(x) = 2x^2 \left(\ln x + \frac{1}{2} \right).$$

Find function f as the general solution of a differential equation and determine a particular function f satisfying the equality $f(1) = 1$.

12.5   Let y be a function of t and $y'(t) = ay(T - t)/t$, where $0 < t \le T$ and a is a positive constant. Find the solution of the differential equation for $a = 1/2$ and $T = 200$.

12.6   Check whether

$$y_1 = x, \qquad y_2 = x \ln x \qquad \text{and} \qquad y_3 = \frac{1}{x}$$

form a fundamental system of the differential equation $x^3 y''' + 2x^2 y'' - xy' + y = 0$. Determine the general solution of the given differential equation.

12.7   Find the solutions of the following differential equations:

(a) $y' - 2y = \sin x$;
(b) $y'' - 2y' + y = 0$ with $y(0) = 1$ and $y'(0) = 2$;
(c) $2y'' + y' - y = 2e^x$;
(d) $y'' - 2y' + 10y = 10x^2 + 18x + 7.6$
with $y(0) = 0$ and $y'(0) = 1.2$.

12.8   Solve the following differential equations:

(a) $y''' - 2y'' + 5y' = 2 \cos x$;   (b) $y''' + 3y'' - 4y = 3e^x$.

12.9   Find the solutions of the following systems of first-order differential equations.

(a) $y_1' = ay_1 \quad -y_2$
    $y_2' = y_1 \quad +ay_2$;
(b) $y_1' = y_1 \quad -2y_2 \quad -y_3$
    $y_2' = -y_1 \quad +y_2 \quad +y_3$
    $y_3' = y_1 \qquad\qquad -y_3$
    with $y_1(0) = 12$, $y_2(0) = 6$ and $y_3(0) = -6$.

12.10   A time-dependent process is described by the linear difference equation

$$y_{t+1} = a(t)y_t + b(t).$$

(a) Find the solution of the difference equation for $a(t) = 2$ and $b(t) = b$ (b is constant) such that $y_0 = 1$.

(b) Investigate how the forcing term b influences the monotonicity property of the solution.

(c) Graph the solution for $b = 1/2$ and for $b = -2$.

12.11 Consider the following cobweb model. In period t, the supply of a product is given by $y_t = -5 + 2p_{t-1}$, and the demand is given by $x_t = 7 - p_t$, where p_t denotes the price of this product at time t. Moreover, equality $y_t = x_t$ holds.

(a) Formulate the problem as a difference equation.

(b) Solve this difference equation with $p_0 = 3$.

(c) Graph the cobweb.

12.12 Find the solutions of the following difference equations:

(a) $y_{t+2} = -2y_{t+1} - 4y_t$;

(b) $y_{t+2} = 2y_{t+1} + 8y_t + 3t + 6$ with $y_0 = -2/3$ and $y_1 = 5$;

(c) $y_{t+2} = -2y_{t+1} - y_t + 4^t$.

12.13 A company intends to modernize and enlarge its production facilities, which is expected to lead to an increase in the production volume. The intended increase in the production volume in year $t + 1$ is 40 per cent of the volume of year t plus 15 per cent of the volume of year $t - 1$. In year t, it is assumed that 440 units are produced, and in year $t - 1$, 400 units have been produced.

(a) Formulate a difference equation for this problem.

(b) Find the general solution of this difference equation.

(c) What is the production volume after five years?

12.14 Given is the difference equation $y_{t+1} = 3y_t + 2t + 1$.

(a) Find a solution by using the solution formula.

(b) Find the solution as the sum of the general solution of the corresponding homogeneous difference equation and a particular solution of the non-homogeneous difference equation. Proceed as in the case of a linear difference equation of second order with constant coefficients.

(c) Prove by induction that for $t \in \mathbb{N}$, the solutions found in (a) and (b) coincide.

Selected solutions

1 INTRODUCTION

1.1   (a) false; (d) true; (e) false

1.2 $D : A \wedge B \wedge C, \quad E : A \wedge \overline{B} \wedge \overline{C}, \quad F : A \vee B \vee C,$
$G : (A \wedge \overline{B} \wedge \overline{C}) \vee (\overline{A} \wedge B \wedge \overline{C}) \vee (\overline{A} \wedge \overline{B} \wedge C), \quad H : \overline{F}, \quad I : \overline{D}$

1.4   (a) $A \Leftrightarrow B$; (b) $A \Rightarrow B$; (c) $B \Rightarrow A$; (d) $A \Leftrightarrow B$; (e) $A \Rightarrow B$

1.5   (a) true, negation $\bigvee\limits_{x} x^2 - 5x + 10 \leq 0$ is false;

     (b) false, negation $\bigvee\limits_{x} x^2 - 2x \leq 0$ is true

1.8   (a) T, F, F, T; (b) T, F, T, T; (c) T, T, F, F; (d) F, T, T, T

1.9 $A \cup B = \{1, 2, 3, 5, 7, 8, 9, 11\}, \quad |A \cup B| = 8, \quad A \cap B = \{1, 3, 7, 9\}, \quad |A \cap B| = 4,$
$|A| = 6, \quad |B| = 6, \quad A \setminus B = \{5, 11\}, \quad |A \setminus B| = 2$

1.10   subsets: $\emptyset, \{1\}, \{2\}, \{1, 2\}; \quad |P(P(A))| = 16$

1.13   4

1.14   110 students have a car and a PC, 440 have a car but no PC, 290 have a PC but no car and 840 students have a car or a PC

1.15 $A \times A = \{(1, 1), (1, 2), (2, 1), (2, 2)\}, \quad A \times B = \{(1, 2), (1, 3), (2, 2), (2, 3)\},$
$B \times A = \{(2, 1), (3, 1), (2, 2), (3, 2)\}, \quad A \times C = \{(1, 0), (2, 0)\},$
$A \times C \times B = \{(1, 0, 2), (1, 0, 3), (2, 0, 2), (2, 0, 3)\},$
$A \times A \times B = \{(1, 1, 2), (1, 2, 2), (2, 1, 2), (2, 2, 2), (1, 1, 3), (1, 2, 3), (2, 1, 3), (2, 2, 3)\}$

1.16 $M_1 \times M_2 \times M_3 = \{(x, y, z) \mid (1 \leq x \leq 4) \wedge (-2 \leq y \leq 3) \wedge (0 \leq z \leq 5)\}$

1.17   479,001,600 and 21,772,800

1.18   56

1.19   125 and 60

1.20 $n^2 - n$

1.21   5

1.22   (a) 1,128

1.23   (a) $\dfrac{30y - 7x}{6xy + 12y^2};$ (b) $\dfrac{a^2 + b^2}{2a^2}$

1.24   (a) $\{x \in \mathbb{R} \mid x \geq -1\};$ (b) $\{x \in \mathbb{R} \mid (x < 2) \vee (x \geq 11/4)\};$
      (c) $\{x \in \mathbb{R} \mid (x < -1) \vee (0 < x \leq 2)\};$ (d) $\{x \in \mathbb{R} \mid (x < -2) \vee (2 < x < 4)\}$

1.25   (a) $(-2, 2);$ (b) $(1, 5);$ (c) $(1/2, 5/2);$ (d) $(0, \infty);$ (e) $(-1, -1/3]$

1.26   (a) $-19/7$;   (b) $9c/b^6$

1.27   (a) $x = 15$;   (b) $x_1 = \sqrt{2}$, $x_2 = -\sqrt{2}$, $x_3 = 1$, $x_4 = -1$;
       (c) $x_1 = -2$, $x_2 \approx 19.36$, $(x_3 \approx -4.99)$

1.28   (a) $a = 2$;   (b) $x = 1$

1.29   (a) $x = 6$;   (b) $x_1 = 1$, $x_2 = 10$, $x_3 = 0.001$;   (c) $x = 9$

1.30   (b) $x_1 = 2i$, $x_2 = -2i$, $x_3 = 3i$, $x_4 = -3i$

1.31

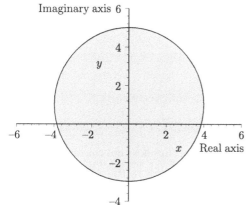

1.32 $z_1 + z_2 = -1 + 5i$, $z_1 - z_2 = 3 + 3i$, $z_1 z_2 = -6 - 7i$, $\dfrac{z_1}{z_2} = \dfrac{2}{5} - \dfrac{9}{5}i$

1.33   (a) $\Re(z) = 0$, $\Im(z) = -1$, $z = -i$;
       (b) $\Re(z) = -1$, $\Im(z) = 0$, $z = -1$;
       (c) $\Re(z) = -\sqrt{3}$, $\Im(z) = -3$, $z = -\sqrt{3} - 3i$

1.34   (a) $z = -i = 1(\cos \frac{3}{2}\pi + i \sin \frac{3}{2}\pi) = e^{3\pi i/2}$;
       (b) $z = -1 = 1(\cos \pi + i \sin \pi) = e^{\pi i}$;
       (c) $z = 3(\cos 60° + i \sin 60°) = 3e^{\pi i/3}$;
       (d) $z = \sqrt{5}(\cos 243° + i \sin 243°) = \sqrt{5} \cdot e^{4.25i}$

1.37   (a) $z_1 = \sqrt{3} + i$, $z_2 = -1 + \sqrt{3}\,i$, $z_3 = -\sqrt{3} - i$, $z_4 = 1 - \sqrt{3}\,i$;
       (b) $z_1 = \frac{5}{2}i$, $z_2 = \frac{5}{4}(-\sqrt{3} - i)$, $z_3 = \frac{5}{4}(\sqrt{3} - i)$

1.38 $a_1 = -32$, $a_2 = 1$

2   SEQUENCES; SERIES; FINANCE

2.1   (a) $a_{101} = 815$;   (b) $d = 2$, $a_1 = 7$, $a_n = 2n + 5$

2.2   (a)   strictly increasing   (b)   unbounded;   (c) $a_{n+1} = a_n + 1/2$

2.3   (a) $a_1 = 3$, $n = 16$;   (b) $a_1 = 18$, $q = 1/3$

2.4 $\{a_n\}$ is strictly decreasing and bounded, $\displaystyle\lim_{n\to\infty} a_n = -5$;
       $\{b_n\}$ is not monotone but bounded, $\displaystyle\lim_{n\to\infty} b_n = 0$;
       $\{c_n\}$ is decreasing and bounded, $\displaystyle\lim_{n\to\infty} c_n = 0$

2.5 (a) $\lim\limits_{n\to\infty} a_n = 2/e$;

(b) $\lim\limits_{n\to\infty} b_n = \begin{cases} -\infty & \text{for} \quad a < 0 \\ 1/3 & \text{for} \quad a = 0 \\ \infty & \text{for} \quad a > 0 \end{cases}$;

(c) $\lim\limits_{n\to\infty} c_n = 0$ (both for $c_1 = 1$ and $c_1 = 4$)

2.6 (a) shirts: $a_1 = 2,000$; $a_2 = 2,500$; $a_3 = 3,000$; $a_{10} = 6,500$;
trousers: $a_1 = 1,000$; $a_2 = 1,200$; $a_3 = 1,440$; $a_{10} = 5,159.78$;

(b) shirts: $s_{15} = 82,500$; trousers: $s_{15} = 72,035.11$

2.7 (a) $s = -10/7$;   (b) $s = 1$

2.8 (a) series converges for $-1 \le x < 1$;   (b) series converges for $|x| > 1$

2.9 (a) $s_1 = -12$, $s_2 = -8$, $s_3 = -\dfrac{92}{9}$, $s_4 = -\dfrac{157}{18}$,   series converges;

(b) $s_1 = 1$, $s_2 = 3$, $s_3 = \dfrac{9}{2}$, $s_4 = \dfrac{31}{6}$,   series converges;

(c) $s_1 = \dfrac{1}{2}$, $s_2 = \dfrac{113}{162}$, $s_3 = \dfrac{113}{162} + \dfrac{3^9}{4^9}$, $s_4 = s_3 + \dfrac{4^{16}}{5^{16}}$,   series converges;

(d) $s_1 = 0$, $s_2 = \dfrac{3}{2}$, $s_3 = \dfrac{5}{6}$, $s_4 = \dfrac{25}{12}$,   series does not converge.

2.10 (a) 17,908.47 EUR; (b) 17,900.51 EUR;   (c) 18,073.83 EUR; (c) is the best.

2.11 49,696.94 EUR

2.12 interest rate $i = 0.06$

2.13 (a) $A_{annually} = 16,288.95$ EUR; $A_{quarterly} = 16,436.19$ EUR;
$A_{monthly} = 16,470.09$ EUR;

(b) $i_{\text{eff}-quarterly} = 0.050945$; $i_{\text{eff}-monthly} = 0.051162$

2.14 5,384.35 EUR

2.15 (a) bank A: $2,466$ EUR;   bank B: $2,464.38$ EUR
(b) bank A: $2,478$ EUR;   bank B: $2,476.21$ EUR

2.16 (a) sinking fund deposit: 478.46 EUR;
(b) sum of interest and deposit: 550.46 EUR

2.17 1,809.75 EUR

2.18 (a) $V_1 = 62,277.95$ EUR;   (b) $V_2 = 63,227.62$ EUR

2.19 (a) $A = 82,735.69$ EUR;   (b) $P_{120} = 58,625.33$ EUR

2.20 (a) $P = 22,861.15$ EUR;   total payment 228,611.50 EUR

Redemption table for answer 2.20(a) (EUR)

Period (year)	Annuity	Amortization instalment	Interest	Amount of the loan at the end
1	22,861.15	10,111.15	12,750.00	139,888.85
2	22,861.15	10,970.59	11,890.55	128,918.25
3	22,861.15	11,903.10	10,958.05	117,015.15
4	22,861.15	12,914.86	9,946.29	104,100.29
5	22,861.15	14,012.62	8,848.52	90,087.66
6	22,861.15	15,203.70	7,657.45	74,883.96
7	22,861.15	16,496.01	6,365.13	58,387.95
8	22,861.15	17,898.17	4,962.97	40,489.77
9	22,861.15	19,419.52	3,441.63	21,070.25
10	22,861.15	21,070.18	1,790.97	0.07

(b) total payment: 220,125.00 EUR

Redemption table for answer 2.20(b) (EUR)

Period (year)	Annuity	Amortization instalment	Interest	Amount of the loan at the end
1	27,750	15,000	12,750	135,000
2	26,475	15,000	11,475	120,000
3	25,200	15,000	10,200	105,000
4	23,925	15,000	8,925	90,000
5	22,650	15,000	7,650	75,000
6	21,375	15,000	6,375	60,000
7	20,100	15,000	5,100	45,000
8	18,825	15,000	3,825	30,000
9	17,550	15,000	2,550	15,000
10	16,275	15,000	1,275	0

(c) 15.1 years

2.21 (a) the project should not go ahead; (b) 1 per cent, 2 per cent; (c) yes

2.22 (a) depreciation amount in year 1: 7,000 EUR, in year 8: 7,000 EUR
(b) depreciation amount in year 1: 12,444.44 EUR, in year 8: 1,555.55 EUR
(c) depreciation amount in year 1: 9,753.24 EUR, in year 8: 4,755.21 EUR

3 RELATIONS; MAPPINGS; FUNCTIONS OF A REAL VARIABLE

3.1 (a) R:

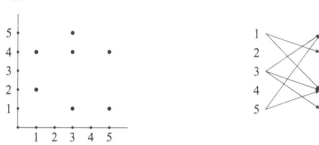

(b) T, F, F, F;
(c) $R^{-1} = \{(2,1), (4,1), (1,3), (4,3), (5,3), (1,5), (4,5)\}$;
(d) R: no mapping; S: mapping

3.2 (a) bijective mapping, (b) surjective mapping,
(c) mapping, (d) injective mapping

3.3 (a) Illustration by graphs:

f: g: $f \circ g$:

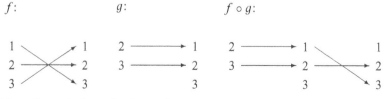

(b) $D_f = A$, $R_f = B$; $D_g = C$, $R_g = \{1,2\} \subseteq A$; $D_{f \circ g} = R_{f \circ g} = C \subseteq B$;
(c) f: bijective, g: injective, $f \circ g$: injective, $g \circ f$: no mapping

3.4 F^{-1} is a mapping with domain $D_{F^{-1}} = \mathbb{R}$ and range $R_{F^{-1}} = [-2, \infty)$.

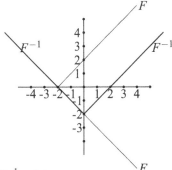

3.5 F is a function, F^{-1} exists.
G is not a function, $y = \pm\sqrt{9 - 4.5x^2}$ is an ellipse with midpoint $(0, 0)$ and $a = \sqrt{2}$,
$b = 3$.

3.6 $(g \circ f)(x) = 4x^2 + 4x - 1$ $(f \circ g)(x) = 2x^2 - 3$

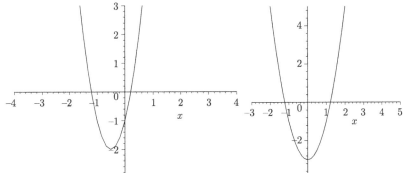

3.7   (a) f and g bijective;   (b) $f^{-1}(x) = \ln x$, $g^{-1}(x) = -x$;
      (c) $f \circ g = e^{-x}$, $g \circ f = -e^x$

3.8 $a = -1$, $R_f = \{x \in \mathbb{R} \mid -4 \le y < \infty\}$, $f^{-1}(x) = -1 + \sqrt{4 + x}$

3.9   (a) $D_f = \{x \in \mathbb{R} \mid x \ge 0\}$, $R_f = \{y \in \mathbb{R} \mid -1 \le y < 1\}$;

            $f^{-1}:$ $y = 16 \cdot \dfrac{(x + 1)^2}{(x - 1)^2}$, $-1 \le x < 1$;

      (b) $D_f = \mathbb{R}$, $R_f = \mathbb{R}$, $f^{-1}:$ $y = \sqrt[3]{x} + 2$

3.10 (a) $(P_5/P_2)(x) = 2x^3 - 2x^2 - 12x$;   (b) $P_5(x) = (x - 1)(x - 1)(x + 2)(x - 3)2x$
      (d)

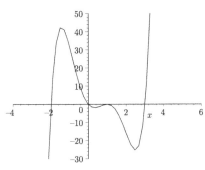

3.11 x_1, x_2, x_4 are zeroes; x_3 is not a zero;

$$f(x) = (x - 1)(x + 1)(x + 1)(x + 2)\left[x - \tfrac{1}{2}(1 + \sqrt{3}i)\right]\left[x - \tfrac{1}{2}(1 - \sqrt{3}i)\right]$$

3.12   (a) $f_1 : D_{f_1} = \mathbb{R}$, $R_{f_1} = [-1, 1]$, odd; $f_2 : D_{f_2} = \mathbb{R}$, $R_{f_2} = [-2, 2]$, odd;

   $f_3 : D_{f_3} = \mathbb{R}$, $R_{f_3} = [-1, 1]$, odd; $f_4 : D_{f_4} = \mathbb{R}$, $R_{f_4} = [1, 3]$;

   $f_5 : D_{f_5} = \mathbb{R}$, $R_{f_5} = [-1, 1]$;

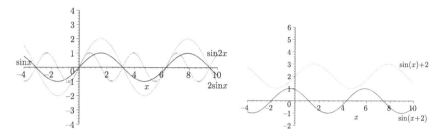

   (b) $f_1 : D_{f_1} = \mathbb{R}$, $R_{f_1} = \{x \in \mathbb{R} : x > 0\}$; $f_2 : D_{f_2} = \mathbb{R}$, $R_{f_2} = \{x \in \mathbb{R} : x > 0\}$;

   $f_3 : D_{f_3} = \mathbb{R}$, $R_{f_3} = \{x \in \mathbb{R} : x > 0\}$; $f_4 : D_{f_4} = \mathbb{R}$, $R_{f_3} = \{x \in \mathbb{R} : x > 2\}$;

   $f_5 : D_{f_5} = \mathbb{R}$, $R_{f_5} = \{x \in \mathbb{R} : x > 0\}$

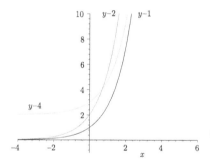

3.13   (a) $D_f = \{x \in \mathbb{R} \mid x \neq 0\}$, $R_f = \mathbb{R}$,

   f unbounded and even,

   f strictly decreasing for $x < 0$,

   f strictly increasing for $x > 0$

   (b) $D_f = \{x \in \mathbb{R} \mid x > 0\}$, $R_f = \mathbb{R}$,

   f unbounded,

   f strictly increasing

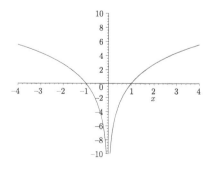

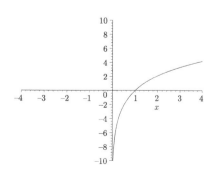

(c) $D_f = \mathbb{R}, R_f = \{y \in \mathbb{R} \mid y \geq 5\}$,
f bounded from below and even,
f strictly decreasing for $x \leq 0$,
f strictly increasing for $x \geq 0$

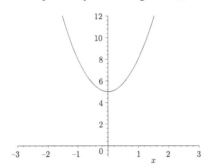

(d) $D_f = \{x \in \mathbb{R} \mid |x| \leq 2\}$,
$R_f = \{y \in \mathbb{R} \mid 0 \leq y \leq 2\}$,
f bounded and even,
f strictly increasing for $x \leq 0$,
f strictly decreasing for $x \geq 0$

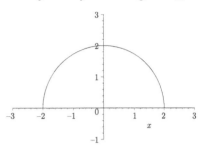

(e) $D_f = \mathbb{R}, R_f = \{y \in \mathbb{R} \mid y > 1\}$,
f bounded from below,
f strictly decreasing

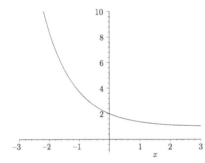

(f) $D_f = \mathbb{R}, R_f = \{y \in \mathbb{R} \mid y \geq 0\}$,
f bounded from below,
f decreasing

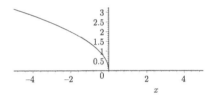

4 DIFFERENTIATION

4.1 (a) $\lim_{x \to x_0} f(x) = a$; (b) $\lim_{x \to 1} f(x) = 1$;

(c) $\lim_{x \to 0} f(x) = 0$; (d) $\lim_{x \to 1+0} f(x) = 2$, $\lim_{x \to 1-0} f(x) = 1$

4.2 (a) $\lim_{x \to 2} \dfrac{x^3 - 3x^2 + 2x}{x - 2} = 2$, gap;

(b) $\lim_{x \to 2-0} \dfrac{x^3 - 3x^2}{x - 2} = \infty$, $\lim_{x \to 2+0} \dfrac{x^3 - 3x^2}{(x - 2)} = -\infty$, pole;

(c) $\lim_{x \to 2} \dfrac{x^3 - 3x^2}{(x - 2)^2} = -\infty$, pole

4.3 (a) $f(2)$ not defined, $x_0 = 2$ gap; (b) continuous; (c) jump; (d) jump

4.4 (a) not differentiable at $x_0 = 5$, differentiable at $x_1 = 0$;

(b) differentiable at $x_0 = 0$, not differentiable at $x_1 = 2$

4.5 (a) $y' = 6x^2 - 5 - 3\cos x$; (b) $y' = (4x^3 + 4)\sin x + (x^4 + 4x)\cos x$;

(c) $y' = \dfrac{4x + 2x\sin x + 2\sin x + \sin^2 x - x^2 \cos x + \cos^2 x}{(2 + \sin x)^2}$;

(d) $y' = 4(2x^3 - 3x + \ln x)^3 (6x^2 - 3 + \frac{1}{x})$;

(e) $y' = -4(x^3 + 3x^2 - 8)^3(3x^2 + 6x)\sin(x^3 + 3x^2 - 8)^4$;

(f) $y' = -4(3x^2 + 6x)\cos^3(x^3 + 3x^2 - 8)\sin(x^3 + 3x^2 - 8)$;

(g) $y' = \dfrac{e^x \cos e^x}{2\sqrt{\sin e^x}}$;   (h) $y' = \dfrac{2x}{x^2 + 1}$

4.6   (a) $y' = \sin x + \cos x$;   (b) $y' = \dfrac{1}{\cos x}$;   (c) $y' = \frac{2}{3}(x^{-2/3} + x^{-1/3})$

4.7   (a) $f'(x) = \left[\ln(\tan x) + \dfrac{x}{\sin x \cos x}\right](\tan x)^x$;

(b) $f'(x) = \left(\ln x + 1 - \dfrac{1}{x}\right)x^{x-1}\cos(x^{x-1})$;

(c) $f'(x) = \dfrac{(x+2)\sqrt{x-1}}{x^3(x-2)^2}\left(\dfrac{1}{x+2} + \dfrac{1}{2(x-1)} - \dfrac{3}{x} - \dfrac{2}{x-2}\right)$

4.8   (a) $f'''(x) = 6\cos x - 6x\sin x - x^2\cos x$;   (b) $f'''(x) = \dfrac{4}{x^3}$;

(c) $f'''(x) = -12 \cdot \dfrac{x^2 + 12x + 12}{(x-2)^6}$;   (d) $f'''(x) = (x+4)e^x$

4.9   exact change: 28;   approximation: 22

4.10 -10

4.11 $\varrho_f(x) = \frac{1}{2}$; $\varrho_g(x) = \dfrac{1}{2x}$; $\varepsilon_f(x) = \frac{1}{2}x$; $\varepsilon_g(x) = \frac{1}{2}$;

$\varrho_f(1) = \varrho_f(100) = \dfrac{1}{2}$; $\varrho_g(1) = \dfrac{1}{2}$; $\varrho_g(100) = \dfrac{1}{200}$.

Function f is inelastic for $x \in (0, 2)$ and elastic for $x > 2$;   function g is inelastic for $x > 2$.

When f changes from $x_0 = 1$ to 1.01, the function value changes by 0.5 per cent; when f changes from 100 to 101, the function value changes by 50 per cent.

When g changes from some x_0 by 1 per cent, the function value always changes by 0.5 per cent.

4.12 $\varepsilon_D(p) = -4(p-1)p$;

demand D is elastic for $p > (1 + \sqrt{2})/2$ and inelastic for $0 < p < (1 + \sqrt{2})/2$; $p \neq 1/2$ $(\varepsilon_D(1/2) = 1)$

4.13   (a)   local minima at $P_1 : (0, -5)$ and $P_2 : (2, -9)$,
local maximum at $P_3 : (0.25; -4.98)$,   global maximum at $P_4 : (-5; 1,020)$;

(b)   global maximum at $P_1 : (3, 4)$,   global minimum at $P_2 : (-5, -4)$;

(c)   local and global maximum at $P_1 : (0, 1)$;

(d)   local minimum at $P_1 : (4, 8)$,   local maximum at $P_2 : (0; 0)$, global maximum does not exist;

(e)   global minimum at $P_0 : (0, 0)$,
global maximum at the endpoint of I, i.e. $P_1 : (5, 0.69)$

4.14   local minimum at $P_1 : (1, 5/6)$,   local maximum at $P_2 : (2, 2(2 - \ln 2)/3$
for $a = -2/3$, $b = -1/6$

4.15   (a)   1;   (b)   1/4;   (c)   1;   (d)   1;   (e)   1/6;   (f) $-1/6$

4.16   (a) $D_f = \mathbb{R}\backslash\{2\}$,   no zero, discontinuity at $x_0 = 2$, local minimum at $P_1 : (-1/2, 1/5)$, inflection point at $P_2 : (-7/4, 13/45)$, $\lim\limits_{x \to \pm\infty} f(x) = 1$, f strictly decreasing on $(-\infty, -1/2] \cup (2, \infty), f$ strictly convex for $x \geq -7/4$

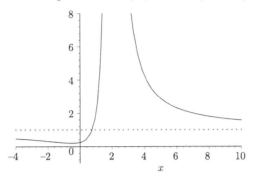

(b) $D_f = \mathbb{R} \setminus \{0, 1/2\}$, zero: $x_0 = 4/3$, discontinuities: $x_1 = 0$ and $x_2 = 1/2$, $\lim_{x \to \pm\infty} f(x) = -3/2, f$ strictly decreasing on D_f, f strictly convex for $x > 1/2$

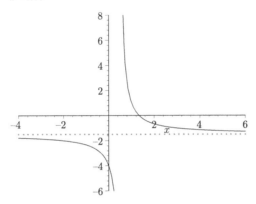

(c) $D_f = \mathbb{R} \setminus \{0; 1\}$, zero: $x_0 = -1$, discontinuities: $x_1 = 0$ and $x_2 = 1$, local minimum at $P_3 : (3.56, 8.82)$, local maximum at $P_4 : (-0.56, 0.06)$, inflection point at $P_5 : (-0.20, 0.02)$, $\lim_{x \to -\infty} f(x) = -\infty$, $\lim_{x \to \infty} f(x) = \infty$, f strictly decreasing on $[-0.56, 0) \cup (1, 3.56], f$ strictly convex for $x \geq -0.2$

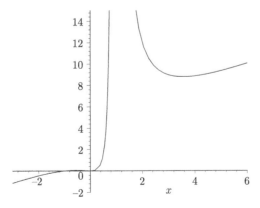

(d) $D_f(x) = \mathbb{R}$, no zeroes, local maximum at $P_1 : (2, 1)$, inflection points at $P_2 : (1, e^{-1}), P_3 : (3, e^{-1})$, $\lim_{x \to \pm\infty} f(x) = 0, f$ strictly increasing on $(-\infty, 2]$,

f strictly concave for $1 \leq x \leq 3$

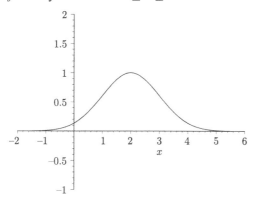

(e) $D_f = \{x \in \mathbb{R} \mid x > 2\}$,   no zero, local maximum at P_1 : $(4, -3 \ln 2)$, inflection point at P_2 : $(6.83, -2.27)$, $\lim\limits_{x \to \infty} f(x) = -\infty$, $\lim\limits_{x \to 2+0} f(x) = -\infty$, f strictly decreasing for $x \geq 4$, f strictly convex for $x \geq 6.83$

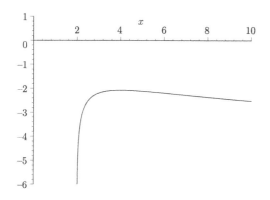

(f) $D_f(x) = \mathbb{R}$,   zeroes: $x_0 = 0$, $x_1 = 2$, local minimum at P_2 : $(0,0)$, local maximum at P_3 : $(1.33, 1.06)$, inflection point at P_4 : $(2,0)$, $\lim\limits_{x \to -\infty} f(x) = \infty$, $\lim\limits_{x \to \infty} f(x) = -\infty$, f strictly decreasing on $(-\infty, 0] \cup (4/3, \infty)$, f strictly convex for $x \geq 2$

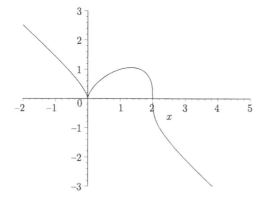

4.17 (a) $f(x) = 1 - \frac{\pi^2}{32}(x-2)^2 + \frac{\pi^4}{6144}(x-2)^4 + R_5,$

$R_5 = -\frac{\pi^6}{4^6} \cdot \sin\left[\frac{\pi}{4}(2 + \lambda(x-2))\right]\frac{(x-2)^6}{6!}, \quad 0 < \lambda < 1;$

(b) $f(x) = x - \frac{x^2}{2} + \frac{x^3}{3} - \frac{x^4}{4} + - \ldots + (-1)^{n-1} \cdot \frac{x^n}{n} + R_n,$

$R_n = \frac{(-1)^n x^{n+1}}{(\lambda x + 1)^{n+1}(n+1)}, \quad 0 < \lambda < 1;$

(c) $f(x) = 2x - 2x^2 - \frac{1}{3}x^3 + x^4 + R_4,$

$R_4 = \frac{x^5}{5!} \cdot e^{-\lambda x}[-41\sin(2\lambda x) - 38\cos(2\lambda x)], \quad 0 < \lambda < 1$

4.18 $T_4(-\frac{1}{5}) = 1 + (-\frac{1}{5}) + \frac{1}{2!}(-\frac{1}{5})^2 + \frac{1}{3!}(-\frac{1}{5})^3 + \frac{1}{4!}(-\frac{1}{5})^4 = 0.81873$

4.19 (a) Newton's method

x_n	$f(x_n)$	$f'(x_n)$
1	-3	-3
0	$+2$	-6
0.333333	$+0.037037$	-5.66667
0.339869	$+4.29153 \cdot 10^{-5}$	-5.65347
0.339877	$+4.29153 \cdot 10^{-5}$	-5.65347

(b) regula falsi

x_n	$f(x_n)$
0	$+2$
1	-3
0.4	-0.336
0.342466	-0.146291
0.339979	-0.000577
0.339881	-0.000023
0.339877	$-9.536 \cdot 10^{-7}$

4.20 Newton's method

x_n	$f(x_n)$
5	0.3905620
4.5117973	0.0051017
4.5052428	0.0000010
4.505241496	$1.6 \cdot 10^{-10}$

5 INTEGRATION

5.1 (a) $e^{\sin x} + C;$

(b) $\frac{1}{2}(\ln x)^2 + C;$

(c) $-\frac{5}{4}\ln|1 - 4x| + C;$

(d) $\frac{1}{2}e^{-3+2x} + C;$

(e) $\sqrt{x^2 + 1} + C;$

(f) $\frac{1}{3}\sqrt{x^2 + 1}(x^2 - 2) + C;$

(g) $\frac{1}{2}\ln(e^{2x} + 1) - 2\arctan e^x + C;$ (h) $\frac{1}{3}\arcsin\frac{3}{\sqrt{2}}x + C;$

(i) $-\dfrac{1}{\sin x} - \sin x + C$; (j) $-\dfrac{1}{\tan \frac{x}{2}} + C$;

(k) $\dfrac{1}{4}\left[\ln\left|\tan\dfrac{x}{2}\right| + \dfrac{1}{2}\tan^2\dfrac{x}{2}\right] + C$

5.2 (a) $e^x(x^2 - 2x + 2) + C$; (b) $\dfrac{1}{2}e^x(\sin x + \cos x) + C$;

(c) $x\tan x + \ln|\cos x| + C$; (d) $\dfrac{1}{2}(\cos x \sin x + x) + C$;

(e) $\dfrac{x^3}{3}\ln x - \dfrac{x^3}{9} + C$; (f) $x\ln(x^2 + 1) - 2x + 2\arctan x + C$

5.3 (a) 3; (b) $4 - 2\ln 3$; (c) 2/3; (d) 10/3;

(e) $\dfrac{1}{2}\ln|2t - 1|$; (f) 1/3; (g) $\pi/4$

5.4 (a) total cost $C = 9,349.99$; total sales $S = 15,237.93$;
total profit $P = 5,887.94$

(b) average sales $S_a = 3,809.48$; average cost $C_a = 2,337.49$;

(c) $P(t) = 10,000e^{-t}(-t^2 - 2t - 2) + 20,000 - 1,000(4t - 2\ln(e^t + 1)$
$+2,000\ln 2$

5.5 (a) 4; (b) $A = 32.75$

5.6 (a) graph for $q_0 = 8$, $t_0 = 4$:

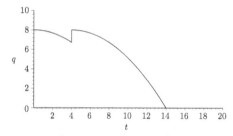

(b) $q_0 T\left(1 - \dfrac{T^2 - 12T + 48}{300}\right)$; (c) $x = T/2$

5.7 (a) 0.784981; (b) 0.783333; (c) 0.785398; exact value: $\pi/4$

5.8 (a) 1; (b) 1/2; (c) 1; $\lambda > 0$; (d) $2/\lambda^2$; $\lambda > 0$
  0; $\lambda = 0$ 0; $\lambda = 0$
  $-\infty$; $\lambda < 0$ $-\infty$; $\lambda < 0$

(e) 4; (f) does not exist

5.9 1,374.13

5.10 $PS = 7$; $CS = 4$.

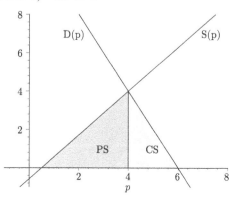

6   VECTORS

6.1 (a) $\mathbf{a} + \mathbf{b} - \mathbf{c} = \begin{pmatrix} 1 \\ -5 \\ -9 \end{pmatrix}$, $\mathbf{a} + 3\mathbf{b} = \begin{pmatrix} 5 \\ -11 \\ -7 \end{pmatrix}$,

$\mathbf{b} - 4\mathbf{a} + 2\mathbf{c} = \begin{pmatrix} -3 \\ -4 \\ 14 \end{pmatrix}$, $\mathbf{a} + 3(\mathbf{b} - 2\mathbf{c}) = \begin{pmatrix} -7 \\ -23 \\ -43 \end{pmatrix}$;

(b) $\mathbf{a} > \mathbf{b}$, $\mathbf{c} \geq \mathbf{a}$, $\mathbf{c} > \mathbf{b}$;

(c) $\mathbf{a}^{\mathrm{T}} \cdot \mathbf{b} = 0$, $\mathbf{a}^{\mathrm{T}} \cdot \mathbf{c} = 0$, $\mathbf{b}^{\mathrm{T}} \cdot \mathbf{c} = -18$;
    $\mathbf{a},\mathbf{b}$ orthogonal; $\mathbf{a},\mathbf{c}$ orthogonal; $\angle(\mathbf{b}, \mathbf{c}) \approx 126.3°$

(d) $(\mathbf{a}^{\mathrm{T}} \cdot \mathbf{b})\mathbf{c} = \begin{pmatrix} 0 \\ 0 \\ 0 \end{pmatrix}$, $\mathbf{a}(\mathbf{b}^{\mathrm{T}} \cdot \mathbf{c}) = \begin{pmatrix} -36 \\ -18 \\ 18 \end{pmatrix}$;

(e) $|\mathbf{b} + \mathbf{c}| = \sqrt{29}$, $|\mathbf{b}| + |\mathbf{c}| = \sqrt{21} + \sqrt{44}$, $|\mathbf{b}^{\mathrm{T}} \cdot \mathbf{c}| = 18$, $|\mathbf{b}||\mathbf{c}| = \sqrt{21}\sqrt{44}$

6.2 $\alpha = \beta - 2$, $\beta \in \mathbb{R}$

6.3 (a) $\sqrt{19}$;

(b) $\mathbf{a} \geq \mathbf{b}$: $M = \{(a_1, a_2) \in \mathbb{R}^2 \mid a_1 \geq b_1 \wedge a_2 \geq b_2\}$;

$|\mathbf{a}| \geq |\mathbf{b}| : M = \{(a_1, a_2) \in \mathbb{R}^2 \mid \sqrt{a_1^2 + a_2^2} \geq \sqrt{b_1^2 + b_2^2}\}$

6.4   all the vectors are linear combinations of $\mathbf{a}^1$ and $\mathbf{a}^2$;
    $(0, 0.5)^{\mathrm{T}}$ is a convex combination

6.5 $\mathbf{a}^4 = \frac{1}{2}\mathbf{a}^1 + \frac{1}{4}\mathbf{a}^2 + \frac{1}{4}\mathbf{a}^3$.

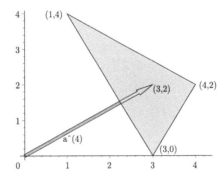

6.6   no

6.7   no; no basis

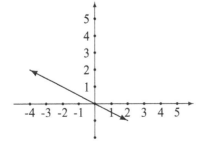

6.8   yes

6.9 (a) $\begin{pmatrix} 3 \\ 3 \\ -3 \end{pmatrix} = 0 \begin{pmatrix} 1 \\ 0 \\ 3 \end{pmatrix} + 3 \begin{pmatrix} 0 \\ 1 \\ 0 \end{pmatrix} + 3 \begin{pmatrix} 1 \\ 0 \\ -1 \end{pmatrix}$;

(b)　bases: $\mathbf{a}, \mathbf{a}^1, \mathbf{a}^2$ and $\mathbf{a}, \mathbf{a}^1, \mathbf{a}^3$;

(c)　$\mathbf{b} = 2\mathbf{a}^1 - \mathbf{a}^2 + \mathbf{a}$

7　MATRICES AND DETERMINANTS

7.1 (a) $A^T = \begin{pmatrix} 3 & 1 \\ 4 & 2 \\ 1 & -2 \end{pmatrix}$, no equal matrices;

(b) $A + D = \begin{pmatrix} 6 & 5 & 1 \\ 5 & 4 & 0 \end{pmatrix}$, $A - D = \begin{pmatrix} 0 & 3 & 1 \\ -3 & 0 & -4 \end{pmatrix}$,

$A^T - B = \begin{pmatrix} 1 & 0 \\ 3 & 0 \\ 2 & -2 \end{pmatrix}$, $C - D = \begin{pmatrix} -2 & -1 & -1 \\ -5 & -1 & -1 \end{pmatrix}$

(c) $A + 3(B^T - 2D) = \begin{pmatrix} -9 & 1 & -2 \\ -20 & -4 & -14 \end{pmatrix}$

7.2 $A = \begin{pmatrix} 2 & 0 & 1.5 \\ 0 & 1 & -0.5 \\ 1.5 & -0.5 & -2 \end{pmatrix} + \begin{pmatrix} 0 & -1 & -1.5 \\ 1 & 0 & -0.5 \\ 1.5 & 0.5 & 0 \end{pmatrix}$

7.3 (a) $AB = \begin{pmatrix} 26 & 32 & 0 \\ 23 & 27 & 1 \end{pmatrix}$; (b) $AB = \begin{pmatrix} 45 & 51 & 58 \\ 25 & 18 & 32 \end{pmatrix}$;

(c) $AB = 10$; $\qquad BA = \begin{pmatrix} -6 & -9 & -12 & -15 \\ 12 & 18 & 24 & 30 \\ -6 & -9 & -12 & -15 \\ 4 & 6 & 8 & 10 \end{pmatrix}$;

(d) $AB = \begin{pmatrix} 0 & 0 \\ 0 & 0 \end{pmatrix}$; $\qquad BA = \begin{pmatrix} -8 & 4 & -12 \\ 2 & -1 & 3 \\ 6 & -3 & 9 \end{pmatrix}$;

(e) $AB = \begin{pmatrix} 2x+3y+ z \\ x+5y+2z \end{pmatrix}$

7.4 $A^T B^T = \begin{pmatrix} 2 & -4 \\ -1 & 2 \\ 3 & -6 \end{pmatrix} \cdot \begin{pmatrix} 2 & 1 & -1 \\ 3 & 0 & -2 \end{pmatrix} = \begin{pmatrix} -8 & 2 & 6 \\ 4 & -1 & -3 \\ -12 & 3 & 9 \end{pmatrix} = (BA)^T$,

$B^T A^T = \begin{pmatrix} 2 & 1 & -1 \\ 3 & 0 & -2 \end{pmatrix} \cdot \begin{pmatrix} 2 & -4 \\ -1 & 2 \\ 3 & -6 \end{pmatrix} = \begin{pmatrix} 0 & 0 \\ 0 & 0 \end{pmatrix} = (AB)^T$

7.5 (a) $ACB = D_{(3,1)}$,

(b) $(AC)B = \begin{pmatrix} -2 & 0 & 1 & 0 \\ -14 & -12 & 7 & -24 \\ -10 & 9 & 5 & 18 \end{pmatrix} \cdot \begin{pmatrix} -1 \\ 2 \\ 0 \\ 7 \end{pmatrix} = \begin{pmatrix} 2 \\ -178 \\ 154 \end{pmatrix}$,

$A(CB) = \begin{pmatrix} 1 & 0 \\ 7 & -4 \\ 5 & 3 \end{pmatrix} \cdot \begin{pmatrix} 2 \\ 48 \end{pmatrix} = \begin{pmatrix} 2 \\ -178 \\ 154 \end{pmatrix}$

7.6 (a) $A^2 = \begin{pmatrix} 0 & 0 & 14 & 46 \\ 0 & 0 & 0 & 35 \\ 0 & 0 & 0 & 0 \\ 0 & 0 & 0 & 0 \end{pmatrix}$, $A^3 = \begin{pmatrix} 0 & 0 & 0 & 70 \\ 0 & 0 & 0 & 0 \\ 0 & 0 & 0 & 0 \\ 0 & 0 & 0 & 0 \end{pmatrix}$, $A^k = 0,\ k \geq 4$;

(b) $B^{2k} = I,\quad k \in N;\quad B^{2k-1} = B;\quad k \in N$

7.7 (a) R_1: 74,000 units and R_2: 73,000 units;

(b) 31 EUR each unit of S_1, 29 EUR each unit of S_2 and 20 EUR each unit of S_3; 275 EUR for F_1 and 156 EUR for F_2.

7.8 (a) $A_{12} = \begin{pmatrix} -1 & 4 \\ 0 & 1 \end{pmatrix}$, $A_{22} = \begin{pmatrix} 2 & 0 \\ 0 & 1 \end{pmatrix}$;

(b) $|A_{12}| = -1,\quad |A_{22}| = 2$;

(c) cofactor of a_{11}: -18, cofactor of a_{21}: -3, cofactor of a_{31}: 12;

(d) $|A| = -33$

7.9 (a) -6, (b) 5, (c) 70, (d) 8, (e) 0, (f) $(-1)^{n-1}3^n$

7.10 (a) $x = 2$; (b) $x_1 = -2,\quad x_2 = 0$

7.11 $x_1 = 2,\quad x_2 = 3,\quad x_3 = -5$

7.12 the zero vector is the kernel of the mapping

7.13 $\begin{pmatrix} x_1 \\ x_2 \\ x_3 \end{pmatrix} \in \mathbb{R}^3 \mapsto \begin{pmatrix} u_1 \\ u_2 \\ u_3 \end{pmatrix} = \begin{pmatrix} -1 & 4 & -1 \\ 10 & 9 & -1 \\ -14 & -10 & 4 \end{pmatrix} \begin{pmatrix} x_1 \\ x_2 \\ x_3 \end{pmatrix} \in \mathbb{R}^3$

7.14 (a) $A^{-1} = \begin{pmatrix} -1/11 & 3/11 & -3/11 \\ -4/11 & 1/11 & 10/11 \\ 4/11 & -1/11 & 1/11 \end{pmatrix}$; (c) $C^{-1} = \begin{pmatrix} 1 & -3/2 & -8 \\ 0 & 1/2 & 2 \\ 0 & 0 & -1 \end{pmatrix}$;

(d) $D^{-1} = \begin{pmatrix} 2 & -1 & -1 & 1 \\ 0 & 1/2 & 1 & -1/2 \\ 5 & -4 & -3 & 2 \\ 2 & -3/2 & -1 & 1/2 \end{pmatrix}$

7.15 $A^{-1} = \begin{pmatrix} 1 & -2 & 5 & 21 & -41 \\ 0 & 1 & -2 & -9 & 16 \\ 0 & 0 & 1 & 3 & -5 \\ 0 & 0 & 0 & 1 & -2 \\ 0 & 0 & 0 & 0 & 1 \end{pmatrix}$

7.16 $(AB)^{-1} = B^{-1}A^{-1} = \frac{1}{17} \begin{pmatrix} -23 & -37 \\ -7 & -12 \end{pmatrix}$

7.17 (a) $X = B^{\mathrm{T}}A^{-1}$; (b) $X = B(A + 2I)^{-1}$; (c) $X = A^{-1}CB^{-1}$;

(d) $X = C^{-1}AB^{-1}$; (e) $X = [(A + 4I)C^{\mathrm{T}}]^{-1}$

7.18 (b) $x \longmapsto (I - A)x = y$; (c) no; (d) $y \in \mathbb{R}^4 \longmapsto x = (I - A)^{-1}y \in \mathbb{R}^4$

7.19 (a) $q = (I - A)p$

with $A = (a_{ij}) = \begin{pmatrix} 0 & 3 & 0 & 0 & 1 \\ 0 & 0 & 0 & 0 & 5 \\ 0 & 0 & 0 & 2 & 0 \\ 0 & 0 & 0 & 0 & 2 \\ 0 & 0 & 0 & 0 & 0 \end{pmatrix}$;

(b) $\mathbf{p} = (I - A)^{-1}\mathbf{q}$;

(c) $\mathbf{r} = B\mathbf{p} = B(I - A)^{-1}\mathbf{q}$

$$\text{with } B = (b_{ij}) = \begin{pmatrix} 1 & 0 & 3 & 0 & 0 \\ 5 & 1 & 0 & 0 & 0 \\ 0 & 0 & 0 & 0 & 7 \end{pmatrix}, \quad (I-A)^{-1} = \begin{pmatrix} 1 & 3 & 0 & 0 & 16 \\ 0 & 1 & 0 & 0 & 5 \\ 0 & 0 & 1 & 2 & 4 \\ 0 & 0 & 0 & 1 & 2 \\ 0 & 0 & 0 & 0 & 1 \end{pmatrix},$$

$\mathbf{r}^\mathrm{T} = (\ 660,\quad 1740,\quad 70\ )$

8 LINEAR EQUATIONS AND INEQUALITIES

8.1 (a) $x_1 = 1$, $x_2 = 2$, $x_3 = 0$; (b) no solution;

   (c) no solution; (d) $x_1 = 5$, $x_2 = 4$, $x_3 = 1$

8.2 (a) (i) $x_1 = x_2 = x_3 = x_4 = 0$;

   (ii) $x_1 = -9t$, $x_2 = 11t$, $x_3 = -3t$, $x_4 = t$, $t \in \mathbb{R}$;

   (b) (i) $x_1 = 3$, $x_2 = -3$, $x_3 = 1$, $x_4 = -1$;

   (ii) $x_1 = -6 - 9t$, $x_2 = 8 + 11t$, $x_3 = -2 - 3t$, $x_4 = t$, $t \in \mathbb{R}$

8.3 (a) $x_1 = 5 + 3t$, $x_2 = \frac{5}{2} - \frac{5}{2}t$, $x_3 = \frac{13}{2} + \frac{13}{2}t$, $x_4 = t$, $t \in \mathbb{R}$;

   $x_1 = 2 + \frac{6}{13}t$, $x_2 = 5 - \frac{5}{13}t$, $x_3 = t$, $x_4 = -1 + \frac{2}{13}t$, $t \in \mathbb{R}$

   (b) $x_1 = 4 - 3t_1 - 2t_2$, $x_2 = -1 + 2t_1 - t_2$, $x_3 = t_1$, $x_4 = 1 + 3t_2$, $x_5 = t_2$;

   $x_1 = 6 + 2t_1 - 7t_2$, $x_2 = t_1$, $x_3 = t_2$, $x_4 = -2 - 3t_1 + 6t_2$,

   $x_5 = -1 - t_1 + 2t_2$;

   $t_1, t_2 \in \mathbb{R}$

8.4 $a \neq \frac{31}{6}$: $x = \frac{8a - 46}{6a - 31}$, $y = \frac{-3a + 12}{6a - 31}$, $z = \frac{7}{6a - 31}$

8.5 the cases 'no solution' and 'unique solution' do not exist for any λ.

   $\lambda = 2$: $x = -\frac{1}{3}z - \frac{2}{3}y$, $y, z \in \mathbb{R}$;

   $\lambda \neq 2$: $x = -\frac{2}{3}y$, $y \in \mathbb{R}$; $z = 0$

8.6 (a) no solution for: $a = 0$, $b \in \mathbb{R}$ or $a \neq -1/3$, $b = 1$;

   unique solution for: $a \neq 0$, $b \neq 1$;

   general solution for: $a = -1/3$, $b = 1$

   (b) $x_1 = 2 + t$, $x_2 = -t$, $x_3 = -3$, $x_4 = t$, $t \in \mathbb{R}$;

   (c) $x_1 = 6$, $x_2 = 0$, $x_3 = 1$, $x_4 = 4$

8.7 $x = 1$, $y = 2$, $z = 3$ for $a \neq 0$, $|a| \neq |b|$;

   $x = 1$, $y = t$, $z = 5 - t$, $t \in \mathbb{R}$, for $a = 0$, $b \neq 0$;

   $x = 4 - t$, $y = 2$, $z = t$, $t \in \mathbb{R}$, for $a \neq 0$, $a = b$;

   $x = -2 + t$, $y = 2$, $z = t$, $t \in \mathbb{R}$, for $a \neq 0$, $a = -b$

8.8 $\ker(A) = (-3t/2,\ -t/2,\ t)^\mathrm{T}$, $t \in \mathbb{R}$

8.9 system 1: $x_1 = 1$, $x_2 = 2$, $x_3 = 0$;

   system 2: $x_1 = 5$, $x_2 = 0$, $x_3 = 3$

8.10 (a) $A^{-1} = \begin{pmatrix} -4 & 4 & 1 \\ 1 & -2 & -1 \\ 1 & 1 & 1 \end{pmatrix}$;   (c) $C^{-1} = \begin{pmatrix} 2 & -1 & -1 & 1 \\ 0 & 1/2 & 1 & -1/2 \\ 5 & -4 & -3 & 2 \\ 2 & -3/2 & -1 & 1/2 \end{pmatrix}$

8.11   (a) $X = \frac{1}{9} \begin{pmatrix} -6 & 78 & 51 \\ -11 & -52 & -40 \end{pmatrix}$;   (b) $X = \frac{1}{5} \begin{pmatrix} -11 & 8 \\ 2 & -1 \end{pmatrix}$;

(c) $X = \begin{pmatrix} -10 & 10 & -2 \\ 15 & -11 & 1 \end{pmatrix}$

8.12   (b) $X = \begin{pmatrix} 5.79 & 17.80 & 14.19 & 10.18 \\ 28.99 & 29.67 & 9.46 & 30.54 \\ 11.59 & 23.73 & 18.52 & 20.36 \\ 11.59 & 35.60 & 14.19 & 25.46 \end{pmatrix}$

8.13   **a** belongs to the set; **b** does not belong to it.

8.14   (b)

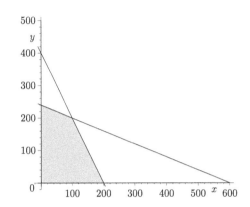

(c) $\begin{pmatrix} x \\ y \end{pmatrix} = \lambda_1 \begin{pmatrix} 0 \\ 0 \end{pmatrix} + \lambda_2 \begin{pmatrix} 200 \\ 0 \end{pmatrix} + \lambda_3 \begin{pmatrix} 0 \\ 240 \end{pmatrix} + \lambda_4 \begin{pmatrix} 100 \\ 200 \end{pmatrix},$

$\sum_{i=1}^{4} \lambda_i = 1, \qquad \lambda_i \geq 0, \qquad i = 1, 2, 3, 4;$

8.15

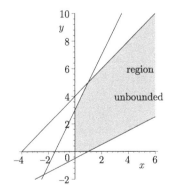

$\begin{pmatrix} x_1 \\ x_2 \end{pmatrix} = \lambda_1 \begin{pmatrix} 0 \\ 0 \end{pmatrix} + \lambda_2 \begin{pmatrix} 0 \\ 3 \end{pmatrix} + \lambda_3 \begin{pmatrix} 1 \\ 5 \end{pmatrix} + \lambda_4 \begin{pmatrix} 1 \\ 0 \end{pmatrix} + \mu_1 \begin{pmatrix} 1 \\ 1 \end{pmatrix} + \mu_2 \begin{pmatrix} 2 \\ 1 \end{pmatrix},$

$\sum_{i=1}^{4} \lambda_i = 1, \qquad \lambda_i \geq 0, \qquad i = 1, 2, 3, 4, \qquad \mu_1, \mu_2 \geq 0$

8.16 (a)
$$\begin{pmatrix} x_1 \\ x_2 \\ x_3 \end{pmatrix} = \lambda_1 \begin{pmatrix} 0 \\ 0 \\ 0 \end{pmatrix} + \lambda_2 \begin{pmatrix} 0 \\ 0 \\ 3 \end{pmatrix} + \lambda_3 \begin{pmatrix} 3/4 \\ 0 \\ 0 \end{pmatrix} + \lambda_4 \begin{pmatrix} 5/9 \\ 7/9 \\ 0 \end{pmatrix}$$

$$+ \lambda_5 \begin{pmatrix} 0 \\ 11/7 \\ 10/7 \end{pmatrix} + \lambda_6 \begin{pmatrix} 0 \\ 1 \\ 0 \end{pmatrix}, \quad \sum_{i=1}^{6} \lambda_i = 1, \quad \lambda_i \geq 0, \quad i = 1, 2, \ldots, 6;$$

(b)
$$\begin{pmatrix} x_1 \\ x_2 \\ x_3 \end{pmatrix} = \lambda_1 \begin{pmatrix} 3 \\ 0 \\ 0 \end{pmatrix} + \lambda_2 \begin{pmatrix} 0 \\ 0 \\ 0 \end{pmatrix} + \lambda_3 \begin{pmatrix} 0 \\ 3 \\ 0 \end{pmatrix} + \lambda_4 \begin{pmatrix} 0 \\ 12/5 \\ 3/5 \end{pmatrix}$$

$$+ \lambda_5 \begin{pmatrix} 0 \\ 0 \\ 3/5 \end{pmatrix}, \quad \sum_{i=1}^{5} \lambda_i = 1, \quad \lambda_i \geq 0, \quad i = 1, 2, \ldots, 5$$

9 LINEAR PROGRAMMING

9.1 (a)

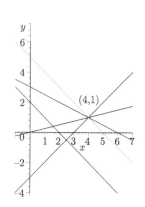

(b)

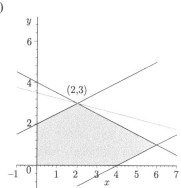

(c) infinitely many solutions

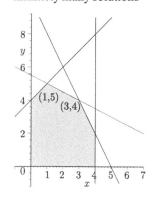

(d) no optimal solution

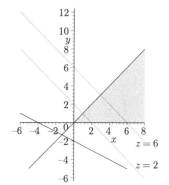

9.2
$$\begin{array}{rcrcl} z = & 10x_1 & + & 6x_2 & \to & \text{max!} \\ & x_1 & & & \leq & 100 \\ & & & x_2 & \geq & 30 \\ & -3x_1 & + & x_2 & \leq & 0 \\ & x_1 & + & 4x_2 & \leq & 200 \\ & & & x_1, x_2 & \geq & 0 \end{array}$$

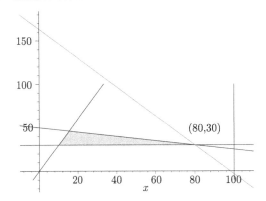

optimal solution: $x_1 = 80$, $x_2 = 30$; $z = 980$

9.3 (a) $\bar{z} = -z = -x_1 + 2x_2 - x_3 \rightarrow$ max!

s.t.
$$
\begin{aligned}
x_1 &+ x_2 + x_3 + x_4 &&= 7\\
-3x_1 &+ x_2 - x_3 &+ x_5 &= 4
\end{aligned}
$$
$$x_j \geq 0, \quad j = 1,2,3,4,5$$

(b) $\bar{z} = -z = -x_1^* - 2x_2 + 3x_3 - x_4^* + x_4^{**} \rightarrow$ max!

s.t.
$$
\begin{aligned}
-x_1^* + x_2 &- \tfrac{1}{2}x_3 + \tfrac{3}{2}x_4^* - \tfrac{3}{2}x_4^{**} &&= 4\\
-x_1^* &+ 2x_3 + x_4^* - x_4^{**} + x_5 &&= 10\\
-2x_1^* &- 2x_3 + 3x_4^* - 3x_4^{**} + x_6 &&= 0
\end{aligned}
$$
$$x_1^*, x_2, x_3, x_4^*, x_4^{**}, x_5, x_6, x_7 \geq 0$$

9.4 (a) $x_1 = 8/5$, $x_2 = 3/5$; $z = 11/5$;

(b) infinitely many optimal solutions: $z = -11$;

$$\mathbf{x} = \lambda \begin{pmatrix} 1 \\ 5 \end{pmatrix} + (1 - \lambda) \begin{pmatrix} 3 \\ 4 \end{pmatrix}, \quad 0 \leq \lambda \leq 1$$

9.5 (a) $x_1 = 0$, $x_2 = 10$, $x_3 = 5$, $x_4 = 15$; $z = 155$;

(b) a (finite) optimal solution does not exist;

9.6 $x_1 = 80$, $x_2 = 30$, $x_3 = 20$, $x_4 = 0$, $x_5 = 210$, $x_6 = 0$; $z = 980$;

9.7 (a) a (finite) optimal solution does not exist;

(b) $z = 24$;

$$\begin{pmatrix} x_1 \\ x_2 \\ x_3 \end{pmatrix} = \lambda \begin{pmatrix} 5.5 \\ 4.5 \\ 9.5 \end{pmatrix} + (1 - \lambda) \begin{pmatrix} 10 \\ 0 \\ 14 \end{pmatrix}, \quad 0 \leq \lambda \leq 1$$

(c) the problem does not have a feasible solution.

9.8 dual problem of problem 9.5 (a):   dual problem of problem 9.7 (a):

$w = 400u_5 + 30u_6 + 5u_7 \rightarrow$ min!

s.t.
$$
\begin{aligned}
20u_5 &+ u_6 &&\geq 7\\
10u_5 &+ u_6 &&\geq 4\\
12u_5 &+ u_6 + u_7 &&\geq 5\\
16u_5 &+ u_6 &&\geq 6
\end{aligned}
$$
$$u_j \geq 0; \quad j = 5,6,7$$

$w = 10u_5 + 4u_6 \rightarrow$ max!

s.t
$$
\begin{aligned}
-3u_4 &+ u_5 &&\leq 1\\
u_4 &+ 2u_5 + u_6 &&\leq 1\\
4u_4 &+ 3u_6 &&\leq -1
\end{aligned}
$$
$$u_j \geq 0; \quad j = 4,5,6$$

dual problem of 9.7(c):

$$w = 4u_5 + 9u_6 + 3u_7 \to \min!$$

s.t.
$$
\begin{array}{rcrcrcr}
u_5 & + & 2u_6 & & & \geq & 2 \\
u_5 & + & 3u_6 & + & u_7 & \geq & -1 \\
-u_5 & - & u_6 & + & 2u_7 & \geq & 1 \\
-u_5 & - & 2u_6 & & & \geq & 0
\end{array}
$$

$$u_5 \in \mathbb{R};\ u_6 \in \mathbb{R};\ u_7 \leq 0$$

9.9 (a) primal problem: $x_1 = 0,\ x_2 = 9/2,\ x_3 = 1/2,\ x_4 = 17/2;\ z = -18;$
   dual problem: $u_5 = 1,\ u_6 = 4,\ u_7 = 1;\ w = -18;$
   (b) primal problem: $x_1 = 17,\ x_2 = 0,\ x_3 = 9,\ x_4 = 3;\ z = 63;$
   dual problem: $u_5 = 25/4,\ u_6 = 3/4,\ u_7 = 35/4;\ w = 63$

9.10 optimal solution of problem (P): $x_1 = 4,\quad x_2 = 2;\quad z = 14;$
   optimal solution of problem (D): $u_3 = 2/3,\quad u_1 = 1/3,\quad u_5 = 0;\quad w = 14$

9.11 (c) optimal solutions:
   (1) $x_1 = 0,\quad x_2 = 100,\quad x_3 = 50,\quad x_4 = 40;\quad z = 190;$
   (2) $x_1 = 50,\quad x_2 = 0,\quad x_3 = 100,\quad x_4 = 40;\quad z = 190$

10 EIGENVALUE PROBLEMS AND QUADRATIC FORMS

10.1 A: $\lambda_1 = 3,\quad \lambda_2 = -2,\quad \mathbf{x}^1 = t_1 \begin{pmatrix} 1 \\ 1 \end{pmatrix},\quad \mathbf{x}^2 = t_2 \begin{pmatrix} 1 \\ -4 \end{pmatrix};$

B: $\lambda_1 = 3 + i,\quad \lambda_2 = 3 - i,\quad \mathbf{x}^1 = t_1 \begin{pmatrix} 1 \\ 1 + i \end{pmatrix},\quad \mathbf{x}^2 = t_2 \begin{pmatrix} 1 \\ 1 - i \end{pmatrix};$

C: $\lambda_1 = 1,\quad \lambda_2 = 2,\quad \lambda_3 = -1,$

$$\mathbf{x}^1 = t_1 \begin{pmatrix} 1 \\ 0 \\ 1 \end{pmatrix},\quad \mathbf{x}^2 = t_2 \begin{pmatrix} 2 \\ -1 \\ 2 \end{pmatrix},\quad \mathbf{x}^3 = t_3 \begin{pmatrix} 3 \\ 1 \\ 2 \end{pmatrix}$$

D: $\lambda_1 = 1,\quad \lambda_2 = \lambda_3 = 2,\quad \mathbf{x}^1 = t_1 \begin{pmatrix} 0 \\ 1 \\ 0 \end{pmatrix},\quad \mathbf{x}^2 = t_2 \begin{pmatrix} 0 \\ -2 \\ 1 \end{pmatrix};$

$t_1, t_2, t_3 \in \mathbb{R};\ t_1, t_2, t_3 \neq 0$

10.2 (a) $\lambda_1 = 1.1$ greatest eigenvalue with eigenvector $\mathbf{x}^{\mathrm{T}} = (3a/2, a),\ a \neq 0;$
   (b) based on a production level $x_1 = 3a/2$ and $x_2 = a\ (a > 0)$ in period t, it follows a proportionate growth by 10 per cent for period $t+1$: $x_1 = 1.65a,\quad x_2 = 1.1a;$
   (c) $x_1 = 6,000;\ x_2 = 4,000;$ subsequent period: $x_1 = 6,600;\ x_2 = 4,400;$
   two periods later: $x_1 = 7,260;\ x_2 = 4,840$

10.3 (a) A: $\lambda_1 = 1,\quad \lambda_2 = 3,\quad \lambda_3 = -2,\quad \lambda_4 = 5,$

$$\mathbf{x}^1 = t_1 \begin{pmatrix} 1 \\ 0 \\ 0 \\ 0 \end{pmatrix},\quad \mathbf{x}^2 = t_2 \begin{pmatrix} 1 \\ 1 \\ 0 \\ 0 \end{pmatrix},\quad \mathbf{x}^3 = t_3 \begin{pmatrix} 4/15 \\ -2/5 \\ 1 \\ 0 \end{pmatrix},\quad \mathbf{x}^4 = t_4 \begin{pmatrix} 1/4 \\ 0 \\ 0 \\ 1 \end{pmatrix};$$

B : $\lambda_1 = \lambda_2 = 4, \quad \lambda_3 = -2,$

$$\mathbf{x}^{1,2} = t_1 \begin{pmatrix} 1 \\ 0 \\ 0 \end{pmatrix} + t_2 \begin{pmatrix} 0 \\ 1 \\ 1 \end{pmatrix}, \quad \mathbf{x}^3 = t_3 \begin{pmatrix} 0 \\ 1 \\ -1 \end{pmatrix} ;$$

$t_1, t_2, t_3, t_4 \in \mathbb{R} \setminus \{0\}$

10.4 $\mathbf{x}^{\mathsf{T}} B \mathbf{x} = \mathbf{x}^{\mathsf{T}} B_s \mathbf{x}$ with $B_s = \begin{pmatrix} 2 & -1/2 \\ -1/2 & 4 \end{pmatrix}$, B_s is positive definite

10.5 (a) $A: \lambda_{1,2} = 2 \pm \sqrt{2}, \qquad B: \lambda_{1,2} = (3 \pm \sqrt{17})/2,$
       $C: \lambda_1 = -4, \quad \lambda_2 = 2, \quad \lambda_3 = 3, \qquad D: \lambda_1 = 1, \quad \lambda_{2,3} = 3 \pm \sqrt{8};$
   (b) A and D are positive definite, B and C are indefinite

10.6 (a) $a_1 = 3, \quad a_2 = 1, \quad a_3 \in \mathbb{R};$
   (b) any vector $\mathbf{x}^1 = (t, t, 0)^{\mathsf{T}}$ with $t \in \mathbb{R}, t \neq 0$, is an eigenvector;
   (c) $a_3 > 1;$ (d) no

11   FUNCTIONS OF SEVERAL VARIABLES

11.1 (a) $f(x_1, x_2) = \sqrt{x_1 x_2}$

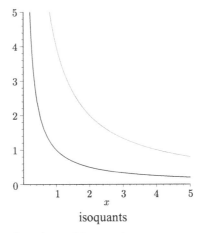

isoquants

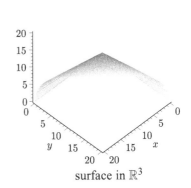

surface in $\mathbb{R}^3$

   (b) domains and isoquants

   (i)

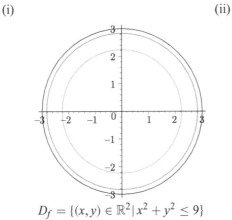

$D_f = \{(x, y) \in \mathbb{R}^2 \mid x^2 + y^2 \leq 9\}$

   (ii)

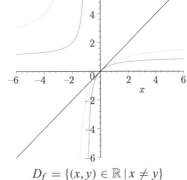

$D_f = \{(x, y) \in \mathbb{R} \mid x \neq y\}$

(iii)

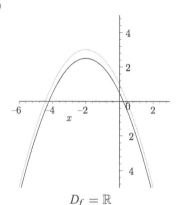

$$D_f = \mathbb{R}$$

11.2 (a) $f_x = 2x \sin^2 y$,

$f_y = 2x^2 \sin y \, \cos y$;

(b) $f_x = y^2 x^{(y^2-1)}$,

$f_y = 2y \, x^{(y^2)} \ln x$;

(c) $f_x = yx^{(y-1)} + y^x \ln y$,

$f_y = x^y \ln x + xy^{(x-1)}$;

(d) $f_x = \dfrac{1}{2x}$,

$f_y = \dfrac{1}{2y}$;

(e) $f_x = e^{x^2+y^2+z^2}(2 + 4x^2)$,

$f_y = 4xye^{x^2+y^2+z^2}$,

$f_z = 4xze^{x^2+y^2+z^2}$;

(f) $f_{x_1} = \dfrac{x_1}{\sqrt{x_1^2 + x_2^2 + x_3^2}}$,

$f_{x_2} = \dfrac{x_2}{\sqrt{x_1^2 + x_2^2 + x_3^2}}$,

$f_{x_3} = \dfrac{x_3}{\sqrt{x_1^2 + x_2^2 + x_3^2}}$

11.3 (a) $C_x = 120 - \dfrac{1,200,000}{x^2}$, $\quad C_y = 800 - \dfrac{32,000,000}{y^2}$;

(b) $C_x(80) = -67.5$, $\quad C_x(120) \approx 36.67$, $\quad C_y(160) = -450$, $\quad C_y(240) \approx 244.5$

11.4 (a) $f_{x_1 x_2} = f_{x_2 x_1} = 6x_2 x_3^3$, $\quad f_{x_1 x_3} = f_{x_3 x_1} = 9x_2^2 x_3^2$, $\quad f_{x_2 x_3} = f_{x_3 x_2} = 18x_1 x_2 x_3^2$,

$f_{x_1 x_1} = 6x_1 - x_1^{-2}$, $\quad f_{x_2 x_2} = 6x_1 x_3^3$, $\quad f_{x_3 x_3} = 18x_1 x_2^2 x_3 - x_3^{-2}$;

(b) $f_{xx} = \dfrac{4y^2}{(1-xy)^3}$, $\quad f_{yy} = \dfrac{4x^2}{(1-xy)^3}$, $\quad f_{xy} = f_{yx} = \dfrac{2xy+2}{(1-xy)^3}$;

(c) $f_{xx} = \dfrac{4xy}{(x^2-y^2)^2}$, $\quad f_{yy} = f_{xx}$, $\quad f_{xy} = f_{yx} = \dfrac{-2(x^2+y^2)}{(x^2-y^2)^2}$

11.5 (a) $\mathbf{grad} f(1,0) = (a,b)^{\mathrm{T}}$, $\quad\quad \mathbf{grad} f(1,2) = (a,b)^{\mathrm{T}}$;

(b) $\mathbf{grad} f(1,0) = (2,1)^{\mathrm{T}}$, $\quad\quad \mathbf{grad} f(1,2) = (3,3.6)^{\mathrm{T}}$;

(c) $\mathbf{grad} f(1,0) = (-1/2, 0)^{\mathrm{T}}$, $\quad\quad \mathbf{grad} f(1,2) = (-1/2, -1)^{\mathrm{T}}$

11.6 the direction of movement is $-\mathbf{grad}\, f(1,1) = (-1.416, -0.909)^{\mathrm{T}}$

11.7 (a) $dz = \dfrac{1}{y} \cos \dfrac{x}{y} \, dx - \dfrac{x}{y^2} \cos \dfrac{x}{y} \, dy$;

(b) $dz = (2x + y^2)\, dx + (2xy + \cos y)\, dy$;

(c) $dz = (2x\, dx + 2y\, dy)e^{x^2+y^2}$;

(d) $dz = \dfrac{1}{x}\, dx + \dfrac{1}{y}\, dy$

11.8 surface: $S = 28\pi$, $\quad$ absolute error: 4.08, $\quad$ relative error: 4.6 per cent

11.9 (a) $\frac{dz}{dt} = 2x_1e^{x_2}\frac{dx_1}{dt} + x_1^2e^{x_2}\frac{dx_2}{dt}$;

(b) (i) $z' = 2x_1e^{x_2}2t + x_1^2e^{x_2}\frac{2}{t} = 6t^5$,

(ii) $z' = 2x_1e^{x_2}\frac{2}{t} + x_1^2e^{x_2}2t = 8e^{t^2}\left(\frac{1}{t} + t\ln t\right)\ln t$;

(c) (i) $z = t^6$; $z' = 6t^5$,

(ii) $z = (\ln t^2)^2e^{t^2}$, $z' = 8e^{t^2}(\frac{1}{t} + t\ln t)\ln t$

11.10 $\frac{\partial f}{\partial r^1}(\mathbf{a}) = -\frac{1}{2}$; $\frac{\partial f}{\partial r^2}(\mathbf{a}) = -\frac{3}{4}\sqrt{2}$; $\frac{\partial f}{\partial r^3}(\mathbf{a}) = \frac{\sqrt{5}}{2}$

11.11 (a) $\mathbf{grad}\,C(3,2,1) = (8,\ 6,\ 10)^T$, $\frac{\partial C}{\partial r_1}(P_0) = 13.36$;

percentage rate of cost reduction: 5.52 per cent;

(b) The first ratio is better or equal

11.12 $\rho_{f,x_1} = 0.002$; $\rho_{f,x_2} = 0.00053$; $\varepsilon_{f,x_1} = 0.2$; $\varepsilon_{f,x_2} = 0.079$

11.13 (a) partial elasticities:

$\varepsilon_{f,x_1} = \frac{x_1(3x_1^2 + 2x_2^2)}{2(x_1^3 + 2x_1x_2^2 + x_2^3)}$, $\varepsilon_{f,x_2} = \frac{x_2(4x_1x_2 + 3x_2^2)}{2(x_1^3 + 2x_1x_2^2 + x_2^3)}$

f homogeneous of degree $r = 3/2$, $r > 1$;

(b) f is not homogeneous

11.14 (a) $y' = \frac{bx}{a\sqrt{x^2 - a^2}}$; (b) $y' = \frac{3x\cos 3x - \sin 3x}{x^2}$; (c) $y' = -\frac{y(yx^y - 1 + 2x^2y)}{x(yx^y\ln x - 1 + yx^2)}$

11.15 $|J| = r$; for $r \neq 0 : r = x^2 + y^2$, $\varphi = \arctan\frac{y}{x}$ $\left(\text{resp. } \varphi = \arccos\frac{x}{x^2 + y^2}\right)$

11.16 local maximum at $(x_1, y_1) = (1/2, 1/3)$; no local extremum at $(x_2, y_2) = (1/7, 1/7)$

11.17 (a) local minimum at $(x_1, y_1) = (0, 1/2)$ with $z_1 = -1/4$;

(b) local minimum at $(x_1, y_1) = (1, \ln\frac{4}{3})$ with $z_1 = 1 - \ln\frac{4}{3}$

11.18 stationary point: $(x^0, y^0) = (100, 200)$;

local minimum point with $C(100, 200) = 824,000$

11.19 local minimum point: $\mathbf{x}^1 = (1,\ 0,\ 0)$,

$\mathbf{x}^2 = (1,\ 1,\ 1)$, $\mathbf{x}^3 = (1,\ -1,\ -1)$ are not local extreme points

11.20 no local extremum

11.21 stationary point $\mathbf{x} = (30,\ 30,\ 15)$ is a local maximum point with $P(30, 30, 15) = 26,777.5$

11.22 (a) $y = 10.05x - 28.25$; (b) $y(18) = 152.64$, $y(36) = 333.5$

11.23 P_1: maximum; P_2: minimum

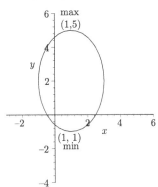

11.24 (a) local maximum at $(x_1, y_1) = (4, 0)$ with $z_1 = 16$;

    (b) local minimum at $(x_1^1, x_2^1, x_3^1) = (-1/12,\ 37/12,\ -30)$ with $z_1 = -73/48$

11.25 local maximum point; values of the Lagrangian multipliers: $\lambda_1 = -13$ and $\lambda_2 = -16$

11.26 length = breadth = 12.6 cm, height = 18.9 cm

11.27 local and global maximum of distance $D_{max} = 84/9$

    at $(x_1, y_1) = (-\frac{2}{3}, \frac{2}{3}\sqrt{5})$, $(x_2, y_2) = (-\frac{2}{3}, -\frac{2}{3}\sqrt{5})$;

    local minimum of distance $D_{1\,min} = 3$ at point $(x_3, y_3) = (-1, 0)$;

    local and global minimum of distance $D_{2\,min} = 1$ at point $(x_4, y_4) = (1, 0)$

11.28 stationary point $(x_1, x_2, x_3; \lambda) = (25,\ 7.5,\ 15;\ 5)$;

    local minimum point with $C(25, 7.5, 15) = 187.5$

11.29 $136/3$

11.30 $32/3$

11.31 $3/8$

12 DIFFERENTIAL EQUATIONS AND DIFFERENCE EQUATIONS

12.1 (a), (c)

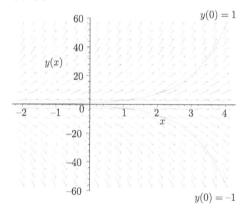

    (b) $y^P = e^x$

12.2 (a) $y = \ln|e^x + C|$;   (b) $y^2 = 1 + 2\ln\dfrac{1 + e^x}{2}$

12.3 $ky \cdot (x - 1) - y + 1 = 0$

12.4 general solution $y = Cx^{x^2}$,   particular solution $y^P = x^{x^2}$

12.5 $y = Ct^{100}e^{-t/2}$

12.6 The functions y_1, y_2, y_3 form a fundamental system;

    $y = C_1 x + C_2 x \ln x + C_3 \dfrac{1}{x}$

12.7 (a) $y = Ce^{2x} - \frac{1}{5}\cos x - \frac{2}{5}\sin x$;   (b) $y^P = e^x(1 + x)$;

    (c) $y = C_1 e^{-x} + C_2 e^{x/2} + e^x$;   (d) $y = -e^x \cos 3x + x^2 + 2.2x + 1$

12.8 (a) $y = C_1 + e^{-x}(C_2 \cos 2x + C_3 \sin 2x) + \dfrac{1}{5}\cos x + \dfrac{2}{5}\sin x$;

    (b) $y = C_1 e^x + C_2 e^{-2x} + C_3 x e^{-2x} + \dfrac{1}{3}x e^x$

12.9 (a) $y_1 = C_1 e^{ax} \cos x + C_2 e^{ax} \sin x$ (b) $y_1^P = 3e^{2x} + 9$

$y_2 = C_1 e^{ax} \sin x + C_2 e^{ax} \cos x$

$y_2^P = -2e^{2x} + 8e^{-x}$

$y_3^P = e^{2x} - 16e^{-x} + 9$

12.10 (a) $y_t = (1+b)2^t - b$; (b) strictly increasing for $b > -1$;

(c)

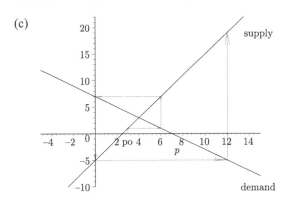

12.11 (a) $p_{t+1} = -2p_t + 12$; (b) $p_t = -(-2)^t + 4$ with $p_0 = 3, p_1 = 6, p_2 = 0, \ldots$;

(c)

12.12 (a) $y_t = 2^t \left(C_1 \cos \frac{2}{3}\pi t + C_2 \sin \frac{2}{3}\pi t \right)$; (b) $y_t^P = 4^t - (-2)^t - \frac{1}{3}t - \frac{2}{3}$;

(c) $y_t = C_1(-1)^t + C_2 t(-1)^t + \frac{1}{100} \cdot 4^t$

12.13 (a) $y_{t+2} = 1.4y_{t+1} + 0.15y_t$; (b) $y_t = C_1(1.5)^t + C_2(-0.1)^t$;

(c) 1,518.76 units

12.14 (a) $y_t = 3^t y_0 + \sum_{i=1}^{t-1} (2i+1)3^{t-i-1}$; (b) $y_t = 3^t y_0 + 3^t - t - 1$

Literature

Anthony, M. and Biggs, N., *Mathematics for Economics and Finance*, Cambridge: Cambridge University Press, 1996.

Bronstein, I.N. and Semandjajew, K.A., *Taschenbuch der Mathematik*, twenty-fifth edition, Stuttgart: Teubner, 1991 (in German).

Chiang, A.C., *Fundamental Methods of Mathematical Economics*, third edition, New York: McGraw-Hill, 1984.

Dück, W., Körth, H., Runge, W. and Wunderlich, L. (eds), *Mathematik für Ökonomen*, Berlin: Verlag Die Wirtschaft, 1979 (in German).

Eichholz, W. and Vilkner, E., *Taschenbuch der Wirtschaftsmathematik*, second edition, Leipzig: Fachbuchverlag, 2000 (in German).

Kalischnigg, G., Kockelkorn, U. and Dinge, A., *Mathematik für Volks- und Betriebswirte*, third edition, Munich: Oldenbourg, 1998 (in German).

Luderer, B. and Würker, U., *Einstieg in die Wirtschaftsmathematik*, Stuttgart: Teubner, 1995 (in German).

Mizrahi, A. and Sullivan, M., *Mathematics. An Applied Approach*, sixth edition, New York: Wiley, 1996.

Mizrahi, A. and Sullivan, M., *Finite Mathematics. An Applied Approach*, seventh edition, New York: Wiley, 1996.

Nollau, V., *Mathematik für Wirtschaftswissenschaftler*, third edition, Stuttgart and Leipzig: Teubner, 1999 (in German).

Ohse, D., *Mathematik für Wirtschaftswissenschaftler* I–II, third edition, Munich: Vahlen, 1994 (in German).

Opitz, O., *Mathematik. Lehrbuch für Ökonomen*, Munich: Oldenbourg, 1990 (in German).

Rommelfanger, H., *Mathematik für Wirtschaftswissenschaftler* I–II, third edition, Hochschultaschen-bücher 680/681, Mannheim: B.I. Wissenschaftsverlag, 1994 (in German).

Rosser, M., *Basic Mathematics for Economists*, London: Routledge, 1993.

Schmidt, V., *Mathematik. Grundlagen für Wirtschaftswissenschaftler*, second edition, Berlin and Heidelberg: Springer, 2000 (in German).

Schulz, G., *Mathematik für wirtschaftswissenschaftliche Studiengänge*, Magdeburg: Otto-von-Guericke-Universität, Fakultät für Mathematik, 1997 (in German).

Simon, C.P. and Blume, L., *Mathematics for Economists*, New York and London: Norton, 1994.

Sydsaeter, K. and Hammond, P.J., *Mathematics for Economic Analysis*, Englewood Cliffs, NJ: Prentice-Hall, 1995.

Varian, H.R., *Intermediate Microeconomics. A Modern Approach*, fifth edition, New York: Norton, 1999.

Werner, F., *Mathematics for Students of Economics and Management*, sixth edition, Magdeburg: Otto-von-Guericke-Universität, Fakultät für Mathematik, 2004.

Index

Advanced Mathematical Economics

Rakesh V. Vohra, Northwestern University, USA

As the intersection between economics and mathematics continues to grow in both theory and practice, a solid grounding in mathematical concepts is essential for all serious students of economic theory.

In this clear and entertaining volume, Rakesh V. Vohra sets out the basic concepts of mathematics as they relate to economics. The book divides the mathematical problems that arise in economic theory into three types: feasibility problems, optimality problems and fixed-point problems.

Of particular salience to modern economic thought are the sections on lattices, supermodularity, matroids and their applications. In a departure from the prevailing fashion, much greater attention is devoted to linear programming and its applications.

Of interest to advanced students of economics as well as those seeking a greater understanding of the influence of mathematics on 'the dismal science', *Advanced Mathematical Economics* follows a long and celebrated tradition of the application of mathematical concepts to the social and physical sciences.

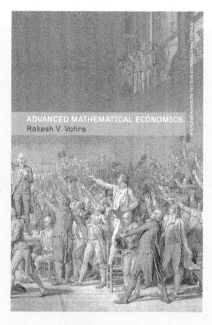

Series: Routledge Advanced Texts in Economics and Finance

November 2004: 208pp
Hb: 0-415-70007-8: £65.00
Pb: 0-415-70008-6: £22.99
eB: 0-203-79995-X: £22.99
Available as an inspection copy

Routledge books are available from all good bookshops, or may be ordered by calling
Taylor & Francis Direct Sales on **+44 (0)1264 343071** (credit card orders). For more information please contact David Armstrong on **020 7017 6028** or email
david.armstrong@tandf.co.uk
2 Park Square, Milton Park, Abingdon, Oxon OX14 4RN

Routledge
Taylor & Francis Group

www.routledge.com/economics

available from all good bookshops

Printed and bound by CPI Group (UK) Ltd, Croydon, CR0 4YY

01/11/2024

01782610-0006